IONIC LIQUIDS IN BIOTRANSFORMATIONS AND ORGANOCATALYSIS

IONIC LIQUIDS IN BIOTRANSFORMATIONS AND ORGANOCATALYSIS

Solvents and Beyond

Edited by
Pablo Domínguez de María

WILEY
A JOHN WILEY & SONS, INC., PUBLICATION

Published by John Wiley & Sons, Inc., Hoboken, New Jersey.
Published simultaneously in Canada.

Library of Congress Cataloging-in-Publication Data:
Ionic liquids in biotransformations and organocatalysis : solvents and beyond / edited by Pablo Dominguez de Maria.
p. cm.
Includes index.
ISBN 978-0-470-56904-7 (hardback)
1. Proteins–Biotechnology. 2. Ionic solutions. 3. Catalysis. I. Dominguez de Maria, Pablo, 1974-
TP248.65.P76I59 2012
572.6–dc23

2011038032

Printed in the United States of America.

10 9 8 7 6 5 4 3 2 1

CONTENTS

FOREWORD

When Pablo Domínguez de María invited me to contribute a chapter to a book that he proposed on biocatalysis and organocatalysis in ionic liquids, I had to decline the offer owing to other pressing commitments, but I agreed to write a foreword to the book. Now that I see the impressive result of his endeavors I am rather sorry that I am not a contributor.

My introduction to the subject dates back to June 1996 when I attended an inspiring lecture on ionic liquids presented by Ken ("Mr. Ionic Liquids") Seddon of the Queen's University Belfast at the Clean Tech '96 conference in London. I was immediately hooked. I was fascinated by the possible benefits to be gained by using ionic liquids as reaction media for catalytic processes. While listening to the lecture it occurred to me that it would be very interesting to try ionic liquids as solvents for conducting biocatalytic processes. I was motivated by the notion that ionic liquids, by virtue of their anticipated compatibility with enzymes, could possibly exert a rate enhancing and/or stabilizing effect, resulting in an improved operational performance compared with that observed in organic solvents. Afterward I asked Ken if he knew whether anybody had tried to use an enzyme in an ionic liquid. His answer was: "No, but why don't we try it? I can supply the ionic liquids." So we decided to try reactions with *Candida antarctica* lipase B (CaLB), as this robust enzyme was known to be thermally very stable and tolerant toward organic solvents under essentially anhydrous conditions. It took a while to find a Ph.D. student, Rute Madeira Lau, to perform the experiments, but the results were gratifying. We observed that Novozyme 435 (an immobilized form of CaLB) was able to catalyze various reactions—esterification, amidation, and perhydrolysis—under anhydrous conditions in the second-generation ionic liquids, [bmim][BF_4] and [bmim] [PF_6], with rates at least as high as those in organic solvents.

Following the publication of our results, in *Organic Letters* in 2000, the use of ionic liquids as reaction media for catalysis in general and biocatalysis in particular has undergone exponential growth. It was soon recognized that the use of second-generation ionic liquids on a large scale was seriously hampered by their high price coupled with ecotoxicity and poor biodegradability. Consequently, attention was devoted to the development of a third generation of ionic liquids that are greener, more sustainable, and less expensive than the second generation. In particular, ionic liquids derived from natural raw materials, such as carbohydrates and amino acids, are emerging as green solvents potentially suitable for large-scale applications. Furthermore, it is possible to design task specific ionic liquids, for example, bio-based chiral ionic liquids, that meet not only environmental requirements but are also eminently suited to particular tasks.

Pablo is to be complimented on bringing this group of knowledgeable authors together to review the state of the art in biocatalysis and organocatalysis in ionic liquids. The subjects covered are wide-ranging, from fundamental aspects of interactions between proteins and ionic liquids to their use as reaction media with both hydrolytic and nonhydrolytic enzymes, whole cell bioconversions, and, as a bonus, organocatalytic reactions. Importantly, practical aspects are highlighted, including process development issues such as downstream processing. Why use ionic liquids as reaction media in the first place? An important motivation for their use as reaction media was that, based on their negligible vapor pressure, they would be environmentally acceptable alternatives to volatile organic solvents. However, the question still remained of how to separate the product from the ionic liquid. An elegant solution to this problem was found in continuous product extraction with supercritical carbon dioxide.

In addition to the enhanced operational performance through increased stability and/or selectivity another important motivation for using ionic liquids as reaction media was their ability to dissolve large amounts of highly polar substrates, such as carbohydrates and nucleosides. In particular, their ability to readily dissolve biopolymers such as cellulose and lignin has become an important asset in the current drive toward the bio-based economy, in which there is a need for effective and sustainable methods for the primary conversion of renewable lignocellulosic raw materials. The challenges of using ionic liquids as reaction media for biotransformations are also addressed. In order to be sustainable they must meet stringent requirements regarding the greenness and economic viability of their synthesis and their environmental footprint, which is governed by properties such as bioaccumulation, biodegradability, and ecotoxicity. In addition to the various chapters on the use of ionic liquids as reaction media, there is an extra treat for the reader: two chapters on nonsolvent applications, in biotransformations and organocatalytic conversions, respectively. This includes interesting concepts such as the use of ionic liquid–coated enzymes and the anchoring of organocatalysts to ionic liquids.

In short, I believe that this book is an important addition to the literature on ionic liquids as reaction media for biocatalytic and organocatalytic processes. In addition to its obvious value to practicing organic chemists in both industry and academia, its educational value should not be underestimated. It should prove to be of great value for advanced undergraduate and graduate students. Finally, I would like to thank Pablo for giving me the opportunity to air my views on the merits of this book. I wish him all the success that he has surely earned.

Roger A. Sheldon
Emeritus Professor of Biocatalysis and Organic Chemistry
Delft University of Technology

January 2012

PREFACE

When I was approached by Wiley to edit a book on ionic liquids in biotransformations and organocatalysis, the spontaneous question that quickly came to my mind was, *is there a gap for such a book*? The field of ionic liquids applied to biotransformations and organocatalysis has developed enormously during the last two decades. Therefore, the realization of a book that could gather, categorize, and provide an updated and complete state-of-the-art in these areas was clearly a demand. There are obviously several comprehensive reviews in the field, but I think none of them can cover the topic(s) in their widest extent. Thanks to the outstanding chapters of many world-class experts in the area, this book is now a reality that I hope will be a useful contribution for researchers in the field, both in academia and industry.

Since the beginning of my work as editor, I have made it clear in my mind that I do not want a book just covering uses of ionic liquids as solvents, albeit, of course, this topic is broad and very important (Chapters 3–7 and 9), nor a book regarding ionic liquids as "green solvents"—an unfortunate label that has surely brought more problems than advantages to ionic liquids (Chapters 1 and 7). In fact, many ionic liquids are not green, but their versatility and tunability makes us optimistic that it will be possible to combine greenness with the acquired know-how on advantages that ionic liquids may bring, for example, leading to the third generation of ionic liquids. Likewise, emerging deep-eutectic-solvents represent a promising option, and first uses in biotransformations are briefly discussed herein (Chapters 1, 5, and 7). In addition, an extensive updated state-of-the-art on toxicity and (bio)degradability of commonly used ionic liquids, together with protocols and rules applied for assessing these parameters, is provided in Chapter 7.

Quite remarkably, ionic liquids are more than mere solvents. They represent a fantastic academic tool for studying and understanding interactions with proteins, enzymatic mechanisms, and so on (Chapters 2 and 5); there are also a number of "non-solvent" approaches for practical applications, such as catalyst immobilization or activation, downstream processing, and catalyst grafting or coating (Chapters 8 and 10). Ionic liquids can also be smartly combined with other nonconventional solvents, such as supercritical fluids, or with innovative process design concepts (Chapter 4). Finally, some ionic liquids can be employed in whole-cell biotransformations, providing novel and promising approaches, including proof-of-principle for deep-eutectic-solvents and whole cells (Chapter 7).

I want to acknowledge a number of people who have made this book a reality. First of all, the greatest credits go obviously to the authors of this book, the actual and unique protagonists of this work (together with the ionic liquids!). Without their outstanding efforts, professionalism, and excellent and readily updated chapters, this project would have never been possible. Furthermore, I wish to thank Dr. Daniela Gamenara, Dr. Fabrizio Sibilla, and Dr. Andreas Buthe for many fruitful and stimulating discussions. Likewise, thanks are given to Prof. Dr. Roger Sheldon for writing the Foreword of this book. I am also indebted to Ms. Anita Lekhwani, Senior Acquisitions Editor at Wiley, for the interest and patience she has had and the hard work she has done throughout the editing process. And my thanks go as well to Dr. Edmund H. Immergut, Consulting Editor for Wiley and Wiley-VCH, for inviting me to edit this book and his trust and support during this time.

I must say that, overall, this project has been for me a fascinating and unforgettable adventure. I really hope that readers will find this book an attractive and useful tool for working in the field of ionic liquids, biotransformations, and organocatalysis. Suggestions for further improvements, data treatment, new topics, and so on are of course welcome for future editions of this work.

Pablo Domínguez de María
Aachen, Germany, January 2012

CONTRIBUTORS

Paola Agrigento, Dipartimento Chimica Organica "E. Paternò", Università di Palermo, Viale delle Scienze, Palermo, Italy

Danielle Dennewald, Lehrstuhl für Bioverfahrenstechnik, Technische Universität München, Garching, Germany

Pablo Domínguez de María, Institute of Technical and Macromolecular Chemistry (ITMC), RWTH Aachen University, Aachen, Germany

Daniela Gamenara, Physical-Organic Chemistry and Bioprocesses Group, Organic Chemistry Department, Facultad de Química, Universidad de la República (UdelaR), Montevideo, Uruguay

Eduardo García-Verdugo, Instituto de Catálisis y Petroleoquímica, CSIC, Campus de la UAM, Cantoblanco Madrid, Spain

Francesco Giacalone, Dipartimento Chimica Organica "E. Paternò", Università di Palermo, Palermo, Italy

Lasse Greiner, Institute of Technical and Macromolecular Chemistry (ITMC), RWTH Aachen University, Aachen, Germany and DECHEMA e.V. Karl-Winnacker-Institut, Frankfurt am Main, Germany

Michelangelo Gruttadauria, Dipartimento Chimica Organica "E. Paternò", Università di Palermo, Palermo, Italy

Christina Kohlmann, Institute of Technical and Macromolecular Chemistry (ITMC), RWTH Aachen University, Aachen, Germany and Cognis GmbH, Düsseldorf, Germany

Pedro Lozano, Departamento de Bioquímica y Biología Molecular "B" e Inmunología, Facultad de Química, Universidad de Murcia, Murcia, Spain

Patricia Saenz Méndez, Physical-Organic Chemistry and Bioprocesses Group, Organic Chemistry Department, Facultad de Química, Universidad de la República (UdelaR), Montevideo, Uruguay and Computational Chemistry and Biology Group, DETEMA, Facultad de Química, Universidad de la República (UdelaR), Montevideo, Uruguay

Renato Noto, Dipartimento Chimica Organica "E. Paternò", Università di Palermo, Palermo, Italy

Radovan Šebesta, Faculty of Natural Sciences, Comenius, University Bratislava, Bratislava, Slovakia

Gustavo Seoane, Physical-Organic Chemistry and Bioprocesses Group, Organic Chemistry Department, Facultad de Química, Universidad de la República (UdelaR), Montevideo, Uruguay

Štefan Toma, Faculty of Natural Sciences, Comenius, University Bratislava, Bratislava, Slovakia

Dirk Weuster-Botz, Lehrstuhl für Bioverfahrenstechnik, Technische Universität München, Garching, Germany

Zhen Yang, College of Life Sciences, Shenzhen University, Shenzhen, Guangdong, China

Hua Zhao, Chemistry Program, Savannah State University, Savannah, GA, USA

PART I

FUNDAMENTALS

1

IONIC LIQUIDS: DEFINITION, APPLICATIONS, AND CONTEXT FOR BIOTRANSFORMATIONS AND ORGANOCATALYSIS

Pablo Domínguez de María

1.1 IONIC LIQUIDS: DEFINITION, DEVELOPMENT, AND OVERVIEW OF CURRENT MAIN APPLICATIONS

Ionic liquids (ILs) (low-temperature molten salts) are *simply* mixtures of cations and anions that do not pack well among them, and therefore remain liquid at low to moderate temperatures. The low melting points are often achieved by incorporating bulky asymmetric cations into the structure, together with weakly coordinating anions. Arbitrarily it has been established that ILs that melt below 100°C fall into the category of "ionic liquids." On the other hand, those that are liquid at room temperature are often regarded as "room-temperature ionic liquids" (RTILs). Although some IL compositions have been known for a long time, it has been in the last decades when an impressive development in the field has emerged, providing innovative applications in many areas of chemistry. This interest is driven by the fact that by changing the cation or the anion of a certain IL, the physicochemical properties of that IL can be finely tuned. Thus, novel solvents can be defined and used for a specific tailored application. Obviously this wide tunability cannot be reached with conventional organic solvents. For instance, IL polarities can be modulated to design ILs that are immiscible with either low-polarity

Ionic Liquids in Biotransformations and Organocatalysis: Solvents and Beyond, First Edition. Edited by Pablo Domínguez de María.

organic solvents or high-polarity solvents. This facilitates conventional extraction methods to be employed in product separation and furthermore provides promising entries in many areas of chemistry. Likewise, combinations of ILs with other solvent systems, for example, supercritical fluids, have provided promising synergies for chemical process setups since properties of both systems can be easily modulated.

Not without discussion, it is usually assumed that the first "true" IL was described in 1914 by Walden. The IL was ethylammonium nitrate (EAN), $[EtNH_3][NO_3]$, with a melting point (m.p.) of 12.5°C. However, the finding attracted rather little attention at that time.[1] Two decades later, in 1934, a patent reported on some pyridinium-based molten salts that were able to dissolve certain amounts of cellulose.[2] Again, in this case, the importance, potential, or utility of this finding was underestimated. Remarkably, nowadays cellulosic biomass pretreatment by means of a wide number of ILs is an important topic of research since some ILs enable the dissolution of different lignocellulosic materials. Once dissolved in these ILs, cellulose can be subsequently depolymerized by, for instance, different hydrolytic enzymes (see also Chapter 5, Sections 5.4 and 5.5).[3]

In a broad sense, ILs started to attract interest in the 1960s. During several decades on (1960–1990), the first generation of ILs appeared and was widely described and chemically characterized. Typical cations for first-generation ILs were dialkylimidazolium and alkylpyridinium derivatives. As anions, chloroaluminate and other metal halide structures were used. As an important drawback for practical applications, first-generation ILs were found to be sensitive to water and air. These features clearly hampered further applications of first-generation ILs in different fields of chemistry. In the 1990s, the second generation of ILs emerged. Herein, anions were substituted for weakly coordinating anions such as BF_4 or PF_6. These new ILs were air- and water-stable and therefore led to much research and efforts in the area, as the enhanced stability of ILs provided a much wider frame for operating with them under many different processing conditions. More recently, the third generation of ILs has emerged. This third generation comprises biodegradable and readily available ions, such as natural bases (e.g., choline), amino acids, and naturally occurring carboxylic acids.[4,5] Together with this third generation of ILs, so-called deep eutectic solvents (DES) represent a promising alternative because they are simple to prepare, biodegradable, and more economical, compared with other ILs. In general, DES are mixtures of a solid salt with a hydrogen-bond donor in different proportions. An example of DES is represented by the combination of choline chloride (solid salt at room temperature, m.p. 302°C) with urea (solid at room temperature, m.p. 132°C), which leads to a DES with a melting point of 12°C.[5–7] Yet it is not clear if DES can be regarded as "IL" since some of the structures (e.g., urea) are not charged and therefore subsequently produced solvents are not entirely ionic. Despite this, it is believed that many properties of ILs can also be more or less extrapolated to DES. In Figure 1.1 some selected milestones in the IL history are depicted.

Until now, second-generation ILs have been the subject of enormous fundamental research, providing interesting and novel applications in many areas of chemistry. Yet their use at commercial scale is still limited to a few cases, presumably due to economic aspects related to ILs. However, along with the development of the third generation of

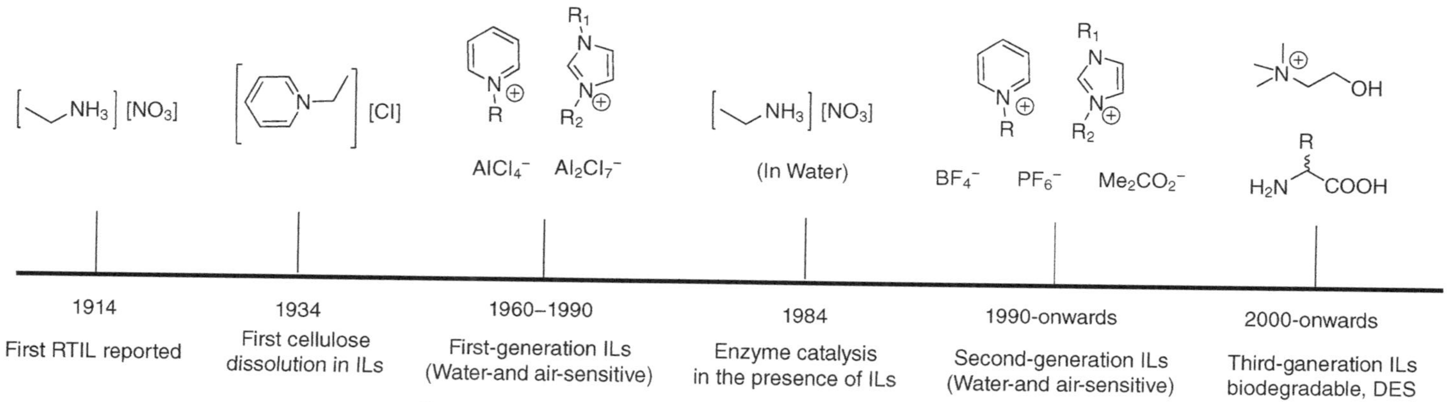

Figure 1.1. Some selected milestones in the history of ILs.[4,5]

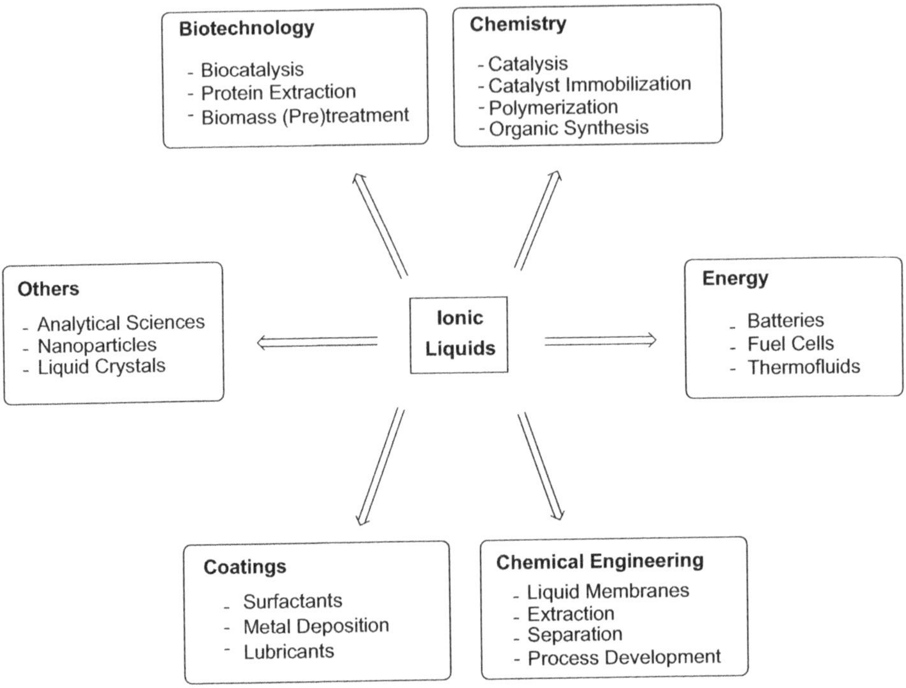

Figure 1.2. Overview of possible applications of ILs in different areas.[8–16]

ILs, which are more sustainable, biodegradable, and cheaper derivatives, it is anticipated that novel IL-based applications will reach the commercial level in the coming years. A significant number of companies already commercialize some ILs and perform R&D-related activities aimed at identifying new market niches and business opportunities.

The enormous potential of ILs is driven by their intrinsic feature (previously mentioned), which is that ILs can be finely tuned by carefully selection of anions and cations. Thus, ILs can be tailored for a specific application, leading to the concept of task-specific ionic liquid (TSIL). By choosing anions and cations, relevant examples of protic ILs, chiral ILs, multifunctional ILs, supported ILs, and so on have been reported. Some general applications for ILs have been put forward (see overview in Figure 1.2). More information on the general applications of ILs can be found in recent reviews and books devoted to various IL areas.[8–16]

1.2 ON THE GREENNESS OF ILs: TOWARD THE THIRD GENERATION OF ILs AND DES

Apart from the ample tunability of ILs (previous section), probably another aspect that has triggered significant interest and research in the field of ILs is the common claim that ILs are "green solvents." This general assumption is based on several important properties commonly attributed to ILs, namely that ILs pose negligible vapor pressure and that they are nonflammable. In this section aspects related to the greenness of ILs will be briefly discussed.

First of all, it is usually reported that ILs do not exert measurable vapor pressure since they are entirely composed of ions. Hence, they cannot be distilled without decomposition and thus are nonvolatile. An obvious conclusion that may be drawn from these postulates is that environmental advantages would be achieved by using ILs instead of volatile organic compounds (VOCs). This nonvolatility statement has been challenged by Seddon and coworkers, who demonstrated that some ILs can in fact be distilled at low pressures.[17] However, at ambient pressures most of the ILs indeed show a negligible vapor pressure, and therefore from that viewpoint they may still be considered environmentally advantageous compared with VOCs.

The second important property to categorize ILs as environmentally-benign solvents is their nonflammability compared again with VOCs. However, this statement has also been challenged by Wilkes, Rogers et al., who showed that a wide number of commonly used ILs were actually combustible since products formed during thermal decomposition of ILs were found to be combustible. Experiments thus showed that it is not safe to operate with ILs close to fire or heat sources.[18] Therefore, although the low flammability of ILs may certainly provide advantages compared with VOCs, it is obvious that ILs should not be regarded as "green solvents" by the mere fact that they are ILs. Traditionally, in publications dealing with ILs, there are not many distinctions among ILs, and usually *all* ILs are generically regarded as "green solvents" or as "nonflammable" solvents. This trend, however, is starting to change.

Apart from the two above-mentioned IL properties (low vapor pressure and nonflammability), there are other aspects—surely more important in assessing the greenness of ILs—that definitely challenge the "green label" of many often used ILs. These aspects are related to the environmental impact that IL syntheses may have (e.g., the *E*-factor of producing ILs), as well as to the eco-toxicity and biodegradability of the ions composing the ILs, and of metabolites formed thereof, when ILs are (accidentally) spoiled in the milieu. These topics have only recently started to receive the attention that they actually deserve.[19–22] In this chapter a brief discussion of these issues is given. Furthermore, a detailed and updated discussion of topics such as biocompatibility, toxicity, and biodegradability of ILs is also available in Chapter 7 (Sections 7.1 and 7.2).

As any other chemical or solvent, the production of ILs clearly involves a synthetic process in which some chemical steps are conducted. Therefore, during IL syntheses some reagents are used and some wastes or by-products are formed together with the IL. These waste and by-product formation is crucial from an environmental viewpoint—green chemistry and green engineering—[23,24] and are often not mentioned or even considered when ILs are claimed as "green solvents." In Table 1.1, the principles labeled "PRODUCTIVELY" (green chemistry) and "IMPROVEMENTS" (green engineering) are summarized.[24]

Despite the importance of these green chemistry principles, it has not been until recently that studies focusing on the environmental concerns of IL syntheses were

TABLE 1.1. Principles of Green Chemistry ("PRODUCTIVELY") and Green Engineering ("IMPROVEMENTS"), as Reported in the Literature[24]

"PRODUCTIVELY"	"IMPROVEMENTS"
Green chemistry	Green engineering
Prevent wastes	Inherently nonhazardous and safe
Renewable materials	Minimize material diversity
Omit derivatization steps	Prevention instead of treatment
Degradable chemical products	Renewable material and energy inputs
Use safe synthetic methods	Output-led design
Catalytic reagents	Very simple
Temperature, pressure ambient	Efficient use of mass, energy, space, and time
In-process monitoring	Meet the need
Very few auxiliary substances	Easy to separate by design
E-factor, maximize feed in product	Networks for exchange of local mass and energy
Low toxicity of chemical products	Test the life cycle of the design
Yes, it is safe	Sustainability throughout the product life cycle

reported.[20] Therein, the widely used alkylimidazolium-based ILs were taken as a model, and critical studies regarding their syntheses (*E*-factor and atom economy), purification steps, discoloration, and source of energy applied were carried out. Overall it was concluded that the production of those ILs is far less green than what is usually claimed in the literature dealing with ILs. At laboratory-scale processes, quaternization synthetic approaches may still provide some green footprints if processes are conducted with either microwave or conduction as the energy source.[24] However, in the other cases, conclusions clearly challenged the environmental label that ILs usually have in the literature. It is clearly expected that more environmental studies on the ILs syntheses will be carried out in the coming years and therefore a better picture will emerge.

Moreover, when assessing the greenness of ILs other important aspects include IL release, eco-toxicity, biodegradability, bioaccumulation, and spatiotemporal range in the milieu.[22] Although it can be expected that environmental release of ILs could be easily controlled compared with VOCs—by virtue of the almost negligible vapor pressure and volatility of ILs —it is clear that sooner or later some appreciable amounts of ILs will reach the environment (e.g., in wastewater effluents). Therefore, it is crucial to assess how these ILs are going to interact with living organisms. To this end, a number of standardized tests and protocols have been established. They include studies on inhibition of acetylcholinesterase enzymes, luminescence inhibition of the marine bacterium *Vibrio fisheri*, growth rate inhibition of the freshwater green alga *Pseudokirchneriella subcapitata*, cell viability of IPC-81 cells, growth inhibition of duckweed, *Lemna minor*, and an acute test with zebrafish, *Danio rerio*.[21] In addition, products formed during the environmental degradation of ILs must also be considered. It has been reported that some of these degradation products may be even more toxic than the original ILs.[21] In Chapter 7 of this book (Sections 7.1 and 7.2), detailed information on the state-of-the-art of these aspects is provided.

A general conclusion that can be set herein is that commonly used ILs are far less green than what they are usually claimed in publications. However, once again it has to be mentioned that the huge versatility of ILs (tunability, tailored properties) might be used to provide greener ILs than the current ones. In general, these environmental concerns are starting to shift research in the field to the production of more sustainable ILs. An envisaged future challenge will be to design ILs that maintain their promising physicochemical properties and potential applications while providing greener footprints, both in terms of *E*-factors and in terms of degradability, toxicity, and so on.[19] In this respect, an alternative that has emerged is the production of ILs starting from natural sources as substrates such as amino acids, carboxylic acids, and sugar-based structures.[25] Herein, natural amino acids have been used extensively[26] because of their interesting tunable chemical properties, greater affordability, and high compatibility for living organisms. In addition, amino acids incorporate chiral centers into the IL, which may add other interesting properties with promising applications. Thus, amino acid-based ILs represent interesting examples of chiral bio-based ILs.[27] Another alternative is to use amino acids as starting materials for the synthesis of ILs. Herein, albeit substrates are obviously environmentally friendly, attention to subsequent synthetic procedures to afford the final derivatives should be taken into account in a case-by-case scenario. In Figure 1.3 some examples of ILs derived from amino acids are depicted.

In the quest for ILs that could deserve the label "green solvents," another important approach is represented by designing ILs that can be not only entirely composed of biomaterials, but also involve derivatization steps that may add limited environmental concerns to IL production. An example of this strategy is the use of available and inexpensive choline hydroxide as staring material. The production of choline-based ILs are conducted simply by substituting choline hydroxide with the correspondent (naturally occurring) carboxylic acid at ambient temperature, producing water as the only by-product of the process (Figure 1.4).[28,29]

Therefore, the design of ILs provides an enormous possibility for tailored ILs, even when environmental concerns are considered, and a useful but green solvent is envisaged. In conclusion, although ILs cannot be generally regarded as safe (green) solvents, it is possible to design ILs that can meet environmental requirements. Given the current trends in environmental processes and green chemistry, it is clear that this will be the most important and sustainable line of development for ILs in the coming years. Furthermore, the design of ILs that can be used for certain applications while maintaining acceptable environmental footprints will be crucial. Some examples of these combinations have just appeared in the field of biocatalysis, including the design of enzyme-friendly choline-based ILs that are able to dissolve cellulose at the same time (AMMOENG 110™, Figure 1.5).[30] This combination provides a promising frame to undertake biocatalytic reactions in reaction media that can be compatible with the substrates/products employed. This approach is discussed more extensively in Chapter 2 (Section 2.6) and Chapter 5 (Sections 5.4 and 5.5).

As stated in the previous section, together with bio-based ILs, another important field in the third generation of ILs are represented by DES.[5–7,9] DES are a combination of a room-temperature salt (e.g., choline chloride) together with a hydrogen-bond-forming molecule (e.g., urea, glycerol, and carboxylic acids). When these structures are

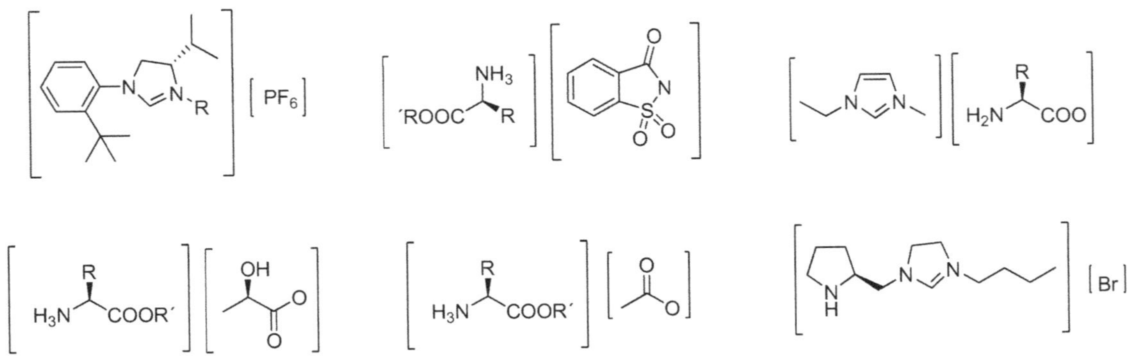

Figure 1.3. Some examples of chiral ionic liquids using amino acids (or derivatives thereof) as substrates.[27]

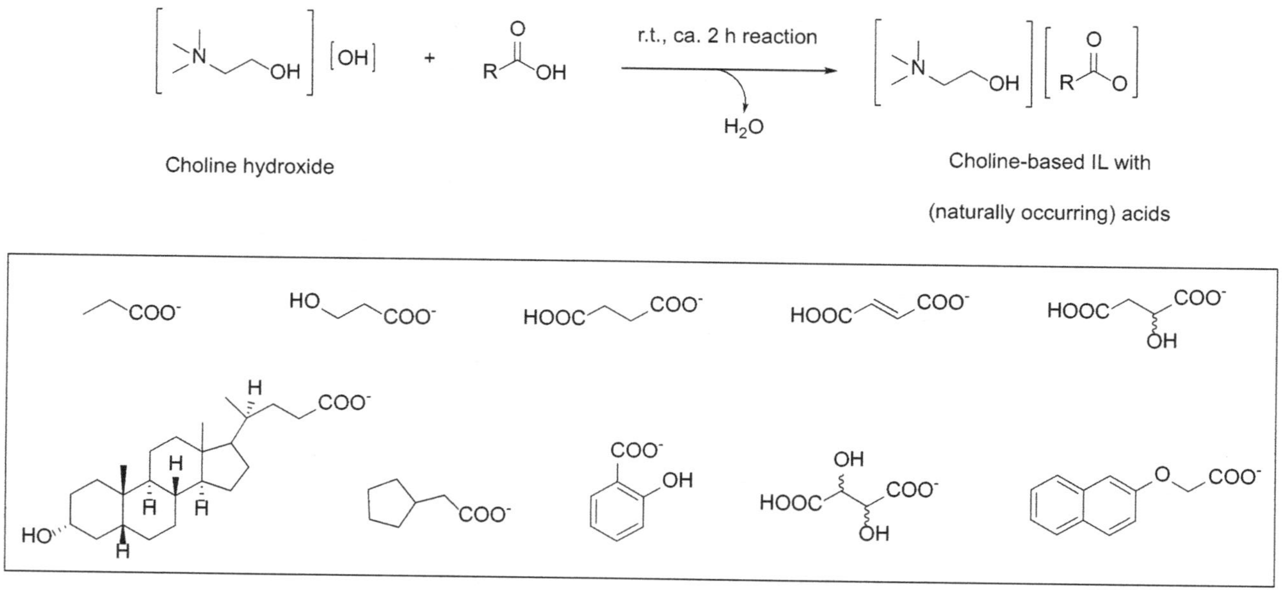

Figure 1.4. Choline-based ionic liquids. Some of the carboxylic acids are naturally occurring, thus providing a 100% bio-based approach for the production of neoteric solvents.[28,29]

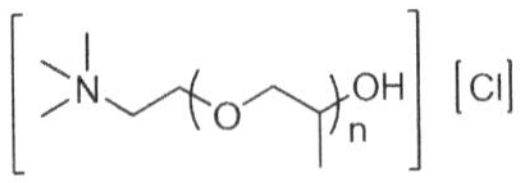

n= 5-15

Figure 1.5. Structure of AMMOENG 110™ recently reported as an enzyme-friendly solvent for cellulose dissolution.[30]

mixed in different proportions, melting points drop, thus providing liquid solvents at different temperatures (even at room temperature). Herein, combinations of choline chloride with urea, malonic acid, ethylene glycol, or glycerol have been successfully reported as DES, showing that a broad range of new solvents can be produced. Because of the ease of their synthesis (a simple mixing of components is often enough) together with the complete biodegradability of their components, it is clear that DES represent a promising entry in the field of tunable green solvents (see also Chapter 5, Section 5.2.3.6, and Chapter 7, Section 7.4). Furthermore, prices for DES are expected to be much lower than those for ILs. Although much more research is needed, it is often considered that many of the observed properties of second-generation ILs will also be valid for both third-generation ILs and DES solvents.

1.3 CONTEXT OF ILs IN BIOTRANSFORMATIONS AND ORGANOCATALYSIS

The fields of biotransformations and organocatalysis have emerged in the last decades as promising alternatives for the environmental-friendly production of different chemicals, not only in the fine-chemical or pharmaceutical arena, but also in the production of other important low-added value commodities.[31–34] Enzymes and many organocatalysts are benign and nontoxic catalysts, which enable the production of relevant compounds, building blocks, and commodities with reduced environmental concerns. These catalysts usually display high enantio-, regio-, or chemoselectivities, operating under mild reaction conditions. As an extension of these fields, ILs have also been largely considered useful alternatives for many applications in biotransformations and organocatalysis.

The first reported combination of enzymes and ILs was published in 1984.[35] In that contribution, the stability and activity of alkaline phosphatase in water solutions, in which different amounts of the *fused salt* EAN, $[EtNH_3][NO_3]$, had been added, was studied. The article showed an unexpectedly high stability of the enzyme under different proportions of EAN. Decades later, the enzyme-catalyzed formation of *Z*-aspartame in $[BMIM][PF_6]$, catalyzed by thermolysin, was reported, demonstrating that ILs may be excellent reaction media for biocatalysis, in terms of enhanced activity, stability, or selectivity.[36] Since then, an impressive number of applications dealing with ILs and biocatalysis have been reported either as patents or as publications. Both types of biocatalysts, free enzymes and whole cells, have been studied. ILs have been used as solvents or co-solvents for biocatalytic reactions, as additives to enhance stability, activity, or selectivity of enzymes, or even as reaction (biphasic) media for whole-cell processes. Most of the applications in this area deal with second-generation ILs. Interestingly, first assessments of biocatalytic reactions using third-generation ILs or DES have started to appear.[5,37–40] Therefore, it is expected that ILs will continue to make an impact in the field of enzyme catalysis since the enormous knowledge being developed with second-generation ILs will be a proper starting point to set up biocatalytic applications with more sustainable and economically accessible third-generation ILs and DES. Chapters 2–8 of this book deal in detail with the use of ILs in biocatalysis, covering in a broad extent all the applications, options, and synergies that the use of ILs may bring the field of enzyme catalysis and biotechnology.

On the other hand, the use of ILs in organocatalysis has also been reported, as documented by several recent reviews.[41,42] The first applications of ILs in organocatalysis were as solvents for organocatalytic reactions. Later on other interesting applications, such as the use of ILs as support for organocatalyst immobilization, were reported. The combination of properties of ILs—finely tuned by selection of cations and anions—with exquisite catalysis that organocatalysts may bring may be a powerful synergy for envisaging new applications. The topics are well covered in Chapters 9 and 10, which deal exclusively with ILs and organocatalysis.

REFERENCES

1 P. Walden, *Bull. Acad. Imper. Sci. (St. Petesburg)* **1914**, *1800*, 405–422.

2 C. Graenacher, US 1943176. **1934**.

3 A. Pinkert, K. N. Marsh, S. Pang, M. P. Staiger, *Chem. Rev.* **2009**, *109*, 6712–6728.

4 J. S. Wilkes, *Green Chem.* **2002**, *4*, 73–80.

5 J. Gorke, F. Srienc, R. J. Kazlauskas, *Biotechnol. Bioprocess Eng.* **2010**, *15*, 40–53.

6 A. P. Abott, D. Boothby, G. Capper, D. L. Davies, R. K. Rasheed, *J. Am. Chem. Soc.* **2004**, *126*, 9142–9147.

7 A. P. Abott, G. Capper, D. L. Davies, R. K. Rasheed, V. Tambyrajah, *Chem. Commun.* **2003**, 70–71.

8 B. Ni, A. D. Headley, *Chemistry* **2010**, *16*, 4426–4436.

9 H. Olivier-Bourbigou, L. Magna, D. Morvan, *Appl. Catal. A* **2010**, *373*, 1–56.

10 V. I. Parvulescu, C. Hardacre, *Chem. Rev.* **2007**, *107*, 2615–2665.

11 S. Chowdhury, R. S. Mohan, J. L. Scott, *Tetrahedron* **2007**, *63*, 2363–2389.

12 Z. C. Zhang, *Adv. Catal.* **2006**, *49*, 153–237.

13 P. Wasserscheid, T. Welton, *Ionic Liquids in Synthesis*, Wiley-VCH, Weinheim, Germany, **2003**.

14 J. Dupont, R. F. de Souza, P. A. Z. Suarez, *Chem. Rev.* **2002**, *102*, 3667–3692.

15 P. Wasserscheid, W. Keim, *Angew. Chem. Int. Ed. Engl.* **2000**, *39*, 3772–3789.

16 C. Baudequin, D. Bregeon, J. Levillain, F. Guillen, J. C. Plaquevent, A. C. Gaumont, *Tetrahedron Asymmetry* **2005**, *16*, 3921–3945.

17 M. J. Earle, J. M. S. S. Esperança, M. A. Gilea, J. N. Canongia Lopes, L. P. N. Rebelo, J. W. Magee, K. R. Seddon, J. A. Widegren, *Nature* **2006**, *439*, 831–834.

18 M. Smiglak, W. M. Reichert, J. D. Holbrey, J. S. Wilkes, L. Sun, J. S. Thrasher, K. Kirichenko, S. Singh, A. R. Katrizky, R. D. Rogers, *Chem. Commun.* **2006**, 2554–2556.

19 N. Wood, G. Stephens, *Phys. Chem. Chem. Phys.* **2010**, *12*, 1670–1674.

20 M. Deetlefs, K. R. Seddon, *Green Chem.* **2010**, *12*, 17–30.

21 T. P. T. Pham, C. W. Cho, Y. S. Yun, *Water Res.* **2010**, *44*, 352–372.

22 J. Ranke, S. Stolte, R. Störmann, J. Arning, B. Jastorff, *Chem. Rev.* **2007**, *107*, 2183–2206.

23 P. T. Anastas, J. B. Zimmerman, *Environ. Sci. Technol.* **2003**, *37*, 94A.

24 S. Tang, R. Bourne, R. Smith, M. Poliakoff, *Green Chem.* **2008**, *10*, 268–269.

25 S. T. Handy, *Chemistry* **2003**, *9*, 2938–2944.

26 K. Fukumoto, M. Yoshizawa, H. Ohno, *J. Am. Chem. Soc.* **2006**, *127*, 2398–2399.

27 X. Chen, X. Li, A. Hu, F. Wang, *Tetrahedron Asymmetry* **2008**, *19*, 1–14.

28 Y. Fukaya, Y. Lizuka, K. Sekikawa, H. Ohno, *Green Chem.* **2007**, *9*, 1155–1157.

29 Y. Yu, X. Lu, Q. Zhou, K. Dong, H. Yao, S. Zhang, *Chemistry* **2008**, *14*, 11174–11182.

30 H. Zhao, G. A. Baker, Z. Song, O. Olubajo, T. Crittle, D. Peters, *Green Chem.* **2008**, *10*, 696–705.

31 A. Liese, K. Seebach, C. Wandrey, *Industrial Biotransformations*, Wiley-VCH, Weinheim, Germany, **2006**.

32 K. Faber, *Biotransformations in Organic Chemistry*, Springer Verlag, Berlin, **2004**.

33 A. Berkessel, H. Gröger, *Asymmetric Organocatalysis*, Wiley-VCH, Weinheim, Germany, **2005**.

34 M. T. Reetz, B. List, S. Jaroch, H. Weinmann, eds., *Organocatalysis*, Springer-Verlag, Berlin, Germany, **2008**.

35 D. K. Magnuson, J. W. Bodley, D. Fernell Evans, *J. Solut. Chem.* **1984**, *13*, 583–587.

36 M. Erbeldinger, A. J. Mesiano, A. J. Rusell, *Biotechnol. Prog.* **2000**, *16*, 1129–1131.

37 M. Moniruzzaman, K. Nakashima, N. Kamiya, M. Goto, *Biochem. Eng. J.* **2010**, *48*, 295–314.

38 P. Lozano, *Green Chem.* **2010**, *12*, 555–569.

39 M. Suershkumar, C. K. Lee, *J. Mol. Catal., B Enzym.* **2009**, *60*, 1–12.

40 F. van Rantwijk, R. A. Sheldon, *Chem. Rev.* **2007**, *107*, 2757–2785.

41 S. Toma, M. Meciarova, R. Sebesta, *Eur. J. Org. Chem.* **2009**, 321–327.

42 P. Domínguez de María, *Angew. Chem. Int. Ed. Engl.* **2008**, *47*, 6960–6968.

2

IONIC LIQUIDS AND PROTEINS: ACADEMIC AND SOME PRACTICAL INTERACTIONS

Zhen Yang

ABBREVIATIONS FOR IONIC LIQUID CATIONS

$[BBIm]^+$	1,2-dibutylimidazolium
$[BdMIm]^+$	1-butyl-2,3-dimethylimidazolium
$[BMIm]^+$	1-butyl-3-methylimidazolium
$[BMPy]^+$	N-butyl-3-methypyridinium
$[Btmsim]^+$	1-butyl-3-trimethylsilylimidazolium
$[C_{1111}N]^+$	tetramethylammonium
$[C_{2222}N]^+$	tetraethylammonium
$[C_{3333}N]^+$	tetrapropylammonium
$[C_{4444}N]^+$	tetrabutylammonium
$[CPMA]^+$	cocosalkyl pentaethoxy methyl ammonium
$[DEA]^+$	diethanolammonium

Ionic Liquids in Biotransformations and Organocatalysis: Solvents and Beyond, First Edition. Edited by Pablo Domínguez de María.

$[EdMim]^+$	1-ethyl-2,3-dimethylimidazolium
$[EMIm]^+$	1-ethyl-3-methylimidazolium
$[EtNH_3]^+$	ethylammonium
$[HMIm]^+$	1-hexyl-3-methylimidazolium
$[MMIm]^+$	1,3-dimethylimidazolium
$[OMIm]^+$	1-octyl-3-methylimidazolium
$[ONIm]^+$	1-octyl-3-nonylimidazolium
$[MEBu_3P]^+$	2-methoxyethyl(tri-n-butyl)phosphonium
$[mmep]^+$	1-methyl-1-(-2-methoxyethyl)pyrrolidinium
$[MTOA]^+$	methyltrioctylammonium
$[N(CH_2CH_2OH)_4]^+$	tetrakis (2-hydroxyethyl) ammonium
$[TBA]^+$	tetrabutylammonium

ABBREVIATIONS FOR IONIC LIQUID ANIONS

$[BF_4]^-$	tetrafluoroborate
$[CB]^{3-}$	Cibacron Blue 3GA
$[dbp]^-$	dibutylphosphate
$[dhp]^-$	dihydrophosphate
$[PF_6]^-$	hexafluorophosphate
$[SbF_6]^-$	hexafluoroantimonate
$[Tf_2N]^-$	bis[(trifluoromethyl)sulfonyl]amide
$[TfO]^-$	trifluoromethanesulfonate
$[TOS]^-$	tosylate
$[MDEGSO_4]^-$	2-(2-methoxyethoxy) ethylsulfate
$[MeSO_3]^-$	methylsulfonate
$[MeSO_4]^-$	methylsulfate
$[OctSO_4]^-$	octylsulfate

ABBREVIATIONS FOR AMMONIUM IONIC LIQUIDS

EAF	ethylammonium formate
EAN	ethylammonium nitrate
EtAF	ethanolammonium formate
MOEAF	2-methoxyethylammonium formate
PAF	propylammonium formate
TEAMS	triethylammonium methylsulfonate

OTHER ABBREVIATIONS

AAIL	amino acid ionic liquid
AIL	aprotic ionic liquid
ABS	aqueous biphasic system
AOT	sodium bis(2-ethylhexyl) sulfosuccinate
ATR-FTIR	attenuated total reflection Fourier transform infrared spectroscopy
a_w	thermodynamic water activity
B_+, B_-	viscosity *B*-coefficient for cation and anion, respectively
BSA	bovine serum albumin
CALB	*Candida antarctica* lipase B
CD	circular dichoism
DLS	dynamic light scattering
DMSO	dimethyl sulfoxide
DSC	differential scanning calometry
FT-IR	Fourier transform infrared spectroscopy
GuHCl	guarnidinium chloride
HRP	horseradish peroxidase
HSA	human serum albumin
IC_{50}	half maximal inhibitory concentration
IL	ionic liquid
LCST	lower critical separation temperature behavior
log *P*	logarithm of the partition coefficient of the solvent in an octanol/water mixture
m.p.	melting point
PA	proton activity
PAGE	polyacrylamide gel electrophoresis
PEG	polyethylene glycol
pI	isoelectric point
PIL	protic ionic liquid
SANS	small angle neutron scattering
THF	tetrahydrofuran
TRIS	tris(hydroxymethyl)aminoethane

2.1 INTRODUCTION

Ionic liquids (ILs) are organic salts remaining as liquids under ambient temperatures. Since the first reports on performing biotransformations in ILs,[1,2] interests in using them as a new type of solvents for biocatalysis, either as a solvent or as a co-solvent added to aqueous solution, have been increasing rapidly because of their unique solvent properties and ability to present excellent enzyme activity, stability, and selectivity.[3–10] One of the major attractions of making ILs an alternative to conventional organic solvents is their "designer solvent" properties; that is, the physical and chemical properties of the ILs, including their polarity, hydrophobicity, viscosity, and solvent miscibility, can be finely tuned by appropriate modification of their cations and anions. This advantage allows an IL to be designed for a specific reaction, such as to solubilize the enzyme for homogeneous catalysis, to increase the substrate solubility, to modify the enzyme selectivity, to enhance the enzyme stability, or to tailor the reaction rate.

In order to take full advantage of these perceived benefits, it is necessary to have a thorough understanding of the fundamental interactions between an IL and an enzyme/protein. The objective of this chapter is to take a comprehensive survey of IL–protein interactions acquired from detailed discussions concerning Hofmeister effects on biocatalysis, impacts of ILs on structure and function of enzymes/proteins, and use of ILs for protein extraction. It is hoped that the information collected here can be used to provide valuable guidelines for designing biocompatible ILs for biotransformations.

2.2 IONIC LIQUIDS, WATER, AND PROTEINS

Biocatalysis in ILs is a complicated system involving ILs, water, and proteins. Our discussion of IL–protein interactions should start from the introduction of each of these three major components, their individual existent states, their specific roles in this biocatalytic system, and their interactions with each other.

2.2.1 Ionic Nature of Ionic Liquids

Perhaps the most distinct feature that differentiates ILs from conventional molecular organic solvents is their ionic nature. Based on a series of experimental data, Dupont[11] proposed that a pure 1,3-dialkylimidazolium IL takes a general structural pattern of $\{[(DAI)_x(X)_{x-n})]^{n+}[(DAI)_{x-n}(X)_x)]^{n-}\}_n$, where DAI is the 1,3-dialkylimidazolium cation and X is the anion: Both its cations and anions are associated with each other via electrostatic and hydrophobic interactions, and more importantly, H-bonding network, to form a polymeric supramolecular structure; one imidazolium hydrogen is bonded to at least three anions and one anion hydrogen bonded to at least three cations, although the number of anions that surround the cation (and vice versa) can change depending upon the anion size and the type of the N-alkyl imidazolium substituents. The strength of the H-bonds between imidazolium cation and its counter anion follows the order of $CF_3CO_2^- > BF_4^- > PF_6^- > BPh_4^-$. This structural pattern is maintained in the solid, liquid, and gas phases. Introduction of other molecules (such as water) and macromolecules (such as proteins) results in a disruption of the H-bond network and in some cases the generation of nanostructures with polar and nonpolar regions. Further addition of other molecules may lead to collapse of the H-bonding supramolecular structure of the IL, thus forming triple ions and even contact ion pairs. Eventually, solvent-separated free ions are formed under infinite dilution in a solvent such as water.

A similar phenomenon was found by Katayanagi et al. when conducting a thermodynamic study on the hydration characteristics of [BMIm][BF_4] and [BMIm][I].[12] On dissolution into water at infinite dilution, both ILs dissociate completely into free ions, although the subsequent hydration is much weaker than for NaCl. As the IL concentration increases, up to a threshold mole fraction ($X_{IL} = 0.015$ for [BMIm][BF_4] and 0.013 for [BMIm][I]), both the IL cations and anions tend to interact with each other. Later on, the IL ions start to organize themselves, either directly or in a H_2O-mediated manner. Finally, at $X_{IL} = 0.5$–0.6, IL clusters are formed with a very similar local arrangement to that in the pure IL state. Zhang et al.[13] investigated the hydration process of [EMIm][BF_4] via two-dimensional vibration spectroscopy, and their results revealed that upon dissolution in water, the three-dimensional structure of the IL is first gradually destroyed to form ionic clusters, which are then further dissociated to ion pairs surrounded by water molecules.

Although studies concerning the hydration process of other IL types are rarely available, it is believed that upon solvation in water they follow a similar pattern to imidazolium ILs.

2.2.2 Protic and Aprotic Ionic Liquids

All the above-mentioned ILs are apolar ionic liquids (AILs). Protic ionic liquids (PILs) are a subset of ILs formed by the equimolar mixing of a Brønsted acid (HA) and a Brønsted base (B) through proton transfer from the acid to the base:

$$B + HA \rightarrow HB^+ + A^- \leftrightarrow [HB][A]$$

In fact, the first IL discovered by Gabriel and Weimer in 1888,[14] ethanolammonium nitrate (m.p. 52–55°C), was a PIL. Ethylammonium nitrate (EAN, [$EtNH_3$][NO_3], m.p. 12.5°C), invented by Walden in 1914,[15] was the first truly room-temperature IL discovered, which has been the focus of most of the PIL investigations because this solvent has water-like characteristics, including a capability for H-bonding, as well as an ionic and hydrophobic character. Figure 2.1 shows the typical cations that are involved in AILs and PILs. PILs are currently under intense study in a variety of applications.[16]

The key property that distinguishes a PIL from an AIL is the presence of both proton-donor and proton-acceptor sites in the PIL, making it ready for the buildup of H-bonding network within itself or with protein molecules. The PIL in its pure state may appear in the form of aggregates (with an aggregation number of 1.1–7.0), neutral species (due to incomplete proton transfer), and dissociated ions. Therefore, as pointed out by MacFarlane and Seddon,[17] PILs challenge our concept of true ILs which are composed entirely of ionic components, and have to be taken as liquid mixtures because a variety of proton transfer and association equilibria are involved in them. As has been discussed in Greaves and Drummond's recent review,[16] PILs possess a number of unique properties compared with AILs, among which the acidity is worth a special attention, as this has to be taken into consideration when selecting a PIL as a solvent. The Brønsted acidity (or proton activity [PA]) of a PIL can be determined by measuring the NH shift ($\delta_{N\text{-}H}$) for the available proton in the PIL by using ^{1}H NMR—a larger shift indicates a lower Brønsted acidity—and can be adjusted by the choice of Brønsted base and Brønsted acid used in PIL formation.[18]

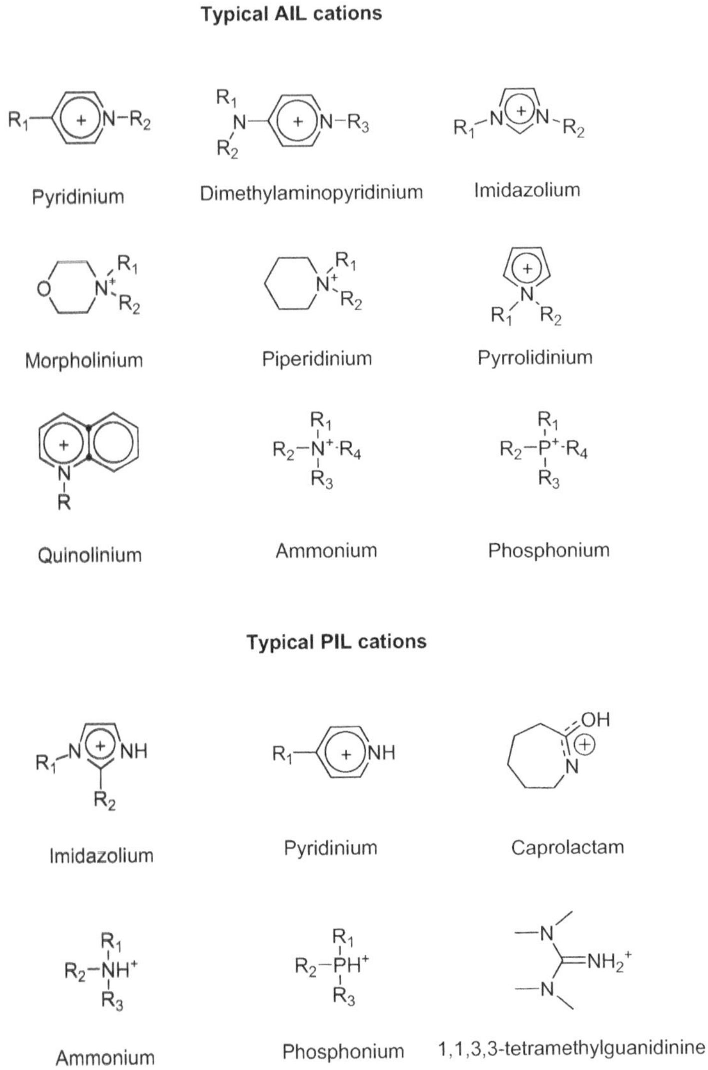

Figure 2.1. Cations typically used in protic and aprotic ionic liquids (PILs and AILs, respectively).

Because of their unique properties, PILs (especially EAN) have found applications in a number of protein studies such as folding/unfolding,[18,19] renaturation,[20] and crystallization,[21] which will be discussed later in Section 2.4. Although there have been numerous publications about investigating biocatalysis in AILs such as imidazolium ILs, very few have been involved in the use of PILs as reaction media for biotransformations. EAN has been reported to be a poor solvent for lipases, due to the dissolution and unfolding of the enzyme caused by the IL anion NO_3^-, which is a strong H-bond acceptor.[126] However, a recent report showed that subtilisin is active in a PIL, diethanolammonium chloride, [DEA][Cl], which can not only dissolve the enzyme but also retain its activity and structure.[22] Mann et al.[23] have studied the effect of four ammonium PILs on both refolding and activity of lysozyme, and have found that the enzyme activity increased in the presence of all the ILs tested, among which ethanolammonium formate (EtAF) behaved the best, not only in presenting the highest enzyme activity but also in stabilizing the protein against unfolding at high temperatures and in promoting protein renaturation upon cooling.

2.2.3 Water Present in the Ionic Liquids

The presence of water in ILs may have significant implications for their properties as solvents such as polarity, viscosity, and conductivity. For instance, an increase in the water mole fraction in [BMIm][PF_6], [BMIm][Tf_2N], and [EMIm][Tf_2N] from 0 to 0.25 is accompanied by a reduction of 65%, 34%, and 30% in the viscosity of the three ILs, respectively.[24]

ILs can easily absorb water from atmosphere, and the amount of water uptake is dependent not only on the humidity and temperature in the environments but also on the nature of the ILs. However, it is not certain yet whether it is the cation or the anion of an IL that is bound by water and is the major determinant for water uptake, and some experiments have presented different, or even contradictory, results. By studying the attenuated total reflection Fourier transform infrared spectroscopy (ATR-FTIR) spectra of water absorbed from air into a series of imidazolium ILs with different anions, Cammarata et al.[25] have concluded that the water molecules are present in the free (not self-associated) state, binding to the IL anions via H-bonding, and that they can form bulk liquid-like associated aggregates in ILs with anions of strong basicity, such as $[NO_3]^-$ and $[CF_3CO_2]^-$. The IL cation, especially the acidic H at C_2 of the imidazolium ring, is not directly involved in water absorption and H-bonding with water molecules. The strength of H-bonding between water molecules and anions increases in the order $[PF_6]^- < [SbF_6]^- < [BF_4]^- < [Tf_2N]^- < [ClO_4]^- < [CF_3SO_3]^- < [NO_3]^- < [CF_3CO_2]^-$. On the other hand, Miki et al.[26] conducted a thermodynamics study to elucidate the effect of [BMIm][Cl] on the molecular organization of H_2O and found that the effect of the IL cation is conspicuous whereas that of the anion is not. Thus, they speculated that in the H_2O-rich composition region the delocalized positive charge of the imidazolium ring attracts some of the water molecules; as a result, the H-bond probability of bulk H_2O is reduced away from solutes.

In fact, we assume that the water molecules present in the IL may bind to either the cation or the anion, depending on their kosmotropicity/chaotropicity, H-bond acidity/basicity, and nucleophilicity.

2.2.4 Interactions of Water and Ionic Liquids with Proteins

It is necessary to have a brief review first about how the interactions of water and organic solvents with proteins affect the enzyme catalysis in organic media. An essential amount of water is required for the enzyme to promote its activity in an organic solvent. The water added to a nonaqueous enzymatic system is actually distributed between enzyme, solvent, substrate, and support (if present), and it is the amount of the water associated with the enzyme that determines the enzyme performance.[27] Different enzymes require different amounts of water, varying from a few to several hundred water molecules per enzyme molecule.[27–29] Thermodynamic water activity (a_w) has been generally considered as a parameter to quantify the hydration level on the enzyme.[30] Each enzyme presents an optimal water activity, which is not solvent-dependent but again varies from enzyme to enzyme.[31] Therefore, in order to avoid the influence of water in the study of solvent effect, it is necessary to compare enzyme activity in different organic solvents when the hydration level of the enzyme (or the water activity) is kept constant. As stated in our previous review,[32] when an enzyme is applied to the solvent system, the solvent generally affects the enzyme performance via interactions with the water, the enzyme, the substrates, and the products that are present in the reaction system.

So far, different types of enzymes have been demonstrated to show catalytic activity in ILs, and it seems evident that enzymes in ILs basically follow the same catalytic mechanism as in water and in organic solvents.[33–35] It is thus reasonable to surmise that an enzyme in an IL, as in an organic solvent, functions with a microaqueous phase associated with the enzyme molecule and that the substrate must penetrate into this aqueous layer to get direct contact with the enzyme for the reaction to occur while the products have to pass out to drive the reaction forward. Accordingly, an essential amount of water is also required for enzyme functioning in ILs, which are expected to play a role similar to that of organic solvents in affecting the enzyme performance. It is therefore necessary to examine the role of water, water activity, and ILs in the biocatalytic IL-based system.

2.2.4.1 Effect of Water and Water Activity. Our recent experiments have confirmed that the hydrolytic activity of *Penicillium expansum* lipase (PEL) in [BMIm][PF_6] is sensitive to the amount of water added.[36] For both solvents [BMIm][PF_6] and hexane, there is a similar bell-shaped relationship between enzyme activity and the amount of water added to the reaction system, although the reaction rates obtained in the IL are much higher (Figure 2.2). The enzyme requires a greater optimal water content in the IL system (2.3%, v/v) than that for the hexane system (0.2%, v/v). Water is a co-substrate for the PEL-catalyzed hydrolysis reaction, and this can partly account for the initial enhancement of the enzyme hydrolytic activity with an increase in the water content. In ILs, water also plays an important role in affecting the enzyme activity, like in organic solvents:[32] An essential amount of water may activate the enzyme by increasing the polarity and structural flexibility of the enzyme active site; but too much water is harmful to the enzyme by facilitating enzyme aggregation, thus diminishing the substrate diffusion and eventually leading to enzyme inactivation.

A previous report deals with the examination of the activity of *Candida rugosa* lipase to catalyze an esterification reaction in three organic solvents and three ILs.[37] The bell-shaped reaction rate versus water content curves confirmed that water is indeed required for optimizing enzyme activity in ILs as in organic solvents. The optimal water

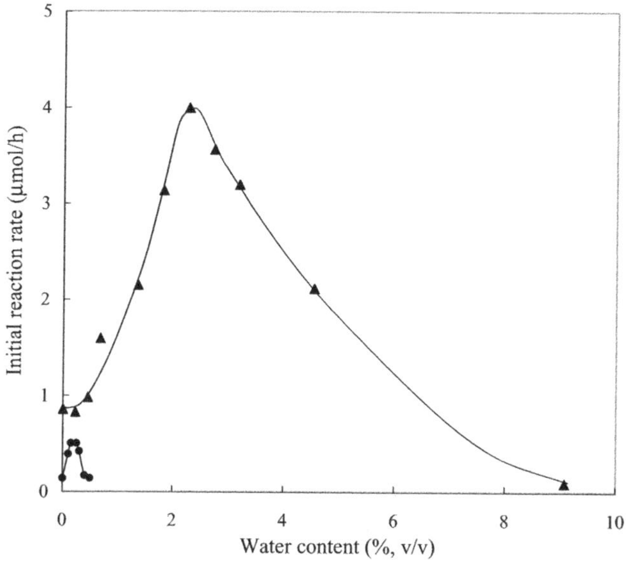

Figure 2.2. Dependence of the hydrolytic activity of *Penicillium expansum* lipase in hexane (•) and in [BMIm][PF_6] (▲) on the amount of water added to the reaction system. (Adapted with permission from Yang et al.[36]).

content varies following the order hexane < toluene < [ONIm][PF_6] < [BMIm][PF_6] ~ THF < [BMIm][BF_4], which is consistent with the hydrophobicity order of the solvents involved. This verifies the high polarity of ILs, which is as strong as lower alcohols and formamide.[5] Among the three ILs examined, [ONIm][PF_6] is the most hydrophobic, with two long alkyl chains attached to the imidazolium ring, and hence requires the least amount of water than [BMIm][PF_6]; and [BMIm][BF_4] requires the maximal amount of water simply because it is water-miscible and has the strongest power to compete with the enzyme for the water.

Kaftzik et al.[38] have determined the water amount in the reaction mixture as a function of water activity for the two ILs, [MMIm][$MeSO_4$] and [BMIm][$OctSO_4$]. At a similar water activity level, [MMIm][$MeSO_4$] always takes up more water than [BMIm][$OctSO_4$]. Because both the cation and the anion of [BMIm][$OctSO_4$] are more hydrophobic than those of the other IL, this experiment confirms that, like organic solvents, a more hydrophobic IL requires less water to achieve the same water activity as compared with a less hydrophobic IL. In fact, ILs are so polar that they have a great ability to retain water. For instance, the water activity of the IL [CPMA][$MeSO_4$] at 40°C was slowly increased from 0.2 to about 0.5 when the water content was raised abruptly from 1% to 8% (v/v).[39] It is therefore easy to understand why as compared with nonpolar organic solvents such as hexane and toluene, an IL system always requires a higher water content for an optimal enzyme activity, as shown in the literature.[36,37]

A number of enzymes have shown their activities responding sensitively upon a change in the water activity. Tyrosinase presented a higher enzyme activity in [BMIm][PF_6] at a higher water activity,[33] consistent with our previous findings about the effect of water activity on the activity of the same enzyme in $CHCl_3$. The activity of chymotrypsin in [BMIm][Tf_2N] and [EMIm][Tf_2N] increased with an increase in the water activity,[40] and the same occurred to *Candida antarctica* lipase B (CALB) in [BMIm][Tf_2N][41] and mandelate racemase in [MMIm][$MeSO_4$].[38] When an immobilized CALB (Novozym® 435) was used to catalyze the esterification between geraniol and acetic acid, the optimal water activity for the enzyme in [BMIm][PF_6] was determined to be 0.6, which is higher than the one obtained in hexane ($a_w = 0.1$).[42]

The above last example reporting different optimal water activity results for the same enzyme in both [BMIm][PF_6] and hexane is actually in contrast with what was observed in another study with organic solvents: The initial rate of esterification catalyzed by an immobilized lipase (lipozyme) showed similar dependence on water activity (with the same optimal $a_w = 0.5$) in different organic solvents ranging in polarity from pentan-3-one to hexane.[31] Our recent investigation has also revealed that while PEL in hexane was most active in catalyzing a hydrolytic reaction at a low water activity of 0.04, the hydrolytic activity of the enzyme in the IL [BMIm][PF_6] did not show a clear trend following a change in water activity.[36] It is also worth noting that the optimal water activity obtained by Novozym 435 (the same immobilized lipase used in Barahona et al.[42]) catalyzing a transesterification reaction in another IL [OMIm][Tf_2N] was 0.2, again different from the value of 0.6 obtained by the same enzyme catalyzing the esterification reaction in [BMIm][PF_6].[42] It appears that the optimal water activity depends on the enzyme, the reaction type, and the IL used. All these different results obtained in terms of the water activity dependence may lead us to speculate that rather than being a simple nonaqueous solvent, an IL may play some other important roles in affecting the biocatalytic performance. Its ionic nature may be one of the major responsible factors. This will be considered in a later section.

In organic systems, a constant water activity can be achieved with two methods: (1) separate pre-equilibration of substrate solution and enzyme preparation with a saturated aqueous solution of a salt; and (2) direct addition of a salt hydrate to the reaction mixture to act as a water buffer. The latter method was discussed by Halling,[43] who considered that the equilibrium of a salt hydrate pair can provide a constant vapor pressure of water at a fixed temperature. Russell's group has successfully demonstrated that salt hydrates can also be used to control water activity in ILs for the synthesis of 2-ethylhexyl methacrylate catalyzed by immobilized *Candida antarctica* lipase.[34]

In our recent study about the activity of PEL in ILs,[36] we tried to use the salt hydrate method for controlling water activity and incidentally found that direct addition of salt hydrates into hexane and [BMIm][PF_6] presented different impacts on the enzyme activity (Figure 2.3). In hexane, as expected, the enzyme activity correlated well with the water activity provided by the salt hydrates, being optimal at $a_w = 0.04$; in the IL, however, instead of showing an obvious V–a_w correlation, the enzyme showed a clear trend of increasing activity upon an increase in the ($B_- - B_+$) value of the salts added (B_+ and B_- are viscosity *B*-coefficients for the cation and anion of the salt hydrate, respectively, which will be introduced in Section 2.3). We suspect that this could be

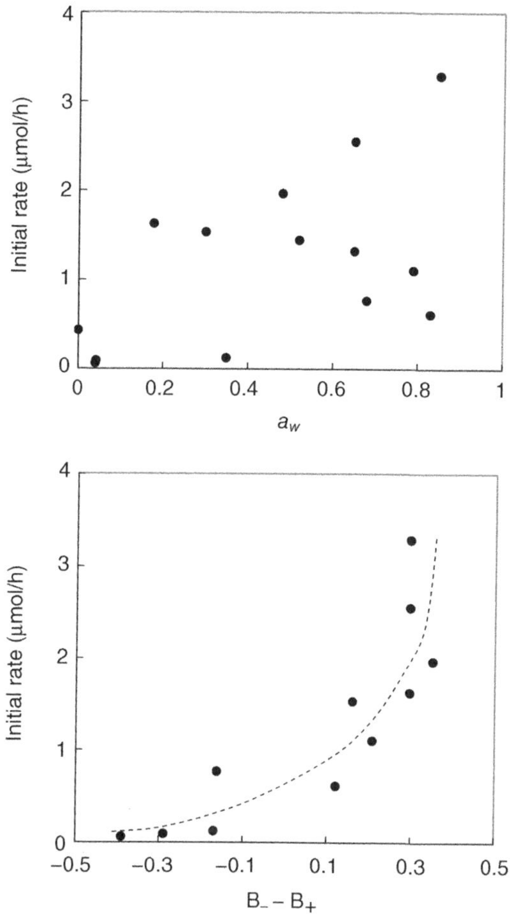

Figure 2.3. Correlation between the hydrolytic activity of *Penicillium expansum* lipase in [BMIm][PF_6] with the water activity (a_w) provided by the salt hydrate added to the reaction system (top) and the ($B_- - B_+$) value of the salt hydrate (bottom). Reproduced with permission from Yang et al.[36].

related to the dual function of the salt hydrates: the water buffering effect and the specific ion effect (Hofmeister effect). In a nonaqueous system, not only can the salt hydrates exist as intact molecules in the bulk solvent phase to control the hydration level of the enzyme (the water activity) via their hydration/dehydration equilibria, but they can also penetrate into the microaqueous phase surrounding the enzyme molecule and dissolve and dissociate into their corresponding cations and anions, exhibiting their specific ion effect (the Hofmeister effect) on the enzyme. In the case of the enzyme used in this study, the latter effect may be dominating in the IL system, because the IL requires a much greater amount of water for enzyme activity (2.3% v/v in the IL vs. 0.2% v/v in hexane). Therefore, care must be taken when trying to use the salt hydrate method to control water activity in IL systems. The concept of Hofmeister effects and viscosity *B*-coefficient will be introduced later in Section 2.3.

2.2.4.2 Effect of Ionic Liquids. When an enzyme is introduced into an IL in the presence of a small amount of essential water, its molecules, with a monomolecular layer of water surrounding outside, are firmly trapped in the H-bonding network produced by the IL.[44] Nanostructures with polar and nonpolar regions are generated where these inclusion compounds are formed. The entrapped molecules are thus stabilized mainly due to the electronic and steric effects provided by the nanostructures.[11] This may be the major reason why enzymes are usually found to be highly stable in ILs.

As nonaqueous solvents, ILs are assumed to affect enzyme performance via the three interactions in the same way as do organic solvents:[6] (1) their great ability of retaining water allows them to strip off the essential water that is associated with the enzyme, leading to enzyme deactivation; (2) they can penetrate into the microaqueous phase surrounding the enzyme molecules and get direct contact with the enzyme, thereby changing the protein dynamics, the protein conformation, and/or the enzyme's active center; and (3) they can also interact with the substrates and products either by direct reactions with them or by altering their partitioning between the bulk IL phase and the microaqueous phase that surrounds the enzyme molecule. Indeed, at least three IL–protein interactions can be used to account for the increased enantioselectivity and remarkable acceleration of lipase-catalyzed transesterification in both organic solvents and ILs, caused by coating the enzyme with a special IL, [BdMIm][cetyl-PEG10-sulfate]: (1) the coating IL may modify the flexibility of the enzyme molecules due to the long poly(oxyethylene)alkyl chain in the anion of the IL; (2) the cationic part of the IL may bind with the enzyme molecule and cause preferential modification of the enzyme conformation; and (3) the hydration situation of the enzyme can be modified as well due to its interaction with the IL.[45–47]

In view of the ionic nature of ILs, which is the major feature that distinguishes ILs from the molecular organic solvents, one can envision that when penetrating into the aqueous phase surrounding the enzyme molecules, the IL would dissolve and at least partially dissociate into individual cations and anions rather than exists as an intact molecule. The presence of these ions may play an important role in affecting the enzyme action in ILs, presumably by direct interaction with the enzyme molecules, especially via H-bonding, electrostatic and hydrophobic interactions, and/or by modifying the microenvironment in and around the enzyme molecules due to their different physicochemical properties. All these specific interactions will be reflected in the forthcoming sections.

2.3 HOFMEISTER EFFECTS ON BIOCATALYSIS

In fact, specific ion effects have been well known in biology for over a century; it was first reported by Hofmeister that inorganic salts and ions showed different abilities of precipitating proteins, following a recurring sequence now known as the Hofmeister series.[48] The fact that enzymatic activity seemed to follow the Hofmeister series has also been recognized for almost half a century.[49] Indeed, the Hofmeister series has been shown to have a universal utility not only in biochemistry, but also in areas such as physical, colloid, polymer, and surface chemistry. Although there has been an intensive effort devoted to explaining Hofmeister effects at the molecular level both through theoretical and molecular dynamics simulations and through experimental approaches,[50–54] an understanding of such effects is still far from complete.[55] As ILs are composed of ions, it is reasonable to surmise that they also elicit Hofmeister effects on enzymes.[10,56] Here the effect of inorganic salt ions on enzyme performance is first examined, followed by our consideration whether the enzyme behavior in IL-based systems also follows the Hofmeister series.

2.3.1 Hofmeister Effects of Inorganic Salts

At low salt concentrations (up to ~0.01 M), ions affect enzyme performance predominantly via electrostatic interactions. The Hofmeister ion effects become important when the electrostatic forces are screened by higher salt concentrations. Numerous studies (see Yang[10], Zhao[56] for literature reviews) have revealed that the effect of ions on enzyme activity and stability usually follows the Hofmeister series and that an enzyme/protein solution is normally stabilized by kosmotropic anions and chaotropic cations but destabilized by chaotropic anions and kosmotropic cations (Figure 2.4).

Ions are regarded as kosmotropic and chaotropic depending on their abilities to interact with water and to change the water structure by shifting the equilibrium of low- and high-density water (which is more and less structured, respectively). With a high charge density, a kosmotropic ion interacts more strongly with water than water with itself and tends to strengthen the water structure by shifting the water equilibrium to low-density water. The situation is reversed in the case of a chaotropic ion. However, in addition to their electrostatic and hydration forces, the ionic excess polarizabilities

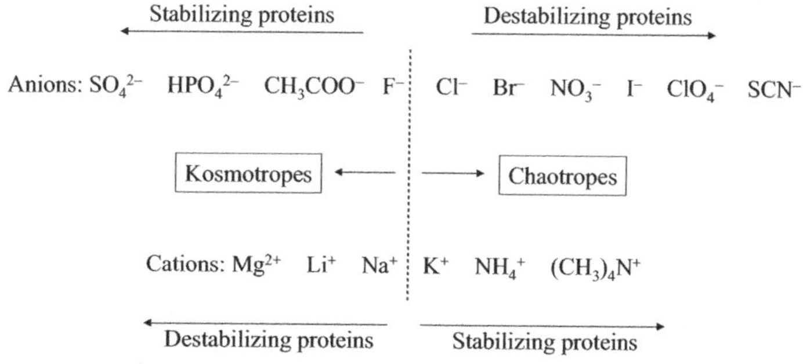

Figure 2.4. The Hofmeister series and classification of ions into kosmotropes and chaotropes.

of ions in solution have also to be considered, because near an interface an ion experiences not only an electrostatic potential, but also a highly specific ionic dispersion potential.[57,58] When salt concentrations are relatively high and the electrostatic contribution is screened, it is this ionic dispersion potential that becomes dominating.[53] Ions show different specificities in water basically because they possess different polarizabilities.

Specifically, an ion may affect the enzyme performance by playing the role of a substrate, a cofactor, or an inhibitor to the enzyme. But more generally, the specific ion effects should be better understood by considering the ability of the ion to alter the bulk water structure, to affect the protein–water interactions, and to directly interact with the enzyme molecules.[10]

2.3.1.1 Quantification of Hofmeister Series. Attempts have been devoted to correlating Hofmeister effects with characteristics of ions and solvents. A number of physicochemical parameters, such as salting out coefficients,[59] lyotropic numbers,[60] melting temperature coefficients,[61] excluded volumes,[62] surface tension increments,[63] and NMR *B′*-coefficients,[64,65] have been proposed for ranking the Hofmeister series. However, the most generally accepted quantitative description of Hofmeister series so far is the viscosity *B*-coefficient of the Jones–Dole equation:[66]

$$\eta/\eta_o = 1 + Ac^{1/2} + Bc + Dc^2$$

Where η and η_o are the viscosities of the salt solution (at the concentration of c) and pure water, respectively. The coefficient A is a constant related to the long-range electrostatic interactions, and its contribution to the viscosity change is predominant only in very dilute solutions. The term Dc^2 is only needed at very high salt concentrations. The coefficient B,[67] characterizing the ion–solvent interactions, is interpreted as a measure of the structure-making and structure-breaking capacity of an ion in solution, thus directly correlating with the kosmotropic/chaotropic properties of the ion. The viscosity *B*-coefficient is normally positive for a kosmotrope and negative for a chaotrope.

Broering and Bommarius considered a number of possible predictors for Hofmeister effects when investigating the deactivation of horse liver alcohol dehydrogenase, α-chymotrypsin, and monomeric red fluorescent protein.[68] They observed that both the deactivation constants (as determined by kinetic studies) and melting temperatures (as determined by differential scanning calorimetry [DSC]) of the three different proteins showed a common relationship with the viscosity *B*-coefficients of the anions in the solution, but not with their surface tension increments. Their further experiments indicated that the protein kinetic deactivation is more dependent on the viscosity *B*-coefficients of the anions rather than on that of the cations.[69]

Such correlation was also observed in nonaqueous enzymology. Ru et al.[70] reported a dramatic activation of subtilisin *Carlsberg* upon co-lyophilization with simple inorganic salts in organic media (from 33-fold activation induced by the chaotropic NaI to 2480-fold by the kosmotropic NaAc) as a function of the viscosity *B*-coefficient.

Lindsay et al.[71] presented a clear trend toward increased activity of penicillin amidase in organic solvents (up to ~35,000-fold activation) by pre-lyophilization with inorganic salts as a function of a greater difference in the viscosity *B*-coefficients of the anions and cations involved in the salts. This strongly supports the comment that enzyme activation is favored by the combination of a highly chaotropic cation and a highly kosmotropic anion.

2.3.1.2 *Effect of Ions on Protein Stability.*

The abilities of the ion to affect the protein–water interaction and to interact directly with the protein molecule can conspire together to decipher the Hofmeister ion effect on protein stability. A protein dissolved in aqueous solution has on its surface many hydrophilic and polar moieties, especially charged groups, responsible for hydration and interactions with the ions in solution. These groups can be divided mainly into two entities: the chaotropic amide and amino groups and the kosmotropic carboxylic groups.[72] According to the "law of matching water affinity,"[73,74] oppositely charged ions tend to form contact ion pairs in solution if they have equal affinities for water but will separate if their water affinities are very different, because the interactions in aqueous salt solution follows the order: kosmotrope–kosmotrope > kosmotrope–water > water–water > chaotrope–water > chaotrope–chaotrope. This suggests that a kosmotropic anion and a kosmotropic cation in aqueous solution will bind together due to the strong interactions between them two, but a cation and an anion with equal chaotropicities also tend to form an ion pair because the relatively strong water–water interactions will keep the two chaotropes together.

The stabilization effect of kosmotropic anions can be well illustrated by preferential hydration.[75] Because of its strong interaction with water and low affinity for the chaotropic amide groups on the protein surface, a kosmotropic anion like SO_4^{2-} competes effectively for the water molecules originally associated with the protein molecule, together being excluded from the protein surface. This encourages the protein molecule to minimize its surface area exposed to the solvent, thereby favoring the acquirement of its native compact state due to the restored driving force of the hydrophobic effect.[54] Unlike its kosmotropic partner, however, a chaotropic anion like SCN^- has not only a low water affinity but also a high polarizability, thereby preferring to bind to the protein–water interface and to interact with the positively charged groups (which are chaotropic) on the protein surface. This may lead to a change in the protein conformation and hence its destabilization.

Cations are usually observed to show a less dominant effect relative to anions of the same charge density because they are less polarizable[76] and hydrate less strongly.[77] Kosmotropic cations are destabilizing because not only can they interact with the enzyme's kosmotropic moieties such as carboxylic groups, but they also have a high tendency of ion pairing with the kosmotropic anions in the solution, thus reducing the abundance of these free anions to play their kosmotropic role.[78] Indeed, a twofold higher affinity of Na^+ (kosmotropic) over K^+ (chaotropic) to the protein surface, especially to the carboxylic groups of aspartate and glutamate, has been quantified by means of molecular dynamics simulations and conductivity measurements.[79] Comparatively, chaotropic cations are more stabilizing than their kosmotropic partners.

The above picture can thus be used to illustrate the generally observed trend of proteins being stabilized by kosmotropic anions and chaotropic cations but destabilized by chaotropic anions and kosmotropic cations.

2.3.1.3 Effect of Ions on Enzyme Activity. The effect of ions on enzyme activity is not so straightforward. As shown in Table 2.1, upon salt addition, many enzymes have their activities varied following the Hofmeister series, that is, presenting higher activities in the presence of kosmotropic anions and chaotropic cations.

There are also some disobeying examples.[80–83] Even for those enzymes following the Hofmeister series, some are inhibited by all the salts tested,[49,80,84] while others are activated.[81,82,85] These complexities may be attributed to the interactions of the ions with the enzyme molecules that give rise to the following changes:

TABLE 2.1. Variation in Activity of Some Enzymes upon Addition of Different Salts in Aqueous Solution (Reproduced with Permission from Yang[10])

Enzymes	Activity order following the Hofmeister series	Effect[a]	Reference
Myosin nucleoside triphosphatase Trypsin Lactate dehydrogenase Estradiol-17β dehydrogenase Fumarase	**Anions:** $CH_3COO^- > Cl^- > NO_3^- > Br^- > I^- > SCN^- > ClO_4^-$ **Cations:** $(CH_3)_4N^+ > Cs^+ > K^+ > Na^+ > Li^+$	Inhibition	49
HIV-1 protease	**Anions:** $SO_4^{2-} > CH_3COO^- > Cl^- > Br^- > I^-$ **Cations:** $NH_4^+ > K^+ > Na^+$		92
Poliovirus picornain-3C protease	**Anions:** citrate^{3-} $> SO_4^{2-} > HPO_4^{2-} > CH_3COO^- > HCO_3^- > Cl^-$	Activation	85
Horseradish peroxidase	**Anions:** $SO_4^{2-} > Cl^- > Br^- \sim NO_3^-$		88
β-glucosidase	**Anions:** $SO_4^{2-} \sim F > CH_3COO^- > Cl^- > NO_3^- \sim Br^- > SCN^- > ClO_4^- > I^-$ **Cations:** $Na^+ > K^+ > Rb^+ > NH_4^+ > Cs^+ > Mg^{2+} > Ca^{2+} > Li^+ > Zn^{2+} > Cu^{2+} \sim Fe^{2+}$	Inhibition	84
Candida rugosa lipase	**Anions:** $SO_4^{2-} > Cl^- > SCN^-$		87
NADH oxidase	**Anions:** Bell-shaped relationship	Inhibition	80
NADH oxidase	**Cations:** V-shaped relationship	Activation	81
Acid sialidases	No obvious relationships	Activation	82
Alkaline phosphatase	Bell-shaped relationship		83

[a] In this column, "Inhibition" and "Activation" refer to whether the enzyme was inhibited or activated by addition of all the salts tested, respectively.

(1) *Surface pH of the enzyme.* The surface pH determines the ionization state of the amino acid residues of the enzyme's active site, which is important in modulating the enzyme's catalytic activity. Boström et al.[86] developed a modified ion-specific double-layer model to demonstrate that the surface pH of proteins is dependent on the salt concentration and on the ionic species following the Hofmeister series. Holding a higher polarizability than Cl^-, SCN^- has a stronger interaction with the protein surface, resulting in the accumulation of H^+ around the protein surface and in turn a reduction in the surface pH. In the case of *Candida rugosa* lipase,[87] this may lead to the protonation of both Glu^{341} and His^{449}, which then cannot assist the oxygen of Ser^{209} in its nucleophilic attack on the carbonyl carbon of the ester substrate, thus prohibiting the enzyme.

(2) *Net charge of the enzyme.* Addition of salts in the buffer solution may also induce a change in the buffer pH following the Hofmeister series, as has been observed.[83,87,88] This would affect the activity and stability of the enzyme in the buffer solution accordingly. Part of the reason for this variation may be the change in the net charge of the enzyme that is caused. However, it is not certain whether the ion-induced change in the net charge of an enzyme is critical in determining the enzyme performance. Sedlák et al.[72] observed a similar Hofmeister effect of ions on the thermal stability of two different model enzymes, the acidic *Desulfovibrio desulfuricans* apoflavodoxin (net charge −19 at pH 7) and the basic horse heart cytochrome *c* (net charge +17 at pH 4.5), suggesting direct interactions between ions and the peptide backbone, rather than the charged groups, of the enzyme. This was supported by Broering and Bommarius,[68] who reported that three enzymes/proteins with different isoelectric points (pI), horse liver alcohol dehydrogenase (pI = 8.52), α-chymotrypsin (pI = 8.31), and monomeric red fluorescent protein (pI = 5.65), followed a similar ion-dependent deactivation pattern at pH 7.0. On the other hand, the Hofmeister series can be direct or reverse depending on the charge and hydrophobicity/hydrophilicity of the surface in a colloidal system.[89] This was supported by two earlier experiments involving the study of Hofmeister effects on protein solubility: Both acidic *Hypoderma lineatum* collagenase (pI = 4.1) at pH 7.2[90] and basic hen egg white lysozyme (pI = 11.1) at pH 4.5[91] had their solubilities in water varied, following the direct and reversed Hofmeister series, respectively.

(3) *Active site and catalytic mechanism of the enzyme.* Ions may have strong interactions with the functional groups on the surface of the enzyme, especially those in the enzyme's active site, which are crucial for catalysis. This will trigger a change in the enzyme's active site both chemically and physically, resulting in a modification in the enzyme's catalytic activity and even its catalytic mechanism. This effect may be more severe than the above-mentioned effects caused by the ion-induced change in both surface pH and net charge of the enzyme (please refer to the following paragraphs for examples and discussions).

Gouvea et al.[85] have observed a significant activation of poliovirus picornain-3C protease (up to 104-fold) by a series of anions following the Hofmeister series:

citrate^{3-} > SO_4^{2-} > HPO_4^{2-} > CH_3COO^- > HCO_3^- > Cl^-, which was reflected by an increase in k_{cat} and no significant change in K_m. This salt-induced activation is associated with conformational changes as demonstrated by structural studies using fluorescence and circular dichroism (CD) spectroscopy. The thermodynamic data also indicated a more organized active site in the presence of the kosmotropic citrate. The kinetic study of HIV-1 protease,[92] however, showed that the ion effect was more on the K_m rather than on the k_{cat} of the enzyme, indicating a higher affinity of the enzyme for the substrate induced by a kosmotropic anion.

On the other hand, Žoldák et al.[80] reported an unusual bell-shaped relationship between the activity of NADH oxidase and the anion position in the Hofmeister series, implying that the enzyme activity is modulated by both kosmotropic and chaotropic anions via different mechanisms: The chaotropic anions decrease the apparent k_{cat} but increase the apparent K_m, whereas the kosmotropic anions decrease both k_{cat} and K_m. This indicates the importance of the flexibility of the enzyme's active site: Both the high rigidity of the active site caused by kosmotropic anions and the high flexibility induced by the chaotropic ones have a decelerating effect on the enzyme activity. This was supported by another study of cation effect on the same enzyme,[81] reinforcing the importance of the accessibility and flexibility of the enzyme's active site in regulating the enzyme activity through the perturbation of the balance between the open and closed conformations of the enzyme's active site. Indeed, a higher flexibility of cytochrome *c* in the presence of chaotropic anions has been confirmed by denaturation study of the protein, as monitored by DSC and CD.[93]

A plausible explanation for the effect of anions on the activity of *Pseudomonas cepacia* lipase involves opposite roles played by both kosmotropic and chaotropic anions on the transition state and active site of the enzyme.[94] A kosmotropic anion tends to stabilize the transition state of the E–S complex (by increasing its polarity) and destabilize the hydrophobic substrate, thus resulting in a lower activation energy. A chaotropic anion may act in the opposite way, and what is more, its strong interactions with the enzyme due to its high polarizability may distort the enzyme's active site, leading to destabilization of the transition state and hence enzyme inactivation. Indeed, the five different enzymes studied by Warren et al.[49] showed a parallelism between general structure-disrupting effectiveness and activity-inhibiting effectiveness caused by a series of Hofmeister ions, implicating that the ions inhibited the enzyme activity by disrupting the protein structure.

Recently, we reported the first investigation of Hofmeister effects on both activity and stability in parallel for the same enzyme, alkaline phosphatase.[83] Both the enzyme activity and stability correlated well with the Hofmeister series (Figure 2.5). The stability study offers another example to support the general trend of more kosmotropic anions and chaotropic cations favoring higher enzyme stability. The activity of alkaline phosphatase showed a bell-shaped relationship with the ($B_- - B_+$) values of the salts present, being optimal in the presence of a salt, such as KNO_3, the cation and anion of which have similar kosmotropic/chaotropic properties. This salt-induced activity change may result from different interactions of the salt ions with the enzyme. Strong kosmotropic anions and cations may affect the coordination of Zn^{2+} and Mg^{2+} at the enzyme's active site by binding to the metal ions and their original ligands (which are kosmotropic

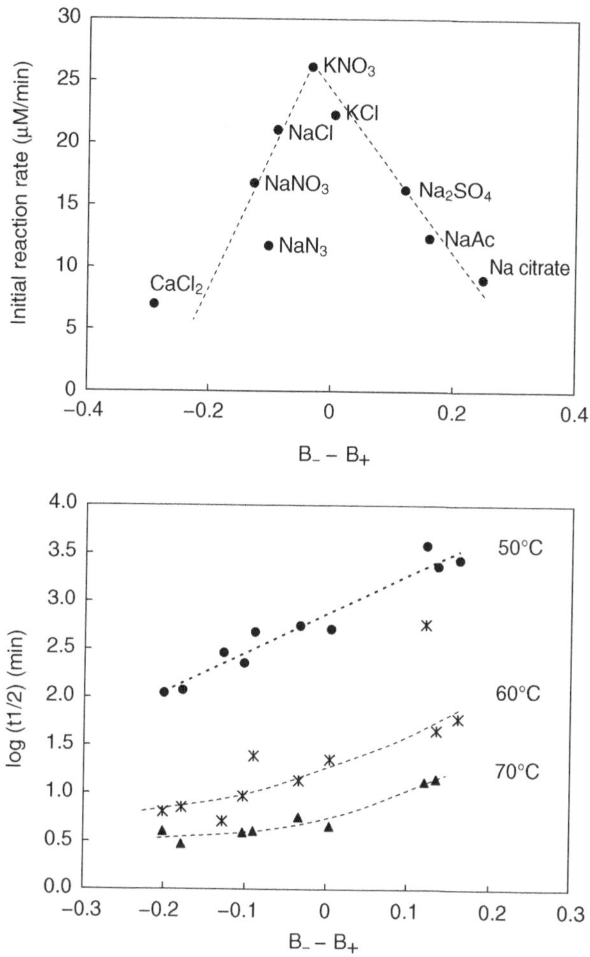

Figure 2.5. Relationship between the enzyme activity (top) and stability (bottom) of alkaline phosphatase in the aqueous solution and the ($B_- - B_+$) value of the salt added. Reproduced with permission from Yang et al.[83].

carboxylic groups of Glu and Asp, located on the enzyme surface), respectively, thus modulating the enzyme's catalytic mechanism. Chaotropic anions tend to bind to the protein–H_2O interface and to interact with the chaotropic cationic moieties (such as amino groups) on the enzyme surface, leading to a change in both the surface pH and the protein conformation of the enzyme. A chaotropic cation can bind to the enzyme surface, due to its high polarizability, to neutralize the net negative charge of the enzyme, or ion-pair with its chaotropic counter-anion in the solution to lessen the deactivating effect caused by the latter.

Therefore, it can be concluded from our study that the effect of salts on enzyme stability is general whereas the one on activity is enzyme-specific, depending on the individual impacts of both anions and cations on the enzyme's surface pH, active site, and catalytic mechanism.

2.3.2 Hofmeister Effects of Ionic Liquids

When ILs are used as enzymatic reaction media, they tend to penetrate into the micro-aqueous phase surrounding the enzyme molecule and dissolve and at least partially dissociate into individual cations and anions. It is reasonable to speculate that these IL ions also exhibit some Hofmeister effects on the enzyme functioning. Instead of studying the IL effects on enzyme performance by placing the enzyme in different ILs, one can place the enzyme in aqueous solution with addition of different ILs and their associated ions. This can provide valuable information for elucidating the importance of the IL's ionic nature in affecting enzyme performance.[10]

2.3.2.1 Effect of Ionic Liquid Ions on Enzyme Performance in Aqueous Solution. Table 2.2 lists some examples showing how the activity and stability of some enzymes in aqueous solution varied upon addition of different ILs. It can be observed that under most circumstances enzymes do follow the general Hofmeister effects, favored by the presence of kosmotropic anions and chaotropic cations.

The cation effect is obvious, as enzymes usually present both higher activity and higher stability following the decreasing order of the cation kosmotropicity. Although the viscosity *B*-coefficients for most IL cations are not available yet, it is generally believed that while $[MMIm]^+$ and $[EMIm]^+$ are chaotropes, larger immidazolium and ammonium cations with longer alkyl chains are more kosmotropic due to their stronger hydrophobic hydration.[56,95] Lange et al.[96] have examined the effect of N′-alkyl-N-methylimidazolium chlorides on the destabilization of lysozyme induced by heat and guarnidinium chloride (GuHCl). All the tested ILs were found to reduce the melting temperature of the enzyme and lower the midpoint concentration for GuHCl-induced protein unfolding, with a higher tendency corresponding to a longer alkyl chain attached to the imidazolium cation. This strongly supports our comment regarding the cation effect noted earlier in Section 2.3.1.2, suggesting that an imidazolium cation with a longer hydrophobic alkyl chain has a higher preference of destabilizing the enzyme due to (1) its strong interaction with the kosmotropic moieties, such as the carboxylic groups, on the enzyme surface, and (2) its hydrophobic interaction with the inner hydrophobic moieties of the enzyme molecule, leading to the disruption of the enzyme's native conformation.

Another typical example for illustrating the cation effect was given by Constantinescu et al.,[97] who used DSC to characterize the thermal denaturation of RNase A with addition of ILs holding Br^- and Cl^- as the common anions. Again almost all ILs rendered the enzyme to have its transition temperature lowered, in the order of $K^+ > Na^+ > [C_{1111}N]^+ > Li^+ > [C_{2222}N]^+ \sim [EMIm]^+ > [BMIm]^+ \sim [C_{3333}N]^+ > [HMIm]^+ \sim [C_{4444}N]^+$, which is rather consistent with the kosmotropicity order of the cations.

As regards the anion effect, there are indeed a number of examples showing that the IL anions follow the Hofmeister series to affect the enzymes and proteins. For instance, Saadeh et al.[98] have examined the effect of a series of tetrabutylammonium

TABLE 2.2. Activity and Stability Order of Some Enzymes in IL-Containing Aqueous Solution (Reproduced with Permission from Yang[10])

Enzymes	Condition (IL%)[a]	Activity/stability order	Relation to Hofmeister series[b]	Reference
Chloroperoxidase	Conversion (30%)	[MMIm][MeSO$_4$] > [BMIm][MeSO$_4$]	Direct	176
Lysozyme	Melting temperature (1 M)	**Cation effect** (with Cl^- as a common anion): $EMIm^+ > BMIm^+ > HMIm^+$ $OH\text{-}EMIm^+ > OH\text{-}PMIm^+ > OH\text{-}HMIm^+$	Direct	96
Lipase	Initiate rate (20%, 25%)	**Cation effect** (with BF_4^- as a common anion): $EMIm^+ > PMIm^+ > BMIm^+$	Direct	99
	Initiate rate (10–20%)	**Anion effect** (with $BMIm^+$ as a common anion): $BF_4^- > Cl^- > Br^- > NO_3^- > HSO_4^-$	Reversed	99
Papain	Activity (15%)	Cation effect (with BF_4^- as a common anion): $C_2MIm^+ > C_3MIm^+ > C_4MIm^+ > C_5MIm^+ > C_6MIm^+$ Anion effect (with $BMIm^+$ as a common anion): $BF_4^- > CH_3COO^- > NO_3^- > Cl^- > HSO_4^-$	Direct	110
D-amino acid oxidase	Initial rate (20%)	[MMIm][MMPO$_4$] > [BMIm][BF$_4$] > [BMIm][OcSO$_4$]		177
	Stability (40%)	[MMIm][MMPO$_4$] > [BMIm][BF$_4$] > [BMIm][OcSO$_4$]		177
Protease P6	Stability (0.7 M)	[EMIm][CH$_3$COO] ~ [EMIm][CF$_3$COO] > [EMIm][Br] > [EMIm][TOS] > [BMIm][CF$_3$COO]	Direct	178
	Activity (0.5 M)	[EMIm][CF$_3$COO] > [BMIm][CF$_3$COO]	Direct	179
3α-Hydrosteroid dehydrogenase	Activity (10%)	[BMIm][lactate] > [EMIm][CF$_3$SO$_3$] > [BMIm][BF$_4$] > [BMIm][CF$_3$SO$_3$]		127
Cytochrome c	Stability (80 wt%)	[choline][H$_2$PO$_4$] > [choline][dbp] > [BMIm][CH$_3$COO] > [BMIm][lactate] ~ [BMIm][MeSO$_4$]	Direct	95

(*Continued*)

TABLE 2.2 (*Continued*)

Enzymes	Condition (IL%)[a]	Activity/stability order	Relation to Hofmeister series[b]	Reference
CYP3A4	Activity (0–100%)	$[BMIm][PF_6] > [BMIm][BF_4] \sim [MMIm][BF_4]$	Reversed	180
RNase A	Melting temperature (0–2 M)	**Cation effect** (with Cl^- and Br^- as common anions): $K^+ > Na^+ > C_{1111}N^+ > Li^+ > C_{2222}N^+ > EMIm^+ > BMIm^+ > C_{3333}N^+ > HMIm^+ \sim C_{4444}N^+$	Direct	97
		Anion effect (with $EMIm^+$ as a common cation): $SO_4^{2-} > HPO_4^{2-} > Cl^- > EtSO_4^- > BF_4^- \sim Br^- > MeSO_4^- > TfO^- > SCN^- \sim N(CN)_2^- > Tf_2N^-$	Direct	97
Alcohol dehydrogenase	Conversion (20%)	$[EMIm][CH_3COO] > [BMIm][CH_3COO] > [EMIm][MeSO_4]$	Direct	166
	Initial rate (10%)	$[BMIm][PF_6] > [BMIm][BF_4] > [EMIm][TOS]$	Reversed	181
	Half-life (10%)	$[BMIm][EtSO_4] > [BMIm][PF_6] > [EMIm][TOS] > [BMIm][BF_4]$		181
Glucose dehydrogenase	Half-life (10%)	$[BMIm][EtSO_4] > [BMIm][BF_4] > [BMIm][PF_6] > [EMIm][TOS]$	Direct	181
Tyrosinase	Half-life (2%)	$[BMIm][BF_4] > [BMIm][PF_6] > [BMIm][MeSO_4]$		109
	Activity (2%)	$[BMIm][PF_6] > [BMIm][BF_4] > [BMIm][MeSO_4]$	Reversed	109
Bovine serum albumin Catalase	Denaturation (0–0.10 mM)	**Anion effect** (with tetrabutylammonium as a common cation): Formate > acetate > propionate > butyrate > linoleate	Direct	98

[a] In this column, the values in the parentheses refer to the IL contents in the aqueous solution; except otherwise stated, they are the volumetric percentages of the IL.

[b] In this column, the "Direct" and "Reversed" relations mean that the activity/stability order of the enzyme follows the Hofmeister series directly or reversely.

(TBA) ILs (with different carboxylic anions) on the structure of catalase and bovine serum albumin (BSA) by following their fluorescence emission spectra. For both proteins in aqueous solution, the maximum fluorescence emission at 350 nm, characteristic of tryptophan residues exposed to the aqueous milieu, increased with an increase in the concentration of each IL added, and increased with an increase in the length of the alkyl chain of the IL anion (indicating a decrease in kosmotropicity). This suggests that more tryptophan residues of the hydrophobic region of the protein become exposed or accessible to the aqueous phase due to the interaction with the TBA salts and that this denaturation becomes more serious in the presence of a more chaotropic IL anion.

However, the anion effect does not always strictly follow the Hofmeister sequence (see Table 2.2). Some enzymes even presented their activities corresponding reversely to the ranking of the anions in the Hofmeister order, and this usually occurs in the presence of a more kosmotropic cation, $[BMIm]^+$. For instance, the initial reaction rate of lipase-catalyzed enantioselective hydrolysis of D,L-phenylglycine methyl ester in the IL-containing aqueous buffer varied in the order of $[BMIm][BF_4]$ > [BMIm][Cl] > [BMIm][Br] > $[BMIm][NO_3]$ > $[BMIm][HSO_4]$.[99] The lower enzyme activity in the presence of a more kosmotropic IL anion in this situation could possibly be explained by the higher tendency of the kosmotropic cation ($[BMIm]^+$) to ion-pair with its kosmotropic counter-anion, thus reducing its abundance in the aqueous bulk solution to play its stabilizing/activating role, as has been proposed in Section 2.3.1.2.

Therefore, it is reasonable to conclude that both the cation and anion of an IL function cooperatively to affect the enzyme activity and stability. An excellent example to support this is given by $[choline][H_2PO_4]$, a perfect IL combination made up of a chaotropic cation and a kosmotropic anion.[95] Cytochrome *c* dissolved in this IL hydrated with 20 wt% water and maintained its activity and structure after 18 months of storage at room temperature. But if stored in aqueous buffer solution such as Tris-HCl or phosphate buffer, the protein will lose its activity after only 1–2 weeks. As another example, the protease enantioselectivity in aqueous IL mixtures has also been found to correlate well with the difference in the viscosity *B*-coefficients of IL anions and cations.[100]

Nevertheless, that the activity or stability of a few enzymes did not follow the Hofmeister series reminds us of the fact that the kosmotropicity/chaotropicity of the IL ions is not the only factor affecting enzyme performance, which in fact may also be governed by a series of other complicated mechanisms. In particular, whether ILs present in their aqueous solution can be totally dissociated into ions has to be considered seriously. As pointed out before,[11] an imidazolium-based IL holds a highly organized H-bonded network structure in its solid/liquid form. This structure may be at least partially maintained when the IL has a concentrated level in the aqueous solution, and under such conditions the Hofmeister series may not be well applicable. This is also true for those ILs composed of cations and anions with similar water affinity; they have a high tendency of ion-pairing with each other and are hence uneasy to dissolve and dissociate into ions. Therefore, one has to be cautious when trying to use the Hofmeister effects to explain the IL impacts on enzymatic catalysis, and some other solvent properties (such as polarity and hydrophobicity) may also have to be taken into account.

TABLE 2.3. Effect of IL Addition on the Kinetic Parameters of Some Enzymes in Aqueous Systems (Reproduced with Permission from Yang[10])

Enzyme	IL (IL content)[a]	K_m[b]	V_{max}[b]	V_{max}/K_m[b]	Reference
Laccase	[BMPy][BF_4] (25%)	↑	↓	↓	101
	[EMIm][[$MeSO_3$] (25%), [EMIm] [$EtSO_4$] (25%), [EMIm] [$MDEGSO_4$] (25%)	↓	↓	↓	102
Alkaline phosphatase	[EMIm][BF_4] (25–75%)	↓	↓	↑	182
Lipase	[BMIm][BF_4] (20%)	↓	↑	↑	99
Papain	[C_nMIm][BF_4] (15%), $n = 2$–6	↓	↑	↑	110
Alcohol dehydrogenase	[BMIm][Cl] (0.05 g/mL)	↑	↑	↓	105
Protease P6	[EMIm][CF_3COO] (0.5 M)	↓	↓	↑	179
	[BMIm][CF_3COO] (0.5 M)	↓	↓	↓	179
Horseradish peroxidase	[BMIm][BF_4] (25%)	↑	↓	↓	103
	[BMPy][BF_4] (0–25%)	↑	↓	↓	104
	[BMIm][$MeSO_4$] (0–25%)	↓	↓	↓	104
Tyrosinase	[BMIm][BF_4] (2%), [BMIm] [$MeSO_4$] (2%), [BMIm][PF_6] (1.7%)	↑	↑	↑	109

[a] All the percentages in this column refer to the volumetric percentages of ILs in aqueous solution, unless otherwise stated.

[b] Symbols ↑ and ↓ in these three columns refer to the increase and decrease in the kinetic parameters obtained by the enzyme in the presence of the IL as compared with the values obtained in the IL-free system.

2.3.2.2 Kinetic Studies of Enzymes in Ionic Liquid-Containing Aqueous Solution. A comparison of kinetic parameters of enzymes in aqueous solution with and without addition of ILs should be informative to our understanding of the IL effects. As can be seen from Table 2.3, IL addition affects the kinetic parameters of different enzymes very differently.

Even for the same enzyme laccase upon addition of ILs, the K_m values for oxidation of syringaldazine[101] and 2,2′-azino-bis(3-ethylbenzthiazoline-6-sulfonic acid) diammonium salt[102] can be increased and decreased, respectively, as compared with the results obtained in the IL-free aqueous buffer. This may largely be associated with the different ILs used: For the former case, [BMPy][BF_4] was added, which was composed of a kosmotropic cation and a chaotropic anion,whereas the latter enzymatic oxidation utilized three ILs, [EMIm][$MeSO_3$], [EMIm][$EtSO_4$], and [EMIm][$MDEGSO_4$], all in a combination of chaotropic cations and kosmotropic anions, the favorite form for enzyme action. This analysis is strongly supported by the similar results for horseradish peroxidase (HRP).[103,104] In addition, Zong's group obtained an enhancement in V_{max} for both lipase[99] and alcohol dehydrogenase[105] by addition of ILs, both in accord with a reduction in the activation energy. Nevertheless, the complexity of the IL-induced variation in the kinetic parameters reinforces the fact that ILs affect enzyme performance depending on a combination of complicated mechanisms rather than on a single simple one such as kosmotropicity.

2.3.2.3 Enzyme Performance in Ionic Liquid-Dominating Reaction Systems. It would be interesting to examine whether the enzyme activity and stability also follow the Hofmeister effect as discussed above when ILs are used as the dominating medium for the enzyme catalysis. This is indeed the case for quite a number of enzymes and proteins. Perhaps the best annotation for this phenomenon was given by Fujita et al.[95] The activity of cytochrome *c*, after 3 weeks of storage at room temperature in different ILs containing 20 wt% H_2O, varied in the order: [choline][dhp] > [BMPy][dhp] > [choline][dbp] > [BMIm][acetate] > [BMIm][lactate] > [BMIm][$MeSO_4$]. It should be noted that the water content present in the system was merely sufficient for hydrating the IL at a ratio of 3 water molecules per IL ion pair, and therefore no free water molecules existed. The protein dissolved in [choline][dhp] retained similar activity to that in fresh buffer solution, whereas no activity was observed in [BMIm][lactate] and [BMIm][$MeSO_4$]. Structural studies have also revealed that the conformation of the protein was well maintained in hydrated [choline][dhp], but the secondary structure and the heme coordination state of the protein was obviously altered when the protein was incubated in [BMPy][dhp] or [BMIm][lactate], as demonstrated by ATR-FTIR and Raman spectra. This is actually not surprising when considering the kosmotropicity orders of the IL anions ([dhp] > [dbp] > [acetate] > [lactate] > [$MeSO_4$]) and cations ([choline] < [BMPy] < [BMIm]) involved.

There are a few other examples showing that the imidazolium ILs with shorter alkyl chains attached to the imidazolium ring favor either enzyme activity[39,106,107] or stability.[96,108] For instance, the activity of three lipases to catalyze a transesterification reaction in ILs under controlled water activity all decreased following the same order:[106] [EMIm][Tf_2N] > [BMIm][Tf_2N] > [HMIm][Tf_2N]. This is in concert with the chaotropicity order of the cations.

We have recently conducted a systematic study on activity and stability of mushroom tyrosinase in three different systems:[33,109] in ILs, in IL-containing aqueous solution, and in aqueous solution with addition of Na^+ or K^+ salts of the IL anions. Three ILs were selected: [BMIm][PF_6], [BMIm][BF_4], and [BMIm][$MeSO_4$], all holding the same cation but different anions. This study provides a good example to elucidate the cooperative functioning of the IL anions and cations following the Hofmeister effect (Figure 2.6).

When tyrosinase was assayed in the three different ILs, both its activity and stability varied in the same order of [BMIm][PF_6] > [BMIm][BF_4] > [BMIm][$MeSO_4$].[33] A similar situation was observed in the aqueous solution containing different ILs (2%, v/v),[109] suggesting that the influence brought from ILs was the same regardless of whether the ILs were used directly as the reaction medium or were introduced into the aqueous solution as an additive. However, when the effect of IL anions was examined by introducing only their Na^+ or K^+ salts into the aqueous solution, the situation was somewhat reversed and the enzyme was most stable and exhibited the highest K_m and

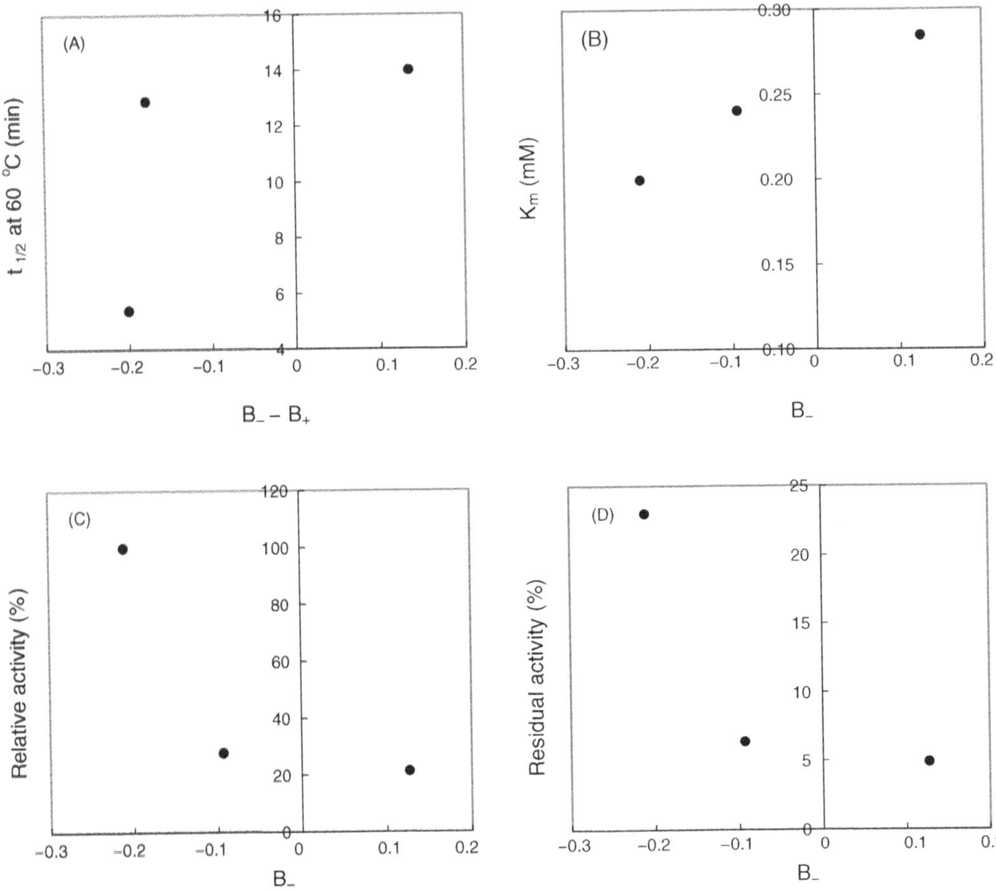

Figure 2.6. Effect of three ILs ([BMIm][PF_6], [BMIm][BF_4], [BMIm][$MeSO_4$]) and their associated anions (PF_6^-, BF_4^-, $MeSO_4^-$) on the activity and stability of mushroom tyrosinase. (A) Half-lives of the enzyme in phosphate buffer (50 mM, pH 6.0) containing 0.3 M of KPF_6, $NaBF_4$, and $KMeSO_4$, respectively, at 60°C. (B) The K_m values of the enzyme in phosphate buffer (50 mM, pH 6.0) containing the three different ILs (2%, v/v of [BMIm][BF_4] and [BMIm][$MeSO_4$], and saturated level (1.7%, v/v) of [BMIm][PF_6]). (C) Relative activity of the enzyme in three different ILs as assayed at 35°C and a water activity of 0.90. (D) Retained activity of the enzyme after being treated with different ILs for 24 hours at 30°C. The residual activity (%) refers to the percentage of the initial reaction rate obtained by the enzyme (as assayed in chloroform) after IL treatment as compared with the one obtained without being treated with any IL. Reproduced with permission from Yang[10].

V_{max} in the presence of $KMeSO_4$ relative to the values obtained in the solution containing $NaBF_4$ or KPF_6.

The kosmotropicity order for both these cations and anions involved are as follows: $[BMIm]^+ > Na^+ > K^+$, and $MeSO_4^- > BF_4^- > PF_6^-$. Thus, it is not surprising that the enzyme was most stable and active in the presence of $KMeSO_4$. This can be explained by the stabilizing effect of the kosmotropic anion, $MeSO_4^-$, which was further strengthened by the chaotropic cation, K^+. However, the situation was reversed when ILs were used directly as the reaction medium or were added into the aqueous solution as an additive, and this can be attributed to the cation effect. The kosmotropic IL cation $[BMIm]^+$ possesses a high propensity for ion-pairing with the kosmotropic anions (such as its counter-anion, $MeSO_4^-$) in the solution to minimize their stabilization/activation effect. It also tends to interact strongly with the kosmotropic moieties, such as the carboxylic groups, of the enzyme molecules, especially when the negative charge of the protein (pI = 4.8) in the pH 6.0 buffer is concerned. This may be a convenient explanation to account for the fact that tyrosinase showed the lowest activity and stability in the presence of $[BMIm][MeSO_4]$, as compared with the results obtained in the systems containing the other two ILs. This work indicates that IL cations and anions work cooperatively to affect the activity and stability of mushroom tyrosinase, which can be well illustrated by the Hofmeister effects.

Undoubtedly, Hofmeister effects have offered a reasonable explanation for the impact of ILs on biocatalysis, especially when ILs are used as a co-solvent or an additive in the aqueous solution. We have to admit that there are a considerable number of reports on enzyme behavior in ILs that do not follow the Hofmeister effects. This reminds us again of the fact that Hofmeister series is not the only determinant for controlling enzyme action, especially when an IL is used as the sole solvent. An IL in its pure liquid form presents a polymeric supramolecular structure with H-bonding network and is hardly dissociated into cations and anions in the presence of the small amount of water added to the system or within the microaqueous phase surrounding the enzyme molecules.[11] This may be the major reason why some ILs do not effectively exhibit the Hofmeister effects on the enzyme function. Under these circumstances, some other factors or interactions may be predominant in regulating the enzyme activity and stability, as will be discussed in the forthcoming section.

2.4 IMPACT OF IONIC LIQUIDS ON ENZYMES AND PROTEINS

2.4.1 Effect of Ionic Liquids on Enzyme Activity and Stability

***2.4.1.1 Hydrophobicity and Log* P.** In the literature, there have been a certain amount of experimental results that cannot be simply explained by the Hofmeister series. Some are even contradictory to it. For instance, imidazolium ILs with a longer alkyl chain on the imidazolium ring (indicating higher kosmotropicity and higher hydrophobicity) are usually found to promote higher enzyme activity[110–114] and higher enzyme stability.[114,115] Hydrophobicity may be a better explanation for this phenomenon, especially when considering that there was no water activity control in most of these studies. Indeed, some other studies have also revealed that enzymes are sometimes more active or more stable in more hydrophobic ILs.[33,112,116–119] Typically, Russell's group[120] observed that free lipase (*Candida rugosa*) showed promising activity in the hydrophobic IL, $[BMIm][PF_6]$, but was inactive in all the other more hydrophilic ones tested, namely $[BMIm][CH_3CO_2]$, $[BMIm][NO_3]$, $[BMIm][CF_3CO_2]$, $[mmep][CH_3CO_2]$,

[mmep][NO_3], [mmep][CH_3SO_3], [mmep][CF_3CO_2], and [mmep][CH_3SO_3]. Lee et al.[106] have also observed that the immobilized *Candida antarctica* lipase (Novozym 435) maintained its full activity after incubation in hydrophobic ILs such as [BMIm][Tf_2N], [BMIm][PF_6], and [EdMIm][Tf_2N], for almost 50 hours, whereas the half-lives for the same enzyme in hydrophilic [BMIm][BF_4] and [BMIm][TfO] were less than 6 hours.

It has been generally accepted that the hydrophobicity of an organic solvent can be quantified in terms of log *P*, the logarithm of the partition coefficient of the solvent in an octanol/water mixture. This has been taken as a key determinant of enzyme activity in organic media. Solvents with a high log *P*, such as hexane (log $P = 3.5$), are usually found to be more hydrophobic and more favorable for enzymatic reactions than those with a low log *P*, such as ethanol (log $P = -0.24$),[121] simply because the more hydrophobic solvents have a lower tendency of stripping off the essential water from the enzyme, thus favoring the maintenance of its native structure. There have been attempts to apply this log *P* concept to IL. Russell's group[120] was the first to determine the log *P* values for ILs, and the extremely low log *P* values (−2.90 to −2.39) they obtained were verified by Ropel et al.,[122] who measured the octanol–water partition coefficients for a series of imidazolium ILs. For the ILs with $[BMIm]^+$ as the cation, the log *P* values decreased in the order $Tf_2N^- > PF_6^- > Cl^- > NO_3^- > Br^- > BF_4^-$, and for the ILs with the most hydrophobic anion, Tf_2N^-, the log *P* values increased with an increase in the length of the alkyl chain on the imidazolium ring of the cation. In spite of their extremely low log *P* values, some ILs, such as [BMIm][Tf_2N] and [BMIm][PF_6], are immiscible with water (hydrophobic) while others are miscible (hydrophilic), such as [BMIm][BF_4] and most ammonium ILs. The direct correlation between hydrophobicity of ILs and their log *P* values still exists.

In view of how solvent hydrophobicity affects enzyme functioning in organic media, it is not difficult to understand the above-mentioned experimental results as regards the higher enzyme activity or stability obtained in more hydrophobic ILs. In fact, a hydrophilic IL can not only strip off the essential water that is originally associated with the protein molecules, but also penetrate into that microaqueous layer and dissolve and dissociate into individual ions, interacting specifically (such as through Hofmeister effects) with the protein; some of these interactions may be deleterious. Indeed, the initial rates of the lipase-catalyzed transesterification reaction[106] showed an obvious trend of progressively rising with an increase in the log *P* values of the ILs used. Earlier, the same group[123] reported a successful use of ILs as additives for sol-gel immobilization of *Candida rugosa* lipase to increase its activity and stability. A linear relationship was also obtained between the log *P* value of the IL and the residual activity of the immobilized lipase after 5 days' incubation in hexane at 50°C, implying that hydrophobic ILs are more effective for stabilizing the enzyme. This suggests that hydrophobic ILs in the sol–gel process can act as a template during gelation and behave as a stabilizer to protect the enzyme from the inactivation by alcohol or heat.

However, one has to keep in mind that IL hydrophobicity itself is not a universal determinant of enzyme performance, taking into account the complicated mechanisms and interactions involved. One exception is that among all the nine hydrophobic and hydrophilic ILs tested, the best synthetic activity of lipase (from *Candida antarctica*) was obtained when the enzyme was assayed in the water-miscible IL, [CPMA]

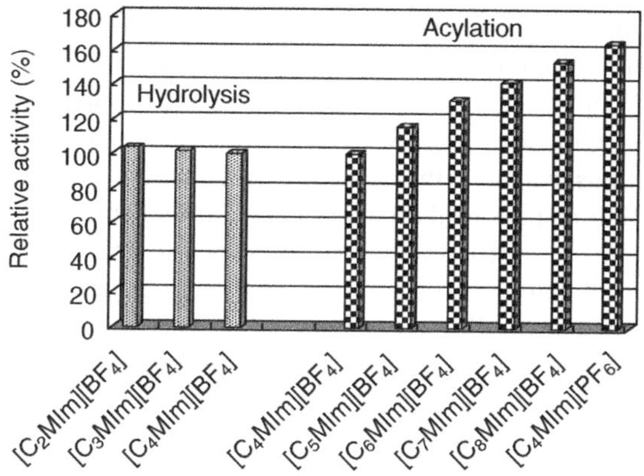

Figure 2.7. Effect of the alkyl side chain of imidazolium ILs ($[C_nMIm][BF_4]$) on the activity of *Candida antarctica* B (Novozym 435) in the hydrolysis of D,L-phenylglycine methyl ester in aqueous solution containing 20% or 25% (v/v) IL[99] and in acylation of (*R,S*)-1-trimethylsilylethanol with vinyl acetate in ILs.[114] The reaction rate in $[C_4MIm][BF_4]$ was taken as 100 in each case.

$[MeSO_4]$.[39] The following two experimental results also cannot be simply explained by using hydrophobicity only.

An interesting phenomenon (Figure 2.7) can be observed when comparing the two opposite results obtained by Zong's group, who tried to examine the effects of different imidazolium ILs ($[C_nMIm][BF_4]$, $n = 2$–8) on the activity of the same enzyme, the immobilized lipase from *Candida antarctica* B (Novozym 435), in two different types of reactions.

The lipase-catalyzed acylation reaction was carried out in ILs under a pre-equilibrated water activity of 0.75, and the initial rate was obviously higher in the IL containing a longer alkyl side chain,[114] following the trend of hydrophobic solvents promoting enzyme activity. However, when the enzyme was used to catalyze the hydrolysis reaction in aqueous buffer solution containing 20% or 25% IL (v/v), a slight decline in the reaction rate was observed as the alkyl chain attached to the imidazolium cation was longer.[99] Presumably, in aqueous solution the Hofmeister effects of ILs are predominant in controling the enzyme activity because of their dissolution and dissociation into ions, as discussed earlier, whereas when ILs are used directly as the reaction medium, the effect of their hydrophobicity may be more determining. In this particular case, the lipase activity is boosted in a more hydrophobic IL not because of its water-stripping effect (as all the reactions were carried out under a controlled water activity) but presumably because of its hydrophobic interactions with the enzyme.

Another interesting result was reported by Zhao et al.[124] While working on the activation effect induced by microwave irradiation on the same enzyme used above, the authors found that the initial rate of the transesterification reaction catalyzed by the enzyme showed a reversed V-shaped relationship with the log *P* values of the IL used as the solvent but did not correlate well with the viscosity or polarity of the IL. The initial increase in the activity upon an increase in log *P* may be related to favorable effects of the higher hydrophobicity of the IL and the lower nucleophilicity and H-bond basicity (H-bond accepting ability) of the IL anion. The former factor can ensure the enzyme hydration, while the latter two can minimize both the nucleophilic and H-bonding interactions between the IL anion and the protein, thus facilitating the enzyme activity. With respect to the later decrease in enzyme activity with a further increase in log *P*, one of the reasons may be the substrate ground-state stabilization,[125] leading to a higher activation energy and hence a lower reaction rate. But more importantly, as log *P* is greater, the IL may become larger and more hydrophobic (in either cation or anion or both), thus inducing stronger hydrophobic interactions with the protein and conformational change in the protein structure.

2.4.1.2 Nucleophilicity and H-bond Basicity. The above discussion draws our attention to two other factors, nucleophilicity and H-bond basicity, which may have impacts on controlling the enzyme action in ILs. In fact, the importance of nucleophilicity has been raised by Kaar et al.[120] They found that *Candida rugosa* lipase was only active in [BMIm][PF_6] but not in all other hydrophilic ILs holding NO_3^-, $CF_3CO_2^-$, $CH_3CO_2^-$, $CF_3SO_3^-$, $CH_3SO_3^-$ as the anion. These IL anions are more nucleophilic than PF_6^- and may coordinate more strongly to the positively charged sites in the enzyme structure, causing conformational changes. Therefore, they proposed that use of ILs with low anion nucleophilicity is essential to enzyme activity.

The importance of another factor, H-bond basicity, was brought into attention by Lau et al.[126] They observed a 10 times lower activity for CALB in ILs containing alkylsulfate, nitrate, and lactate anions, which can dissolve the enzyme, as compared with the activity obtained in [BMIm][PF_6] and [BMIm][BF_4], in which the enzyme was insoluble. By means of FT-IR spectroscopy, it was concluded that the low activity of the enzyme in [BMIm][lactate] was accompanied by its denaturation upon dissolution. It was suspected that the strong ability of lactate to form stable H-bonds with the polypeptide backbone of the enzyme may be responsible for this situation. An IL anion with a strong H-bond basicity may cause dissociation of the hydrogen bonds that maintain the structural integrity of the α-helices and β-sheets, which in turn will cause the protein to unfold.

Undoubtedly, both nucleophilicity and H-bond basicity of the IL anions can be important in affecting enzyme performance. It has appeared to become a consensus that ILs with highly nucleophilic or highly H-bond forming anions are interfering. But whether their effects on enzyme action are positive or negative has to be considered seriously because this may be dependent very much on the enzyme and its catalytic mechanism. Strongly nucleophilic and H-bond forming IL anions may be detrimental to some enzymes because of their potential to perturb the protein conformation by strong interactions with the enzyme, but sometimes they may be beneficial to other enzymes. Lactate seems to be the best example to illustrate this. While [BMIm][lactate] has been well known to dissolve and denature CALB,[126] it is actually the solvent that permitted 3α-hydroxysteroid dehydrogenase with the highest activity:[127] The enzyme activity in aqueous solution was enhanced by addition of 10% [BMIm][lactate] as a co-solvent, whereas addition of the other three ILs, [EMIm][CF_3SO_3], [BMIm][CF_3SO_3], and [BMIm][BF_4], all caused depression to the enzyme activity, and the major reason for this may also be attributed to the strong H-bond forming ability of the lactate anion.

Another issue derived from the topic of nucleophilicity and H-bond basicity is the possible protein dissolution. Some ILs with high nucleophilicity and high H-bond basicity are capable of dissolving the enzyme, implicating strong interactions between the enzyme and the IL. Dissolution per se does not necessarily mean to denature or deactivate the enzyme, just like an enzyme dissolved in water does not necessarily mean to be denatured or deactivated. Whether the enzyme is denatured or deactivated is determined by whether these IL–enzyme interactions are favorable or unfavorable to the enzyme action, again depending on both the enzyme and the IL. Take the study by Lau et al.[126] again as an example. Both [Et_3MeN][$MeSO_4$] and the other four ILs containing alkylsulfate, nitrate, and lactate anions can dissolve the enzyme, CALB. But the dissolved enzyme maintained its activity and native conformation in the former IL, while being deactivated and denatured in all the four latter ILs. As mentioned earlier, cytochrome *c* was able to dissolve in [choline][dhp] (with 20% H_2O) while maintaining both its activity and structure for 18 months, but a loss of activity and structure occurred to the protein when dissolving in other ILs such as [choline][dbp], [BMIm][acetate], [BMIm][lactate], and [BMIm][$MeSO_4$].[95]

Regardless of all these, the consensus that ILs with highly nucleophilic or highly H-bond forming anions are deleterious to enzyme action appears to have been supported by many experiments and accepted by many researchers. However, the potential of H-bonding and perturbing the protein structure of these anions does not seem to be a sufficient reason to explain why among the five ILs capable of dissolving the lipase, only [Et_3MeN][$MeSO_4$] was able to retain the enzyme activity.[126] It is crucial to understand what makes this IL so distinguishable from the other four (i.e., [EMIm][$EtSO_4$], [BMIm][lactate], [$EtNH_3$][NO_3], and [BMIm][NO_3]). Additionally, most of the successful IL designs introduced later in Section 2.6.3, as well as [choline][dhp][95] and [CMPA][$MeSO_4$][39] discussed before, contain $H_2PO_4^-$, $MeSO_4^-$, $CF_3SO_3^-$ as their anions, which are also strong H-bond acceptors. In fact, one direct way to confirm the IL anion–protein interactions is to introduce the anions into the protein solution in the form of their inorganic salts and to undertake some kinetic, thermodynamic, and structural investigations. However, this is actually what has been done for the studies of Hofmeister effects of inorganic salts and ions. These studies have shown that enzymes are usually stabilized by a kosmotropic anion, but no sensitive relation between enzyme performance and the ion's H-bonding ability has been reported. Although there is no direct correlation between them two, a kosmotropic anion is usually a strong H-bond acceptor. Therefore, whether an IL anion with a strong H-bond basicity is definitely detrimental to enzyme action remains an enigma. In our opinion, this should be enzyme-dependent.

2.4.1.3 Viscosity. Viscosity is another solvent property that has to be considered. The viscosity of an IL is usually higher than that of molecular solvents and may control the enzyme activity by affecting the mass-transfer limitations in the reaction systems. This is possibly the reason why α-chymotrypsin was more active in [EMIm][Tf_2N] than in [MTOA][Tf_2N], which have a viscosity of 34 and 574 cP, respectively.[128]

On the other hand, high viscosity of ILs may also offer a stabilizing effect, slowing down the migration of protein domains from the active conformation into the inactive one.[8] Nevertheless, the viscosity effect on biocatalytic transformations does not seem to be significant.[23,119,124,129] For instance, a lipase-catalyzed transesterification reaction proceeded faster in a phosphonium IL, [$MEBu_3P$][Tf_2N], than in an organic solvent, diisopropyl ether, although the former solvent (72 cP at 25°C) is much more viscous than the latter (0.305 cP at 32°C).[47]

Overall, a number of factors possibly contributing to affect the enzyme functioning in ILs have been proposed, including the solvent properties such as hydrophobicity, viscosity, polarity, nucleophilicity, and H-bond basicity. However, none of these seems to be universal. This is understandable when considering the biocatalytic systems containing so many complicated interactions. In addition to the general solvent properties, some factors specific to the enzyme and the components involved in the reaction system have to be considered as well. For instance, the conversion rate of esculin esterification catalyzed by an immobilized lipase (Novozym 435) in 17 different ILs seemed to decrease with an increase in the esculin solubility.[107] This could be explained by the substrate ground-state stabilization in the IL, which has a higher ability to dissolve the substrate. The superior transesterification activity obtained in [BdMIm][BF_4] by all the three lipases, in comparison with that obtained in [BMIm][BF_4], could be attributed to the lack of acidity of the 2-position of the imidazolium cation, which avoids accumulation of the acetaldehyde oligomer that may deactivate the enzymes.[39]

2.4.2 Effect of Ionic Liquids on Protein Structure and Dynamics

Information regarding the impact of ILs on protein structure and dynamics is crucial for our understanding of protein function in ILs. Nevertheless, in comparison with the numerous experimental results collected with respect to activity and stability, so far there have been only a few biophysical characterizations of proteins in the IL-based systems reported in the literature. Spectroscopic techniques, especially fluorescence, CD, and FT-IR analyses, have been employed to conduct structural studies in order to provide information on the conformational change, stabilization, and denaturation of proteins induced by ILs.

In the first report of protein spectroscopy in ILs, Baker et al.[130] used intrinsic tryptophan fluorescence to monitor the unfolding transition for monellin (a single Trp-containing protein) at soluble levels in an IL, [BMPy][Tf_2N]. Use of the IL affords the protein with a significantly enhanced thermostability, as its unfolding temperature was raised from 40°C in water to 105°C in the IL containing 2% (v/v) water. This stabilization is actually entropically driven because the entropy of unfolding for monellin ($\Delta S°$) was reduced from 250 J/K/mol in water to 136 J/K/mol in the IL, a reflection of more rigid structure of the protein within the IL. Moreover, the blue shift of the emission maximum wavelength upon thermal unfolding implicates shielding or isolation of the Trp by neighboring aromatic residues, thus yielding minimal exposure of the Trp residue to the surrounding IL solvent. Therefore, the thermal stabilization of the protein may result from the IL being able to tighten the protein structure and to keep its internal structure unexposed to the bulk solvent.

Pioneering structural studies have also been conducted by Iborra's group to tackle the issue of structure–stability relationship. The remarkable stabilizing power of ILs (such as [EMIm][Tf_2N]) to α-chymotrypsin[35] and *Candida antarctica* lipase B (CALB)[131] is associated with the structural changes of the proteins, as monitored by using the CD and intrinsic fluorescence spectroscopic techniques. The fluorescence result shows a more compact protein conformation in the IL as compared with that in other solvents such as water, 3 M sorbitol, and 1-propanol, supporting what was observed in the above report,[130] and the CD spectra reveal the evolution of α-helix to β-sheet in the secondary structure of the enzyme when the solvent was switched from water to the IL. It must be these two structural changes that contribute to the formation of a compact, flexible, and active enzyme conformation that offers high levels of activity and stability.

Later on, a series of investigations on the structure–activity relationship have followed. Lau et al.[126] reported that among the ILs that can dissolve CALB, only [Et_3MeN][$MeSO_4$] was able to render the enzyme to promote considerable activity (~1/3 of the activity presented in [BMIm][BF_4]), whereas in all others ([EMIm][$EtSO_4$], [BMIm][lactate], [BMIm][NO_3], and [$EtNH_3$][NO_3]), the reaction was at least 10 times slower than in [BMIm][BF_4]. The FT-IR spectroscopic analysis indicates that the enzyme retained its native secondary structure when dissolved in [Et_3MeN][$MeSO_4$], but when dissolved in the other ILs it suffered a loss of α-helix and β-turn structures and an increased presence of random coil structures and H-bonded carbonyls due to protein unfolding. This clearly demonstrates that the loss of enzyme activity in the IL is directly associated with the loss of the enzyme's native structure. The authors further postulated that H-bonding could be the key to understanding this issue. The IL anions, such as lactate used in this study, could easily form stable H-bonds with the polypeptide backbone, thus disrupting the H-bonds that used to maintain the structural integrity of the α-helices and β-sheets and causing the protein to unfold.

Both ATR-FTIR and fluorescence techniques have been employed by Lou et al. to correlate the conformational changes with the observed hydrolytic activity of papain in a series of imidazolium ILs (15% v/v in 50 mM, pH 7.0 phosphate buffer).[110] For the Group I ILs (with a common anion but varying cations: [C_nMIm][BF_4], $n = 2–6$), the enzyme activity increased with an increase in the length of the alkyl chain attached to the imidazolium cation, all higher than the activity obtained in aqueous buffer. For the Group II ILs (with a common cation but varying anions: [BMIm][X], X = HSO_4^-, Cl^-, NO_3^-, CH_3COO^-), the hydrolytic rates obtained by the enzyme were all lower than that obtained in aqueous buffer, except in [BMIm][BF_4]. The second derivative ATR-FTIR spectra in the amide I region of papain in the systems containing the Group I ILs indicate that the enzyme preserved a near native conformation, with a decrease in the α-helix structure and an increase in both β-sheets and β-turns as compared with the spectrum obtained in the control buffer. This may support a more compact and stable enzyme conformation capable of exhibiting catalytic activity. However, the ATR-FTIR spectra obtained by the enzyme in the presence of Group II ILs were greatly different from the control, a reflection of a non-native conformation, which was strongly supported by the fluorescence data. This study clearly indicates that the catalytic performance and conformational structure of the enzyme is profoundly influenced by both IL cations and anions (in particular). A high or low activity presented by the enzyme in the IL systems is corresponding to a native or non-native protein structure, respectively, depending more on the choice of the IL anions. The enzyme exerted a low activity in the presence of the Group II ILs, presumably because all the anions involved in this group have high H-bond accepting capability. BF_4^-, on the other hand, due to its lower H-bond basicity and lower nucleophilicity, shows a lower propensity for triggering a conformational change by disrupting H-bonds and interacting with the positively charged moieties of the enzyme, thus affording a high enzyme activity. As a strong support to the earlier reports, this study manifests that (1) a high enzyme activity or stability presented in ILs is accompanied by a native-like structure; and (2) IL anions with high H-bond basicity may be deleterious to enzyme functioning.

As mentioned earlier, Fujita et al.[95] have also undertaken structural studies for elucidating the remarkable power of hydrated [choline][H_2PO_4] in maintaining the activity of cytochrome *c* after 18 months of storage in the dissolved form at room temperature. Among the various ILs that were tested in the study, only [choline][H_2PO_4] enabled the protein to show excellent thermal stability and long-term stability while also retaining its native secondary structure and conformation, as has been monitored by ATR-FTIR and resonance Raman spectroscopies, respectively. These results support the above-mentioned first notion regarding the correlation between activity/stability and structure, but it has to be noted that $H_2PO_4^-$ in this case seems to be exceptional to the second notion with respect to H-bond basicity because it is a strong H-bond acceptor and yet it dissolved cytochrome *c* without causing denaturation.

With the aid of dynamic light scattering (DLS) and small angle neutron scattering (SANS) techniques, Sate et al.[132] investigated the structure of CALB dissolved in different solvents such as water, dimethyl sulfoxide (DMSO), and three ILs ([EMIm][$N(CN)_2$], [EMIm][NO_3], [EMIm][$EtSO_4$]), in order to compare any conformational changes in the solvents and to correlate these with changes in catalytic activity. Their measurements of the protein dimensions reveal that while the enzyme in water formed cylindrical nanostructures as monomers, it aggregated both in DMSO and in all the three ILs. The enzyme dissolved in [EMIm][$N(CN)_2$] is completely inactive and existed as disc-shaped, "side-by-side" aggregates including roughly 150 CALB molecules, rather than as the "end-to-end" dimeric or trimeric aggregation seen in DMSO. The aggregates in the other two ILs were smaller and elicited very low levels of enzymatic activity. The average size of the CALB aggregates in these ILs increased with the increase in the electron density (and hence H-bonding potential) of the anion, in the order $EtSO_4^- < NO_3^- < N(CN)_2^-$, also corresponding to the decrease in the enzyme activity. Removal of the essential water from the enzyme may be part of the reason for enzyme aggregation and inactivation in the IL, but IL anions with higher nucleophilicity and stronger H-bond accepting capability may play a more important role, as they can interact directly with the positively charged groups on the enzyme surface and disrupt the H-bonds originally present and responsible for the preservation of the protein structure. Therefore, these structural measurements provide further support for the consensus that ILs with strongly H-bonding anions may disfavor biocatalysis.

Micaêlo and Soares presented their molecular dynamics simulation study of a serine protease, cutinase, in two different ILs, [BMIm][PF_6] and [BMIm][NO_3].[133] Their molecular modeling studies have provided very useful information on the protein–IL–

H_2O interactions and their effects on the protein function. The enzyme was preferentially stabilized in [BMIm][PF_6] but destabilized in [BMIm][NO_3] because, as their results showed, the former IL allowed the retention of more water at the enzyme surface and rendered a more native-like enzyme structure than the latter. Only a fraction of the total water present in the system was shown to locate at the enzyme surface plus some internal waters, whereas the excess water molecules were found on bulk solution interacting preferentially with the IL anions via H-bonds. Both IL cations and anions were spread over the enzyme surface, but the anion species dominated the nonbonded interactions with the enzyme; the number of H-bonds between $[NO_3]^-$ and the enzyme was approximately twice the number found for $[PF_6]^-$ at room temperature. This gives strong evidence to support the idea that interactions with strong H-bonding IL anions are responsible for structural unfolding and deactivation. The finding that [BMIm][NO_3] is more destabilizing than [BMIm][PF_6] is in accordance with the previous experimental observations,[120,126] and seems to be against the aforementioned Hofmeister effects, because the anion NO_3^- is more kosmotropic than PF_6^- (viscosity *B*-coefficients for both anions are −0.043 and −0.21, respectively[67]).

Recently, Bright's group started an investigation on how solvation within an IL influences the structure and dynamics of a protein.[134] A large multidomain model protein, human serum albumin (HSA), was labeled at Cys-34 (located in loop 1 of domain I) with the polarity-sensitive fluorescent probe acrylodan (Ac), which provides a convenient means to assess how ILs affect the local microenvironment surrounding a single site within the protein by fluorescence spectroscopic examinations. As the fluorescence emission spectra of the probe, either attached to the protein or free in the solution, are sensitive to the solvent polarity, their results clearly demonstrate that the local microenvironment surrounding the Ac residue within native HSA–Ac in the phosphate buffer is not as polar as the bulk solvent, a reflection of the shielding from aqueous solution afforded by the neighboring amino acid residues in HSA. The situation is different when the protein is solvated in different ILs (with 2% v/v H_2O): The polarity of the local microenvironment surrounding the HSA-attached Ac in [BMIm][BF_4] at room temperature is lower than that in the solvent itself, ruling out the possibility of solvent exposure; when solvated in [BMIm][Tf_2N] and [BMIm][PF_6], however, the local polarity around the Ac residue attached to HSA is higher than, and approximately equal to, the polarity of the specific solvent, respectively. In fact, the local polarity around the Ac residue within HSA–Ac varies according to the change in the solvent used—[BMIm][Tf_2N] > phosphate buffer > [BMIm][PF_6] > [BMIm][BF_4]—although this polarity order is somewhat temperature-sensitive. Coincidently, a few enzymes have been found to show their activities in the presence of different ILs following the same IL order,[106–108,111,123] and the major reason for this may be the IL-induced alteration in the local polarity within the microenvironment around the enzyme's active site. Additionally, it was found that the Ac reporter motion is always coupled to the global HSA protein motion while in the phosphate buffer, but not at all in the three IL media. As the temperature increases, the Ac is more exposed to the IL-rich location and thus more associated with the neighboring nonpolar amino acid residues and/or IL components, an indication of the protein being denatured. Therefore, these results clearly demonstrate that different ILs significantly affect the local polarity of the protein's microenvironment, which may be the key reason for the alteration of enzyme activity, and that the thermal unfolding process of HSA and the dynamic motion of the protein in the three IL systems deviate markedly from the mechanism followed in aqueous buffer.

The same research group furthered their study to explore the HSA–Ac rotational dynamics in aqueous buffer and in the aforementioned IL-based media as a function of temperature and water loading by using time-resolved fluorescence anisotropy and intensity decay measurements.[135] The results of this research show that the behavior of multidomain proteins dissolved in IL/water mixtures can be quite complex, depending on factors such as protein characteristics (e.g., secondary and tertiary structure, pI, disulfide and salt bridges, metal cofactors), the domain(s) in question, temperature, and the solvent composition (choice of IL, water loading).

In summary, several important implications can be derived from the above structural investigations: First, ILs have strong impacts on the protein structure and dynamics by actions such as stripping off essential water from the protein, interacting with the protein via electrostatic, hydrophobic, and H-bond interactions, and altering the physicochemical properties (such as polarity) of the microenvironment in the protein. Second, the IL-induced alteration in the protein structure and dynamics may be the major factor responsible for the change in the catalytic activity and thermal stability of the protein. A high activity or stability presented by an enzyme in an IL-based system corresponds to a compact, native-like protein structure. Finally, the IL–protein interactions may or may not follow the Hofmeister effect, but it is likely that the IL anions with high H-bond basicity have a strong propensity for dissolving and denaturing proteins.

2.4.3 Effect of Ionic Liquids on Protein Refolding and Renaturation

ILs have also been found to be able to improve protein refolding and renaturation. Summers and Flowers were the first to explore the use of ILs as additives for protein refolding.[20] The thermal unfolding of lysozyme in the absence and presence of EAN was followed by using DSC technique. No refolding was observed for the denatured protein without EAN treatment. However, if the protein was denatured in the presence of 5% EAN, it could be refolded simply by dilution, resulting in ~87% refolding and ~90% activity regained, although both the melting temperature and the unfolding enthalpy of the protein were slightly reduced. This suggests that although EAN is a denaturant, it is effective in enhancing protein refolding and renaturation. A supposition is that the ethyl group of EAN interacts with the hydrophobic portion of the protein and protects it from intermolecular association while the charged portion of the salt stabilizes the electrostatic interactions of its secondary structure. Therefore, it is likely that the major contribution of EAN in the protein refolding is to depress the aggregation of the unfolded state rather than to stabilize the refolded state.

The remarkable ability of EAN to improve refolding recovery was convincingly supported by Angell's group.[19] After storage as refrigerated in the EAN solution for 3 years, both the melting temperature and unfolding enthalpy of the lysozyme sample was almost unchanged as compared with that of the freshly prepared samples, exemplifying that the IL has bestowed an extraordinary level of protection on the protein. Interestingly, using triethylammonium methylsulfonate (TEAMS) in place of EAN resulted in a slight increase in both the unfolding temperature and unfolding enthalpy, indicative of a more effective refolding enhancer.

Mann et al.[23] used near-ultraviolet (UV) CD spectroscopy to assess the ability of four ammonium-based ILs (ethylammonium formate [EAF], propylammonium formate [PAF], 2-methoxyethylammonium formate [MOEAF], and ethanolammonium formate [EtAF]) to assist the refolding of lysozyme after it was heated to 90°C. The activity of the enzyme in aqueous solution containing different amounts of these four ILs have also been assayed and compared. The results have shown that all of them were both effective refolding enhancers and activity enhancers. But comparatively, the most significant improvement in both refolding and activity was given by EtAF, followed by MOEAF. It is speculated that the interaction between the IL cations and the protein plays an important role here. The longer alkyl chain in the cation of PAF may be responsible for its lower ability in improving refolding and activity, as adsorption of this hydrophobic cation to the protein's hydrophobic core effectively protects it from the bulk hydrophilic solvent. This can also be explained by the Hofmeister effects, as discussed earlier. The fact that both EtAF and MOEAF are superior can be attributed to the H-bond forming ability of their cations. Between these two ILs, EtAF performed more effectively because the hydroxyl group on its cation side chain can provide both an H-bond donor and an H-bond acceptor site, whereas the ether oxygen in MOEAF can act only as an H-bond acceptor. This issue will be discussed further in Section 2.6.3.

The above ILs effective in promoting refolding are all PILs. Indeed, a number of other PILs have shown the ability to convey surprising stabilization against hydrolysis and aggregation, permitting multiple unfold/refold cycles without loss to aggregation. This stabilization effect can be optimized by selecting a basic PIL or PIL mixtures, because Bryne and Angell[18] have discovered that both the unfolding temperature and unfolding enthalpy of two proteins, hen egg white lysozyme and ribonuclease A, increased with an increase in the $\delta_{N\text{-}H}$ of the PIL solution, which is a sensitive parameter for judging the PA of the PIL, as has been introduced earlier in Section 2.2.2. A higher $\delta_{N\text{-}H}$ indicates a more basic solution. Unfortunately, the knowledge about the correlation between the structure and combination of the PIL or PIL mixture and its PA and about the solvent–protein interactions in these systems is not sufficient to allow people to take full advantage of these unusual and highly tunable solvent media for the study of protein folding/unfolding processes.

On the other hand, Lange et al.[96] have selected aprotic imidazolium ILs to investigate their effects on protein refolding and renaturation. The two groups of ILs, the *N′*-alkyl and *N′*-(ω-hydroxyalkyl)-N-methylimidazolium chlorides, acted as refolding enhancers for the two model proteins, that is, hen egg white lysozyme and the single-chain antibody fragment ScFvOx. Generally, the refolding yield was higher and was reached at a higher IL concentration, when the alkyl chain attached to the imidazolium ring was longer. The protein aggregation was also effectively suppressed by these ILs. However, they were also found to trigger protein destabilization, in terms of decreasing both the unfolding temperature and unfolding enthalpy of lysozyme and lowering the midpoint concentration for guanidinium chloride (GdHCl)-induced protein unfolding. The destabilization was more serious in the presence of the IL with a longer alkyl chain attached to the imidazolium cation. This again can be illustrated by the Hofmeister effect, as has been discussed earlier in Section 2.3.2.1. The authors classified these ILs as preferentially bound, slightly to moderately chaotropic co-solvents for proteins.

The protein refolding mechanism tells us that the refolding of a given protein may be promoted by stabilizing its native state, by accelerating the kinetics of the correct folding reaction, and by suppressing unspecific aggregation of the unfolded polypeptide and/or intermediates on the folding pathway. The results obtained by Lange et al.[96] using AILs appear to be similar to what was observed with EAN, the first PIL used as protein refolding enhancer,[20] thus substantiating the idea that both AILs and PILs can be effective in improving protein refolding, presumably due to their capability of resisting the protein aggregation but not because the protein's native state is stabilized.

2.4.4 Effect of Ionic Liquids on Protein Crystallization and Fibrilization

ILs can also be used for protein crystallization[21,136,137] and fibrilization.[138] However, so far little effort has been devoted to these studies. Although a number of positive results have been reported as shown in the cited references, a detailed understanding of the IL effects and the mechanisms involved is still lacking.

2.5 PROTEIN EXTRACTION BY MEANS OF IONIC LIQUIDS

ILs can work not only as excellent reaction media for biotransformations but also as promising solvents for extraction and separation of biomolecules. Recently, a number of IL-based extraction systems have been successfully developed for biocompatible isolation and extraction of proteins and enzymes. A careful examination of the extraction mechanisms involved may help us achieve a better understanding of the IL–protein interactions.

2.5.1 Aqueous/Ionic Liquid–Liquid Extraction Systems

Aqueous/organic solvent liquid–liquid extraction processes have played an important role in biotechnological applications such as separation and extraction of biomolecules. However, it is always problematic when adopting conventional organic solvents as extraction medium because of their natural toxicity to biomolecules. It has been of increasing interest to utilize ILs in place of organic solvents in these extraction processes.

Cheng et al.[139] presented the first use of ILs for selective isolation of heme proteins. Hemoglobin (100 ng/μL, pH 7) was quantitatively extracted into IL 1-butyl-3-trimethylsilylimidazolium hexafluorophosphate ([Btmsim][PF_6]) without using any coexisting extractants/additives. The extraction efficiency fell to 20% and 93% when two other ILs, [BMIm][PF_6] and [BBIm][PF_6], were used as the extraction solvents, respectively. This implicates that a higher extraction efficiency occurred to an IL with a higher hydrophobicity. The extraction is also very selective: other proteins such as cytochrome *c*, apo-myoglobin, BSA, and transferrin, were not able to be extracted into the above three ILs. The practical applicability of this new extraction system was demonstrated by the achievement of selective separation of hemoglobin from human whole blood.

The direct extraction of hemoglobin into the IL phase can be attributed to the coordination between the ferrous ion in the heme group of hemoglobin and the imidazolium group of the IL cation. Imidazole is a strong covalent coordinating ligand with iron atom in the heme group. For the heme proteins such as hemoglobin and myoglobin, the sixth vacant coordinating position of iron atom is available, and the formation of the $[Btmsim]^+$–heme complex facilitates the transfer of heme proteins into the IL phase. This has been well illustrated by spectroscopic studies such as UV-visible (UV-Vis) spectra, fluorescence spectra, ^{58}Fe Mossbauer spectra, and CD spectra. The selectivity of this protein extraction, specific for hemoglobin and myoglobin, provides another strong support for this extraction mechanism. All those proteins (except cytochrome c) that cannot be transferred to the IL phase do not contain a heme group. Cytochrome *c*, on the other hand, cannot be extracted into the IL phase as well from a pH 7 aqueous solution. The Fe ion in the heme group of this protein is strongly coordinated in the vertical axis by two strong-field amino acid residues, His^{18} and Met^{80},[140] leaving no vacant coordinating positions for the attack from the IL cation. However, a partial extraction of cytochrome *c* can be achieved by adjusting the aqueous solution to pH 1. A plausible reason for this switch may be related to the breakage of the axial coordination bond of Fe with His (rather than Met, as pointed out in the original paper[139]): The protonation of His under a low pH may diminish its coordination capability, facilitating the attack of the IL cation $[Btmsim]^+$ to Fe and in turn the transfer of the protein into the IL phase.

Tzeng et al.[141] have introduced $[BMIm]_3[CB]$ (an organic salt derived from an affinity dye, Cibacron Blue 3GA [CB]) into the IL $[BMIm][PF_6]$ phase for liquid–liquid extraction of lysozyme from aqueous solution. The extraction efficiency decreased with an increase in pH, with a value of higher than 90% obtained at pH 4, and the protein was almost quantitatively recovered from the IL phase to the aqueous solution of 1 M KCl under pH 9–11. The extraction was specific for lysozyme in contrast to cytochrome *c*, ovalbumin, and BSA, and was repeatable for at least eight cycles. It is believed that this selective extraction is driven not only by the dye–protein affinity interaction but also by the electrostatic interaction: At pH 4, which is well under the isoelectric point of lysozyme (pI ~ 11), the protein becomes positively charged and interacts strongly with the three negatively charged sulfonate groups on CB of the organic salt $[BMIm]_3[CB]$ and even with the IL anion PF_6^-, thus facilitating the transfer of the protein into the IL phase.

2.5.2 Ionic Liquid-Based Aqueous Biphasic Systems

Aqueous biphasic systems (ABS) are clean alternatives for traditional aqueous/organic solvent liquid–liquid extraction systems. ABS are formed when aqueous solutions of one polymer (such as polyethylene glycol [PEG]) and one kosmotropic salt, or of two salts (one chaotropic and another kosmotropic), are mixed together at appropriate concentrations to form two distinct phases. By discarding the use of volatile organic solvents, aqueous biphasic extraction systems are more protein benign and environmentally friendly. In 2003, Rogers and his coworkers reported the first study of developing aqueous biphasic extraction systems based on water-miscible ILs (IL-ABS) for recycle, metathesis, and study of the distribution ratios of short chain alcohols.[142] The use of IL-ABS for protein extraction was first achieved by Du et al.,[143] who extracted proteins from human body fluids by employing a [BMIm][Cl]/K_2HPO_4 system.

Dreyer and Kragl[144] reported the first application of IL-ABS for partial purification of enzymes. A special class of ILs, "AMMOENG™," which are acyclic ammonium salts containing cations with oligoethyleneglycol units of different chain lengths, has been found to be highly effective in this application. By using ABS based on the IL AMMOENG110™ and K_2HPO_4, two different alcohol dehydrogenases (from *Lactobacillus brevis* and a thermophilic bacterium) were enriched in the IL-containing upper phase, resulting in an increase of specific activity by a factor of 2.1 and 4.0, respectively. As the IL provides extra beneficial effects such as stabilizing the enzyme and enhancing the solubility of hydrophobic substrates, this IL-ABS offers the opportunity of combining the enzyme purification process with the enzyme catalytic process.

The above IL/K_2HPO_4-ABS was further studied by the same research group.[145] By working on the effect of different protein characteristics (e.g., pH and total charge, surface area, hydrophobicity, molecular weight) on the partitioning behavior of the four model proteins (lysozyme, myoglobin, albumin, and trypsin), the authors suggested that the major driving force for the extraction process is the electrostatic interaction between the negatively charged protein and the positively charged IL cation. This assumption was supported by the SDS-PAGE results: As compared with the bands shown by the four model proteins from aqueous solution, samples from the upper phase of the IL-ABS showed a deceleration of the protein run as well as an enlargement of the bands themselves. This can be easily explained by an increase in protein size due to the attachment of the IL cation to the protein, and a number of 2–30 IL cations attached to each protein molecule can be worked out. It is worth mentioning that in this study no direct relationship was found between protein hydrophobicity and its partitioning behavior, implicating that hydrophobic interactions may not be the major factor responsible for the enrichment of the protein in the IL-containing upper phase.

Another study conducted by Pei et al.,[146] however, has shown that protein extraction in the IL-ABS was driven mainly by hydrophobic interactions, although electrostatic interactions and salting-out effects were also important. These authors have investigated the partitioning of four proteins (BSA, trypsin, cytochrome *c*, and γ-globulins) in IL-ABS based on three imidazolium ILs ([BMIm][[Br], [HMIm][Br], and [OMIm][Br]). Between 75% and 100% of the proteins could be extracted into the IL-rich phase, while their conformations were not altered, as determined by UV-Vis and FT-IR spectroscopy.

In fact, protein extraction in IL-ABS may not be driven by one or two single interactions independently, but should be attributed to cooperative functioning of a combination of IL–protein and salt–protein interactions including H-bonding and electrostatic and hydrophobic interactions. For instance, the efficiency of extracting BSA in the IL-ABS at pH 9.3 increased with an increase in the length of the alkyl chain in the imidazolium ring of the IL used[146]: [BMIm][Br] < [HMIm][Br] < [OMIm][Br]. This is against the Hofmeister effect, but in this case the hydrophobic effect may be predominant. The hydrophobicity of the IL is enhanced because of its longer alkyl chain. This gives rise to stronger hydrophobic interactions between the IL and the protein, thus driving the protein to transfer to the IL phase. On the other hand, electrostatic interactions are also responsible for this protein transfer under the extraction condition. BSA has an isoelectric point of 4.6 and hence is negatively charged at pH 9.3, facilitating the electrostatic interactions between the protein and the IL cation. Particularly, an imidazolium IL with a longer alkyl chain is more kosmotropic and hence has a higher tendency of ion pairing with the kosmotropic carboxylic groups exposed on the protein surface, especially when this protein is negatively charged. Moreover, the salt K_2HPO_4 used in the lower aqueous phase is kosmotropic and its strong salting-out effect may also be partly responsible for pushing the protein to be transferred to the IL-rich phase.

Extraction efficiency of the protein in the IL-ABS may also depend on the protein's molecular weight, as has been verified in several studies.[145,146] The protein transfer to the IL-rich phase requires the breakage of the interacting networks originally present within that phase, so as to create a cavity for the protein to be inserted; energy for this process is obtained from the interactions between proteins and oppositely charged ILs. A protein with a larger size requires a higher energy for this process and hence is more difficult to partition into the IL-rich upper phase.

Cao et al.[147] have investigated the partitioning of horseradish peroxidase (HRP) in the IL-ABS composed of [BMIm][Cl] and K_2HPO_4. Under optimal conditions, about 80% of the enzyme was extracted in the IL-rich upper phase, holding greater than 90% of the original enzyme activity. According to the change in the cation of the IL used, the enzyme activity retained in the IL-rich phase varied in the order of [EMIm][Cl] > [BMIm][Cl] > [HMIm][Cl] > [OMIm][Cl]. Interestingly, this trend is in constrast to the one obtained in the preceding experiment.[146] Presumably in this situation, the Hofmeister effect is dominating, and an imidazolium IL with a longer alkyl chain attached to the cation ring, due to its higher kosmotropicity, shows a higher potential of destabilizing the enzyme.[10] With respect to the enzyme stability, HRP present in the IL-rich phase (containing 42% H_2O) retained a fairly constant activity within 8 hours; the similar was true for the enzyme present in the same IL solvent ([BMIm][Cl]) containing 40% and 50% water, whereas a fast deactivation within 0.5–1 hour was observed for HRP in the IL containing 20% or 30% water. This indicates the crucial role of water in maintaining the HRP stability. The experiment has also revealed that the viscosity of the IL-rich upper phase (4.12 cP) is much lower than that of the PEG 4000-based ABS (79.6 cP). Obviously, this is a significant improvement compared with the traditional PEG-based ABS, making both the extraction process and the further treatments much easier.

2.5.3 Water-in-Ionic Liquid Microemulsion Systems

Water-in-oil (w/o) microemulsions, also called reversed micelles, are a ternary system normally consisting of water, an organic solvent (usually apolar), and a surfactant. The dispersed aqueous phase ("water pools") is formed, surrounded by a monolayer of surfactant molecules with their hydrophilic heads inward and hydrophobic tails outward, dissolved in the surrounding bulk organic phase. Proteins can be entrapped in the water pools with their biological activity and structure fully retained. Therefore, this system has been well accepted as an attractive nonconventional reaction medium for biocatalysis[148–150] and has been employed as a useful tool for protein extraction.[151] Recently, interest has been roused in the development of water-in-IL (w/IL) microemulsion systems for biocatalytic reactions[152,153] and for protein extraction,[154] where ILs have been introduced as a substitute of the conventional organic solvent.

Shu et al.[154] have developed a w/IL microemulsion system, prepared with water, sodium bis(2-ethylhexyl) sulfosuccinate (AOT), and the IL [BMIm][PF_6], for selective extraction of hemoglobin. An extraction efficiency of ca. 96% was achieved for a 100 ng/μL hemoglobin solution (pH 6.3) by using an equal volume of the microemulsions, and 73% of the hemoglobin that was transferred could be rapidly extracted back into an aqueous phase with urea as stripping reagent. This system is only specific for hemoglobin but cannot be used to extract other proteins such as cytochrome *c*, BSA, and transferrin. The protein transferred in the micromulsion system distributed in two states: About 60% dispersed in the bulk IL phase and 30% entrapped in the "water pools" surrounded by the bulk IL entities. The coordination effect may be the major driving force for hemoglobin transfer into the bulk IL phase, as has been demonstrated before;[139] and the protein is drawn into the water pools of the microemulsions mainly due to the electrostatic interactions, which can be verified by the dependency of the extraction efficiency on both pH and salt concentration in the aqueous phase. At a pH lower than the isoelectric point of hemoglobin (i.e., pH < 6.9), a quantitative extraction was achieved simply due to the electrostatic attraction between the positively charged hemoglobin and the negatively charged AOT in the "water pools." When the pH increased, the extraction efficiency significantly declined because of the increase in electrostatic repulsion between the protein and the surfactant, which were both negatively charged. The protein was back-extracted into an aqueous solution with a high salt concentration, implicating that the entrapment of the protein in the water pools was weakened because of the stronger electrostatic interactions in the back-extracted aqueous phase.

The w/IL microemulsion systems are not only suitable for protein extraction, but also promising for performing biocatalytic reactions. When encapsulated in the w/IL microemulsions composed of an anionic surfactant AOT, a hydrophobic IL [OMIm][Tf_2N], and 1-hexanol, HRP was found to be both more active and more stable than in a conventional w/o microemulsion system (composed of AOT/isooctane/water).[152] Similar findings were also reported for three lipases in the w/IL microemulsion system composed of a nonionic surfactant Tween 20 or Triton X-100 and a hydrophobic IL [BMIm][PF_6], and structural studies via FT-IR and CD spectroscopy indicated that the lipases entrapped in the w/IL microemulsions tend to retain their native structure or adapt a more rigid structure in comparison with other reaction media, which correlated well with the higher operational stability obtained in the w/IL system.[153]

2.6 PROPER SELECTION OF IONIC LIQUIDS FOR BIOCATALYSIS

ILs have been regarded as designer solvents because their physical and chemical properties can be tailored by modification of their structures and therefore ILs can be designed deliberately for different reaction conditions. This is one of the major attractions of making ILs an alternative to conventional organic solvents. However, whether this beneficial attraction can be advantaged in biocatalytic processes is determined very much by our understanding of the relationship between IL structure and its properties and of the dependence of enzymatic performance on the IL structure and properties. Although a tremendous amount of work has been carried out on applications of ILs and a reasonable amount of knowledge about their structure and physicochemical properties has been accumulated over recent years, much work is still needed to achieve a comprehensive understanding on this aspect. Meanwhile, there has been a growing demand for approaching green chemistry, while the commonly used ILs are not always green,[155,156] and their toxicity and biodegradability have to be concerned.[157,158] Therefore, it has become of significant importance to explore new biocompatible ILs and to conduct systematic research on the relationship between IL structure and properties and enzymatic performance, so as to work out valuable criteria for IL selection. The amino acid ILs and ammonium and phosphonium ILs, to be introduced in this section, represent this trend.

2.6.1 Amino Acid Ionic Liquids

Because an amino acid has both a carboxylic group and an amino group, both of which have the ability to introduce other functional moieties, and therefore can be used as both a cation and an anion, amino acid ionic liquids (AAILs) prepared from amino acids and their derivatives provide a big group of high-quality functionalized ILs that not only have chiral centers, biodegradable characteristics, and high biocompatibility, but are also a perfect choice for structure–property studies.

Since Fukumoto et al.[159] reported their first synthesis of AAILs from 20 natural amino acids, a large number of AAILs have been prepared by using amino acids or their derivatives as cations[160] or anions,[161,162] and their properties have been characterized. It has been found that the properties of AAILs, such as their melting and decomposition temperatures, viscosity, and ionic conductivity, are highly dependent on the side groups of the amino acids involved, and the symmetry of both the cation and anion also plays some determining roles.[163] A careful examination of the 20 AAILs (with $[EMIm]^+$ as the cation, prepared by Fukumoto et al.[159]) revealed that ILs with carboxylic, amide, hydroxyl, and aromatic groups had a glass transition temperature (T_g) higher than that of other AAILs and that an increase in the alkyl side-chain length in the amino acid anion coincided with a gradual increase in T_g. Obviously, this results from stronger H-bonding and hydrophobic interactions, respectively. Tao et al.[160] have also found that for those AAILs they synthesized (with amino acids as cations), esterification on the carboxylic group (such as Asp, Glu) resulted in a significant decrease in the melting point of the IL. It is reasonable to assume that high melting points or high glass transition temperatures result from strong interactions within the IL complex. H-bonds should be strengthened in the presence of carboxylic, amide, and hydroxyl groups, but weakened by esterification. Meanwhile, hydrophobic interactions would be strong with the help of aromatic groups. Kagimoto et al.[161] have prepared 20 tetraalkylphosphonium-based AAILs, some of which show lower viscosities and higher decomposition temperatures (>300°C) than previously reported ammonium-based AAILs. Jiang et al. have also synthesized four tetraalkylammonium-based AAILs with low viscosities (down to 81 cP).[162]

An interesting feature for AAILs is their "lower critical separation temperature (LCST)" behavior,[163] that is, the solubility of water in the IL increases upon cooling, and below the LCST the mixture of IL and water is miscible in all proportions. This behavior is unusual because solubility of most materials usually increases during heating. Although it has been found in polymer blends,[164] LCST behavior is rare for low molecular weight liquid mixtures. So this behavior is very useful for biocatalytic processes because enzymes are easily denatured under high temperatures. The phase separation temperature can be lowered to ambient temperature by increasing the water content, and the reversible phase change between a homogeneous IL/H_2O mixture and separated phases can be implemented simply by changing the temperature of the solution by a few degrees. Taking into consideration the solubility of the enzyme, the substrates, and the products in both IL and water, unique reaction processes can be designed so that the reaction components can be introduced into different phases, homogeneous reactions can be carried out in the IL/H_2O mixture at a lower temperature, and a slight increase in the temperature will ease the phase separation and recovery of each reaction component involved. Enzymes can even be introduced into the IL phase after some modifications.[117]

Although so many AAILs have been prepared with good solvent properties, their successful applications in biocatalysis have been rarely reported so far. Zhao et al.[165] synthesized some ILs carrying anions of chiral- or ω-amino acids and found that in their aqueous solution, these ILs were capable of stabilizing the protease activity and enhancing the enzyme's enantioselectivity.

2.6.2 Ammonium and Phosphonium Ionic Liquids

Ammonium and phosphonium ILs represent another promising type of biocompatible ILs that can be prepared from natural sources. Because both ammonium and phosphonium salts are commonly found in living creatures, it is reasonable to assume that the ILs that are based on these two cations should be protein benign and thus may provide a good environment for enzymes.

Indeed, CALB has been found to dissolve in [Et_3MeN][$MeSO_4$] while maintaining both its activity and structure,[126] and so did subtilisin in diethanolammonium chloride ([DEA][Cl]).[22] There are other ammonium ILs that promote excellent enzyme performance, as has been reported in the literature (see Mann et al.[23], de Diego et al.[39], Fujita et al.[95], de Gonzalo et al.[166], Das et al.[167] for example). It has to be noted that most of the ammonium-based ILs listed here contain hydroxyl groups, which may be critical for boosting enzyme catalysis. This will be discussed later in Section 2.6.3.

Comparatively, phosphonium-based ILs have drawn less attention in terms of biocatalysis, although a number of literature reports have been given with respect to the synthesis and characterization of these ILs, including those tetraalkylphosphonium-based AAILs as mentioned earlier.[161] Abe et al.[47] have observed a rather high rate of lipase-catalyzed transesterification reaction when 2-methoxyethyl(tri-n-butyl)phosphonium bis(trifluoromethanesulfonyl)imide ([MEBu$_3$P][Tf$_2$N]) was used as a solvent, thus offering an example of a lipase-catalyzed reaction in a phosphonium IL superior to that in diisopropyl ether.

However, ecotoxicity and biodegradation of these ILs have to be concerned. A few results have revealed that some phosphonium ILs showed low levels of biodegradation because they are highly toxic to the microorganisms responsible for biodegradation.[168,169] The anticancer activity and cytotoxicity of some ammonium- and phosphonium-based ILs have also been determined recently.[170] The results showed that the chain length of the alkyl substitution on the cations played a crucial role toward antitumor activity and cytotoxicity of these ILs and that the phosphonium-based ILs were generally more active and less cytotoxic than ammonium ones.

2.6.3 Design of Ionic Liquids for Biocatalysis

Walker and Bruce were the first to describe the idea of IL design for biocatalytic reactions.[171] They synthesized a range of ILs, using [BMIm][PF$_6$] as a template, but with incrementally increased H-bonding capabilities through the introduction of a hydroxyl group to the cation and the use of a more strongly nucleophilic anion (such as Cl$^-$ and glycolate) in place of PF$_6^-$. These ILs were evaluated by using an NADP$^+$-dependent morphine dehydrogenase (MDH) to catalyze the oxidation of codeine to codeinone. For the three ILs with the same hydroxy-functionalized cations ([OH-BMIm]$^+$), the conversion decreased following the order [OH-BMIm][PF$_6$] > [OH-BMIm][glycolate] > [OH-BMIm][Cl], which is consistent with the reversed order of the H-bonding capability of the anions. This again supports the presumption proposed by Lau et al.[126] that IL anions with strong H-bonding capability would elicit enzyme dissolution and denaturation. For the ILs holding the same anion ([PF$_6$]$^-$ or glycolate), the one with a hydroxylated cation always promoted a higher production yield, compared with its nonhydroxylated analogue, at the same water content. Another advantage presented by the hydroxylated ILs is that they were still fairly active at a very low water level (1% and <100 ppm), whereas the nonhydroxylated ILs were almost inactive under this situation. For the first time, this experiment substantiated that enzyme activity can be significantly reduced with increasing H-bonding capability of the anion, but can benefit from an enhancement in the same property of the cation.

Almost simultaneously, the same authors of the above study extended their research and employed one of the new ILs they synthesized, [OH-BMIm][glycolate], in a combined biological and chemical catalysis for the production of opioid oxycodone from codeine.[172] The ability of this functionalized IL to dissolve the substrate, the chemical catalyst, the enzyme (with little denaturation), and its cofactor permits this reaction to be conducted in a single solvent and provides the potential for performing combined biochemical homogeneous catalytic processes in "one pot."

It is interesting that ILs with hydroxylated cations promote a high enzyme activity, especially at extremely low water levels. Holding the essential water for the enzyme may be one explanation for this. But more importantly, these hydroxyl groups can directly H-bond with the hydrolytic entities on the enzyme surface, so that the enzyme resumes its flexible conformation and does not sensitively require water for optimizing its activity. Since the favorite environment for enzymes is water, which is a powerful H-bonding medium, selecting ILs that mimic the molecular structure of water to hold H-bond accepting/donating functionalities would be advantageous in solubilizing the enzyme and facilitating the enzyme action. Hydroxy-functionalized ILs meet the requirement for this water-mimicking property, and therefore it is no wonder that they can generate the beneficial effect to the enzymes. However, it is unexplainable why the enzymes seem to prefer the IL cations possessing H-bond forming capability but not the anions.

The advantage of hydroxylation on the IL cation has been confirmed by Lange et al.[96] and Mann et al.[23] As described by Lange et al.,[96] although both N′-alkyl and N′-(ω-hydroxyalkyl) N-methylimidazolium chlorides destabilized lysozyme, terminal hydroxylation of the alkyl chain in the cation allows the IL to effectively suppress the protein destabilization induced both by heat and by GdHCl. The hydroxylated ILs were also superior to their nonhydroxylated analogues in eliciting higher refolding yields for the two proteins, lysozyme and antibody fragment ScFvOx. As has been exemplified in Section 2.4.3, among the three primary ammonium ILs containg a 2-C side chain, the one with a terminal hydroxyl group is superior to the one containing a methoxy group, both being more effective in promoting protein refolding and protein activity than their nonfunctionalized analogues: EtAF > MOEAF > EAF,[23] substantiating that the IL cation with a higher H-bond forming ability (hydroxyl group > ether oxygen > nonfunctionalized alkyl chain) is more beneficial to the enzyme. More importantly, as this conclusion applies to both imidazolium[96,171] and ammonium[23] ILs, it is reasonable to infer that it must also be applicable to other ILs. Actually, this is further supported by the IL screening test conducted by Arning et al.,[158] which will be discussed later in this section. Figure 2.8 shows the structures of rationally designed ILs with hydroxyl-functionalized and ether-containing cations that have been discussed in this chapter.

Inspired by Walker and Bruce's work regarding cation hydroxylation and also the well-known stabilization effect of TRIS (tris(hydroxymethyl)aminoethane, a hydrophilic lyoprotectant) as an excipient, Das et al.[167] designed an IL possessing four hydroxyethyl moieties, while keeping a structural resemblance to TRIS, for HRP. By using this TRIS-like IL as the solvent, $[N(CH_2CH_2OH)_4][CF_3SO_3]$, the HRP activity was at least 30- to 240-fold enhanced, relative to that obtained in conventional ILs such as $[BMIm][CF_3SO_3]$ and $[BMIm][BF_4]$, and more than 10 times greater than that in methanol, a common organic solvent used for HRP.

A short period later, de Gonzalo et al.[166] also independently designed four hydroxy-functionalized "TRIS-like" ILs, holding TRIS-mimicking cations and $MeSO_4^-$, $EtSO_4^-$,

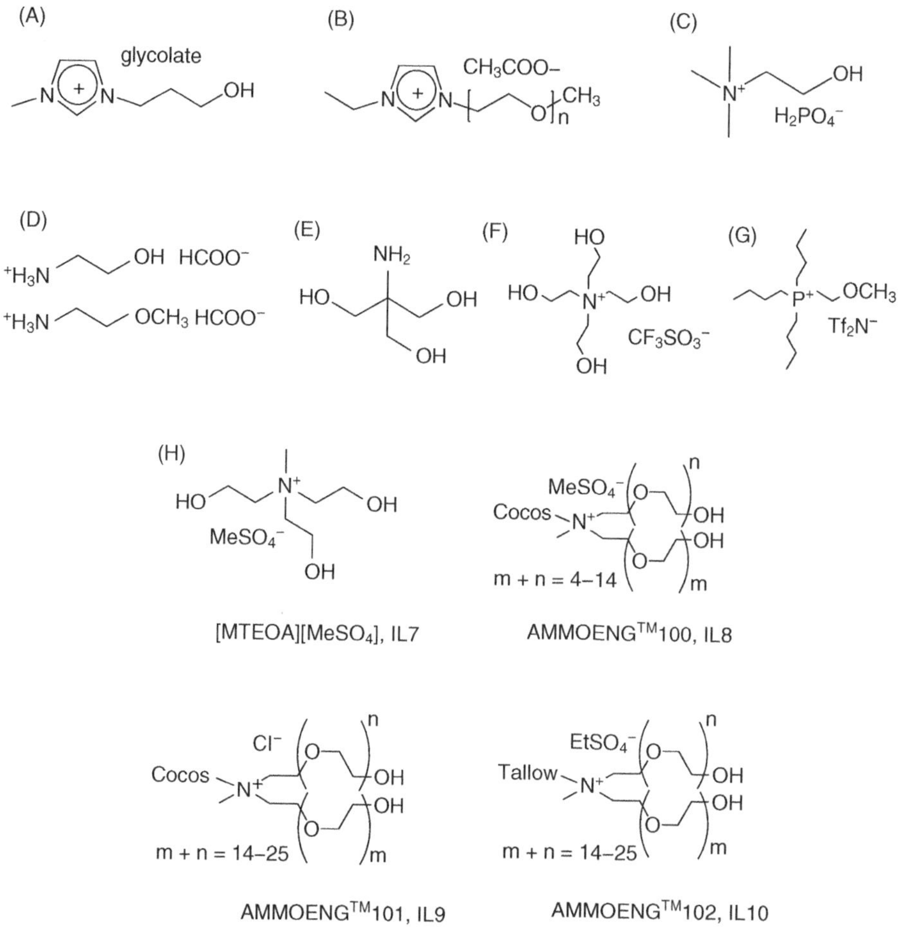

Figure 2.8. Hydroxy-functionalized and ether-containing ionic liquids that are rationally designed. (A) 1-(3-hydroxypropyl)-3-methylimidazolium glycolate[171]; (B) $[Me(OEt)_3\text{-Et-Im}]$ $[CH_3COO]$[173]; (C) [choline]$[H_2PO_4]$[95]; (D) ethanolammonium formate (EtAF) and 2-methoxyethylammonium formate (MOEAF)[23]; (E) tris(hydromethyl) aminomethane (TRIS); (F) tetrakis (2-hydroxyethyl) ammonium triflouromethanesulfonate ($[N(CH_2CH_2OH)_4][CF_3SO_3]$)[167]; (G) 2-methoxyethyl(tri-n-butyl)phosphonium bis(trifluoromethanesulfonyl)imide ($[MEBu_3P]$ $[Tf_2N]$)[47]; (H) four hydroxy-functionalized ILs used by de Gonzalo et al.[166]

and Cl^- as the counter-anions. Alcohol dehydrogenase can be dissolved in aqueous solution containing up to 90% (v/v) of these H_2O-miscible ILs to catalyze ketone reduction with excellent conversions (Figure 2.9, for the IL structures see Figure 2.8H). Enzymes normally suffer a significantly diminishing activity in the presence of a high IL content.[101,158]

The above observations motivated Zhao et al.[173] to design a series of imidazolium ILs that are capable of dissolving carbohydrates but do not considerably inactivate the

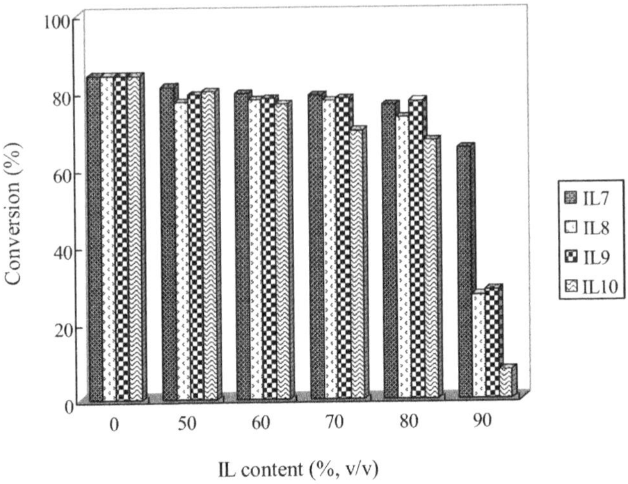

Figure 2.9. Conversions (%) for the ketone reduction catalyzed by the alcohol dehydrogenase ADH-"A" from *Rhodococcus ruber* at various concentrations of different ionic liquids (plotted using data from de Gonzalo et al.[166]).

immobilized lipase B from *Candida antarctica*, so that lipase-catalyzed transesterification of methyl methacrylate with D-glucose and cellulose can be performed while the substrates can be dissolved in the same IL solvent. These rationally designed ILs consist of glycol-substituted cations and acetate anions, in the form of $[Me(OEt)_n\text{-}Et\text{-}Im][CH_3CO_2]$. The H-bond forming anions, oxygen-containing cations, and low cation bulkiness are identified to be beneficial to the carbohydrate dissolution. The cellulose solubility in the IL was lowered in the presence of the imidazolium cations with a longer alkyloxyalkyl chain or with a hydroxyl group at the end of the side chain. The oxygen atom in the alkyloxyalkyl chain of the cation is believed to act as a Lewis-base/H-bond acceptor. Although a long oxygenated chain may offer more IL cation–cellulose interactions in favor of the cellulose dissolution, it may also render the IL to be more bulky and more hydrophobic, which is a problem for solvating the cellulose. Interestingly, in disagreement with the results from the preceding studies, terminal hydroxylation of the alkyloxyalkyl chain of the imidazolium cation did not favor the enzyme activity, nor the cellulose dissolution. The authors' explanation is that the terminal hydroxyl group on the cation may form H-bonds with the H-bond-forming IL anion, acetate, thus reducing the dissolution power of the overall IL, and that the possibility of interaction between this terminal hydroxyl group and the co-substrate of the transestesterification reaction, 1-propanol, may be responsible for the low activity presented by the enzyme. As far as the effect of IL structure on lipase activity is concerned, the activity increased with an increase in the length of the alkyloxyalkyl chain on the imidazolium cation. It is reasonable to speculate that a longer oxygenated chain permits more H-bonding interactions with the enzyme, hence stabilizing it; on the other hand, as the side chain of the cation is longer, the fraction of the anion in the entire IL molecule becomes smaller, and in turn its detrimental effect in terms of H-bonding to the enzyme to cause enzyme denaturation, an issue put forward by Lau et al.,[126] will be weakened. Another interesting finding from this study is that high lipase activity was usually accompanied by low cellulose solubility. For instance, among all the 36 ILs tested, [choline][Tf_2N] was the one that offered the greatest reaction rate and yet the cellulose solubility in it was too low to be determined. This appears to be coincident with what was observed by Hu et al.,[107] reinforcing the effect of substrate ground-state stabilization.

On the other hand, a recent study conducted by Arning et al.[158] deserves a special attention. In order to contribute a deeper insight into the (eco)toxicological hazard potential of ILs to humans and the environment, a set of 79 ILs of different types were used in an acetylcholinesterase (AchE) inhibition screening assay to identify the specific effect of IL cations and anions on enzyme activity. This study has provided valuable information on the IL structure–activity relationships. The inhibitory effect of an IL was evaluated by using its half maximal inhibitory concentration (IC_{50}), which represents the concentration of the IL that is required for 50% inhibition of the enzyme, acetylcholinesterase. Among the broad set of anion species tested (inorganic, organic, and complex borate anions), the vast majority exhibited no effect on the enzyme, and only F^-, PF_6^-, and SbF_6^- can be identified as AchE inhibitors. The cations elicited a more profound effect as follows, in terms of their head groups and side chains:

(1) The most striking inhibitory effect could be detected for the N-dimethylaminopyridinium and the quinolinium head groups, which are both highly aromatic, whereas the polar and nonaromatic morpholinium head group, as well as the sterically bulky tetrabutyl-ammonium and phosphonium cation, exhibited the lowest inhibitory potential to the enzyme. The remaining cations, such as the aromatic pyridinium and imidazolium head groups, as well as the heterocyclic but nonaromatic piperidinium and pyrrolidinium moieties, showed intermediate inhibitory effect. The strong inhibitory effect exhibited by the highly aromatic head groups may have been triggered by their strong hydrophoblic interactions with the enzyme, thereby perturbing the protein conformation and yielding deactivation.

(2) The strong effect on the enzyme activity for the dimethylaminopyridinium and the quinolinium cations is dominated by the cationic core structure, whereas for the less active head groups the lipophilicity of the side chain is the dominating factor mediating the inhibitory potential. Indeed, for a series of imidazolium head groups connected to different alkyl and functionalized side chains, a quantitative structure–activity relationship was derived by the linear regression of the log IC_{50} values versus the logarithms of the HPLC-derived lipophilicity parameter k_0 of the IL cations: A decrease in the IC_{50} value corresponded to an increase in the lipophilicity of the side chain. For instance, the longer side chain in the imidazolium, pyridinium, and ammonium side chains elicited a lower IC_{50} value. This is consistent with the previous findings[39,96,97,106] and can be explained by the Hofmeister effects, as has been discussed in Section 2.3.2.1.

(3) Another discovery derived from the above IC_{50}–lipophilicity relationship is that the more polar functionalized side chains of the IL cations exhibited a

lower inhibitory potential than their lipophilic alkyl references, in concert with what Walker and Bruce have suggested.[171] Obviously, the hydroxy-functionalized side chains (both an H-bond donor and acceptor) showed the weakest inhibitory potential compared with the less polar ether analogues (only an H-bond acceptor), which again were less inhibitory as compared with the head groups containing only nonfunctionalized nonpolar alkyl chains. This strongly supports the preceding observations with respect to the favorable effect of using hydroxy-functionalized[23,96,166,167,171,174,175] and ether-containing[47,173,174] IL cations.

(4) The fact that [OMIm][Cl] was identified as a potent inhibitor while its uncharged analogue, 1-octyl-imidazolium, was not implicates that the positive charge of the IL cation is crucial for the imidazolium IL to interact with the enzyme causing inhibition.

Taking into account all the above results, the authors concluded that while IL anions do not have a significant impact on the inhibition of acetylcholinesterase, the positively charged imidazolium ring, a widely delocalized aromatic system, and the lipophilicity of the side chains connected to the cationic head groups can be identified as the key structural elements responsible for inducing enzyme inhibition. This has been well explained by the active site structure and catalytic mechanism of acetylcholinesterase, and it is believed that these conclusions should be generally applicable to some extent. Therefore, when aiming at the design of nontoxic ILs, the dimethylaminopyridinium and the quinolinium cations should be avoided, and the inhibition potential of the IL can be lowered by choosing polar, nonaromatic head groups or by incorporating polar hydroxy, ether, or nitrile functions into the side chains connected to the cationic core structure.

2.6.4 Proposed Guidelines for Selecting/Designing Biocompatible Ionic Liquids

How to select or design a biocompatible IL is indeed a challenging job to face, as there are so many issues to be considered, including the environmental problems (such as eco-toxicology and biodegradability), solvent properties, and biocatalytic performance. Most studies in the past were exploratory. To our delight is, as the knowledge about the relationships between IL structures and enzyme performance is accumulating, our use of ILs for biocatalytic processes has been switching from random selection to rational design.

Based on our discussion of the Hofmeister effects, we have proposed that designing water-mimicking ILs composed of chaotropic cations and kosmotropic anions may be the trend to be followed, as these ILs have shown the capability of maintaining a high level of enzyme activity and stability as well as a potential of dissolving the enzyme for the purpose of homogeneous catalysis.[10] Incorporating hydroxyl or ether functionalities into the cationic side chain is definitely a good way to tune the IL to become water-mimicking, and has proven to be effective in making the IL more enzyme-friendly. So far, the best illustration for these ideas was given by [choline][H_2PO_4], an

IL with remarkable stabilizing effect as mentioned earlier. This is a new biocompatible IL consisting of a chaotropic choline cation and a kosmotropic dihydrophosphate anion, both being active H-bond formers. Actually, a considerable number of the successful IL designs introduced above also share this similar structural pattern.[23,47,166,167] Additionally, from a closer look at these examples one can also tell that most of these ILs were made up of hydroxy-functionalized ammonium cations; ILs with a TRIS-like cation have shown remarkable potential in promoting biocatalytic reactions.[166,167] In fact, our very recent study[183] on both activity and stability of *Penicillium expansum* lipase and mushroom tyrosinase has given a direct demonstration for this. Our study has also revealed that the IL cations play a more crucial role than their counter anions in affecting the enzyme performance, in support of what Arning et al. have observed[158] and that ammonium ILs composed of chaotropic cations (favorably with H-bonding capability) and kosmotropic anions are favored for enzyme catalysis.

Among all the different IL types, ammonium and phosphonium ILs are more natural, as has been discussed in Section 2.6.2, and more importantly, they have shown to be capable of promoting excellent biocatalytic performance, such as activity[22,23,47,107] and stability,[95] exhibiting a very weak inhibitory potential toward enzymes,[158] and enhancing protein refolding and renaturation.[18,20] Therefore, according to the observations and suggestions from Arning et al.,[158] use of ammonium- and phosphonium-based ILs should be recommended, while IL cations with highly aromatic head groups such as dimethylaminopyridinium and quinolinium should be avoided. Of course one has to keep in mind that considerable care should be exercised in the choice of these ILs concerning their ecotoxicity and biodegradability.

Nevertheless, more work has to be done before coming up with a firm solution to this IL design issue. For example, why are those highly H-bond-forming anions are detrimental to some enzymes but beneficial to others? Why do some enzymes become more active or more stable when the side chain of the IL cation is shorter while other enzymes act in the opposite way? Why do enzymes usually prefer IL cations to be H-bond forming but not IL anions? As discussed before, we believe that answers to questions like these may vary, depending on the conformation, dynamics, active site structure, and catalytic mechanism of each specific enzyme.

2.7 CONCLUDING REMARKS

We have described what is currently known about the IL–protein interactions, which are reflected via the Hofmeister effects of ILs on proteins and enzymes, the roles that ILs play in affecting the biological functions of these biomolecules, such as their activity and stability, structure and dynamics, refolding and renaturation, and the use of ILs in protein extraction and purification. All the impacts of ILs on protein functions can be explained basically by their H-bonding and electrostatic and hydrophobic interactions with the protein, depending on their different physicochemical properties such as hydrophobicity, nucleophilicity, and H-bond acidity/basicity. A thorough understanding of this issue is still far from complete, and the following relationships must be investigated further: (1) the correlation between the structure of an IL and its physicochemical properties; (2) the correlation between the IL structure and its interactions with proteins; and (3) the correlation between the IL structure and protein functions. Undoubtedly, a thorough understanding of all these relationships will stimulate our development of instructive guidelines for designing biocompatible ILs to facilitate their applications in biotransformations.

REFERENCES

1 M. Erbeldinger, A. J. Mesian, A. J. Russell, *Biotechnol. Prog.* **2000**, *16*, 1129–1131.

2 R. M. Lau, F. van Rantwijk, K. R. Seddon, R. A. Sheldon, *Org. Lett.* **2000**, *2*, 4189–4191.

3 U. Kragl, M. Eckstein, N. Kaftzik, *Curr. Opin. Biotechnol.* **2002**, *13*, 565–571.

4 S. Park, R. Kazlauskas, *Curr. Opin. Biotechnol.* **2003**, *14*, 432–437.

5 F. van Rantwijk, R. M. Lau, R. A. Sheldon, *Trends Biotechnol.* **2003**, *21*, 131–138.

6 Z. Yang, W. Pan, *Enzyme Microb. Technol.* **2005**, *37*, 19–28.

7 Y. H. Moon, S. M. Lee, S. H. Ha, Y.-M. Koo, *Korean J. Chem. Eng.* **2006**, *23*, 247–263.

8 F. van Rantwijk, R. A. Sheldon, *Chem. Rev.* **2007**, *107*, 2757–2785.

9 P. Domínguez de María, *Angew. Chem. Int. Ed. Engl.* **2008**, *47*, 6960–6968.

10 Z. Yang, *J. Biotechnol.* **2009**, *144*, 12–22.

11 J. Dupont, *J. Braz. Chem. Soc.* **2004**, *15*, 341–350.

12 H. Katayanagi, K. Nishikawa, H. Shimozaki, K. Miki, P. Westh, Y. Koga, *J. Phys. Chem. B* **2004**, *108*, 19451–19457.

13 L. Zhang, Z. Xu, Y. Wang, H. Li, *J. Phys. Chem. B* **2008**, *112*, 6411–6419.

14 S. Gabriel, J. Weiner, *Ber.* **1888**, *21*, 2669–2679.

15 P. Walden, *Bull. Acad. Imper. Sci.* **1914**, *8*, 405–422.

16 T. L. Greaves, C. J. Drummond, *Chem. Rev.* **2008**, *108*, 206–237.

17 D. R. MacFarlane, K. R. Seddon, *Aust. J. Chem.* **2007**, *60*, 3–5.

18 N. Byrne, C. A. Angell, *J. Mol. Biol.* **2008**, *378*, 707–714.

19 N. Byrne, L.-M. Wang, J.-P. Belieres, C. A. Angell, *Chem. Commun.* **2007**, 2714–2716.

20 C. A. Summers, R. A. Flowers II, *Protein Sci.* **2000**, *9*, 2001–2008.

21 J. A. Garlitz, C. A. Summers, R. A. Flowers II, G. E. O. Borgstahl, *Acta Crystallogr.* **1999**, *D55*, 2037–2038.

22 F. Falcioni, H. R. Housden, Z. Ling, S. Shimizu, A. J. Walker, N. C. Bruce, *Chem. Commun.* **2010**, *46*, 749–751.

23 J. P. Mann, A. McCluskey, R. Atkin, *Green Chem.* **2009**, *11*, 785–792.

24 S. Pandey, K. A. Fletcher, S. N. Baker, G. A. Baker, *Analyst* **2004**, *129*, 569–573.

25 L. Cammarata, S. G. Kazarian, P. A. Salter, T. Welton, *Phys. Chem. Chem. Phys.* **2001**, *3*, 5192–5200.

26 K. Miki, P. Westh, K. Nishikawa, Y. Koga, *J. Phys. Chem. B* **2005**, *109*, 9014–9019.

27 A. Zaks, A. M. Klibanov, *J. Biol. Chem.* **1988**, *263*, 8017–8021.

28 R. H. Valivety, P. J. Halling, A. R. Macrae, *FEBS Lett.* **1992**, *301*, 258–260.

29 A. Zaks, A. M. Klibanov, *J. Biol. Chem.* **1988**, *263*, 3194–3201.

30 P. J. Halling, *Trends Biotechnol.* **1989**, *7*, 50–51.

31 R. H. Valivety, P. J. Halling, A. R. Macrae, *Biochim. Biophys. Acta* **1992**, *1118*, 218–222.

32 Z. Yang, A. J. Russell, in *Enzymatic Reactions in Organic Media* (Eds. A. M. P. Koskinen, A. M. Klibanov), Blackie Academic & Professional, New York, **1996**, pp. 43–69.

33 Z. Yang, Y.-J. Yue, M. Xing, *Biotechnol. Lett.* **2008**, *30*, 153–158.

34 J. A. Berberich, J. L. Kaar, A. J. Russell, *Biotechnol. Prog.* **2003**, *19*, 1029–1032.

35 T. de Diego, P. Lozano, S. Gmouh, M. Vaultier, J. L. Iborra, *Biotechnol. Bioeng.* **2004**, *88*, 916–924.

36 Z. Yang, K.-P. Zhang, Y. Huang, Z. Wang, *J. Mol. Catal., B Enzym.* **2010**, *63*, 23–30.

37 O. Ulbert, T. Fráter, K. Bélafi-Bakó, L. Gubicza, *J. Mol. Catal., B Enzym.* **2004**, *31*, 39–45.

38 N. Kaftzik, W. Kroutil, K. Faber, U. Kragl, *J. Mol. Catal., A Chem.* **2004**, *214*, 107–112.

39 T. de Diego, P. Lozano, M. A. Abad, K. Steffensky, M. Vaultier, J. L. Iborra, *J. Biotechnol.* **2009**, *140*, 234–241.

40 M. Eckstein, M. Sesing, U. Kragl, P. Adlercreutz, *Biotechnol. Lett.* **2002**, *24*, 867–872.

41 M. Noël, P. Lozano, M. Vaultier, J. L. Iborra, *Biotechnol. Lett.* **2004**, *26*, 301–306.

42 D. Barahona, P. H. Pfromm, M. E. Rezac, *Biotechnol. Bioeng.* **2006**, *93*, 318–324.

43 P. J. Halling, *Biotechnol. Tech.* **1992**, *6*, 271–276.

44 E. Fehér, B. Major, K. Bélafi-Bakó, L. Gubicza, *Biochem. Soc. Trans.* **2007**, *35*, 1624–1627.

45 T. Itoh, S. Han, Y. Matsushita, S. Hayase, *Green Chem.* **2004**, *6*, 437–439.

46 T. Itoh, Y. Matsushita, Y. Abe, S.-H. Han, S. Wada, S. Hayase, M. Kawatsura, S. Takai, M. Morimoto, Y. Hirose, *Chemistry* **2006**, *12*, 9228–9237.

47 Y. Abe, K. Kude, S. Hayase, M. Kawatsura, K. Tsunashima, T. Itoh, *J. Mol. Catal., B Enzym.* **2008**, *51*, 81–85.

48 F. Hofmeister, *Arch. Exp. Pathol. Pharmakol.* **1888**, *24*, 247–260. Translated and republished by W. Kunz, J. Henle, B. W. Ninham, *Curr. Opin. Colloid Interface Sci.* **2004**, *9*, 19–37.

49 J. C. Warren, L. Stowring, M. F. Morales, *J. Biol. Chem.* **1966**, *241*, 309–316.

50 R. L. Baldwin, *Biophys. J.* **1996**, *71*, 2056–2063.

51 B. Hribar, N. T. Southall, V. Vlachy, K. A. Dill, *J. Am. Chem. Soc.* **2002**, *124*, 12302–12311.

52 P. Bauduin, A. Renoncourt, D. Touraud, W. Kunz, B. W. Ninham, *Curr. Opin. Colloid Interface Sci.* **2004**, *9*, 43–47.

53 M. Boström, D. R. M. Williams, B. W. Ninham, *Curr. Opin. Colloid Interface Sci.* **2004**, *9*, 48–52.

54 K. D. Collins, G. W. Neilson, J. E. Enderby, *Biophys. Chem.* **2007**, *128*, 95–104.

55 W. Kunz, P. Lo Nostro, B. W. Ninham, *Curr. Opin. Colloid Interface Sci.* **2004**, *9*, 1–18.

56 H. Zhao, *J. Mol. Catal., B Enzym.* **2005**, *37*, 16–25.

57 B. W. Ninham, *Adv. Colloid Interface Sci.* **1999**, *83*, 1–7.

58 M. Boström, D. R. M. Williams, P. R. Stewart, B. W. Ninham, *Phys. Rev. E* **2003**, *68*, 041902.

59 J. Setschenow, *Z. Phys. Chem.* **1889**, *4*, 117–125.

60 A. Voet, *Chem. Rev.* **1937**, *20*, 169–179.

61 P. H. von Hippel, K. Y. Wong, *Science* **1964**, *145*, 577–580.
62 D. Hall, A. P. Milton, *Biochim. Biophys. Acta* **2003**, *1649*, 127–139.
63 W. Melander, C. Horvath, *Arch. Biochem. Biophys.* **1977**, *183*, 200–215.
64 H. Zhao, *J. Chem. Technol. Biotechnol.* **2006**, *81*, 877–891.
65 H. Zhao, Z. Song, *J. Chem. Technol. Biotechnol.* **2007**, *82*, 304–312.
66 G. Jones, M. Dole, *J. Am. Chem. Soc.* **1929**, *51*, 2950–2964.
67 H. D. W. Jenkins, Y. Marcus, *Chem. Rev.* **1995**, *95*, 2695–2724.
68 J. M. Broering, A. S. Bommarius, *J. Phys. Chem. B* **2005**, *109*, 20612–20619.
69 J. M. Broering, A. S. Bommarius, *Biochem. Soc. Trans.* **2007**, *36* (Part 6), 1602–1605.
70 M. T. Ru, S. Y. Hirokane, A. S. Lo, J. S. Dordick, J. A. Reimer, D. S. Clark, *J. Am. Chem. Soc.* **2000**, *122*, 1565–1571.
71 J. P. Lindsay, D. S. Clark, J. S. Dordick, *Biotechnol. Bioeng.* **2004**, *85*, 553–560.
72 E. Sedlák, L. Stagg, P. Wittung-Stafshede, *Arch. Biochem. Biophys.* **2008**, *479*, 69–73.
73 K. D. Collins, *Biophys. J.* **1997**, *72*, 65–76.
74 K. D. Collins, *Methods* **2004**, *34*, 300–311.
75 T. Arakawa, S. N. Timasheff, *Biochemistry* **1982**, *21*, 6545–6552.
76 A. Grossfield, P. Ren, J. W. Ponder, *J. Am. Chem. Soc.* **2003**, *125*, 15671–15682.
77 J. E. Combariza, N. R. Kestner, J. Jortner, *J. Chem. Phys.* **1994**, *100*, 2851–2864.
78 D. K. Eggers, J. S. Valentine, *J. Mol. Biol.* **2001**, *314*, 911–922.
79 L. Vrbka, J. Vondrášek, B. Jagoda-Cwiklik, R. Vácha, P. Jungwirth, *Proc. Natl. Acad. Sci. U.S.A.* **2006**, *103*, 15440–15444.
80 G. Žoldák, M. Sprinzl, E. Sedlák, *Eur. J. Biochem.* **2004**, *271*, 48–57.
81 K. Tóth, E. Sedlák, M. Sprinzl, G. Žoldák, *Biochim. Biophys. Acta* **2008**, *1784*, 789–795.
82 M. Nagaoka, T. Shiraishi, F. Furuhata, Y. Uda, *Biol. Pharm. Bull.* **2003**, *26*, 295–298.
83 Z. Yang, X.-J. Liu, C. Chen, P. J. Halling, *Biochim. Biophys. Acta* **2010**, *1804*, 821–828.
84 E. M. Bowers, L. O. Ragland, L. D. Byers, *Biochim. Biophys. Acta* **2007**, *1774*, 1500–1507.
85 I. E. Gouvea, W. A. S. Judice, M. H. S. Cezari, M. A. Juliano, T. Juhász, Z. Szeltner, L. Polgár, L. Juliano, *Biochemistry* **2006**, *45*, 12083–12089.
86 M. Boström, D. R. M. Williams, B. W. Ninham, *Biophys. J.* **2003**, *85*, 686–694.
87 A. Salis, D. Bilaničová, B. W. Ninham, M. Monduzzi, *J. Phys. Chem. B* **2007**, *111*, 1149–1156.
88 P. Bauduin, F. Nohmie, D. Touraud, R. Neueder, W. Kunz, B. M. Ninham, *J. Mol. Liq.* **2006**, *123*, 14–19.
89 T. López-León, M. J. Santander-Ortega, J. L. Ortega-Vinuesa, D. Bastos-González, *J. Phys. Chem. C* **2008**, *112*, 16060–16069.
90 C. Carbonnaux, M. Ries-Kautt, A. Ducruix, *Protein Sci.* **1995**, *4*, 2123–2128.
91 M. M. Ries-Kautt, A. F. Ducruix, *J. Biol. Chem.* **1989**, *264*, 745–748.
92 E. M. Wondrak, J. M. Louis, S. Oroszlan, *FEBS Lett.* **1991**, *280*, 344–346.
93 N. Tomášková, R. Varhač, G. Žoldák, D. Sedláková, E. Sedlá, *J. Biol. Inorg. Chem.* **2007**, *12*, 257–266.
94 D. Bilaničová, A. Salis, B. W. Ninham, M. Monduzzi, *J. Phys. Chem. B* **2008**, *112*, 12066–12072.

95 K. Fujita, D. R. MacFarlane, M. Forsyth, M. Yoshizawa-Fujita, K. Murata, N. Nakamura, H. Ohno, *Biomacromolecules* **2007**, *8*, 2080–2086.

96 C. Lange, C. Patil, R. Rudolph, *Protein Sci.* **2005**, *14*, 2693–2701.

97 D. Constantinescu, H. Weingärtner, C. Herrmann, *Angew. Chem. Int. Ed. Engl.* **2007**, *46*, 8887–8889.

98 S. M. Saadeh, Z. Yasseen, F. A. Sharif, H. M. A. Shawish, *Ecotoxicol. Environ. Saf.* **2009**, *72*, 1805–1809.

99 W.-Y. Lou, M.-H. Zong, Y.-Y. Liu, J.-F. Wang, *J. Biotechnol.* **2006**, *125*, 64–74.

100 H. Zhao, S. M. Campbell, L. Jackson, Z. Song, O. Olubajo, *Tetrahedron Asymmetry* **2006**, *17*, 377–383.

101 G. Hinckley, V. V. Mozhaev, C. Budde, Y. L. Khmelnitsky, *Biotechnol. Lett.* **2002**, *24*, 2083–2087.

102 A. P. M. Tavares, O. Rodriguez, E. A. Macedo, *Biotechnol. Bioeng.* **2008**, *101*, 201–207.

103 E. S. Hong, O. Y. Kwon, K. Ryu, *Biotechnol. Lett.* **2008**, *30*, 529–533.

104 E.-S. Hong, J.-H. Park, I.-K. Yoo, K.-G. Ryu, *J. Microbiol. Biotechnol.* **2009**, *19*, 713–717.

105 X. Shi, M. Zong, C. Meng, Y. Guo, *Chin. J. Catal.* **2005**, *26*, 982–986.

106 S. H. Lee, Y.-M. Koo, S. H. Ha, *Korean J. Chem. Eng.* **2008**, *25*, 1456–1462.

107 Y. Hu, Z. Guo, B.-M. Lue, X. Xu, *J. Agric. Food Chem.* **2009**, *57*, 3845–3852.

108 A. P. de los Ríos, F. J. Hernández-Fernández, M. Rubio, D. Gómez, G. Víllora, *J. Chem. Technol. Biotechnol.* **2007**, *82*, 190–195.

109 Z. Yang, Y.-J. Yue, W.-C. Huang, X.-M. Zhuang, M. Xing, *J. Biochem.* **2009**, *145*, 355–364.

110 W.-Y. Lou, M.-H. Zong, T. J. Smith, H. Wu, J.-F. Wang, *Green Chem.* **2006**, *8*, 509–512.

111 F. J. Hernández-Fernández, A. P. de los Ríos, M. Rubio, D. Gímez, G. Víllora, *J. Chem. Technol. Biotechnol.* **2007**, *82*, 882–887.

112 A. P. de los Ríos, F. J. Hernández-Fernández, F. Tomás-Alonso, D. Gómez, G. Víllora, *Process Biochem.* **2008**, *43*, 892–895.

113 N. Iwai, T. Tanaka, T. Kitazume, *J. Mol. Catal., B Enzym.* **2009**, *59*, 131–133.

114 W.-Y. Lou, M.-H. Zong, *Chirality* **2006**, *18*, 814–821.

115 M. Katsoura, M. Patila, I. Pavlidis, H. Stamatis, *New Biotechnol.* **2009**, *25S*, S131.

116 S. J. Nara, J. R. Harjani, M. M. Salunkhe, *Tetrahedron Lett.* **2002**, *43*, 2979–2982.

117 K. Nakashima, J. Okada, T. Maruyama, N. Kamiya, M. Goto, *Sci. Technol. Adv. Mater.* **2006**, *7*, 692–698.

118 W.-G. Zhang, D.-Z. Wei, X.-P. Yang, Q.-X. Song, *Bioprocess Biosyst. Eng.* **2006**, *29*, 379–383.

119 E. Husson, C. Humeau, F. Blanchard, X. Framboisier, I. Marc, I. Chevalot, *J. Mol. Catal., B Enzym.* **2008**, *55*, 110–117.

120 J. L. Kaar, A. M. Jesionowski, J. A. Berberich, R. Moulton, A. J. Russell, *J. Am. Chem. Soc.* **2003**, *125*, 4125–4131.

121 C. Laane, S. Boeren, K. Vos, C. Veeger, *Biotechnol. Bioeng.* **1987**, *30*, 81–87.

122 L. Ropel, L. S. Belvèze, S. N. V. K. Aki, M. A. Stadtherr, J. F. Brennecke, *Green Chem.* **2005**, *7*, 83–90.

123 S. H. Lee, T. T. N. Doan, S. H. Ha, Y.-M. Koo, *J. Mol. Catal., B Enzym.* **2007**, *45*, 57–61.

124 H. Zhao, G. A. Baker, Z. Song, O. Olubajo, L. Zanders, S. M. Campbell, *J. Mol. Catal., B Enzym.* **2009**, *57*, 149–157.

125 K. Ryu, J. S. Dordick, *Biochemistry* **1992**, *31*, 2588–2598.

126 R. M. Lau, M. J. Sorgedrager, G. Carrea, F. van Rantwijk, F. Secundo, R. A. Sheldon, *Green Chem.* **2004**, *6*, 483–487.

127 M. Okochi, I. Nakagawa, T. Kobayashi, S. Hayashi, S. Furusaki, H. Honda, *J. Biotechnol.* **2007**, *128*, 376–382.

128 P. Lozano, T. de Diego, J.-P. Guegan, M. Vaultier, J. L. Iborra, *Biotechnol. Bioeng.* **2001**, *75*, 563–569.

129 A. Basso, S. Cantone, P. Linda, C. Ebert, *Green Chem.* **2005**, *7*, 671–676.

130 S. N. Baker, T. M. McCleskey, S. Pandey, G. A. Baker, *Chem. Commun.* **2004**, 940–941.

131 T. de Diego, P. Lozano, S. Gmouh, M. Vaultier, J. L. Iborra, *Biomacromolecules* **2005**, *6*, 1457–1464.

132 D. Sate, M. H. A. Janssen, G. Stephens, R. A. Sheldon, K. R. Seddon, J. R. Lu, *Green Chem.* **2007**, *9*, 859–867.

133 N. M. Micaêlo, C. M. Soares, *J. Phys. Chem. B* **2008**, *112*, 2566–2572.

134 T. A. McCarty, P. M. Page, G. A. Baker, F. V. Bright, *Ind. Eng. Chem. Res.* **2008**, *47*, 560–569.

135 T. A. Page, N. D. Kraut, P. M. Page, G. A. Baker, F. V. Bright, *J. Phys. Chem. B* **2009**, *113*, 12825–12830.

136 M. L. Pusey, M. S. Paley, M. B. Turner, R. D. Rogers, *Cryst. Growth Des.* **2007**, *7*, 787–793.

137 R. A. Judge, S. Takahashi, K. L. Longenecker, E. H. Fry, C. Abad-Zapatero, M. L. Chiu, *Cryst. Growth Des.* **2009**, *9*, 3463–3469.

138 N. Byrne, C. A. Angell, *Chem. Commun.* **2009**, 1046–1048.

139 D.-H. Cheng, X.-W. Chen, Y. Shu, J.-H. Wang, *Talanta* **2008**, *75*, 1270–1278.

140 T. Y. Tsong, *Biochemistry* **1975**, *14*, 1542–1547.

141 Y.-P. Tzeng, C.-W. Shen, T. Yu, *J. Chromatogr. A* **2008**, *1193*, 1–6.

142 K. E. Gutowski, G. A. Broker, H. D. Willauer, J. G. Huddleston, R. P. Swatloski, J. D. Holbrey, R. D. Rogers, *J. Am. Chem. Soc.* **2003**, *125*, 6632–6633.

143 Z. Du, Y. L. Yu, J. H. Wang, *Chemistry* **2007**, *13*, 2130–2137.

144 S. Dreyer, U. Kragl, *Biotechnol. Bioeng.* **2008**, *99*, 1416–1424.

145 S. Dreyer, P. Salim, U. Kragl, *Biochem. Eng. J.* **2009**, *46*, 176–185.

146 Y. Pei, J. Wang, K. Wu, X. Xuan, X. Lu, *Sep. Purif. Technol.* **2009**, *64*, 288–295.

147 Q. Cao, L. Quan, C. He, N. Li, K. Li, F. Liu, *Talanta* **2008**, *77*, 160–165.

148 P. L. Luisi, L. J. Magid, *CRC Crit. Rev. Biochem.* **1986**, *20*, 409–474.

149 K. Martinek, N. L. Klyachko, A. V. Kabanov, Y. L. Khmelnitsky, A. V. Levashov, *Biochim. Biophys. Acta* **1989**, *981*, 161–172.

150 Z. Yang, D. A. Robb, *Biocatal. Biotransformation* **2005**, *23*, 423–430.

151 C. W. Shen, T. Yu, *J. Chromatogr.* **2007**, *A1151*, 164–168.

152 M. Moniruzzaman, N. Kamiya, M. Goto, *Langmuir* **2009**, *25*, 977–982.

153 I. V. Pavlidis, D. Gournis, G. K. Papadopoulos, H. Stamatis, *J. Mol. Catal., B Enzym.* **2009**, *60*, 50–56.

154 Y. Shu, D. Cheng, X. Chen, J. Wang, *Sep. Purif. Technol.* **2008**, *64*, 154–159.

155 R. P. Swatloski, J. D. Holbrey, R. D. Rogers, *Green Chem.* **2003**, *5*, 361–363.

156 J. Dupont, J. Spencer, *Angew. Chem. Int. Ed. Engl.* **2004**, *43*, 5296–5297.

157 A. Romero, A. Santos, J. Tojo, A. Rodríguez, *J. Hazard. Mater.* **2008**, *151*, 268–273.

158 J. Arning, S. Stolte, A. Böschen, F. Stock, W.-R. Pitner, U. Welz-Biermann, B. Jastorffa, J. Ranke, *Green Chem.* **2008**, *10*, 47–58.

159 K. Fukumoto, M. Yoshizawa, H. Ohno, *J. Am. Chem. Soc.* **2005**, *127*, 2398–2399.

160 G.-H. Tao, L. He, N. Sun, Y. Kou, *Chem. Commun.* **2005**, 3562–3564.

161 J. Kagimoto, K. Fukumoto, H. Ohno, *Chem. Commun.* **2006**, 2254–2256.

162 Y.-Y. Jiang, G.-N. Wang, Z. Zhou, Y.-T. Wu, J. Geng, Z.-B. Zhang, *Chem. Commun.* **2008**, 505–507.

163 H. Ohno, K. Fukumoto, *Acc. Chem. Res.* **2007**, *40*, 1122–1129.

164 T. Tanaka, *Phys. Rev. Lett.* **1978**, *40*, 820–823.

165 H. Zhao, L. Jackson, Z. Song, O. Olubajo, *Tetrahedron Asymmetry* **2006**, *17*, 1549–1553.

166 G. de Gonzalo, I. Lavandera, K. Durchschein, D. Wurm, K. Faber, W. Kroutil, *Tetrahedron Asymmetry* **2007**, *18*, 2541–2546.

167 D. Das, A. Dasgupta, P. K. Das, *Tetrahedron Lett.* **2007**, *48*, 5635–5639.

168 A. S. Wells, V. T. Coombe, *Org. Process Res. Dev.* **2006**, *10*, 794–798.

169 F. Atefi, M. T. Garcia, R. D. Singer, P. J. Scammells, *Green Chem.* **2009**, *11*, 1595–1604.

170 V. Kumar, S. V. Malhotra, *Bioorg. Med. Chem. Lett.* **2009**, *19*, 4643–4646.

171 A. J. Walker, N. C. Bruce, *Chem. Commun.* **2004**, 2570–2571.

172 A. J. Walker, N. C. Bruce, *Tetrahedron* **2004**, *60*, 561–568.

173 H. Zhao, G. A. Baker, Z. Song, O. Olubajo, T. Crittlea, D. Peters, *Green Chem.* **2008**, *10*, 696–705.

174 G. Guo, B. Chen, R. L. Murillo, T. Tan, X. Xu, *Org. Biomol. Chem.* **2006**, *4*, 2772–2776.

175 C. Vafiadi, E. Topakas, V. R. Nahmias, C. B. Faulds, P. Christakopoulos, *J. Biotechnol.* **2009**, *139*, 124–129.

176 S. Sanfilippo, N. D'Antona, G. Nicolosi, *Biotechnol. Lett.* **2004**, *26*, 1815–1819.

177 S. Lutz-Wahl, E.-M. Trost, B. Wagner, A. Manns, L. Fischer, *J. Biotechnol.* **2006**, *124*, 163–171.

178 H. Zhao, O. Olubajo, Z. Song, A. L. Sims, T. E. Person, R. A. Lawal, L. A. Holley, *Bioorg. Chem.* **2006**, *34*, 15–25.

179 H. Zhao, S. Campbell, J. Solomon, Z. Song, O. Olubajo, *Chin. J. Chem.* **2006**, *24*, 580–584.

180 A. Chefson, K. Auclair, *ChemBioChem* **2007**, *8*, 1189–1197.

181 W. Hussain, D. J. Pollard, M. Truppo, G. J. Lye, *J. Mol. Catal., B Enzym.* **2008**, *55*, 19–29.

182 M. López-Pastor, A. Domínguez-Vidal, M. J. Ayora-Cañada, B. Lendl, M. Valcárcel, *Microchem. J.* **2007**, *87*, 93–98.

183 J.-Q. Lai, Z. Li, Y.-H. Lü, Z. Yang, *Green Chem.* **2011**, *13*, 1860–1868.

PART II

IONIC LIQUIDS IN BIOTRANSFORMATIONS

3

IONIC LIQUIDS IN BIOTRANSFORMATIONS: MOTIVATION AND DEVELOPMENT

Christina Kohlmann and Lasse Greiner

3.1 FIRST USES OF IONIC LIQUIDS IN BIOTRANSFORMATIONS

As the term "ionic liquid" (IL) for salts that are liquid at temperatures below 100°C was not established until the 1990s, it is difficult to determine which publication can be referred to as the first to use ILs in biotransformations. In our opinion, the report on the influence of triethylammonium nitrate on the activity and stability of the isolated enzyme alkaline phosphatase by Magnuson et al. in 1984 is the first publication.[1] Earlier examples of applying a biocatalyst in a medium containing a "fused" or "liquid salt" might exist. Arguably, an IL diluted in water may not be considered to be an IL. The pioneering work was motivated by similarities in the behavior of the IL and water, as the authors were interested in the nature of protein structure in solution. Within the study, the activity of the enzyme was assayed spectrophotometrically, monitoring the catalytic activity in the hydrolysis of *p*-nitrophenyl phosphate. The dependency of the *p*-nitrophenol extinction coefficient on IL contents was taken into account. In the range of 10–80 vol% IL in TRIS buffer, an activation of alkaline phosphatase of maximal activation 1.6 at 10 vol% IL was observed. All higher additions led to decreased activities and deactivation. Kinetics suggested a noncompetitive inhibition of the

Ionic Liquids in Biotransformations and Organocatalysis: Solvents and Beyond, First Edition. Edited by Pablo Domínguez de María.

enzyme. Incubation of the alkaline phosphatase with addition of different IL contents for 10 minutes showed reduced activities with rising IL content.

Even though these early investigations hinted toward promising results and effects, the research concerning the application of ILs in biotransformations gained momentum not before the year 2000. Within the first years, research focused mainly on hydrolases in IL-enriched mixtures, triggered by the incentive of alternatives to conventional organic solvents for biotransformations. In the following a summary of the different publications from the first two years of research concerning the use of ILs for biotransformations is given.

Cull et al. presented the first example of whole-cell biocatalysis in an aqueous organic two-phase system in 2000.[2] The reaction under investigation was the *Rhodococcus* R312-catalyzed biotransformation of 1,3-dicyanobenzene to 3-cyanobenzamide, since the substrate is hardly water-soluble and the desired product 3-cyanobenzamide undergoes further hydration to the corresponding 3-cyanobenzoic acid by an amidase from the nitrile metabolism pathways (Figure 3.1).

As common for the early research on IL and biotransformations, ILs were prepared by the researchers. Here, [BMIM][PF_6] was synthesized by anion metathesis from [BMIM][Cl] with hexafluorophosphoric acid. Even though the storage stability of the cells in IL was higher than in toluene, the initial rates for product formation were slightly lower. The authors attributed this to the reduced mass-transfer rate between IL and aqueous phase out of other possibilities. Additionally, as phase separation for the water–[BMIM][PF_6] two-phase system was improved, future benefits for downstream processing were predicted.

Also in 2000, Erbeldinger et al. used [BMIM][PF_6] with 5 vol% water as solvent in the thermolysin-catalyzed condensation of carbobenzoxy-L-aspartate and L-phenylalanine methyl ester hydrochloride yielding Z-aspartame (Figure 3.2).[3] They observed excellent enzyme stabilities, as well as activities and yields similar to con-

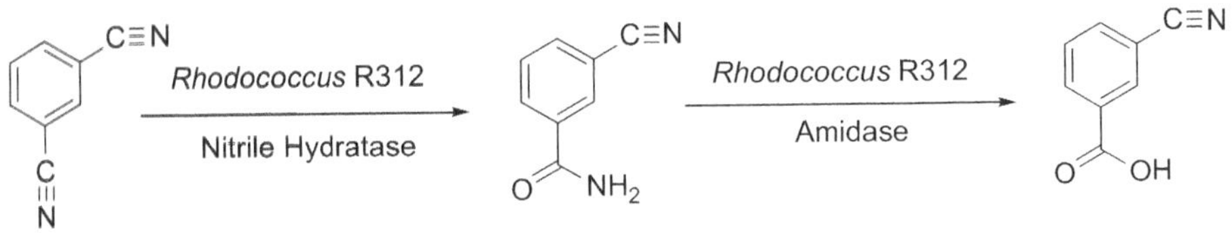

Figure 3.1. *Rhodococcus* R312-catalyzed biotransformation of 1,3-dicyanobenzene to 3-cyanobenzamide and successive transformation to 3-cyanobenzoic acid.

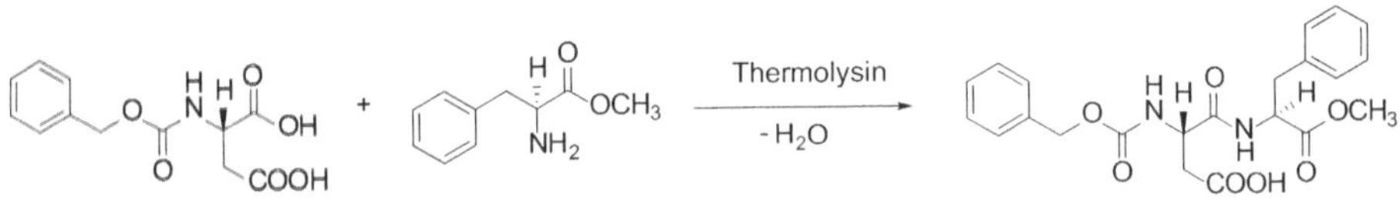

Figure 3.2. Thermolysin-catalyzed condensation of carbobenzoxy-L-aspartate and L-phenylalanine methyl ester hydrochloride yielding *Z*-aspartame.

ventional organic solvents with low water content. Furthermore, they were able to recycle the IL and reuse it with no apparent loss in conversion rates compared with fresh IL.

In succession, different preparations of *Candida antarctica* lipase B (CALB) were applied in the common reactions alcoholysis, ammonolysis, and perhydrolysis with anhydrous [BMIM][PF_6] and [BMIM][BF_4] as reaction media.[4] Generally, reaction rates similar to and higher than in conventional organic solvents were found depending on the reaction, biocatalyst preparation, and solvent used.

Subsequently, within the next year, various publications dealing with different hydrolase (mainly lipases) catalyzed reactions with IL as solvents or additives followed. For example Husum et al. tested the influence of neat [BMIM][PF_6] on four different lipases as well as the potential of [BMIM][BF_4] as a co-solvent for β-galactosidase from *Escherichia coli* and subtilisin protease Savinase™-catalyzed hydrolysis reactions.[5] Even though conversion rates for lipase-catalyzed ester syntheses in IL are comparable to those in tetrahydrofuran (THF) and butanone, superior rates were found for hexane. However, the authors reported unquantified high stabilities for CALB in the IL. Additionally, in 50 vol% solvent water mixtures, they found higher activities for Savinase™ in [BMIM][BF_4] compared with acetonitrile or ethanol, whereas β-galactosidase exhibited highest activities in ethanol. Noteworthy, in pure buffer, the activities were much better than with addition of any solvent.

Itoh et al. compared five different [BMIM]-IL as solvents for the lipase-catalyzed enantioselective acylation of 5-phenyl-1-penten-3-ol and found a dependency on the counter-anion for the achieved yields.[6] While contrasting yields, conversions, and rates obtained in IL and 2-propanol, only [BMIM][PF_6] and [BMIM][BF_4] showed comparable values. Furthermore, enantioselectivity as given by E values were reduced in the presence of IL. Nevertheless, after extraction of product and remaining substrate with ether, four reuses of *Novozym 435* in [BMIM][PF_6] with similar conversions and selectivities were possible.

In contrast, Kragl and coworkers screened activities of nine lipases and two esterases in different commercial ILs and found improved enantioselectivities for some of the biocatalysts.[7] The reaction under investigation was the kinetic resolution of 1-phenylethanol by transesterification. Unfortunately, as the commercial production of IL was still in its infancy, some of the ILs were strongly colored and therefore analysis via HPLC with UV detection was not feasible. While both esterases were inactivated by all tested IL, the performance of the lipases varied. Compared with methyl *tert*-butyl ether (MTBE), in particular *Candida antarctica* lipase A showed improved conversions for a variety of ILs, while for the lipases from *Pseudomonas* sp. and *Alcaligenes* sp. increased enantiomeric excesses were found. Furthermore, in contrast to MTBE, the enantioselectivity of the lipase is less influenced by both temperature and water content.[8] However, an appropriate water content is required to maintain biocompatibility. Additionally, Kragle et al. also tested the recycling of CALB in [BMIM][$(CF_3SO_2)_2N$] after removal of the product via distillation at reduced pressure. It was possible to reuse the lipase with 10% loss in activity per cycle for three times. Besides that, the β-galactosidase-catalyzed synthesis of *N*-acetyl-lactosamine in buffer containing IL as a co-solvent was also a target of investigation. By adding 25 vol% [BMIM][$MeSO_4$] to

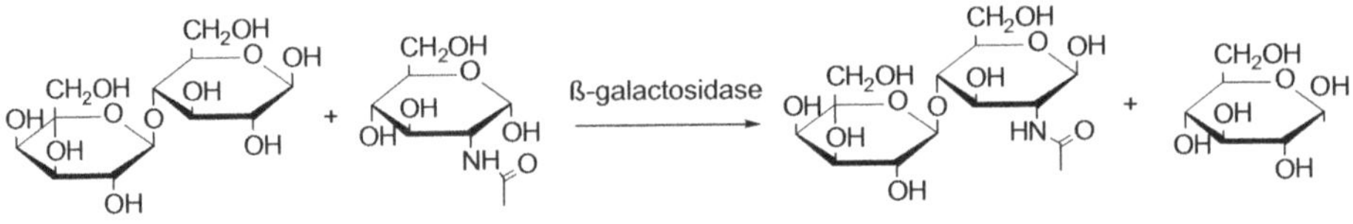

Figure 3.3. β-Galactosidase-catalyzed transglycosylation of *N*-acetylglycosamine and lactose.

the reaction medium, secondary hydrolysis of the product by the biocatalyst could be prevented (Figure 3.3). In general, the authors hint toward shifts in pH to be correlated with impurities in the IL.

Moreover, improved enantioselectivities for different transesterifications catalyzed by two lipases in [BMIM][PF_6] and [EMIM][BF_4] were also reported by Kim et al.[9] Immobilized CALB and native *Pseudomonas cepacia* lipase were tested in two transesterifications and the results were compared with THF and toluene as solvents. For all reactions in ILs, up to 25 times improved enantioselectivities were observed. Furthermore, it was possible to reuse CALB in [BMIM][PF_6] with almost full residual activity.

Iborra and coworkers studied both CALB and α-chymotrypsin three commercial ILs with 2 vol% water content, as α-chymotrypsin was not active in the absence of water.[10,11] Their results suggest coherences between polarity of the solvent and activity of the biocatalyst. CALB showed improved activities in the following order [EMIM][$(CF_3SO_2)_2N$] > [BMIM][PF_6] > [BMIM][$(CF_3SO_2)_2N$]. In addition, with the exception of [EMIM][BF_4], α-chymotrypsin activity increased in a similar order ([EMIM][$(CF_3SO_2)_2N$] > [BMIM][BF_4] > [BMIM][PF_6]); thus, the activity increases with increasing polarity of the solvent. Moreover, the tested ILs were able to stabilize both enzymes in solution. However, no trend concerning structure–activity relation or mechanism could be derived.

The performance of α-chymotrypsin in ILs was also determined by investigating the transesterification of N-acetyl-L-phenylalanine ethyl ester with 1-propanol in [BMIM][PF_6] and [OMIM][PF_6].[12] Laszlo et al. studied the influence of different water contents on the transesterification rate. In line with the results from Iborra and coworkers, α-chymotrypsin was inactive in neat IL. However, the observed reaction rates for IL with an appropriate amount of water were only in the order of magnitude of those observed in nonpolar organic solvents. The authors supposed that IL—similar to organic solvents—might strip off essential water from the biocatalyst. Furthermore, the use of supercritical carbon dioxide ($scCO_2$) as a co-solvent was investigated, as it is highly soluble in both ILs, whereas the ILs are virtually insoluble in $scCO_2$. Interestingly, the enzyme was active in neat IL–$scCO_2$ mixture and additional water did not improve reaction rates, suggesting less potential to remove water from the enzyme.

Park et al. addressed the question of influences of impurities in IL on isolated enzymes.[13] Therefore, they tested two different methods for removing halides from the solvents (Figure 3.4).

For the first method, they washed IL with $AgBF_4$, filtered the formed suspension through celite, and concentrated the mixture via evaporation, before dissolving in

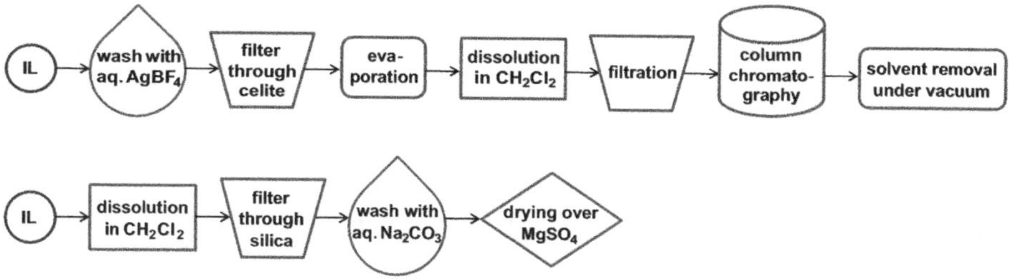

Figure 3.4. IL purification procedures described by Park et al.[13]

dichloromethane and a second filtration step. Final purification was achieved via column chromatography and solvent removal. The other purification method included dilution of IL with dichloromethane and filtering through silica gel, before washing with aqueous sodium carbonate, followed by subsequent drying over anhydrous magnesium sulfate and removal of solvent under reduced pressure. These methods where tested for 10 different imidazolium- or pyridinium-based IL with tetrafluoroborate or hexafluorophosphate anions. These were applied in the acetylation of 1-phenylethanol catalyzed by lipase from *Pseudomonas cepacia* and of glucose catalyzed by CALB. Thereby, the authors found a consistent behavior of the IL purified by the second method. Enzymes that were not active in unpurified IL showed activity after treatment. The observed reaction rates varied only slightly when applying IL with minor structural differences. Differences in the results obtained for both purifications were associated with silver ions caused by the first procedure. In general, reaction rates comparable to those in acetone or toluene for the acylation of 1-phenylethanol and improved regioselectivities for the acylation of glucose were found. The authors attributed this to the higher solubility of the substrate.

Howarth et al. were the first to report on a biocatalytic redox reaction in the presence of IL, and their work is the only example for a nonhydrolytic biocatalyst in 2000 and 2001.[14] They investigated the reduction of several ketones by immobilized baker's yeast in an aqueous [BMIM][PF_6] biphasic system and found enantioselectivities comparable to those achieved in alternative media. The obtained yields varied with substrate. Furthermore, the IL could be reused after distillation of some of the alcohols. Additionally, the biocatalyst was also applied in pure IL; however, in these setups, yields and enantiomeric excesses were extremely poor.

Mostly, research was limited to hydrolytic enzymes, which were tested mainly in IL as pure solvent, where—depending on the work—small amounts of water were diluted. Furthermore, besides the existence of various different ILs, almost all focused on the ILs [BMIM][PF_6] and [BMIM][BF_4]. In most studies, the ILs were synthesized by the researchers themselves, and only in rare cases were quality controls for impurities and even water contents carried out, even though some authors already noted their impact on enzymes. Only few findings of these early contributions hinted toward the potential of IL for biotransformations.

In summary, ILs were applied successfully in diverse transformations, showing improved biocatalysts activities, stabilities, as well as reactant solubilities. Furthermore, the utilization of ILs enabled promising reaction concepts and simplification of downstream processing. Thus, ILs are promising reaction media for biotransformations. Nevertheless, each application has distinct requirements on the medium to choose; hence, there is not only one beneficial IL but a variety of possible candidates.

3.2 MOTIVATION TO USE IL IN BIOTRANSFORMATIONS

Even though biocatalysts show superior selectivities when compared to chemical catalysts in many cases, the natural aqueous reaction medium limits the number of possible applications. Only a small number of industrially interesting substrates are sufficiently water-soluble. Furthermore, aqueous media represent a challenge for reaction engineering in terms of downstream processing and integration into a process chain. With the discovery that some biocatalysts are active and stable in the presence of conventional organic solvents, the field of possible applications was extended. However, not all problems of solubility, stability, and activity could be overcome by the application of organic solvents. Additionally, the application of organic solvents itself has intrinsic challenges and hazards. Therefore, the need for alternative media is unbroken in general and in biocatalysis. One opportunity discussed is the use of ILs as solvents in biotransformations. So, in the beginnings of this research field, attention was focused mainly on finding biocompatible substitutes to organic solvents for reactions catalyzed by hydrolytic enzymes. By now, the topics addressed are more diverse, as the ongoing research brings about new interesting results broadening the field of advantageous applications.

An important driver for using ILs in biotransformations is their potential as solubilizers for various substances. In these applications the IL may be denoted as a performance additive and is the minor component in the reaction media. ILs are able to either dissolve nonpolar substrates in polar aqueous medium or polar substances in nonpolar organic solvents. For example, the application of different water-miscible ILs as solubilizers for thioanisol as substrates for the *Caldariomyces fumago* chloroperoxidase-catalyzed synthesis of (*R*)-phenylmethylsulfoxide was tested (Figure 3.5).[15]

Depending on the IL, even small amounts of less than 10 vol% were sufficient to drastically increase the solubility of the substrate. Best results were obtained with [BMIM][$MeSO_4$]; merely 5 vol% of this IL resulted in a fivefold increase, resulting in 20 mmol/L instead of 3.8 mmol/L. For batch reactions the biocatalytic sulfoxidation was combined with the electrochemical generation of the oxidant H_2O_2. The tested ILs all showed a significant increase in both biocatalyst utilization and achievable productivity in contrast to reaction in pure buffer. 2 vol% of IL altered not only the substrate

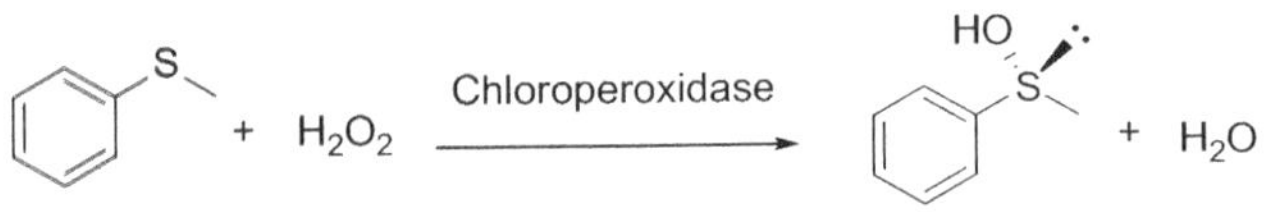

Figure 3.5. Chloroperoxidase-catalyzed sulfoxidation of thioanisol.

solubility, but also biocatalyst stability and activity as well as electrochemical properties of the reaction medium. All ILs showed different influences and a structure–activity relationship was not obvious and/or could not be identified. Productivity increased more than fourfold when using the most effective solubilizer, [BMIM][$MeSO_4$], thereby underlining the potential of ILs as performance additives for reactions limited by solubility.

The ability of IL to act as a biocompatible solubilizer was also tested for different alcohol dehydrogenase-catalyzed reactions. For example, Dreyer et al. reported an exponential increase in acetophenone solubility with increasing amounts of AMMOENG™ 110 up to 70 vol% IL.[16] Applying 10 vol% of this IL in batch syntheses resulted in almost doubled space time yields of 0.8 mol/L d. Moreover, the solubility of different aliphatic ketones could be enhanced by application of AMMOENG™ 101.[17] Depending on the substrate used, for addition of 100 g IL per liter reaction medium up to 86 times increased solubility could be found and space time yields improved by more than factor 10 were possible.

ILs were also effectively applied as solvents for polar substrates in nonaqueous biocatalytic reactions. For example, various authors reported on the enormous potential of dicyanamide-containing ILs to dissolve carbohydrates.[18,19] For instance, the solubility of D-glucose in [BMIM][DCA] is 145 g/L at 25°C, whereas with other commonly applied ILs such as [BMIM][PF_6] and [BMIM][BF_4], solubilities of less than 0.5 g/L comparable to organic solvents, for example, 0.3 g/L in *tert*-butyl alcohol, were achieved.[18] However, van Rantwijk et al. reported fast irreversible deactivation of CALB in the presence of these ILs.[20]

Lee et al. reported transesterification and direct esterification of glucose catalyzed by *Novozym 435* with vinyl laurate and lauric acid, respectively, in ILs (Figure 3.6).[21]

As these reactions are limited by the solubility of glucose in the reaction medium, Lee et al. applied three different methods to dissolve the substrate in IL. For the simplest procedure (method A), glucose in excess and IL were mixed and stirred for 12 hours. Alternatively, an excess of glucose was stirred with IL at 333K for 12 hours and afterwards at 298K for 2 hours (method B). In the third method (method C), an aqueous glucose solution was mixed with IL at room temperature. This mixture was stirred until a clear solution was obtained. Then, water was removed under vacuum for 12 hours at 60°C. The obtained suspension was cooled to 25°C and incubated for a further 2 hours. The efficient removal of water was controlled via Karl Fischer titration. To determine

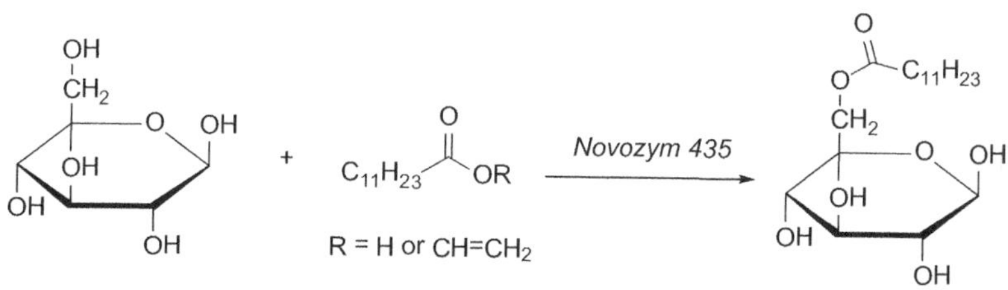

Figure 3.6. Transesterification and direct esterification of glucose catalyzed by *Novozym 435* with vinyl laurate and lauric acid.

the content of dissolved glucose for the different methods, samples were centrifuged and the supernatant was directly analyzed via the dinitrosalicylic acid method.[22] Furthermore, the authors investigated whether the IL might interfere with this assay, and they did not observe any influences. The obtained solubilities increase in the following order: method A < method B < method C. When applying method C, very high solubilities of >500 g/L, 113.4 g/L, and 46.3 g/L could be obtained, especially for [EMIM][Me_2SO_4], [EMIM][$(CF_3SO_2)_2N$], and [BMIM][$(CF_3SO_2)_2N$], respectively. Nevertheless, as *Novozym 435* was not active in the presence of [EMIM][Me_2SO_4], syntheses could not be carried out with this IL. Transesterifications of 40 g/L partially or fully dissolved glucose in [BMIM][$(CF_3SO_2)_2N$] depending on the procedure used showed the following results: In contrast to methods A and B, reaction rates were improved by a factor of 88 and 18, respectively, when applying method C. Additionally, with ~96% the overall conversion was much higher than the 20% observed for methods A and B. Furthermore, the authors found increasing reaction rates for decreasing water contents in the reaction medium. Direct esterification was carried out with different ILs as reaction media. Again, the best results were obtained when dissolving glucose according to method C. Achieved reaction rates and conversions are according to the following order: [BMIM][$(CF_3SO_2)_2N$] > [EMIM][$(CF_3SO_2)_2N$] > [BMIM][BF_4]. Therefore, the authors addressed the hydrophobicity of the IL to exhibit the strongest influence on the activity of the biocatalyst.

Recently, ILs are also discussed as solvents for macromolecules such as biogenic polymers. In particular, the dissolution of cellulose as well as lignin or wooden material is of great interest in academia and industry as renewable resources.[23–32] Nevertheless, the dissolution mechanisms are still not well understood even though the dissolution of cellulose in IL is claimed in a patent dating back to 1934.[33] Finding suitable biocatalysts for the transformation of biogenic material dissolved in ILs is a target of current research (see also Chapter 5, Sections 5.4 and 5.5).[34–36] Another incentive for IL is their potential to stabilize biocatalysts in solution. Already in 2000, improved storage stability of thermolysin in [BMIM][PF_6] compared with ethyl acetate was reported.[3] The enzyme showed no loss in activity even after 144 hours, which compares well with stabilization by immobilization.

In addition, Lozano et al. gave account for both CALB and α-chymotrypsin stabilization by a range of ILs.[10,11] CALB showed a 2.8-fold increase in half-life when stored in [EMIM][BF_4] compared with 1-butanol or hexane. Moreover, in recycling experiments with product extraction and subsequent addition of substrate, the lipase showed even better operational stability. Half-lives increased at least by a factor of 310 in contrast to storage stability in the same IL. Best results were observed with [BMIM][PF_6]: More than 2300-fold improved stabilities were found. Similar results were obtained for α-chymotrypsin; with [MO_3AM][$(CF_3SO_2)_2N$] compared with 1-propanol, storage stabilities improved by a factor of 18 were possible. Best results for stability in the presence of the substrates *N*-acetyl-L-tyrosine ethyl ester and 1-propanol were again obtained with [BMIM][PF_6], even though the effect was less pronounced than for CALB. In contrast to the absence of the substrates, 19-fold increased half-lives were obtained which correspond to an almost 200-fold increase in half-lives compared with 1-propanol.

Moreover, improved stability of D-amino acid oxidase from *Trigonopsis variabilis* and chloroperoxidase from *Caldariomyces fumago* was reported when using IL as additive.[15] Experiments were carried out with 2 to 10 vol% of IL in buffer. In the presence of increasing amounts of three of the tested ILs ([EMIM][Et_2PO_4], [MMIM][Me_2PO_4], [MMIM][$MeSO_4$]), the amino acid oxidase showed increasing half-lives, in the other three ILs ([BMIM][$MDEGSO_4$], [EMIM][$MDEGSO_4$], [BMIM][$MeSO_4$]) decreasing half-lives. [EMIM][Et_2PO_4] exhibited the greatest potential as enzyme stabilizer. With addition of 10 vol% of this IL, half-life was 2.5-fold increased. The influence of [EMIM][Et_2PO_4] on the stability of the chloroperoxidase is also striking. While with the exception of this IL all other ILs show the same overall trend of increased stability with increasing amounts of ILs, still the best half-life with a fivefold increase was measured for 2 vol% of [EMIM][Et_2PO_4].

Hussain et al. screened the stability of an alcohol dehydrogenase from *Rhodococcus erythropolis* and glucose dehydrogenase 103 in the presence 10 vol% of 11 ILs and three organic solvents and compared the obtained half-lives with the half-life observed in pure buffer.[37] While the half-lives of the glucose dehydrogenase could be more than just doubled with butyl-methyl-pyrrolidinium, [$(CF_3SO_2)_2N$], [EMIM][$EtSO_4$], as well as AMMOENG™ 110 and 102, the organic solvents THF, DMSO, and toluene led to drastically decreased half-lives. In contrast, the stability of the alcohol dehydrogenase could be almost threefold increased with addition of DMSO. Nevertheless, at least with butyl-methyl-pyrrolidinium, [$(CF_3SO_2)_2N$], the stability of the enzyme could be further increased, resulting in an improved half-life 3.4 times higher than in pure buffer.

The ability to stabilize enzymes in solution of AMMOENG™ 110 (see Figure 3.8, Chapter 2) was also described by Dreyer et al.[16] Storage stabilities of *Lactobacillus brevis* alcohol dehydrogenase were tested with different volumetric contents from 1% to 30% of the IL in buffer and showed an increased stability with increasing amount of IL. AMMOENG™ was also applied in the synthesis of (*S*)-phenylethanol by an alcohol dehydrogenase from a thermophilic bacterium. While in buffer a conversion of only 40% could be reached, with addition of 10 vol% IL a conversion higher than 90% was possible. The authors addressed this fact to improved the stability of the biocatalyst in the presence of IL, as the half-life of the enzyme in pure buffer was as low as 30 minutes.

Weuster-Botz and coworkers presented various studies concerning cofactor depended whole-cell-catalyzed reductions in aqueous IL two-phase systems.[38–40] To determine possible toxic effects of the IL on different organisms, the integrity of their membranes was tested via the LIVE/DEAD Baclight assay by Invitrogen (Karlsruhe, Germany). The results of the tests can be correlated to the stability of the biocatalyst in the reaction medium. Screening of different ILs did show increased membrane integrities in contrast not only to organic solvents but also to aqueous media. Furthermore, ILs were also able to circumvent toxic effects exhibited by substrates and products, indicating their potential as solvents for whole-cell biotransformations. The topic of whole cells and ILs is broadly treated in Chapter 7.

The potential to stabilize biocatalysts in solution motivated the use of IL as an additive in biocatalyst immobilization. ILs themselves or supported on host matrices especially for applications in biosensors have been reported.[41–47] For example, horseradish peroxidase entrapped in an [BMIM][BF_4] containing silica gel displayed highly increased activity and excellent thermal stability.[44] In contrast to the conventional gel, horseradish peroxidase entrapped in the IL-containing gel exhibited a 30-fold increased specific activity. Furthermore, this gel was effectively applied in an amperometric biosensor.[45]

Jiang et al. immobilized *Candida rugosa* lipase on magnetic nanoparticles supported by different covalently bound ILs, and applied the modified nanoparticles in the esterification of oleic acid and butanol.[46] Therefore, 1-methyl-3-(triethoxysilylpropyl)-imidazolium IL with Cl^-, BF_4^-, and PF_6^- anions, as well as 3-(triethoxysilylpropyl)-imidazolium [PF_6] with methyl, butyl, or octyl functional groups at position one of the cation, were prepared. In comparison with the free enzyme, the immobilized lipase showed superior activities, following the trend $Cl^- < BF_4^- < PF_6^-$. Additionally, activity increased with increasing chain length at the cation. Moreover, investigations concerning thermal stability demonstrated enormous stability of the immobilized biocatalyst. Whereas above 40°C drastically decreased activities were found for the native lipase, the immobilized enzyme still exhibited a residual activity of 60% at 80°C. Furthermore, when reused the immobilisate retained 65% residual activity equaling 5% loss per cycle, whereas the native enzyme lost 15% of its activity per cycle.

Solid supports are not mandatory to effectively immobilize biocatalysts with IL. For instance, a straightforward method was reported by Lee and Kim.[47] They mixed [1-(3′-phenylpropyl)-3-methylimidazolium)][PF_6] with a lyophilisate of lipase from *Pseudomonas cepacia* at temperatures above the IL melting point of 53°C. While cooling to room temperature, the suspension solidified. The so-called ionic liquid-coated enzyme (ILCE) was then tested in transesterification of five different secondary alcohols in toluene. All but one reaction proceeded with 1.5 to 2.0 times increased selectivity as compared with the lyophilisate. Only for 2-chloro-1-phenylethanol was the selectivity slightly decreased. While the first application of the ILCE exhibited a residual activity of only 65%, in the second run full activity was observed. This fact was attributed to diffusion limitations within the first experiment. Another three reuses resulted in 93% of the native enzyme activity, which equals 5% loss of activity per cycle (see also Chapter 8).

In addition to improved biocatalyst stabilities, improved stabilities of reduced nicotinamide cofactors are reported as a special advantage of using ILs in biotransformations where these cofactors are involved.[15,48] The addition of all tested ILs resulted in enhanced stabilities of NADPH and NADH. The possible stabilization increased with increasing amounts of the IL in the reaction medium. [EMPY][$EtSO_4$] showed the highest potential to stabilize the cofactors. In a medium containing 10 vol% of this IL, 2.9 and 2.5 times increased half lives were measured for NADPH and NADH, respectively.

The ability to support enzyme activity is reported not only for immobilized biocatalysts, but also for biocatalysts in solution. This ability to improve the activity and therefore the obtainable reaction rates of a biocatalyst represents another reason for working with IL. However, to identify an effective IL is challenging since for most ILs only comparable and quite often also reduced activities are reported. Nevertheless, some examples exist. Even long before the main research concerning IL as a solvent for biocatalysis started, Magnuson et al. reported improved activities for the alkaline phosphatase-catalyzed hydrolysis of *p*-nitrophenyl phosphate in the presence of up to 20 vol% of triethylammonium nitrate in TRIS buffer.[1] As best result, an increased activity by 60% was possible with 10 vol% IL. In 2006, Fischer and coworkers presented a study concerning the performance of D-amino acid oxidase in the presence of ILs as co-solvents.[49] As activity assay, the conversion of D-phenyl alanine to the corresponding imine was investigated with a coupled peroxidase-o-dianisidin assay. While testing [MMIM][$MePO_4$], a relative activity of 1.45 for 20 vol% IL and a relative activity of 1.29 for 40 vol% of IL were found in contrast to initial activities in pure buffer. However, with 60 vol% of IL, the enzyme was almost inactive (2% residual activity) and when testing the enzyme in immobilized form, no activation could be measured.

Moreover, improved regio- and enantioselectivities of biocatalyst applied in IL have been observed. As improved selectivity may lead to higher product purities, reduced amounts of waste, and might be advantageous for downstream processing, they are one important reason for applying IL in biotransformations. For example, Park et al. investigated the regioselective 6-*O*-acetylation of glucose catalyzed by CALB in different immidazolium- or pyridinium-based ILs and compared the results with those obtained for acetone and THF.[13] Even though quite high conversions of glucose are obtained in both organic solvents (72% for acetone and 99% for THF) not only the wanted product 6-O-acetyl D-glucose, but also diacetylated 3,6-O-diacetyl D-glucose, was obtained with selectivity toward the monoacetylation over diacetylation of 75% and 66%, respectively. However, in ILs, selectivities for the monoacetylation of 88% to 99% were possible. The authors explained the improved selectivity by higher solubility of D-glucose in ILs. In organic solvents, less than 0.04 mg m/L substrate can be dissolved and the reaction is conducted with solid substrate. As the monoacetylated D-glucose shows a higher solubility in the organic medium, it is preferably further acetylated by CALB. Best results were obtained with 3-(2-methoxyethyl)-1-methylimidazolium tetrafluoroborate; this IL shows an improved solubility by a factor of 100, resulting in 99% conversion of D-glucose and a yield of 93% for the desired product. The lipase-catalyzed regioselective synthesis of 4,6-di-O-acetyl-d-glucal in the presence of [BMIM][PF_6] or [BMIM][BF_4] and THF was addressed by Nara et al. (Figure 3.7).[50]

Pseudomonas cepacia lipase supported on celite was applied in the hydrolysis of 3,4,6-tri-O-acetyl-D-glucal with aqueous solvent mixtures and in a nonaqueous approach for the alcoholysis of the same substrate. With 84% conversion and a selectivity of >98%, hydrolysis reactions showed best results in the aqueous [BMIM][PF_6] two-phase system. With THF a conversion of 60% and a selectivity of 62% were possible. [BMIM] [BF_4] had a selectivity comparable to THF. However, a conversion of merely 13% was

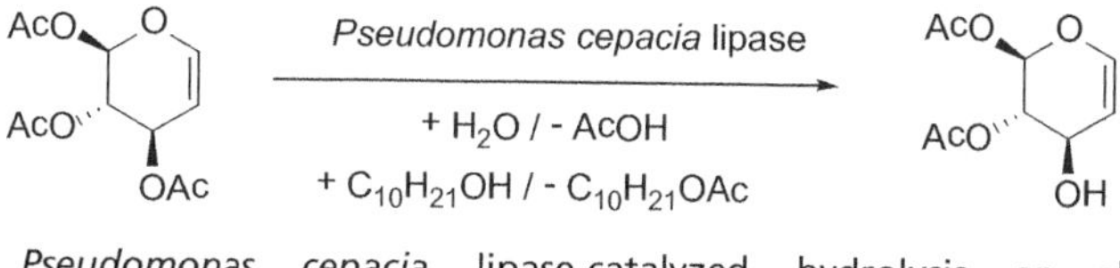

Figure 3.7. *Pseudomonas cepacia* lipase-catalyzed hydrolysis or alcoholysis of 3,4,6-tri-O-acetyl-D-glucal.

possible for the same reaction time. Additionally, nonaqueous alcoholysis also proceeded with very poor conversion and selectivity in [BMIM][BF_4]. While after 2 hours' reaction time a conversion of only 13% was possible, the product reacted further, yielding only traces of the desired product for a reaction time longer than 6 hours. Nevertheless, with a selectivity of 98% and conversion of 55% and 48%, respectively, THF and [BMIM][PF_6] showed comparable results.

Improved enantioselectivities were found for the *Alcaligenes* sp. lipases-catalyzed dynamic kinetic resolution of 1-phenylethanol.[7] Whereas in MTBE the enzyme showed no enantioselectivity, an enantiomeric excess of up to >98% was obtained in [HMIM][BF_4], [OMIM][BF_4], and [BMPY][BF_4]. Furthermore, a reduced influence by both temperature and water content on the enantioselectivity is reported.[8] Moreover, different transesterifications catalyzed by immobilized CALB and *Pseudomonas cepacia* lipase showed improved enantioselectivity when carried out in [BMIM][PF_6] or [BMIM][BF_4].[9] In particular, the kinetic resolution of 1-chloro-3-phenoxypropan-2-ol by *Pseudomonas cepacia* lipase showed promising results: In contrast to toluene and THF, the E value (enantioselectivity) could be improved from 85 and 150, respectively, to more than 1000 with an enantiomeric excess higher than 99.5% for the product when using [BMIM][PF_6]. Furthermore, Mohile et al. reported improved enantioselectivity for the hydrolysis of butyl 2-(4-chlorophenoxy) propionate by *Candida rugosa* lipase.[51] While in buffer an enantiomeric excess of only 47% was observed, with the addition of both water-miscible and water-immiscible ILs ([BMIM][PF_6], [HMIM][BF_4] and [BMIM][BF_4]), an increase to >99% was possible. The overall conversions for buffer and buffer IL mixtures were comparable; however, the reaction rates for all ILs (especially for [BMIM][BF_4]) were much lower.

All these advantages of biotransformations in ILs indicate that their application might also be beneficial in engineering and process development. For example, isolation and purification of the target protein represents a challenging task. Aqueous two-phase systems (ATPS) proved to be an interesting strategy to ease this processes; however, some enzymes might be sensitive to the thereby applied polymers. ILs from the AMMOENG™ series (see Figure 2.8, Chapter 2) hold ammonium cations with functional groups similar to ethylene glycol and have therefore been effectively used for the formation of ATPS. Furthermore, a range of accounts on their ability to stabilize proteins in solution have been published. Thus, it was possible to effectively apply an AMMOENG™ 110-based ATPS in the purification of two alcohol dehydrogenases from *Lactobacillus brevis* and a thermophilic bacterium, expressed in *Escherichia coli*.[16] Optimal conditions for the purification were obtained via experimental design according to Box and Wilson.[52] Applying the optimum ATPS conditions calculated for each protein resulted in an activity recovery of 95% for *Lactobacillus brevis* alcohol dehydrogenase and an activity recovery of 200% for the alcohol dehydrogenase from the thermophilic bacterium. According to the authors, the increase in activity for alcohol dehydrogenase from the thermophilic bacterium was due to the excellent biocompatibility of the IL in contrast to glycerol, in which the unpurified protein was stored.

Furthermore, aqueous IL biphasic systems were effectively applied in synthesis reactions. Figure 3.8 depicts the principle of those systems.

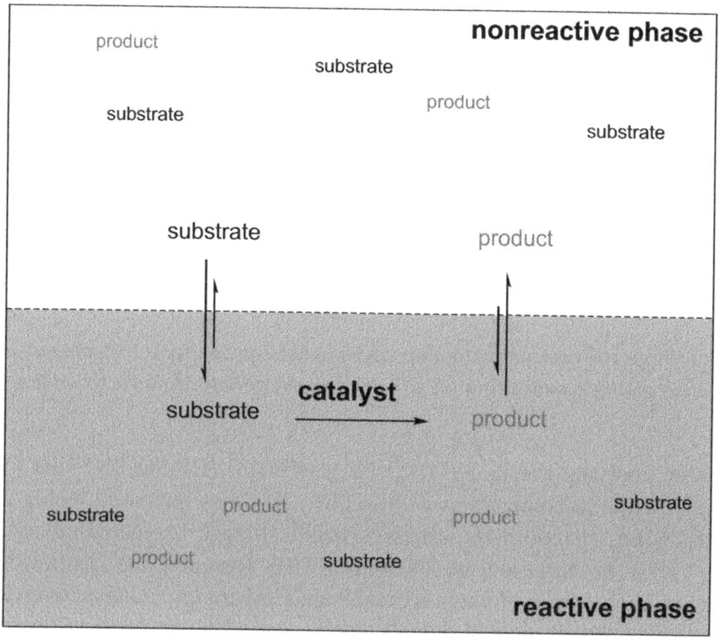

Figure 3.8. Universal demonstration of a biphasic system.

Eckstein et al. investigated the *Lactobacillus brevis* alcohol dehydrogenase-catalyzed reduction of 2-octanone with MTBE and [BMIM][$(CF_3SO_2)_2N$] as second phase.[53] The cofactor NADPH was regenerated substrate coupled via oxidation of 2-propanol to acetone. While the partition coefficients of both co-substrate and co-product were comparable in MTBE, improved values were found for [BMIM][$(CF_3SO_2)_2N$]. In the aqueous IL two-phase system, the partitioning coefficients for acetone were about double as that for 2-propanol, which shifts the equilibrium position toward the product. Furthermore, in line with these findings, conversions of up to 88% were reached, whereas with MTBE, a conversion of 61% was observed. Also, as the cofactor regeneration represents the rate-limiting step of the reaction, in the IL-containing system improved overall rates were obtained. Furthermore, conversions of 88% were reached, while in the presence of MTBE a conversion of merely 61% was observed. A similar approach was used for whole-cell biotransformation of 1-phenyl-2-propanone to 1-phenyl-2-propanol (Figure 3.9).[54]

Recombinant *Escherichia coli* cells as expression host for an alcohol dehydrogenase from *Lactobacillus brevis* were used. 2-Propanol was used as reducing agent. Whereas in the presence of MTBE a yield of 24% was obtained, with application of [BMIM][$(CF_3SO_2)_2N$] it was possible to increase the yield to >95% and accelerate the overall reaction rate. Also, Weuster-Botz and coworkers reported different cofactor-dependent whole-cell-catalyzed reactions in aqueous IL two-phase systems.[38–40] One of their first examples was the reduction of 4-chloroacetophenone to (*R*)-1-(4-chlorophenyl) ethanol with *Lactobacillus kefir.* Compared with reactions in pure buffer, yields could

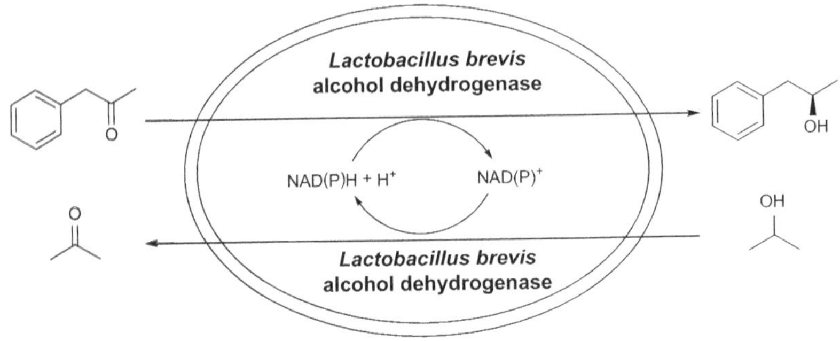

Figure 3.9. Whole-cell biotransformation of 1-phenyl-2-propanone to 1-phenyl-2-propanol with *Escherichia coli* cells expressing an alcohol dehydrogenase from *Lactobacillus brevis*.

be doubled by applying the IL [BMIM][PF_6], [BMIM][$(CF_3SO_2)_2N$], and [MO_3AM][$(CF_3SO_2)_2N$], while in comparison with MTBE, 20 times increased yields were possible. Furthermore, enantiomeric excesses could slightly be increased to >99% as opposed to 98.1% in buffer and 96.3% with MTBE. Recently, the synthesis of (*R*)-2-octanol was addressed by applying a recombinant *Escherichia coli* host overexpressing both *Lactobacillus brevis* alcohol dehydrogenase and *Candida boidinii* formate dehydrogenase. With 1-hexyl-1-methylpyrrolidinium [$(CF_3SO_2)_2N$] as a nonreactive phase, 180 g/L d of the chiral alcohol with an enantiomeric excess of 99.7% was produced (see also Chapter 7).

Despite the previously discussed poor solubilities of glucose in [BMIM][BF_4] and *tert*-butyl alcohol, Ganske and Bornscheuer successfully applied both solvents in a two-phase system for CALB-catalyzed synthesis of glucose fatty esters.[55,56] As acyl donors, lauric acid vinyl ester, myristic acid vinyl ester, and palmitic acid were used, resulting in conversions of 90%, 89%, and 64%, respectively.

Moreover, IL accounted for establishing novel or optimized synthesis routes and new concepts for downstream processing. For example, due to the application of AMMOENG™ 101 (Figure 2.8, Chapter 2) as a co-solvent in the *Lactobacillus brevis* alcohol dehydrogenase-catalyzed continuous synthesis of (*R*)-2-octanol, increased substrate solubilities were feasible, causing the following improvements:[57] First, the turnover number of both biocatalysts is slightly increased and an excellent enantiomeric excess of >99.9% (instead of 95.6% in the pure buffer system) was possible. Second, the space time yield (STY) was more than tripled. Third, cofactor regeneration was 6.1 times more effective. Finally, 80% less waste is generated while producing the same amount of product. Reactions were carried out in an enzyme membrane reactor and cofactor regeneration was realized by glucose dehydrogenase-catalyzed oxidation of D-glucose. Furthermore, integrated product separation was realized by combination of solid phase extraction (SPE) and recycling of the SPE material utilized with $scCO_2$ (Figure 3.10), resulting in more than 30 reuses of the SPE phase without apparent loss in capacity.

Neither the reaction medium applied during the reaction, nor the media applied during downstream processing, should contain volatile and toxic organic solvents. In

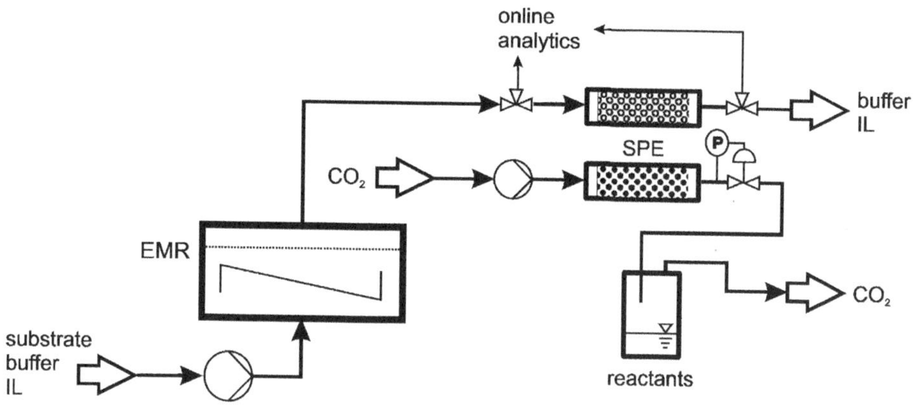

Figure 3.10. Reaction scheme for integrated product separation by SPE and recycling via $scCO_2$.

an attempt to realize integrated product separation, Reetz et al. concentrated on $scCO_2$ extraction.[58] To verify the concept, they investigated CALB-catalyzed acylation of 1-octanol by vinyl acetate in [BMIM][$(CF_3SO_2)_2N$] and extracted the formed product as well as surplus vinyl acetate solution with $scCO_2$. With this approach, a product yield of 92% was possible. The procedure was then repeated three times, yielding 97%, 98%, and 98% product per cycle, indicating an apparently constant biocatalyst performance and effective product separation. Subsequently, a continuous process was established, resulting in a space time yield of 0.1 kg/L h and an integrated yield of 94%. By an optimized workup it was possible to directly separate the products of the lipase-catalyzed kinetic resolution of racemic alcohols by transesterification with vinyl laurate.[59] Separation of the formed (*R*)-ester and the unreacted (*S*)-alcohol was achieved by controlled density reduction via temperature and pressure variation. Thus, the unreacted starting material could be recycled (Figure 3.11) (see also Chapter 4).

In addition to various other topics concerning the use of IL for biotransformations and applicable reactor concepts,[10,11,60–70] Lozano et al. put considerable effort into establishing a strategy for continuous chemo-enzymatic dynamic kinetic resolution in the biphasic system IL and $scCO_2$.[71–74] To this end the CALB-catalyzed enantioselective transesterification of racemic 1-phenylethanol with vinyl propionate was combined with the chemical racemization of the remaining (*S*)-alcohol (Figure 3.12).

Within the first attempts the biotransformation and the chemical reaction step had to be separated due to deactivation of the biocatalyst, resulting in a yield of 70–80% and an enantiomeric excess of 90–97%.[71,72] Recently, the authors reported a concept that allows for both reactions to take place in one pot.[73,74] Therefore, CALB and the acidic zeolite CBV400 were coated with [BM_3AM][$(CF_3SO_2)_2N$]. During 14 days of operation, the continuous reactions carried out at 50°C and 100 bar proceeded without any significant loss in activity and enabled yields up to 98.0% for (*R*)-phenylethylpropionate with an enantiomeric excess of 97.3% (see also Chapter 4).

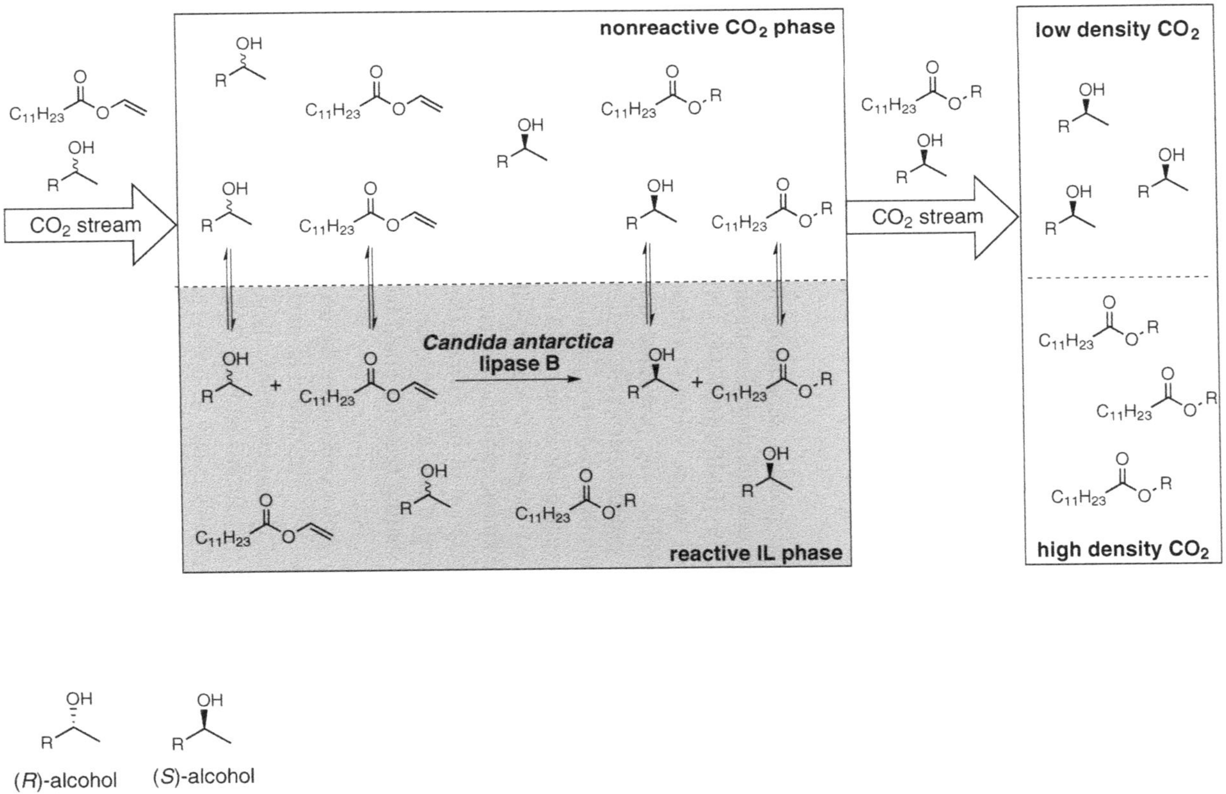

Figure 3.11. *Candida antarctica* lipase B-catalyzed continuous kinetic resolution of racemic alcohols in a biphasic system out of IL and $scCO_2$ with selective separation of the formed products via density reduction.

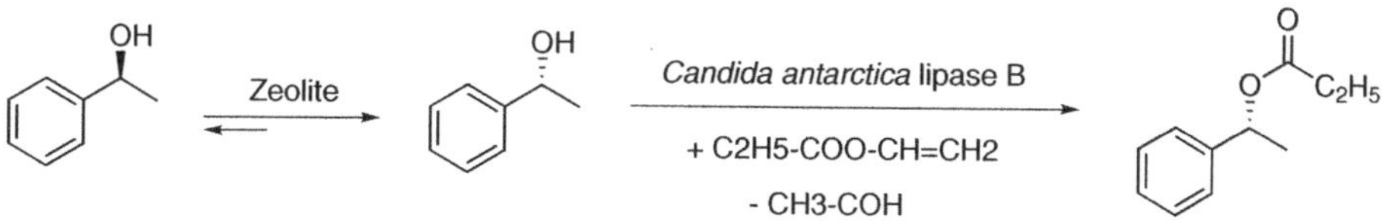

Figure 3.12. Chemo-enzymatic dynamic kinetic resolution of *rac*-1-phenylethanol catalyzed by *Candida antarctica* lipase B and racemization via zeolite as heterogeneous catalyst.

ILs were also successfully applied as supported liquid membranes (SLMs) for enzyme-facilitated resolution of ibuprofen.[75,76] Figure 3.13 depicts the principle used. In a first step, lipase from *Candida rugosa* selectively catalyzed the esterification of (*S*)-ibuprofen from its racemic mixture at interface 1. The obtained ester can dissolve in the IL and is therefore able to diffuse through the membrane. When passing interface 2, the ester hydrolyzes by a lipase from porcine pancreas, yielding the primary (*S*)-ibuprofen. The SLM used was prepared by immersion of a hydrophobic poly(propylene) film in the respective IL. To effectively realize the concept, four ILs and six water-immiscible organic solvents were tested. On average, the selected ILs showed higher initial permeate fluxes of ibuprofen ethyl ester through the resulting SLMs, while [BMIM][$(CF_3SO_2)_2N$] was superior to the other ILs applied. The observed differences were referred to the solubility of the ester in the IL phase and the activities of the biocatalyst at the interfaces. Even though additional lipases from different organisms were tested with an enantiomeric excess of 75% for both enzymes, the combination of the lipases from *Candida rugosa* and porcine pancreas showed the best enantioselectivity possible (see also Chapter 5, Section 5.3.2.4, and Chapter 8).

3.3 CHALLENGES FOR THE USE OF IL IN BIOTRANSFORMATIONS

Since 2000 the interest in ILs as potential solvents for biotransformation has increased rapidly. ILs were considered the designer solvents of the future, being an adequate alternative to organic solvents and a sustainable green high-technology medium. However, only 10 years later, the growth in this field is leveling off as depicted by Figure 3.14.

It is not that results so far achieved by IL research were not promising, but using ILs proved to be challenging in its own right. The most pressing question of which IL to choose cannot be answered *ab initio* so far. Even though progress is made for the prediction of thermodynamic properties, the results can only give trends that have to be tested in reality. The complex interactions with the catalyst and the influence on activity and stability are even less well understood; even if for selected biocatalysts such as the lipase B from *Candida antarctica*, empirical correlations can be made as enough data are accumulated. Furthermore, for two-phase systems the interphase may have further effects on the performance.

For the application of pure ILs as solvents for biotransformations in general the following conclusions can be presented: Hydrophobic IL might be superior to

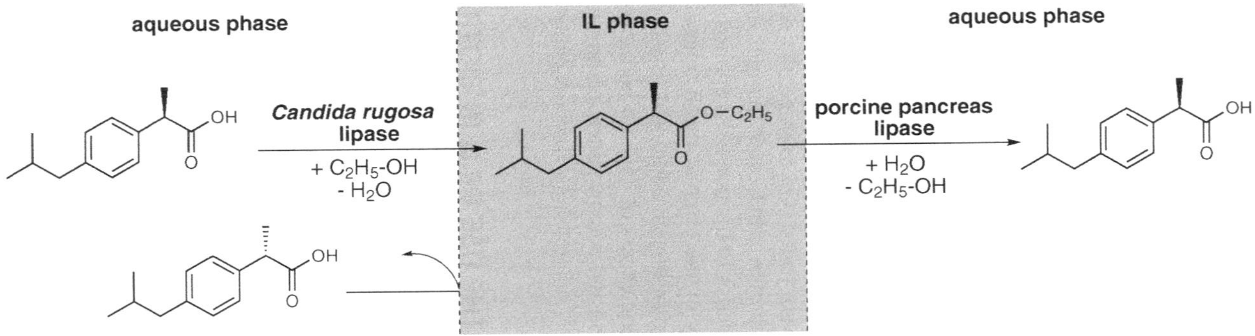

Figure 3.13. IL-supported liquid membranes for enzyme-facilitated resolution of ibuprofen.

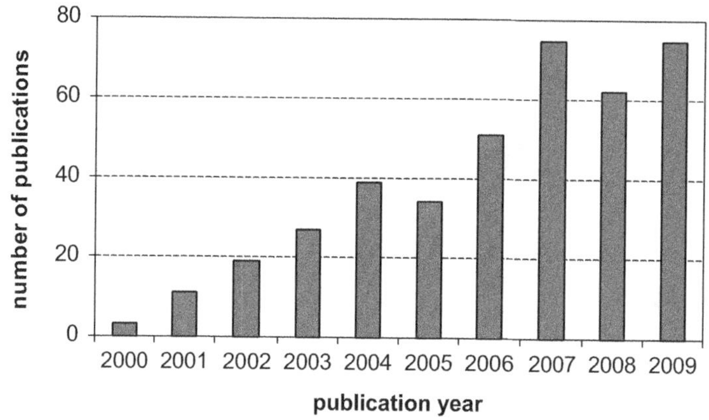

Figure 3.14. Number of articles concerning the use of IL as reaction medium for biotransformations published in journals.

hydrophilic (water-miscible) ones, as they possess a higher tendency to strip off essential water from the enzyme.[12,77] Additionally, the polarity of the solvent is discussed as one factor influencing biocompatibility.[10,77–80] Biocatalysts are supposed to have superior activities in IL with increasing polarity. Nevertheless, the different available methods to predict polarities led to discriminative classification for the same IL,[80] thus rendering predictions difficult.

For aqueous dilutions of hydrophilic IL, very often the Hofmeister series (or more precise viscosity *B*-coefficients) is addressed to explain observed phenomena (for detailed information on the topic see Chapters 2 and 5).[81–85] Due to their ability to support or weaken aqueous hydrogen bonds and therefore influence the structure of water, ions can be classified as kosmotrope or chaotrope. According to common knowledge, a combination of chaotropic cations and kosmotropic anions is able to stabilize proteins in solution, whereas the reverse combination leads to destabilization. While the outcomes presented in some publications are in line with the Hofmeister theory, others do not support its accuracy.[16,85]

Why is it so difficult to forecast if a certain IL and a biocatalyst are compatible? First of all, the rather anthropomorphic or phenomenological definition for a salt being an IL when exhibiting a melting point below 100°C lumps together chemicals with very different physicochemical properties. Consequently, the number of postulated ILs is enormous (ranging from 10^{18} to more conservative figures in the 10^6 range). However, the number of ILs utilized in a biotransformation is well below 100. Cations and anions utilized in biotransformation are presented in Figure 3.15.

In general, almost only imidazolium-based cations (especially 1-butyl-3-methylimidazolium) and anions such as BF_4^-, PF_6^-, and $[(CF_3SO_2)_2N]^-$ have gained attention. Considering the number of possible ILs, the exploration of biocatalysts in these media is by far too low to yield distinct tendencies and consolidated findings.[86]

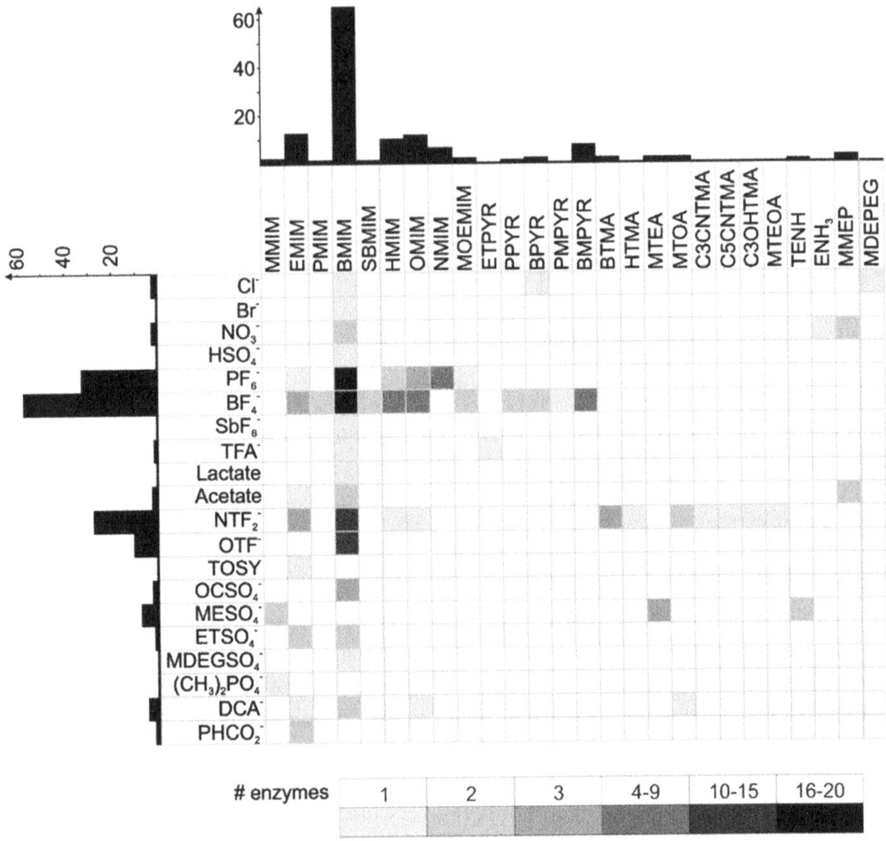

Figure 3.15. Cations and anions utilized in biotransformations. Reproduced with permission of Springer.

Additionally, the appearance of clear trends is hampered by various overlaying effects, which will be discussed consecutively in the following text. Furthermore, research is limited to hydrolases and oxidoreductases. Due to biological diversity, the behavior of these biocatalysts is only comparable to a certain extent; thus, the observed trends cannot be extrapolated to each other and might not be true for other types of biocatalysts. Additionally, it is possible to use IL as a single solvent, as a co-solvent, and in biphasic systems representing the nonreactive or reactive phase. Whereas hydrolases are mainly used in pure solvents, oxidoreductases are preferably applied in aqueous media with ILs as co-solvents or second-phase solvents. This fact further reduces the number of comparable investigations, as the exerted influence of the IL on the biocatalyst strongly depends on the manner of its application.

Noteworthy, experimental results vary and prevent comparability even if the IL and the biocatalyst are investigated under apparently same conditions. For example,

Lutz-Wahl et al. reported an increased activity of isolated D-amino acid oxidase from *Trigonopsis variabilis* for the oxidation of D-phenylalanine in the presence of 20 vol% of [MMIM] [Me_2PO_4] in TRIS buffer.[49] In contrast, an almost linear activity decrease in the same enzyme with increasing amounts of 2–10 vol% of the same IL was reported for the oxidation of D-alanine in potassium phosphate buffer.[16] Furthermore, chloroperoxidase from *Caldariomyces fumago* was first found to be inactivated with [MMIM] [$MeSO_4$] as the co-solvent,[88] while in a more recent publication residual activities of 40% were obtained.[15] Moreover, for the utilization of [BMIM][PF_6] and [BMIM][BF_4] as reaction media for CALB-catalyzed transesterifications, the following results were published independently: First, almost no activity and conversions lower than 5% compared with 40–50% for MTBE and various other ILs (even very similar ones such as [EMIM][BF_4]) were achieved.[7,89] Second, two independent publications quote reaction rates comparable to those obtained for THF, toluene, and 2-propanol.[6,9] Finally, even increased reaction rates were reported.[11]

In almost all cases, discrepancies between published results and lack of reproducibility may be explained by impurities in the ILs used. Since most of the impurities result from IL preparation, it is worth knowing the most common synthesis strategies to be aware of emerging contamination. Figure 3.16 presents the most important routes using the example of imidazolium-based ILs, as these ILs are prevailing in biotransformations.

In the halide-based route, a Lewis base (here 1-methylimidazole) is alkylated by nucleophilic substitution of an alkyl halide, in this case yielding a 1-alkyl-3-methylimidazolium halide. Unreacted starting material is normally removed via extraction by an organic solvent. In a successive step, either the halide may be exchanged by the required anion (upper arrow) or it may be chemically modified to result in the targeted structure (lower arrow). Depending on the miscibility of the formed IL, the by-product acids or metal halides are removed differently. Water-immiscible ILs are purified via extraction with water and phase separation, whereas water-miscible ILs are purged by removal of H_2O under reduced pressure, addition of $CHCl_3$, and repetitive cycles of cooling and filtration of formed precipitates.[90,91] Nevertheless, trace impurities in the form of halides, unreacted organic salts, and acids as well as organic solvents and water remain in the IL according to the preparation route.[91–94]

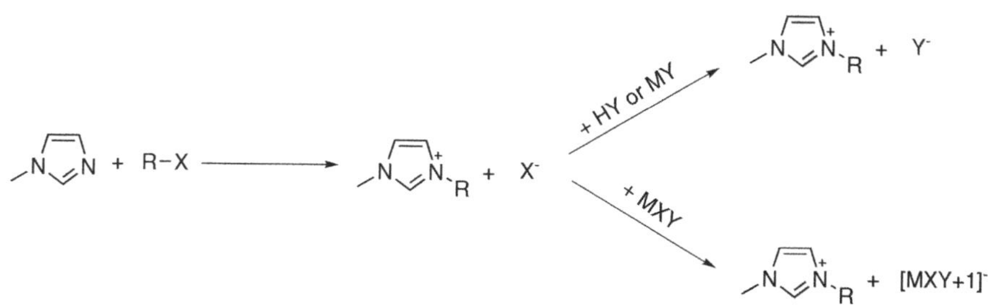

Figure 3.16. Halide-based synthesis routes for the preparation of IL at the example of imidazolium-based IL.

These contaminants are known to influence the solvent behavior as well as the performance of biocatalysts, which accounts for the variety in literature reports.[8,79,80,95] As in the beginning almost only self-made ILs were applied, it is impossible to reassess the impurities, not to mention quantifying them.

One of the first to address the purity issue were Park et al.[13] by comparing conversions obtained for the acetylation of 1-phenylethanol catalyzed by *Pseudomonas cepacia* lipase in unpurified and purified ILs. By removing halides via two different routes, increased conversions were achieved. For example, in unpurified [BMIM][BF_4] a conversion of 8% was measured, whereas for the same IL conversions of 13% and 36% were possible, depending on the purification. Differences in the results obtained for both purifications were associated with silver ion impurities caused by one of the routes. Furthermore, ILs such as [EMIM][BF_4] in which apparently no reaction occurred showed conversions of up to 46% after purification. Nevertheless, in a later work concerning the lipase-catalyzed direct condensation of L-ascorbic acid and fatty acids in the presence of IL, the aqueous sodium carbonate washing step resulted in the accumulation of the latter, accelerating subsequent oxidative decomposition of the formed product.[96] Here, 2.5-fold increased yields of the desired product were possible with the alternative method, thus pointing to the necessity of an adequate purification procedure.

Additionally, the findings of Lee et al. underline the effect of impurities on enzymatic performance.[97] They conducted a study to discover the influence of chloride on enzymatic activity, applying the transesterification of benzyl alcohol and vinyl acetate catalyzed by lipases from *Rhizomucor miehei* and from *Candida antarctica* in [OMIM][$(F_3CSO_2)_2N$]. The content of chloride ions in IL was varied via addition of different amounts [OMIM][Cl]. While for *Candida antarctica* lipase an approximately linear decrease in activity with increasing amount of [OMIM][Cl] became obvious, *Rhizomucor miehei* lipase activity decreased exponentially, resulting in inactive *Candida antarctica* lipase with addition of 20% [OMIM][Cl] and ~ 2% residual activity of *Rhizomucor miehei* lipase for only 2% of the latter IL.

Therefore, the alkylation by alternative halide-free routes, for example, using alkyl sulfates (Figure 3.17), has gained importance. This favorable method is feasible for alkylating agents such as dialkyl sulfate, alkyl trifluoromethansulfate, and alkyl bis(trifluoromethanesulfony)amide, resulting in anions commonly used in biotransformations.[90,92]

Beside halides, water is a common impurity as most ILs are hygroscopic; nevertheless, resulting consequences are frequently underestimated. As known for conventional organic solvents,[98,99] when carrying out biotransformations in pure solvents, the water

Figure 3.17. Halide-free synthesis route for the preparation of [EMIM][$EtSO_4$] with diethyl sulfate as alkylating agent.

content or rather the water activity will affect the behavior of the biocatalyst. Early on, it was noted that most biocatalysts are inactive in water-free ILs. Independently reported publications referring to α-chymotrypsin being inactive in IL without addition water were among the first evidence.[10,12,100] Eckstein et al. investigated the kinetics of N-acetyl-L-phenylalanine ethyl ester transesterification with 1-butanol catalyzed α-chymotrypsin in ILs and conventional organic solvents at fixed water activities.[100] In general the different solvents behaved similarly, since with increasing water content for all solvents, obtained reaction rates increased, whereas the enantioselectivity decreased. Moreover, the enantioselective esterification of (*R,S*)-2-chloropropanoic acid with 1-butanol catalyzed by *Candida rugosa* lipase was also investigated in IL and conventional organic solvents. In addition to reactions carried out at known initial water contents, reactions where the water content was maintained via removal of emerging water by pervaporation were also carried out.[95] While for [BMIM][PF_6] maximum ester yields of 29% were achieved for an initial water content of 0.2%, superior yields of 37% were possible when conducting the synthesis at a constant water content of 0.5%. Accordingly, it is not only required to measure initial water contents or preferably water activities, but it is also necessary to regulate them when aiming for reproducible and comparable results.

Depending on the IL used, accumulated water might not only directly interfere with the biocatalyst, but might also cause hydrolysis of involved ions. Anions such as BF_4^- or PF_6^- tend to hydrolyzed under release of hydrofluoric acid (HF),[101] which is not only able to inhibit enzymes and acidify the reaction medium, but might also be hazardous for the experimenter. In particular, for aqueous mixtures with ILs, control of the resulting pH is recommended.[8,78] Otherwise, obtained results might only result from pH shifts triggered by IL or its impurities. For instance, the addition of 30 vol% of [MMIM][Me_2PO_4] to a potassium phosphate buffer at pH 2.7 resulted in a pH of 3.7.[88] When conducting the chlorination of monochlorodimedone by chloroperoxidase form *Caldariomyces fumago* in this reaction medium, a residual biocatalyst activity of only 30% became obvious, whereas a residual activity of 90% is maintained when the reaction is carried out after adjusting the required pH.

In addition, under basic conditions, decomposition of imidazolium-based cations can take place via carbene formation and subsequent disproportionation.[102] As N-heterocyclic carbenes have successfully been applied as catalysts for transesterifications,[103] even traces of those nucleophiles have to be avoided to exclude background reactions in biocatalytic transesterifications. Furthermore, deactivation of biocatalysts by accumulating decomposition products might also be possible.

Altogether, adequate analytics for the determination of impurities in ILs should not only be delivered by the suppliers, but also need to be addressed by the scientists themselves, when aiming for reproducibility and comparability. Furthermore, appropriate purification steps need to be established. In addition, especially for the application of ILs as co-solvents, the way of stating the amount of ILs used should be questioned. In contrast to the common designation of vol%, concentrations, mass, or mole fractions might lead to more accurate conclusions.

In addition to impurities, mass transport limitations through comparatively high viscosities of ILs might also affect the results. For instance, de Gonzalo et al. investigated the reduction of 2-octanone by *Escherichia coli* cells with alcohol dehydrogenase from *Rhodococcus ruber* in a rather viscous reaction medium with up to 90 vol% of water-miscible IL.[104] To discover probable mass-transfer limitations, they carried out the reactions at 30 and 37°C, which was feasible as the biocatalyst activity is identical at these temperatures. Actually, increased reaction temperature resulted in improved conversions, which the authors attributed to decreasing viscosity at higher temperatures. For AMMOENG™ 100, 101, and 102 (see Figure 2.8 and Chapter 2), conversions were improved by 149%, 47%, and even 265%, respectively. This underlines the importance of discriminating between possibly overlaying effects, when aiming for the discovery of clear trends.

Moreover, when using a photometric measurement, the extinction coefficients may be altered by additives. Even though this fact was considered by Magnuson et al. long before the main interest in IL as a solvent arose,[1] unfortunately this is not reflected in most current research.

In addition to the challenge to select a biocompatible IL itself, frequently a biocompatible solubilizer for the associated reactants is required. Despite the fact that ILs are discussed to be designer solvents, by now even properties such as melting point, miscibility, and solubility cannot be predicted reliably,[79] thus complicating the selection of a task-specific IL.

In conclusion, when aiming for reliable and precise outcomes, the investigator has to fulfill high demands while working with ILs in biotransformations. Special care has to be taken on very likely occurring impurities to ensure that the impact of these contaminations and the IL itself is taken into account. Furthermore, results should be compared on the basis of significant physicochemical properties rather than on arbitrary classifications.

REFERENCES

1 D. K. Magnuson, J. W. Bodley, D. F. Evans, *J. Solut. Chem.* **1984**, *13*, 583–587.

2 S. G. Cull, J. D. Holbrey, V. Vargas-Mora, K. R. Seddon, G. J. Lye, *Biotechnol. Bioeng.* **2000**, *69*, 227–233.

3 M. Erbeldinger, A. J. Mesiano, A. J. Russell, *Biotechnol. Prog.* **2000**, *16*, 1131–1133.

4 R. M. Lau, F. van Rantwijk, K. R. Seddon, R. A. Sheldon, *Org. Lett.* **2000**, *2*, 4189–4191.

5 T. L. Husum, C. T. Jorgensen, M. W. Christensen, O. Kirk, *Biocatal. Biotransformation* **2001**, *19*, 331–338.

6 T. Itoh, E. Akasaki, K. Kudo, S. Shirakami, *Chem. Lett.* **2001**, *32*, 262–263.

7 S. H. Schöfer, N. Kaftzik, U. Kragl, P. Wasserscheid, *Chem. Commun.* **2001**, 425–426.

8 U. Kragl, N. Kaftzik, S. H. Schofer, M. Eckstein, P. Wasserscheid, C. Hilgers, *Chim. Oggi.* **2001**, *19*, 22–24.

9 K.-W. Kim, B. Song, M.-Y. Choi, M.-J. Kim, *Org. Lett.* **2001**, *3*, 1507–1509.

10 P. Lozano, T. De Diego, J.-P. Guegan, M. Vaultier, J. L. Iborra, *Biotechnol. Bioeng.* **2001**, *75*, 563–569.

11 P. Lozano, T. De Diego, D. Carrie, M. Vaultier, J. L. Iborra, *Biotechnol. Lett.* **2001**, *23*, 1529–1533.

12 J. A. Laszlo, D. L. Compton, *Biotechnol. Bioeng.* **2001**, *75*, 181–186.

13 S. Park, R. J. Kazlauskas, *J. Org. Chem.* **2001**, *66*, 8395–8401.

14 J. Howarth, P. James, J. Dai, *Tetrahedron Lett.* **2001**, *42*, 7517.

15 C. Kohlmann, L. Greiner, W. Leitner, C. Wandrey, S. Lütz, *Chem. Eur. J.* **2009**, *15*, 11692–11790.

16 S. Dreyer, U. Kragl, *Biotechnol. Bioeng.* **2008**, *99*, 1416–1424.

17 C. Kohlmann, N. Robertz, S. Leuchs, Z. Dogan, K. Bitzer, S. Lütz, S. Na'amnieh, L. Greiner, *J. Mol. Catal. B: Enzym.* **2011**, *68*, 147–153.

18 Q. Liu, M. H. A. Janssen, F. van Rantwijk, R. A. Sheldon, *Green Chem.* **2005**, *7*, 39–42.

19 S. A. Forsyth, D. R. MacFarlane, R. J. Thomson, M. von Itzstein, *Chem. Commun.* **2002**, 714–715.

20 F. van Rantwijk, F. Secundo, R. A. Sheldon, *Green Chem.* **2006**, *8*, 282–286.

21 S. H. Lee, T. D. Dung, H. H. Sung, W.-J. Chang, Y.-M. Koo, *Biotechnol. Bioeng.* **2008**, *99*, 1–8.

22 G. L. Miller, *Anal. Chem.* **1959**, *31*, 426–428.

23 R. P. Swatloski, S. K. Spear, J. D. Holbrey, R. D. Rogers, *J. Am. Chem. Soc.* **2002**, *124*, 4974–4975.

24 S. Zhu, Y. Wu, Q. Chen, Z. Yu, C. Wang, S. Jin, Y. Dinga, G. Wuc, *Green Chem.* **2006**, *8*, 325–327.

25 D. A. Fort, R. C. Remsing, R. P. Swatloski, P. Moyna, G. Moyna, R. D. Rogers, *Green Chem.* **2007**, *9*, 63–69.

26 I. Kilpeläinen, H. Xie, A. King, M. Granstrom, S. Heikkinen, D. S. Argyropoulos, *J. Agric. Food Chem.* **2007**, *55*, 9142–9148.

27 C. Cuissinat, P. Navard, T. Heinze, *Cellulose* **2008**, *15*, 75–80.

28 Y. Fukaya, K. Hayashi, M. Wada, H. Ohno, *Green Chem.* **2008**, *10*, 44–46.

29 B. Kosan, C. Michels, F. Meister, *Cellulose* **2008**, *15*, 59–66.

30 A. Pinkert, K. N. Marsh, S. Pang, M. P. Staiger, *Chem. Rev.* **2009**, *109*, 6712–6728.

31 J. Vitz, T. Erdmenger, C. Haensch, U. S. Schubert, *Green Chem.* **2009**, *11*, 417–424.

32 M. Zavrel, D. Bross, M. Funke, J. Büchs, A. C. Spiess, *Bioresour. Technol.* **2009**, *100*, 2580–2587.

33 C. Graenacher, Chem, Ind Basel, US 1943176, **1934**.

34 H. Zhao, G. A. Baker, J. V. Cowins, *Biotechnol. Prog.* **2010**, *26*, 127–133.

35 I. P. Samayam, C. A. Schall, *Bioresour. Technol.* **2010**, *101*, 3561–3566.

36 S. Datta, B. Holmes, J. I. Park, Z. Chen, D. C. Dibble, M. Hadi, H. W. Blanch, B. A. Simmons, R. Sapra, *Green Chem.* **2010**, *12*, 338–345.

37 W. Hussain, D. J. Pollard, M. Truppo, G. J. Lye, *J. Mol. Catal., B Enzym.* **2008**, *55*, 19–29.

38 H. Pfruender, M. Amidjojo, U. Kragl, D. Weuster-Botz, *Angew. Chem. Int. Ed. Engl.* **2004**, *43*, 4529–4531.

39 D. Weuster-Botz, *Chem. Rec.* **2007**, *7*, 334–340.

40 S. Bräutigam, D. Dennewald, M. Schuermann, J. Lutje-Spelberg, W.-R. Pitner, D. Weuster-Botz, *Enzyme Microb. Technol.* **2009**, *45*, 310–316.

41 X. Lu, J. Hu, X. Yao, Z. Wang, J. Li, *Biomacromolecules* **2006**, *7*, 975–980.

42 R. Yan, F. Zhao, J. Li, F. Xiao, S. Fan, B. Zeng, *Electrochim. Acta* **2007**, *52*, 7425–7431.

43 K. Nakashima, N. Kamiya, D. Koda, T. Maruyama, M. Goto, *Org. Biomol. Chem.* **2009**, *7*, 2353–2358.

44 Y. Liu, M. Wang, J. Li, Z. Li, P. He, H. Liu, J. Li, *Chem. Commun.* **2005**, 1778–1780.

45 Y. Liu, L. Shi, M. Wang, Z. Li, H. Liu, J. Li, *Green Chem.* **2005**, *7*, 655–658.

46 Y. Jiang, C. Guo, H. Xia, I. Mahmood, C. Liu, H. Liu, *J. Mol. Catal., B Enzym.* **2009**, *58*, 103–109.

47 J. K. Lee, M.-J. Kim, *J. Org. Chem.* **2002**, *67*, 6845–6847.

48 C. Kohlmann, L. Greiner, S. Lütz, DE 10 2008 061 866.7, **2008**.

49 S. Lutz-Wahl, E. M. Trost, B. Wagner, A. Manns, L. Fischer, *J. Biotechnol.* **2006**, *124*, 163–171.

50 S. J. Nara, S. S. Mohile, J. R. Harjani, P. U. Naik, M. M. Salunkhe, *J. Mol. Catal., B Enzym.* **2004**, *28*, 39–43.

51 S. S. Mohile, M. K. Potdar, J. R. Harjani, S. J. Nara, M. M. Salunkhe, *J. Mol. Catal., B Enzym.* **2004**, *30*, 185–188.

52 G. E. P. Box, K. B. Wilson, *J. R. Stat. Soc. B Stat. Methodol.* **1951**, *13*, 1–45.

53 M. Eckstein, M. Villela Filho, A. Liese, U. Kragl, *Chem. Commun.* **2004**, 1084–1085.

54 K. Schroer, E. Tacha, S. Lütz, *Org. Process Res. Dev.* **2007**, *11*, 836–841.

55 F. Ganske, U. T. Bornscheuer, *Org. Lett.* **2005**, *7*, 3097–3098.

56 F. Ganske, U. T. Bornscheuer, *J. Mol. Catal., B Enzym.* **2005**, *36*, 40–42.

57 C. Kohlmann, S. Leuchs, L. Greiner, *Green Chem.* **2011**, *13*, 1430–1436.

58 M. T. Reetz, W. Wiesenhofer, G. Francio, W. Leitner, *Chem. Commun.* **2002**, 992–993.

59 M. T. Reetz, W. Wiesenhöfer, G. Franciò, W. Leitner, *Adv. Synth. Catal.* **2003**, *345*, 1221–1228.

60 P. Lozano, T. de Diego, D. Carrie, M. Vaultier, J. L. Iborra, *Chem. Commun.* **2002**, 692–693.

61 P. Lozano, T. De Diego, D. Carrie, M. Vaultier, J. L. Iborra, *J. Mol. Catal., B Enzym.* **2003**, *21*, 9–13.

62 P. Lozano, T. De Diego, D. Carrie, M. Vaultier, J. L. Iborra, *Biotechnol. Prog.* **2003**, *19*, 380–382.

63 P. Lozano, T. De Diego, D. Carrie, M. Vaultier, J. L. Iborra, *J. Mol. Catal., A Chem.* **2004**, *214*, 113–119.

64 P. Lozano, T. de Diego, S. Gmouh, M. Vaultier, J. L. Iborra, *Biotechnol. Prog.* **2004**, *20*, 661–669.

65 P. Lozano, T. De Diego, S. Gmouh, M. Vaultier, J. L. Iborra, *Biocatal. Biotransformation* **2005**, *23*, 169–176.

66 P. Lozano, T. de Diego, J. L. Iborra, *Methods Biotechnol.* **2006**, *22*, 257–268.

67 P. Lozano, T. De Diego, J. L. Iborra, *Chim. Oggi.* **2007**, *25*, 76–79.

68 P. Lozano, R. Piamtongkam, K. Kohns, T. De Diego, M. Vaultier, J. L. Iborra, *Green Chem.* **2007**, *9*, 780–784.

69 P. Lozano, E. Garcia-Verdugo, R. Piamtongkam, N. Karbass, T. De Diego, M. I. Burguete, S. V. Luis, J. L. Iborra, *Adv. Synth. Catal.* **2007**, *349*, 1077–1084.
70 P. Lozano, T. De Diego, T. Sauer, M. Vaultier, S. Gmouh, J. L. Iborra, *J. Supercrit. Fluids* **2007**, *40*, 93–100.
71 P. Lozano, T. Diego, M. Larnicol, M. Vaultier, J. L. Iborra, *Biotechnol. Lett.* **2006**, *28*, 1559–1565.
72 P. Lozano, T. De Diego, S. Gmouh, M. Vaultier, J. L. Iborra, *Int. J. Chem. React. Eng.* **2007**, *5*, A53.
73 P. Lozano, T. De Diego, M. Vaultier, J. L. Iborra, *Int. J. Chem. React. Eng.* **2009**, *7*, A79.
74 P. Lozano, T. De Diego, C. Mira, K. Montague, M. Vaultier, J. L. Iborra, *Green Chem.* **2009**, *11*, 538–542.
75 E. Miyako, T. Maruyama, N. Kamiya, M. Goto, *Biotechnol. Lett.* **2003**, *25*, 805–808.
76 E. Miyako, T. Maruyama, N. Kamiya, M. Goto, *Chem. Commun.* **2003**, 2926–2927.
77 E. Feher, B. Major, K. Belafi-bako, L. Gubicza, *Biochem. Soc. Trans.* **2007**, *35*, 1624–1627.
78 S. Park, R. J. Kazlauskas, *Curr. Opin. Biotechnol.* **2003**, *14*, 432–437.
79 F. van Rantwijk, R. Madeira Lau, R. A. Sheldon, *Trends Biotechnol.* **2003**, *21*, 131–138.
80 F. van Rantwijk, R. A. Sheldon, *Chem. Rev.* **2007**, *107*, 2757–2785.
81 J. M. Broering, A. S. Bommarius, *J. Phys. Chem. B* **2005**, *109*, 20612–20619.
82 F. Hofmeister, *Arch. Exp. Pathol. Pharmakol.* **1888**, *24*, 247–260.
83 W. Kunz, J. Henle, B. W. Ninham, *Curr. Opin. Colloid Interface Sci.* **2004**, *9*, 19–37.
84 Z. Yang, *J. Biotechnol.* **2009**, *144*, 12–22.
85 H. Zhao, S. M. Campbell, L. Jackson, Z. Y. Song, O. Olubajo, *Tetrahedron Asymmetry* **2006**, *17*, 377–383.
86 H. Zhao, *J. Mol. Catal., B Enzym.* **2005**, *37*, 16–25.
87 C. Roosen, P. Müller, L. Greiner, *Appl. Microbiol. Biotechnol.* **2008**, *81*, 607–614.
88 C. Chiappe, L. Neri, D. Pieraccini, *Tetrahedron Lett.* **2006**, *47*, 5089–5093.
89 U. Kragl, M. Eckstein, N. Kaftzik, *Curr. Opin. Biotechnol.* **2002**, *13*, 565–571.
90 A. Große Böwing, A. Jess, P. Wasserscheid, *Chem. Ing. Tech.* **2005**, *77*, 1430–1439.
91 P. Wasserscheid, T. Welton, *Ionic Liquids in Synthesis*, Vol. 1, 2nd ed., Wiley-VCH, Weinheim, **2008**.
92 A. Stark, P. Behrend, O. Braun, A. Müller, J. Ranke, B. Ondruschka, B. Jastorff, *Green Chem.* **2008**, *10*, 1152–1161.
93 K. R. Seddon, A. Stark, M.-J. Torres, *Pure Appl. Chem.* **2000**, *72*, 2275–2287.
94 P. J. Scammells, J. L. Scott, R. D. Singer, *Aust. J. Chem.* **2005**, *58*, 155–169.
95 L. Gubicza, N. Nemestothy, T. Frater, K. Belafi-Bako, *Green Chem.* **2003**, *5*, 236–239.
96 S. Park, F. Viklund, K. Hult, R. J. Kazlauskas, *Green Chem.* **2003**, *5*, 715–719.
97 S. H. Lee, S. H. Ha, S. B. Lee, Y.-M. Koo, *Biotechnol. Lett.* **2006**, *28*, 1335–1339.
98 A. M. Klibanov, *Nature* **2001**, *409*, 241–246.
99 M. N. Gupta, *Eur. J. Biochem.* **1992**, *203*, 25–32.
100 M. Eckstein, M. Sesing, U. Kragl, P. Adlercreutz, *Biotechnol. Lett.* **2002**, *24*, 867–872.
101 R. P. Swatloski, J. D. Holbrey, R. D. Rogers, *Green Chem.* **2003**, *5*, 361–363.

102 M. Maase, in *Ionic Liquids in Synthesis*, Vol. 1, 2nd ed. (Eds. P. Wasserscheid, T. Welton), Wiley-VCH, Weinheim, **2008**.

103 R. Singh, R. M. Kissling, M.-A. Letellier, S. P. Nolan, *J. Org. Chem.* **2004**, *69*, 209–212.

104 G. de Gonzalo, I. Lavandera, K. Durchschein, D. Wurm, K. Faber, W. Kroutil, *Tetrahedron Asymmetry* **2007**, *18*, 2541–2546.

4

IONIC LIQUIDS AND OTHER NONCONVENTIONAL SOLVENTS IN BIOTRANSFORMATIONS: MEDIUM ENGINEERING AND PROCESS DEVELOPMENT

Pedro Lozano and Eduardo García-Verdugo

4.1 INTRODUCTION: TOWARD GREENER CATALYTIC PROCESSES

The reduction and elimination of hazardous substances in the design, manufacture, and application of chemical products is the main goal of green chemistry.[1] In this context, the use of solvents is one of the major concerns to develop more environmentally benign processes.[2] Indeed, solvents often account for the vast majority of mass wasted in syntheses and processes.[3] Furthermore, many conventional solvents are toxic, flammable, and/or corrosive. Their volatility and solubility have contributed to air, water, and land pollution; have increased the risk of workers' exposure; and have led to serious accidents. Moreover, their recovery and reuse, when possible, is often associated with energy-intensive distillation and sometimes cross-contamination.

Solvent elimination and substitution are not easy tasks, as both are key elements in chemical processes. Indeed, solvent provides an environment where reactants meet. Solvent dilutes the reactants and/or the catalysts. Furthermore, by changing its nature (inert, polarizable, protogenic, protophilic, etc.) transition states and catalysts can be either stabilized or destabilized. Finally, the solvent should assist for the homogeneous

Ionic Liquids in Biotransformations and Organocatalysis: Solvents and Beyond, First Edition. Edited by Pablo Domínguez de María.

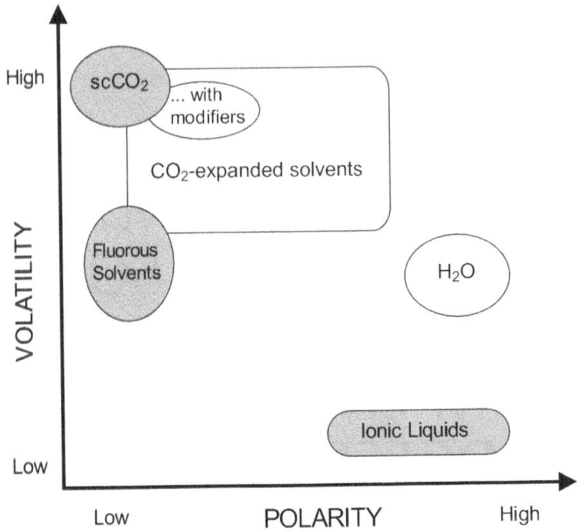

Figure 4.1. Scheme of volatility and polarity characteristic of neoteric solvents.[2a]

distribution of the energy needed for the reaction activation.[4] Thus, new solvents systems should be designed to deliver all these requirements and to be environmentally friendly.

In an effort to address all those shortcomings, chemists started a search for safer new environmentally benign solvents or green solvents, which can easily be recovered/recycled and which can allow catalysts to operate efficiently in them. In this respect, ionic liquids (ILs), supercritical fluids (SCFs), and fluorous solvents are the nonaqueous green solvents, also named neoteric solvents, which have been receiving most attention worldwide in recent years (see Figure 4.1).[5] ILs are a new class of polar liquid solvents and their use has led to a new green chemical revolution because of their unique array of physical–chemical properties, headed by their nonvolatile character, which makes them suitable for numerous industrial applications.[6] SCFs are another class of neoteric solvents with unique tunable properties, which enable them to be applied in green reactions, extraction, and fractionation processes.[7] Supercritical carbon dioxide ($scCO_2$) is much widely used and can be regarded as a safe nontoxic solvent. However, its nonpolar nature renders it unsuitable for most homogeneous catalyses involving polar transition metal complexes. This low-polar character may be modified by addition of small amounts of co-solvents (e.g., ethanol). Alternatively, a gas-expanded liquid (GXL) phase can be also used. GXL is generated by dissolving a compressible gas such as CO_2 or light olefin into the traditional liquid phase at mild pressures (tens of bars). When CO_2 is used as the expansion gas, the resulting liquid phase is termed a CXL. GXLs combine the advantages of compressed gases such as CO_2 and of traditional solvents in an optimal manner.[8] Fluorous solvents are another class of very low polar and easily recyclable green solvents, having a wide range of applications (i.e., refrigerants, lubricants, etc.).[9]

The formation of waste is also linked to the traditional use of a stoichiometric amount of reagents.[10] Switching from stoichiometric methodologies to catalytic processes is perceived as one major way to improve the efficiency of the synthetic toolbox. Thus, the greenness of chemical transformations is closely related to the use of catalytic[11] and engineering approaches.[12] Catalysis can improve the efficiency of a reaction by lowering the energy input required, by avoiding the use of stoichiometric amount of reagents, and by greater product selectivity. In this regard, the tremendous potential of enzymes as "green" catalysts cannot be doubted, since they are able to accelerate stereo-, chemo-, and regioselectively different chemical transformations.[13] Furthermore, a great variety (more than 13,000) enzyme-catalyzed reactions have been successfully demonstrated in laboratory scale, offering clear advantages for the synthesis of enantiopure fine chemicals against any other kind of catalysts.[14] Even the chemical industry is exploring the great potential of biocatalysis to manufacture both bulk and fine chemicals.[15]

Enzymes are proteins, polymeric macromolecules based on amino acid units with unique sequences, which show a high level of three-dimensional (3-D) structural organization. Some of them require the presence of additional nonprotein components (cofactors and coenzymes), before they can function as catalysts. The catalytic activity of an enzyme strongly depends on its unique 3-D structure or native conformation, which is maintained by a high number of weak internal interactions (e.g., hydrogen bonds, and van der Waals), as well as interactions with other molecules, mainly water as the natural solvent of living systems.

In general, enzymes are designed to work in aqueous solutions within a narrow range of environmental conditions (pH, temperature, pressure, etc.), and outside these conditions deactivation occurs, generally, by the unfolding of the native conformation. Therefore, they can be, in principle, considered ideal green catalysts as they operate highly selectively under mild energy requirements and in water as clean solvent. This fact, which is an advantage from the greenness point of view, hampers, to the same extent, a wider applicability of the enzymes due to the very low solubility of most chemicals in water, as well as the noninert character of water, which usually leads to undesired side reactions.[16]

Over the past two decades, strategies based on medium engineering, substrate engineering, and biocatalyst engineering have been developed to demonstrate the suitability of enzymes as catalysts for chemical processes in nonaqueous environments,[17] being now applied into new green nonaqueous solvents (e.g., ionic liquids).[18] In this chapter, an overview of the different strategies recently developed on medium engineering in neoteric solvents to meet the increasing demand for greener biocatalytical processes is given. Figure 4.2 summarizes the different approaches used in the development of biocatalytic processes in aqueous systems and neoteric solvents. The examples deal mainly with use of ILs for biphasic or monophasic systems, but brief insights into other systems such as supercritical or perfluorinated solvents, or their combination with each other or with ILs, are also mentioned.

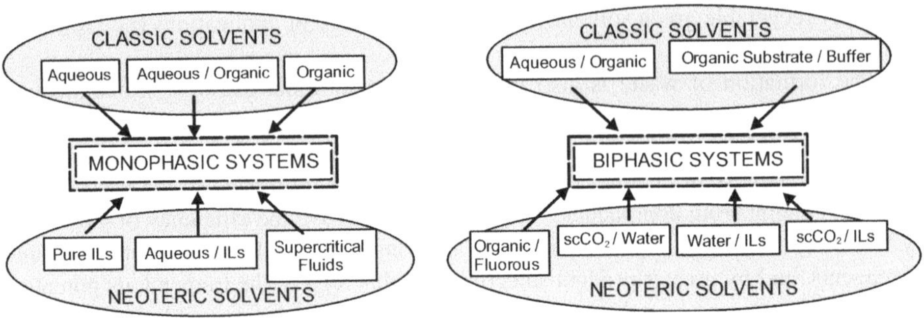

Figure 4.2. Medium engineering to develop biocatalytic processes in either classic or neoteric.

4.2 THE IMPORTANCE OF THE MEDIUM ENGINEERING IN BIOTRANSFORMATIONS

4.2.1 Enzymes in Nonaqueous Environments

Applications of biotransformations in synthetic routes have been hampered by the general idea that biocatalysts can be only used in aqueous solution and under mild conditions. Fortunately, it has been proved that biocatalysis are not so sensitive as thought and under the control conditions can operate under "non-natural" conditions (for example: extreme pH, temperature and pressures, high salt concentration or in presence of other additives).[19] They are also found to be active in different sort of nonconventional solvents ranging from organic solvents to ILs or SCFs. This fact has dramatically increased their use for different chemical processes.[20] Indeed, medium engineering for enzymatic reactions in nonaqueous environments is imperative approach to develop chemical processes based on biotransformations. Medium engineering, in this context, involves the modification of the immediate vicinity of the biocatalyst. Along with it, biocatalyst engineering to provide long-term active and stable enzymes under nonconventional conditions has been a milestone in the development of applied biocatalysis.[21] Thus, developments in genomics, directed evolution, and our exploitation of the natural biodiversity, together with experimental approaches for enzyme immobilization, have led to improvements in the activity, stability, and specificity of enzymes, accompanied by a huge increase in the number and variety of their applications.[22] Furthermore, the catalytic promiscuity of enzymes has been described in nonaqueous environments, related with the ability of a single active site to catalyze more than one chemical transformation (e.g., lipase B from *Candida antarctica* [CALB] is able to catalyze aldol additions, Michael-type additions, and the formation of Si-O-Si bonds).[23] Other enzymes (i.e., peroxidases, laccases, monooxigenases, dehydrogenases, etc.) have also been reported as being excellent tools for organic synthesis in nonaqueous environments.[16,24,25]

There are numerous potential advantages in employing enzymes in nonaqueous environments or media with reduced water content (i.e., in the presence of organic

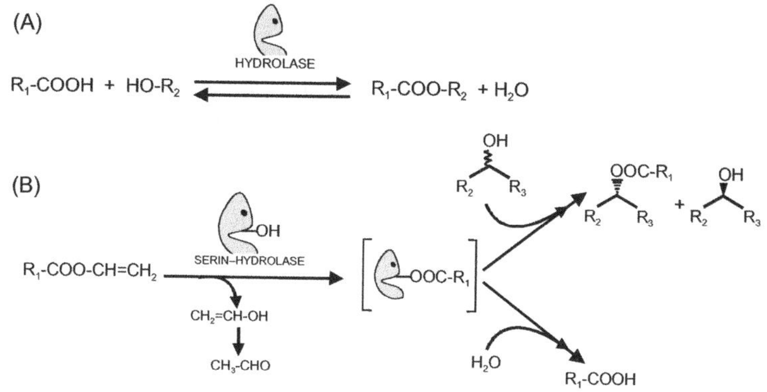

Figure 4.3. (A) Hydrolase-catalyzed ester synthesis by thermodynamic control. (B) Serin–hydrolase-catalyzed ester synthesis by kinetic control.

solvents and/or additives).[24] These advantages include the dramatically higher solubility of hydrophobic substrates, the insolubility of enzymes facilitating their easy reuse, and the elimination of microbial contamination in reactors. Probably the most interesting advantage of using nonaqueous environments for enzyme catalysis arises when hydrolytic enzymes (e.g., lipases, esterases, proteases, and glycosidases) are applied because of their ability to catalyze synthetic reactions different from hydrolysis. Thus, the biocatalyst, as a result of a drastic decrease in the water content of the reaction medium, is able to shift the chemical reversible equilibrium toward the condensation product required for the synthetic mode of action (Figure 4.3A). The success of the approach is directly correlated with the reaction medium engineering used to eliminate the synthetic product or water molecules synthesized as by-products (e.g., by adding water adsorbents).

Substrate engineering provides clear improvements in enzyme-catalyzed synthetic processes. As example, some hydrolases show a catalytic mechanism through a covalent acyl-enzyme intermediate (lipases, α-chymotrypsin [CT], subtilisin, trypsin, papain, etc.). In this context, synthetic reactions can be carried out easily by a kinetically controlled approach using activated substrates (e.g., vinyl esters) in nonaqueous environments, because the alcohol released tautomerizes to a carbonyl compound (e.g., acetaldehyde), which cannot act as a substrate for the enzyme (see Figure 4.3B). The presence of an alternative nucleophile (e.g., alcohol and amine) might involve the formation of a synthetic product, rather than the hydrolytic reaction of the substrate that occurs in the presence of water. For both synthetic approaches, enzymes are able to stereoespecifically recognize only one isomer from a racemic mixture of nucleophiles, resulting in an asymmetric synthetic reaction, also named kinetic resolution (KR).[25]

Switching from water to nonaqueous solvents, as a reaction medium for enzyme-catalyzed reactions, is not always a simple answer. Indeed, the native structure of the enzyme can easily be destroyed, resulting in enzyme deactivation. Water is the key component of all nonconventional media because of the importance that enzyme–water interactions have in maintaining the active conformation of the enzyme.[16] Few clusters of water molecules, presumably bound to charged groups on the surface of enzyme molecules, are required for the catalytic function. It has recently been reported that proteins only achieve full biological activity when the surrounding water has approximately the same mass as the protein in question.[26] In this context, hydrophobic solvents typically afford higher enzymatic activity than hydrophilic ones because the latter have a tendency to strip off some of these essential water molecules from the enzyme molecules.

Two different parameters can be used for determining the correct degree of hydration of enzymes in nonconventional media: water activity (a_w) and water content. In water-miscible solvents, the role of water is mainly defined in terms of thermodynamic water activity rather than water content, due to the high ability of these solvents to reduce the concentration of "free" water molecules by direct solvent–water interactions. For water-immiscible solvents, where the a_w values are close to 1, the most important parameter is the overall amount of water into the media.[27] A "lubricating effect" of water on protein structure and a consequent rise in conformational flexibility has also been suggested to explain enzyme denaturation in aqueous organic mixtures.[16,24]

Enzymes typically fold in such a way that nonpolar residues are buried in a hydrophobic core, while polar residues tend to move to the surface, where they are hydrated. The common hypothesis is that when an enzyme is placed in a dry hydrophobic system, it is trapped in the native state, partly due to the low dielectric constant that greatly intensifies electrostatic forces, enabling it to maintain its catalytic activity. This is one of the most intriguing aspects of nonaqueous biocatalysis, the phenomenon of the so-called memory in anhydrous media.[28] Several factors must be taken into account when choosing a nonaqueous solvent for a given reaction: first, the compatibility of the solvent with the reaction of interest, which is as a function of the solubility of the substrates and products; second, theinertness of the solvent toward the reaction; and finally, the correlation of the enzyme function (activity and stability) with the nature of the medium (e.g., polarity and hydrophobicity). The log P parameter, defined as the ratio between the concentrations of the organic solvent in 1-octanol and in water at equilibrium,

$$P = [\text{solvent}]_{\text{octanol}} / [\text{solvent}]_{\text{water}},$$

has usually been applied to select nonaqueous reaction media: that is, the higher log P values (more hydrophobic solvents) are associated with better enzyme performance in nonaqueous media. In general, enzyme activity in organic solvents appears to be as follows: low in relatively hydrophilic solvents in which log $P < 2$; moderate in solvents where log P is between 2 and 4; and high in hydrophobic solvents with log $P > 4$.[29]

In this context, it is nowadays well known that activity, stability, activity, and selectivity displayed by enzymes in the absence of bulk water, can be altered, modulated, and tailored.[16,22] However, the rates of enzyme-catalyzed reactions in nonaqueous environments are lower than those observed in aqueous media. This diminution has been attributed to several factors such as the decrease in the polarity of the microenvi-

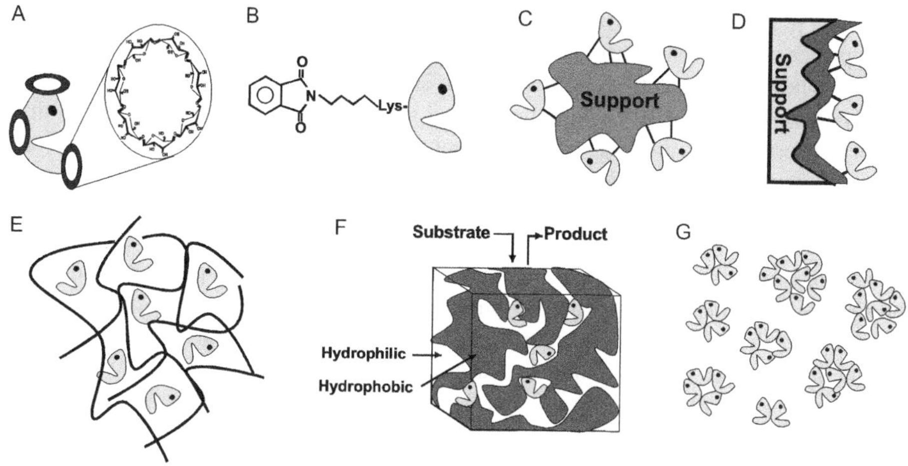

Figure 4.4. Schematic representation of some enzyme preparations for use in nonaqueous environments. (A) Co-lyophilizated with cyclodextrins. (B) Chemically modified with phthalic anhydride. (C) Multipoint attachment to solid support. (D) Covalent attachment to hydrophilic polymers coating inorganic materials. (E) Entrapment into a polymeric matrix. (F) Entrapment into a nanostructured amphiphilic support matrix. (G) Cross-linked enzyme aggregates (CLEAs).

ronment of the enzyme, or the loss in activity during the preparation of enzymes for use in nonaqueous media (e.g., lyophilization).[27]

Medium engineering not only involves the substitution of the aqueous reaction media to nonconventional one, but also implies, outside the context of protein engineering, to stabilize the biocatalysts by immobilization or by other means. Thus, a large variety of techniques have been described for generating highly active enzyme preparations to be used in nonaqueous reaction media, and demonstrating the malleability of biocatalysis by manipulating or engineering the enzyme microenvironment (see Figure 4.4). The simple lyophilization of an aqueous enzyme solution in the presence of organic or inorganic molecules as excipients (e.g., cyclodextrins,[30] polyethylene glycol [PEG][31] or nonbuffer salts such as KCl[32]) can dramatically improve enzyme activity in anhydrous media. In the same way, the chemical modification of enzyme surfaces by covalent attachment of hydrophobic moieties to amino acid residues increases the stability of enzymes in nonaqueous conditions.[33]

Enzyme immobilization onto solid supports, including nanostructures, is another strategy that has been used to improve catalytic rates and, in some cases, to enhance the enzyme stability in nonaqueous media.[34] The binding of enzymes on prefabricated hydrophilic supports (i.e., Eupergit C®),[35a] hydrogels,[35b] or inorganic materials coated with hydrophilic polymers (e.g., α-alumina coated with polyethyleneimine [PEI])[35c] resulted in an enhancement of the operational stability of the biocatalysts used. Furthermore, enzymes can be immobilized by entrapment in sol–gel matrices formed by hydrolytic polymerization of metal alkoxydes (i.e., polymerization of tetraethoxysilane in the presence of the enzyme).[36] The activity and stability of these immobilized derivatives have been improved by co-entrapment of enzymes with hydrophilic additives (i.e., PEG, polyvinyl alcohol [PVA], albumin, etc.) into the silica sol–gel matrix.[37] In this respect, the nature of the support may affect the partitioning of water and substrate molecules with the bulk medium and so favor mass-transfer phenomena and preserve the hydration shell of the enzyme. The use of molecular hydrogels based on a nanostructured enzyme support matrix has recently been reported as being an efficient system for biocatalysis in organic solvents, allowing substrates and products to diffuse through hydrophobic domain to a hydrophilic phase in which the enzyme is entrapped.[38] The amphiphilic nature of the matrix causes the hydrophilic phase to swell upon enzyme loading and the hydrophobic phase to swell in nonpolar solvents, resulting in a great increase in the turnover number of the enzyme. On the other hand, enzyme immobilization onto a support leads to "dilution of activity" because of the large portion of noncatalytic mass. The use of carrier-free immobilized enzymes, such as cross-linked enzyme crystals (CLECs), cross-liked dissolved enzyme (CLEs), and cross-linked enzyme aggregates (CLEAs), has also been reported as a way to obtain active and stable biocatalysts suitable for biotransformations in nonaqueous media.[22,39]

4.3 BIOCATALYSIS IN MONOPHASIC ILs SYSTEMS

4.3.1 Medium Engineering in Monophasic ILs System

Since the beginning of this century, ILs have emerged as exceptionally interesting nonaqueous reaction media for catalytic processes.[40] Both chemo-catalytic and biocatalytic processes have been developed using ILs as alternative solvents.[41] These neoteric solvents are salts, which are liquids below 100°C or typically close to room temperature (RTILs).[6] Typical ILs are based on organic cations, for example, 1,3-dialkylimidazolium, tetraalkylammonium, paired with a variety of anions that have a strongly delocalized negative charge (e.g., BF_4, PF_6, bis((trifluoromethyl)sulfonyl)imide, or NTf_2), resulting in colorless easily handled materials of low viscosity. Their interest as green chemicals resides in their high thermal stability and very low vapor pressure, which can be used to mitigate the problem of volatile organic solvents emission in atmosphere. Moreover, the physical properties of ILs (e.g., density, viscosity, melting points, and polarity) can be finely tuned by the appropriate selection of anions and/or cations. The polarity of a series of ILs has been determined by different techniques. Solvatochromic spectroscopic studies, mainly based on Reichardt's dye method, suggest that these solvents have a polarity comparable with that of the lower alcohols (e.g., methanol and ethanol).[42] Nevertheless, the physical interpretation of the results is complicated. Several reports have appeared recently showing large deviations between the polarity predicted by the Reichardt's dye method and other methods, such as the microwave dielectric constant,[43] other probe dyes, or by means of the vibrational frequencies for single water molecules associated with ILs obtained by Fourier transform infrared (FT-IR) spectroscopy. Reichardt himself observes the limitations of this probe in his own report.[42b] It is clear,

TABLE 4.1. Miscibility of Some Commonly Used ILs in Biocatalyst with Some Molecular Solvents

Cation	Anion	Notation	H_2O	AcCN	Hexane
1-Ethyl-3-methylimidazolium	Tetrafluoroborate	[Emim] [BF_4]	+	+	–
1-Butyl-3-methylimidazolium	Tetrafluoroborate	[Bmim] [BF_4]	+	+	–
1-Butyl-2,3-dimethylimidazolium	Tetrafluoroborate	[Bdmim] [BF_4]	+	+	
1-Hexyl-3-methylimidazolium	Tetrafluoroborate	[Hmim] [BF_4]	+/–	+	–
1-Octyl-3-methylimidazolium	Tetrafluoroborate	[Omim] [BF_4]	+/–	+	–
1-Ethyl-3-methylimidazolium	Hexafluorophosphate	[Emim] [PF_6]	+	+	–
N-Ethylpyridinium	Trifouoroacetate	[Epy] [TFA]	+	+	–
1-Butyl-3-methylimidazolium	Hexafluorophosphate	[Bmim] [PF_6]	–	+	–
1-Butyl-3-methylimidazolium	Bromide	[Bmim] [Br]	+	+	–
1-Butyl-2,3-dimethylimidazolium	Hexafluorophosphate	[Bdmim] [PF_6]	–	+	
1-Hexyl-3-methylimidazolium	Hexafluorophosphate	[Hmim] [PF_6]	–	+	–
1-Octyl-3-methylimidazolium	Hexafluorophosphate	[Omim] [PF_6]	–	+	–
1-Ethyl-3-methylimidazolium	Bis[(trifluoromethyl) sulfonyl]imide	[Emim] [NTf_2]	–	+	–
1-Butyl-3-methylimidazolium	Bis[(trifluoromethyl) sulfonyl]imide	[Bmim] [NTf_2]	–	+	–
Butyltrimethylammonium	Bis[(trifluoromethyl) sulfonyl]imide	[Btma] [NTf_2]	–	+	–
Methyltrioctylammonium	Bis[(trifluoromethyl) sulfonyl]imide	[MTOA] [NTf_2]	–	+	–

+, totally miscible; –, immiscible; +/–, partially miscible.

in any case, that ILs can be designed with tunable polarity to be miscible or immiscible with most hydrophophic organic solvents (e.g., hexane, toluene, and ether), thus providing a nonaqueous polar alternative for two-phase systems that has been widely applied to extracting products from reaction mixtures (see Table 4.1). The ionic nature of ILs contrasts with the nonmiscibility with water of some of them, which is an interesting property for enzymatic catalysis, because water molecules are essential for maintaining enzymes active in nonaqueous environments. All of these properties, including the fact that they are recyclable, make ILs potentially ideal solvents for green chemistry.[6]

A large number of enzymes (e.g., lipases, proteases, peroxidases, dehydrogenases, and glycosidases) and reactions (e.g., esterification, KR, reductions, and oxidations hydrolysis) have been tested in monophasic liquid systems based on ILs,[44] due to their great ability to dissolve both polar and nonpolar compounds.[45]

Although aqueous solutions of ILs do not fit the concept of ILs (liquids composed entirely of ions), these monophasic media are often used in biotransformations to increase the solubility of polar substrates and products with hydrophobic moieties (e.g., amino acids derivatives). Besides, the use of ILs as cosolvents in aqueous solutions may have the similar effects to that described for the addition of certain salts. The ions from water-miscible ILs have the ability to modify the water structure, thus varying the hydration environment of the protein, leading to either their stabilization or their denaturalization.[46] However, the kosmotropic or chaotropic character of ions related with the Hofmeister series may not be the only solid basis for predicting the compatibility of enzymes and water-miscible ILs (see also Chapters 2 and 5).[45b] Organic or inorganic ions present in the bulk solvent phase may interact with water molecules of the first hydration shell surrounding the enzyme. Such interactions could involve the disruption of the protein hydration and/or essential weak interactions (i.e., hydrogen bond, van der Waals, etc.) with the enzyme, to modify the native conformation and so the affinity toward the substrate. Furthermore, a complicated combination of mechanisms including the effects of hydrophobicity, nucleophilicity, and H-bond basicity on IL–water interactions should be also considered. In any case, even the small addition of water-miscible ILs to the system has drastic effects on the reaction outcome. For instance, in the desymmetrization of prochiral malonate diesters using pig liver esterase (PLE), the combination of water, organic solvent, and ILs as co-solvent or additive (concentration $< 1\%$) led to a significant increase of both reaction rate and enantioselctivity (from an ee of 78% to 97%, 0.1–1% of ILs in quaternary ammonium salt form).[47]

The assayed IL concentration, in related water-ILs systems, seems to be a key parameter to develop efficient biocatalytic transformation avoiding deactivation of the enzymes. As a matter of fact, the best enzyme activity and enantioselectivity of the subtilisin-catalyzed resolution of *N*-acetyl amino acid esters in an aqueous solution of [Epy][TFA] was obtained at a 10% v/v IL concentration, above which the enantioselectivity fell drastically.[48a] In another example, the best activities for the laccase-catalyzed oxidation of catechol in [Bmim][Br] and [Bmim][$N(CN)_2$] were observed at concentrations between 10–20% and 50–60% (v/v) in water, respectively, while the activity was decreased at higher and lower concentrations.[48b] A similar behavior was observed for other enzymes, such as chloroperoxidase, hydroxynitrile lyase, formate dehydrogenase, and β-galactosidase, which has been attributed to the ability of ions from water-miscible ILs to strongly interact with proteins, producing deactivation by unfolding of the 3-D structure. However, some water-miscible ILs (e.g., those based on BF_4 anions or those with long alkyl chains in cations, such as cocosalkyl pentaethoxy methyl ammonium methosulfate ([CPMA][MS]), have been shown to be excellent reaction media for lipase-catalyzed transesterifications and glycerolysis at low water content (<5% v/v).[49]

The remarkable results obtained for enzymatic reactions in water-immiscible ILs (e.g., [Bmim][NTf_2], and [Omim][PF_6]) at low water content (<2%) have underlined the suitability of these solvents as a clear alternative to molecular organic solvents for organic synthesis in monophasic systems. Furthermore, as enzymes require a certain degree of hydration to be active, the hygroscopic character of these ILs (e.g., [Bmim][NTf_2] is able to absorb up 1.4% w/w water content) could be regarded as an additional advantage of these neoteric solvents.[50] In this context, the direct addition of enzymes, (e.g., aqueous solution, lyophilized powder of free enzyme, enzyme immobilized onto solid supports, or cross-linked aggregates; see Figure 4.3) into the IL medium containing substrates always provided biotransformations.

Lipases are by far the most used biocatalysts in water-immiscible ILs, such that they are used for the synthesis of aliphatic and aromatic esters, chiral esters by KR of racemic alcohols, carbohydrate esters, polymers, and so on (see Chapter 5).[45] Additionally, the excellent stability displayed by enzymes in water-immiscible ILs for reuse and under high temperatures has been widely described.[51] In the case of free enzymes, the ability of these neoteric solvents to maintain the secondary structure and the native conformation of the protein toward the usual unfolding that occurs in nonaqueous environments has been demonstrated by spectroscopic techniques (e.g., fluorescence, circular dichroism [CD], and FT-IR; see Chapter 2), underlining their excellent ability to protect enzymes against the usual unfolding that occurs in nonaqueous environments (e.g., the half-life time of CALB in [Btma][NTf_2] at 50°C is 1660 times higher than that obtained in hexane).[51c]

In this context, ILs have been applied in biocatalyst medium engineering to improve the performance of immobilized enzymes under nonaqueous environments. For instance, an efficient enzyme derivative based on IL-coated enzymes (ILCEs) can be prepared using a solid IL at room temperature, which melts in the range 50–100°C. In this way, the enzyme is stabilized by mixing the melt ILs with the enzyme, and then cooling the mixture and cutting the solid ILCEs into small pieces (see also Chapter 8).[52] As an example, the IL [1-(3′-phenylpropyl)-3-methylimidazolium)][PF_6], which is solid at room temperature and becomes liquid above 53°C, was used to coat free *Pseudomonas cepacia* lipase in liquid phase. Afterwards, the mixture was cooled to provide a useful immobilized enzyme derivative with markedly enhanced enantioselectivity and without losing any significant activity by reuse.[52] In the same context, the protective effect of ILs during immobilization of *Candida rugosa* lipase (CRL) by sol–gel process has been demonstrated. The highest hydrolytic activity of immobilized lipase was obtained when the hydrophilic IL, [Emim][BF_4], was used as an additive, while the highest stability of immobilized lipase was obtained by using the hydrophobic IL, [C_{16}mim][Tf_2N]. Therefore, the use of binary mixture of 1:1 at the molar ratio of both ILs was used to obtain the optimal immobilized lipase, showing a synthetic activity 14-fold greater than in silica gel without IL and a 84% residual activity in *n*-hexane after 5 days' incubation at 50°C.[53]

Concerning medium engineering for biotransformations in ILs, it has been reported how the use of ILs improves the activity (e.g., synthesis of aliphatic esters,[54a] synthesis of acyl L-canitine)[54b] or the selectivity (e.g., KR of *rac*-menthol,[54c] KR of *rac*-3-phenyllactic acid)[54d] displayed by the enzyme with respect to that observed in organic solvents, although there seem to be no rules for predicting the outcome. Classical parameters in medium engineering for biotransformation in nonaqueous environments (e.g., a_w control and pH memory) have been taken into account to improve results. In this way, salt hydrate pairs were used to control water activity for CALB-catalyzed biotransformations (e.g., synthesis of of 2-ethylhexyl methacrylate[55a] and synthesis of ascorbyl oleate[55b]) in ILs (e.g., [Bmim][PF_6] and [Bmim][BF_4]), resulting in clear improvements in synthetic yields. In another example, a "pH memory" phenomenon, as it is reported in conventional organic solvent at the step of lipase immobilization, was key to explaining the observed conversion and enantioselectivity in the ILs. Magnetic microspheres containing immobilized CRL prepared via suspension polymerization were used to catalyze the resolution of *rac*-menthol in [Bmim][PF_6]. The best enantioselectivity (ee > 90%) was obtained at pH 5.0, then decreasing gradually with increasing pH.[56]

Medium engineering based on the nature of ions involved in ILs has been widely studied to improve the activity and enantioselectivity of enzymes, but no clear rules have been established so far.[45] In this regard, the activity and enantioselectivity of lipase from *Burkholderia cepacia* (BCL) in its free form, immobilized in a sol–gel (BCLxero), and as a CLEA were tested for the acylation of different *sec*-alcohols in dry organic solvents, ILs, and their mixtures. BCL-CLEA displays higher activity than BCLxero for all substrates assayed in the ILs but loses its activity rapidly. BCLxero was suitable for KR in the assayed ILs, but it was impossible to label one IL as being better than another without taking the nature of the substrate into account.[57] However, in another example, several novel phophonium ILs were designed to evaluate the activity and enantioselectivity of the lipase-catalyzed KR of 4-phenylbut-3-en-2-ol. The best results were obtained for the IL 2-methoxyethyl(tri-n-butyl)phosphonium bistriflymide ([$MEBu_3P$][NTf_2]), which represents the first example where this reaction proceeded faster in IL than in a conventional organic solvent, such as i-Pr_2O. However, when a different substrate was used (5-phenylpent-1-en-3-ol), an unexpected decrease in enantioselectivity was observed, which was attributed to a dependence of the lipase activity on the cationic part of the IL.[58]

Another interesting approach of medium engineering to improve biotransformations was carried out by the use of binary ILs systems. As example, the enzymatic production of diglycerides (DG) by glycerolysis of triglycerides (TG) was enhanced by combining a water-immiscible IL, such as [MTOA][NTf_2]), with a water-miscible IL having long alkyl chains, such as ethyloctanodecanoyl oligoethyleneglycol ammonium ethyl sulphate (AMMOENG 102®), as reaction media. This work reported a novel concept to improve DG yield by applying a binary IL system that consisted of one IL with better DG production selectivity and another IL that is able to achieve higher conversion of TGs. The authors thus resort to binary IL systems based on the observations of the characteristics of enzymatic glycerolysis in single IL systems. The best results (85% mol TG conversion and DG yield up to 90% mol) were obtained for the [MTOA][NTf_2]/AMMOENG 102 binary IL system, which are remarkably higher than any organic solvent-based glycerolysis systems ever reported. The tunable property of ILs offers tremendous opportunity to rethink the strategy of current efforts to resolve technical challenges that occurred in many production approaches.[59]

The proper design of ILs is another way to enhance the efficiency of biotransformations, for instance, designing ILs to improve the enzymes solubility in the reaction media. Usually, active enzymes in ILs do not dissolve, but the appropriated nature of ILs allows enzymes to become fully soluble in these neoteric solvents. As example, ILs bearing a combination of triethylmethyl ammonium with alkylsulfate, nitrate, or lactate anions dissolve CALB. However, only $[Et_3MeN][MeSO_4]$ maintains its activity after enzyme dissolution(dissolving enzyme > 1.2 mg/mL). By means of FT-IR spectroscopy, the unfolding of CALB was observed upon dissolution in ILs in which the activity is low, whereas the enzyme conformation in $[Et_3MeN][MeSO_4]$ medium closely resembles the native one.[60] Another common approach to dissolving the enzyme is the modification of the enzyme with crown ether or PEG. This was found to provide high stability and dispersibility of enzyme molecules into the IL, even without immobilization of enzyme or addition of small amounts of water.[31,61]

Alternatively, ILs can be designed to dissolve enzyme through introduction of specific functionalities, which can interact with the enzymes (e.g., moieties forming H-bonding). Turner and coworkers[62] reported that incorporation of a hydroxyl-functionality in the cation of an imidazolium-based IL caused some modifications in the behavior of the ILs that may be advantageous for stabilizing proteins. The potential of 1-(3-hydroxypropyl)-3-methylimidazolium salts for dissolving enzymes without affecting their catalytic activity was also demonstrated.[12] More recently, Zhao and coworkers have found that ether-functionalized ILs are able not only to dissolve but also to stabilize the free enzyme (CALB).[63] Besides, these task-specific ionic liquids (TSILs) increase the dissolution ability toward high hydrophilic substrates such as sugars, cellulose, betulinic acid, or amino acid, which may improve the catalytic efficiency of biocatalytic processes involving these types of reactants (see also Chapter 2, Section 2.6, and Chapter 5, Sections 5.4 and 5.5).

The nature of ILs and presence of the additional functional group seem to be of great importance to solubilize and stabilize the enzyme.[64] Thus, a series of protic hydroxylalkylammonium ILs containing chloride ions, which in traditional ILs are deleterious to enzyme activity,[61a] was used to dissolve chymotrypsin and subtilisin and to measure their protease activity against the model substrate *N*-acetyl-L-phenylalanine ethyl ester. Only subtilisin dissolved in diethanolammonium chloride [DEA][Cl] showed significant activity. The authors speculate that the two hydroxyl groups of the cation are able to coordinate the chloride, thereby overcoming its denaturing capability.

4.3.2 Isolation and Recyclability Issues in Monophasic ILs System

Despite all the excellent catalytic properties shown by enzymes in IL systems, the reuse of biocatalysts and the recovery of products are also key concepts in medium engineering and process development of green chemical bioprocesses. The separation of products from enzyme–IL phase can be facilitated by the inherently low vapor pressure of IL allowing for easy removal of volatile substances. Thus, isolation of the products can be carried out by sublimation or distillation.

As a pioneering example of operational stability of enzymes in ILs, it was reported how CALB-catalyzed KR of 5-phenyl-1-penten-3-ol with phenylthioacetate in [Bmim] $[PF_6]$ obtained an ester yield >99% and enantioselectivities (*E*) of >540.[65] Reaction

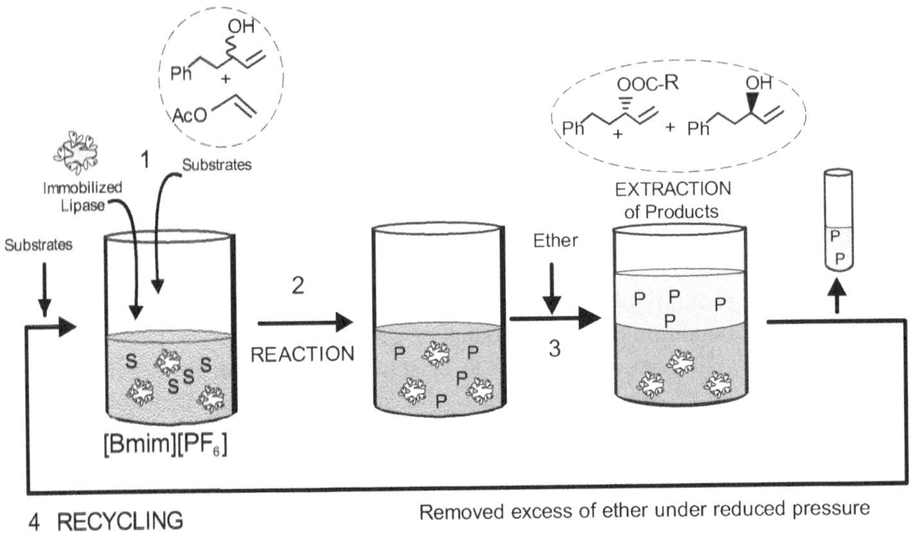

Figure 4.5. Operational strategy for lipase-catalyzed reactions in a monophasic IL system, including recycling and product steps.[65]

products were extracted under reduced pressure (27 hPa), and the enzyme remained anchored to the IL phase, which permits the reuse of the enzyme–IL system by addition of fresh substrates and without any loss in activity (see Figure 4.5).

However, liquid–liquid extraction with organic solvents is the most commonly used approach for the recovery of products after biotransformations performed in monophasic ILs systems. The use of these additional solvents must be considered a clear step back as regards the greenness of the process (see Figure 4.5).[41,45] To overcome this limitation, alternative strategies for product recovery based on either membrane technology (e.g., pervaporation),[66–68] or IL-anchored substrastes approaches[69] have been applied.

The immobilized CALB-catalyzed esterification of acetic acid and ethanol in [Bmim][PF_6] was carried out in a membrane reactor able to remove both the ethyl acetate and water produced by a double pervaporation system using hydrophobic and hydrophilic membranes, respectively, in continuous operation for 72 hours without any loss in the enzyme activity (see Figure 4.6).[66b]

The use of TSILs to anchor substrates and/or products during an enzymatic KR process is another approach to facilitating the separation of enantiomeric products. The key feature of this strategy is the combination of two ILs, one is used as solvent and the other as ionic acylating agent.[69] For this approach, the acylating agent contains both an ionic moiety and an ester moiety in its structure. The latter is recognized by an enzyme to allow the selective resolution and separation of an ionically anchored ester from the unreacted alcohol (see Figure 4.7).

The major advantage of this process is the possibility of separating both free enantiomers of a racemic mixture only by enzymatic resolution in a one-pot reaction using

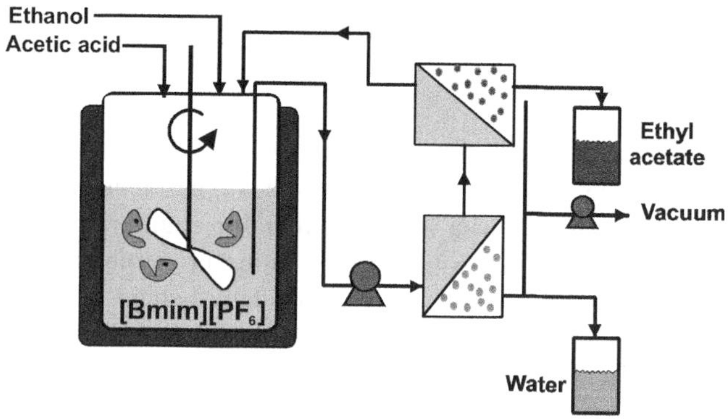

Figure 4.6. Continuous membrane reactor for the enzymatic synthesis of ethyl acetate in $[Bmim][PF_6]$.[66b]

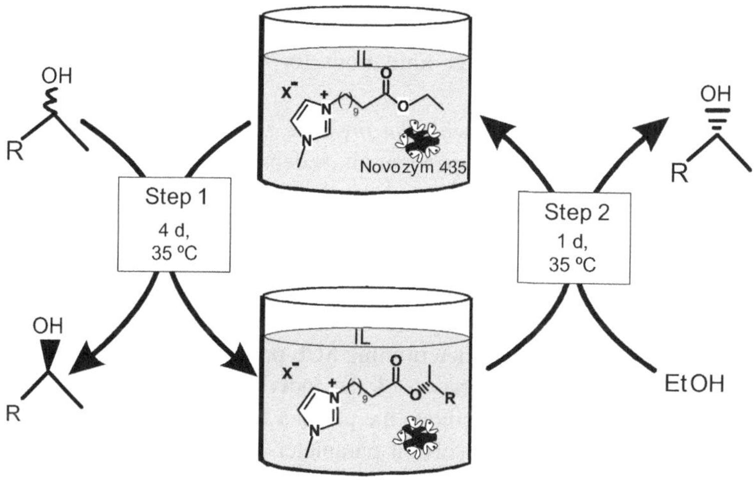

Figure 4.7. Enzymatic resolution of *sec*-alcohols by using an IL as acylating agent in IL.[69]

one equivalent of acylating agent.[69] As shown in Figure 4.7, the anchored enantiomer that results as the product of step 1 can be removed in step 2 by a second enzymatic transesterification (reversible reaction) using a primary alcohol, such as EtOH. The best results were obtained when the ionic acylating agent contains an imidazoliun moiety separated from the ester moiety by a long alkyl chain. For the first enzymatic transformation, the unreacted substrate ((*S*)-1-phenylethanol) was isolated in 51% yield (80.9% ee) after 96 hours. For the second step, the (*R*)-1-phenylethanol was isolated in 41.3% yield (99.3% ee) after 24 hours.[69]

4.4 (BIO)CATALYTIC PROCESSES IN SCFs

4.4.1 Properties of SFCs

A supercritical fluid is defined as a substance above its critical temperature (T_C) and critical pressure (P_C). The critical point represents the highest temperature and pressure at which the substance can exist as a vapor and liquid in equilibrium. Therefore, SCFs are characterized by gas-like viscosities and solvating properties of a wide range of various organic solvents. In particular, environmentally benign carbon dioxide in its supercritical fluid state ($scCO_2$) has great potential to develop cleaner alternative processes, which allow for the total or partial substitution of the most commonly used volatile organic compound (VOC) solvents.

The SCFs have been widely used as solvents in natural products extraction[70] and in the synthesis and processing of materials.[71] In general, the properties suiting them for their use in these processes also render them as attractive media for the following chemo- or biocatalytic reactions:

- *Reduced mass transfer.* The high diffusivity and low surface tension lead to reduced internal mass-transfer limitations for both chemical and biochemical heterogeneous catalysts.[72]
- *Tunable solvent properties with the pressure and temperature.* The solvating power of an SCF is highly dependent on its density. By controlling pressure and temperature it is possible to control the density of the fluid allowing selective manipulation of the solvating power of the fluid. Because of the solubilization power of SCFs, it is claimed that it is possible to design conditions where the fluid can extract a particular solute by varying the solvent power of the fluid. It is reported that solute solubility studies of individual components present in a mixture provide a means of determining SCF process conditions that will afford selective extraction or separation of the individual solutes. Thus, the solvent power can be adjusted to facilitate the product isolation leading to *process benefits.*[73] Furthermore, typical solvent parameters such as the dielectric constant, density, diffusivity, and viscosity of SCF may cause a considerable tuning effect on the chemo-, regio-, and stereoselectivity of the chemical providing *chemical benefits.*[74]
- *Moderate supercritical parameters.* The critical pressures and temperatures associated with most SCFs such as CO_2 (31°C, 73.8 bar), ethane (32°C, 48.8 bar) propane (97°C, 42.4), and fluoroform (25.9°C, 46.9 bar) are not so high as to damage biopolymers or labile biomolecules.
- *Products free from solvent traces.* The use of SCF is not accompanied by the problem of solvent residues in the reaction product because common SCF solvents are generally gases under atmospheric conditions. In particular, the low toxicity makes SCF attractive as an alternative reaction medium for food and pharmaceutical products.

In the view of these exceptional tunable properties, SCF has emerged as replacement for traditional solvents toward the development of greener chemical processes. SCF reactions offer unique process opportunities not only to replace conventional hazardous organic solvents, but also to optimize and control the reaction environment by controlling temperature, pressure and fluid density (solvent properties). These facts permit an easy manipulation of the reaction rate, the elimination of transport limitations, as well as integration of both transformation and product separation steps. Hence, a great number of chemical processes have been developed at both laboratory and industrial levels using SCF as reaction media.[75]

4.4.2 Medium Engineering in Supercritical Biocatalysis

In general, the stability and activity of the enzymes in SCFs are affected by water content, working pressure, and temperature of the system. Thus, under extreme conditions the use of SCFs can lead to enzyme denaturalization and consequent loss of activity by disruption of its 3-D structure.[76] However, under the selection of the appropriate set of conditions, it has been found that enzymes in SCFs can retain their biocatalytic activity, encouraging their application in biocatalytic processes under supercritical conditions.[77]

Since 1985, the use of SCFs as nonaqueous reaction media for enzyme catalysis has been an active area of research. The unique feature of these solvents is the easy way to manipulate their physical properties by simply changing the pressure or temperature, which emphasizes their abilities for extraction, fractionation, and analysis processes, is also underlined in biocatalytic processes. Taking into account that proteins show low solubility in supercritical fluid phase or that they are usually immobilized onto a support, biocatalytic processes in SCFs permits an easy recovery and reuse of the enzyme, while the gas-like diffusivities and low viscosities of SCFs enhance the mass-transport rates of reactants to the active site of the enzymes improving the activity.

Only fluids with low critical temperature (i.e., CO_2, ethane, propane, butane, SF_6, CF_3, etc.) have been assayed for biocatalysis in supercritical conditions because of the tendency of proteins to denature at high temperatures. In the same context, the low polar character of these solvents determined its use in biotransformations of hydrophobic compounds dissolved inside. The most popular SCFs for biocatalysis is $scCO_2$ because it is chemically inert, nontoxic, nonflammable, cheap, and readily available; it also exhibits relatively low critical parameters (e.g., $P_c = 73.8$ bar; $T_c = 31.0°C$) that are suitable for enzymes catalysis, and is considered a green solvent. Other SCFs are less attractive because of their flammability (e.g., ethane, propane), high cost (e.g., CHF_3), or poor solvent power (e.g., SF_6), and have therefore been used in few cases.[78]

At low water content, SCFs can be used as reaction media in which the enzyme kinetics can be correlated with solvent properties such as dielectric constant and hydrophobicity.[79] In this context, the intrinsic activity of enzymes is affected by the nature of the SCF, especially as regards preservation of the essential water layer around proteins. As example, for the lipase-catalyzed transesterification of methylmethacrylates in supercritical ethane, ethylene, fluoroform, sulfurhexafluoride and near-critical propane, it was observed how reaction rates achieved in all the SCFs were significantly greater than those obtained in the conventional solvents examined, being increased with increasing hydrophobicity of the SCFs, as follows: SF_6 > propane > ethane > ethene > CHF_3 > CO_2.[79a] Although scCO_2 is considered a hydrophobic solvent (e.g., 0.31% w/w water content at 344.8 bar and 50°C), the capability of scCO_2 to strip off the essential water molecules from the enzyme microenvironment was pointed out as responsible for this lower activity. This phenomenon was also observed for the lipase-catalyzed ethyl butyrate synthesis by esterification, where the activity is higher in near-critical propane than in scCO_2.[79b]

However, the advantages of scCO_2 to extract, dissolve, and transport chemicals are tarnished in the case of biocatalytic processes because of the detrimental effect of the CO_2 on the enzymes. The ability of CO_2 to form carbamates with ε-amino groups of lysine residues placed on the enzyme surface, as well as to decrease the pH of the aqueous layer around the enzyme, water reacts with CO_2 to form carbonic acid, have been correlated with the usual observed enzyme deactivation.[80] Indeed, a study, in compressed CO_2 and propane, on the influence of temperature, exposure time and pressure on the secondary structure of HRP by far ultraviolet-CD, showed how treating aqueous solutions of enzyme with propane did not induce changes in the secondary structure content of HRP, resulting in good stability. On the contrary, incubation with CO_2 led to a significant loss of the HRP secondary structure, which was accompanied with by significant decrease in the enzyme activity. However, the solid commercial HRP showed no decrease in its activity after treatment with pressurized CO_2 or propane, although treatment in both solvents provoked a loss in the secondary structure.[81] In this context, treatment with scCO_2 has been used to deactivate pectin esterase in orange juice, polyphenol oxidase in grape juice, and α-amylase in liquid foods model systems.[82]

Lipases in scCO_2 are the most widely studied systems because of the catalytic promiscuity of these enzymes toward hydrophobic substrates, and the excellent ability of this fluid to dissolve and transport hydrophobic compounds.[77] Thus, the synthesis of esters by esterification or transesterification (see Figure 4.2) is the most popular enzymatic process in scCO_2 (see Table 4.2), although the use of lipases for the asymmetric synthesis of esters is probably the most active area of research.[83] The excellent enantioselectivity shown by several lipases (e.g., CALB) combined with the unique properties of scCO_2 has successfully permitted the chiral resolution of a large number of racemates (1-phenylethanol, glycidol, ibuprofen, etc.).[84]

Pressure, temperature, and water content are the most important environmental factors affecting enzymatic catalysis in SCFs, particularly their activity, enantioselectivity, and stability. A general finding indicates that enzymatic reactions become less efficient as SCF pressure increases and SCF temperature decreases since these physical changes decrease the diffusion constants of the substrate to the enzyme's active sites.[85]

Additionally to the direct effect of CO_2 on proteins, the high pressure may also have a negative impact on enzyme conformation. The rapid release of CO_2 dissolved in the bound water of the enzyme during depressurization has been claimed to produce structural changes in the enzyme and to cause its inactivation.[86] However, through

TABLE 4.2. Some Bioproducts Obtained by the Lipase Action in $scCO_2$

Products	Reaction conditions	Yield, %	Reference
R-2-Phenyl-1-propyl butyrate	40°C, 100 bar	n.d.	39a
R-1-Phenylethyl acetate	50°C, 100 bar	44 (ee > 97)	39b
Ethyl butyrate	40°C, 100 bar, 24 hours	30	79b
Geranyl acetate	40°C, 140 bar, 72 hours	30	83a
Ethyl oletate	40°C, 150 bar, 3 hours	95	83b
Butyl butyrate	50°C, 90 bar, 3 hours	100	83c
Isoamyl acetate	40°C, 100 bar, 2 hours	100	83d
Structured lipids	55°C, 241 bar, 6 hours	62.2	83e
Fatty acid methyl esters	55°C, 136 bar, 4 hours	26.4	83f
	40°C, 73 bar, 4 hours	63.2	
R-Citronellyl oleate	31.1°C, 84.1 bar	3.6 (ee > 98.9)	83g
3-Hydroxy esters	40°C, 120 bar, 68 hours	63	84a
R-1-Phenylethyl acetate	42°C, 130 bar	48 (ee > 99)	84b
R-1-Phenylethyl acetate	50°C, 200 bar, 6 hours	48 (ee > 99)	84c
S-Glycidyl butyrate	35°C, 140 bar, 10 hours	30 (ee 83)	84d
S-Propyl ester of ibuprofen	50°C, 100 bar, 23 hours	75 (ee 70)	84e

changes in pressure in the vicinity of the critical points (e.g., from 77 to 90 bar), a clear improvement in lipase-catalyzed ester synthesis has been observed. As example, the activity of immobilized CALB-catalyzed butyl butyrate synthesis was exponentially increased by the decrease in $scCO_2$ density for different combinations of pressure and temperature.[83c] In another example, the lipase-catalyzed esterification of *rac*-citronellol with oleic acid in $scCO_2$, the (*R*)-ester product (ee > 99.9%) can be obtained simply by manipulating pressure and temperature around the critical point.[83g]

The effect of pressure on enantioselectivity is indeed noteworthy, although the reason for this is not clear. In another example, the effect of pressure (from 80 to 190 bar) on immobilized CALB-catalyzed enantioselective acetylation of *rac*-1-(*p*-chlorophenyl)-2,2,2-trifluoroethanol with vinyl acetate in $scCO_2$ was studied. At 55°C, the enantioselectivity of the enzyme (*E*-value) was gradually decreased from 50 to 10 when the pressure was increased from 80 to 190 bar, regardless of the reaction time, which was related with changes in $scCO_2$ density.[87] However, for the KR of *rac*-1-phenylethanol catalyzed by the same immobilized lipase, it was demonstrated that changes in pressure did not greatly affect conversion or *E*-values, even when the pressure was increased to 500 bar.[84b]

The effect of temperature on enzyme activity is much more relevant than pressure as a result of enzyme deactivation processes.[77,85] The optimal temperature of enzymatic processes in SCFs is related with pressure because both control solvent properties. The negative effect of temperature on enzymes in supercritical conditions has been related with changes in the hydration level of the enzyme. Water concentration is a key factor in supercritical biocatalysis because of its great influence on enzyme activity and stability. It should be noted that $scCO_2$ may dissolve up to 0.3–0.5% (w/w) water, depending on the temperature and pressure, and produce enzyme deactivation by dehydration in continuous operation. On the other hand, if the water content in the supercritical medium is too high or if the water molecules are produced in the reaction, the increased humidity may also lead to enzyme deactivation. In this context, the appropriate selection of support for the immobilized enzyme, or the addition of salt hydrates (e.g., $Na_4P_2O_7$. 10 H_2O), have been successfully applied to preserve the essential water layer around the enzyme.[88]

4.4.3 Processes Design for SCF Biocatalysis

In general, reactor design is an important feature in supercritical processes. The appropriate reactor design not only helps to facilitate the mass-transfer, to control the reaction conditions (pressure and temperature), and to facilitate the product isolation and recovery, but it also should permit the process to be technically and economically feasible toward a possible scaling-up. In all cases, reactors are made of stainless-steel and contain many different control and safety devices. In some cases, the supercritical reactor also contains elements to simultaneously perform both the biocatalytic and product separation steps (i.e., by producing a back-pressure cascade in a series of coupled high-pressure separator vessels).[83b]

Basic screening and studies aimed at gaining insights into SCF biocatalytic processes are usually performed in high-pressure batch vessels. However, continuous flow processes are preferred because a series of practical and technical advantages are associated with them. These include the improvement in mass and heat transfer, a significant intensification of the process, making available systems work 24 hours a day, 7 days a week, or their easier optimization through the adjustments of simple parameters such as flow, pressure, or temperature. The experimental variables can be easily automated or controlled, leading to a purer and more reproducible product, and a far greater productivity from a fixed amount of catalyst than the one achieved in the batch process. Additionally, the scale-up of flow processes is in general more easily attainable than for batch processes, via different approaches such as the scale-out or the number-up.[89] Therefore, packed-bed continuous flow reactors have been widely known for industrial-scale applications involving heterogeneous catalysts.

Regarding processes in $scCO_2$, Poliakoff and coworkers have proved that laboratory conditions for reactions such as hydrogenation and Friedel–Crafts using catalytic packed-bed continuous flow reactors can be used with only minor modifications to carry out reactions at 100 kg/h on the full-scale plant.[90]

The use of continuous processes for biocatalysis presents some district advantages.[91] For instance, the flow processes allow the reactor to be continuously fed with the substrates, the easy separation of products, and the continuous reutilization of the enzyme in consecutive cycles without the requirement of depressurization.[92] This is quite an important fact to avoid downtime, and consequent loss of productivity, but also favors the long-term stability of the enzyme, as pressurization/depressurization steps, needed in a batch process, cause a major impact on enzyme activity, which is decreased with increasing number of depressurizations.[86c] The productivity is also favored by instant ratio between enzyme and substrate given in a flow system, which is significantly higher in fixed-bed continuous reactors than in conventional batch ones.

Furthermore, as some enzymatic reactions are inhibited by the products, the constant removal of the inhibitory substance clearly improves the efficiency of the process.[93]

In this context, the continuous syntheses of long-chain fatty acid, which are products of interest in the cosmetic, pharmaceutical, and lubricant industries, were studied in pressurized CO_2. As example, the esterification of oleic acid with 1-octanol in dense CO_2 catalyzed by *Rhizomucor miehei* lipase (Lipozyme RM) was performed in both batch and packed-bed bioreactors. Temperature, pressure, reactant dilution, and flow rates were adjusted to optimize the reaction performance. The esterification reaction, under the optimized conditions, took place at the highest conversion, of about 93%, and it was maintained without any significant reduction for a period of 50 days. A comparison with experimental results obtained batch-wise showed that the yield for continuous processes is 10% higher than those observed in the batch process. Furthermore, the authors claimed that the results were better than those reported using conventional organic solvents.[94]

Packed bed reactors containing supported reagents, scavengers, and (bio)catalysts can be easily set up in flow sequential assembling opening the way to produce multistage chemical reactions.[89] In this regard, a two-stage reaction can be performed by combining the metal (Pd)-catalyzed hydrogenation of acetophenone with the lipase-catalyzed KR of the resulting *sec*-alcohol in $scCO_2$.[39b] This process has the advantage that decompression of CO_2 is not required between consecutive steps. The decompression of CO_2 not only affects the stability of the enzyme, but also it is the single largest energy factor contributing to the cost of an $scCO_2$ process. Thus, performing reactions in series has a considerable advantage over performing the reactions separately because there is no requirement for the SCF to be depressurized between reactions. Therefore, the economic productivity of the overall process should be increased when a metal-catalyzed reaction is combined with a selective biocatalytic reaction in a multiple-step synthesis in $scCO_2$.

Immobilized resting cells of *Geotrichum candidum* can also be used in continuous flow reactors for the reduction of ketones under supercritical conditions.[95] The reduction of cyclohexanone was successfully performed, and the biocatalyst was recycled up to four times with only a slight loss in activity. Recycling was not possible using the corresponding batch system because the biocatalysts cannot tolerate repeated depressurization at a very low temperature and separation of the product from the biocatalysts using organic solvents. This method was also applied for the asymmetric reduction of *o*-fluoroacetophenone, achieving excellent enantioselectivity (ee > 99%) and a higher space–time yield than the corresponding batch process (0.24 μmol/min vs. 0.13 μmol/min, at 35°C and 100 bar).

An alternative to performed biocatalytic processes in SCF is the use of high-pressure continuous membrane reactors. Different types of assemblies can be designed.[96] Thus, reactors based on a high-pressure continuous enzymatic flat-shape membrane have been used. Membranes were applied as separation units to retain the biocatalyst in the system (10,000 Da MW cut-off) without the need of depressurizations. In this type of reactor the hydrolysis of oleyl oleate in propane or the hydrolysis of sun flower oil in $scCO_2$, catalyzed by nonimmobilized lipases from *Candida cylindracea* and *Aspergillus niger* respectively, was successfully performed.

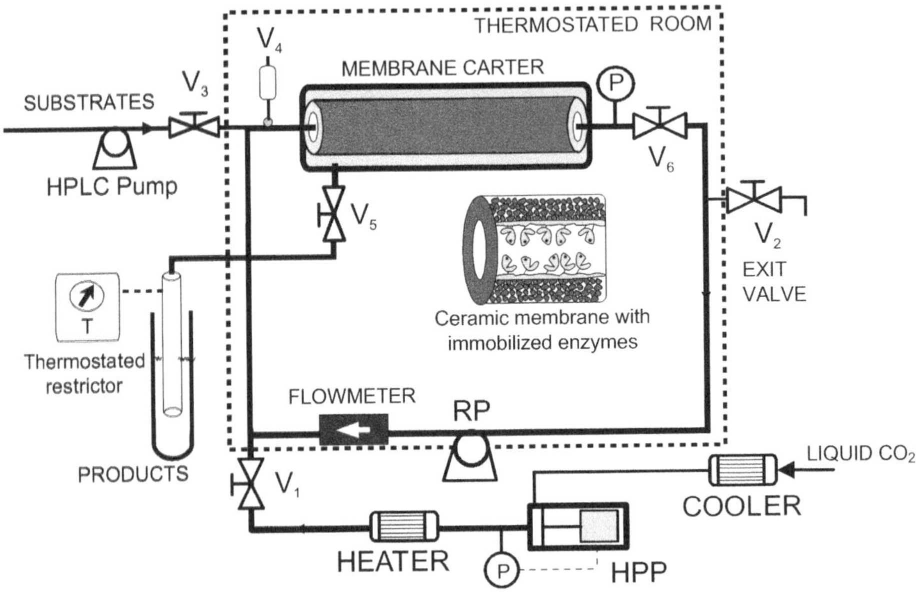

Figure 4.8. Scheme of a high-pressure membrane reactor with recirculation for enzyme-catalyzed transformations in $scCO_2$. HPP, high-pressure pump; RP, recirculation pump.[39c,83c]

Tubular membrane reactors can also be considered in order to design continuous biocatalytic processes in SCF. As a matter of fact, the hydrolysis of carboxymethylcellulose in $scCO_2$ was assayed using a high-pressure continuous enzymatic tubular membrane reactor. Cellulase from *Humicola insolens* was covalently attached on a ceramic tubular membrane, which was previously coated with a hydrophilic inert polymer (see Figure 4.8).[96]

Another example of the use of membrane reactors for the continuous processes in SCF was reported for CALB-catalyzed butyl butyrate synthesis showing an excellent operational behavior, without practically any loss in activity during the assayed time (half-life time higher than 360 cycles). The reactor operated in daily cycles (6 hours of continuous synthetic process in the selected conditions, and 18 hours of storage in the reactor at room temperature). The better enzymatic activity exhibited by the dynamic membrane in $scCO_2$ with respect to the organic solvents clearly showed the suitability of this reaction medium through the appropriate selection of the set of experimental conditions and reactor design to avoid the possible adverse effects of CO_2 on enzyme activity.[83c]

4.5 MULTIPHASE BIOTRANSFORMATIONS

Although a great number of monophasic biocatalytic processes (in aqueous phase, organic solvent, or nonconventional solvents) have been developed, the use of single solvent could present some limitations related to enzyme stability, solubility of substrates, isolation and separation of the products, recyclability of the biocatalysts, and so on. The right combination of two immiscible solvents helps to overcome some of these issues. In a biphasic system the biocatalysts are usually located in the aqueous phase, the second phase being a nonreactive organic solvent. The type of biphasic system used can dramatically affect the activity/productivity or selectivity of the biocatalyzed reaction and influence the stability of the enzyme. The development of nonconventional organic solvents has opened up new opportunities for the design of multiphasic biocatalytic processes, some of which include fluorous solvents, water, IL, and dense CO_2, are discussed in the following sections.

4.5.1 Biocatalytic Processes in Biphasic Fluorous Solvents

Fluorous solvents are another nonconventional reaction medium that can be used in medium engineering for biocatalytic processes. The unique characteristics of the fluorous solvents, such as temperature-dependent miscibility with organic phases, solvophobicity with aqueous and organic solvents, and fluorophilicity with fluorous media, allow the design of both reaction and separation processes.[97]

The early development of fluorous synthetic processes was focused on biphasic catalysis.[98] Thus, for a mixture containing organic and FSs, homogenous fluorous catalytic reactions were carried out at a high temperature, and biphasic separation of the catalyst was conducted at a low temperature.[9,99] This is an innovative concept, and the biphasic recycling system has a potential for different applications.[100,101] However, only few examples of biocatalysis in fluorous biphasic systems have been described to date.[102] The lack of solubility of the proteins in fluorous solvents hindered the full potential of thermosensitive system because of difficulties for a simple catalyst recovery. Therefore, the enzyme in these solvents necessitates a filtration step for its separation and possible reuse. Alternatively, fluorous bi-, or triphasic solvent systems have also been employed to facilitate product separation from a biocatalytic reaction at the work-up stage.[103]

The enantioselective esterification of *rac*-2-methylpentanoic acid with a highly fluorinated decanol catalyzed by lipase from *Candida rugosa* in a perflurohexane/hexane biphasic system is one of the few examples of enzymatic catalysis that have been reported (see Figure 4.9).[104] The acid substrate is dissolved in hexane, while the

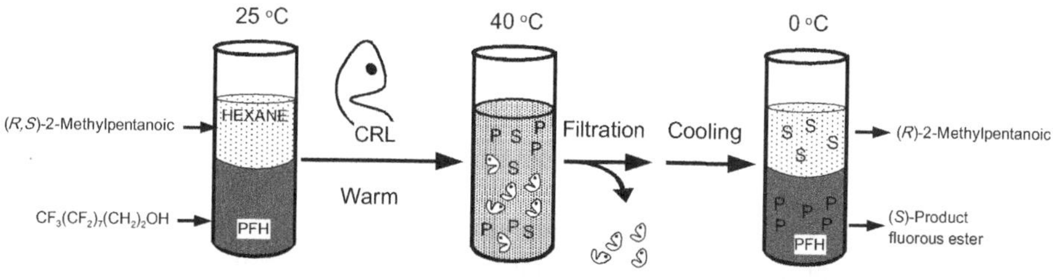

Figure 4.9. General scheme for enantioselective partitioning in lipase-mediated esterification using fluorous biphase system (FBS). CRL, *Candida rugosa* lipase; PFH, perfluorohexane.[104]

fluorinated alcohol is dissolved in the fluorous phase (FS). When the resulting biphasic mixture is warmed, becoming one phase, the lyophilized enzyme is added to perform the reaction. The lipase selectively catalyzes the esterification of (*S*)-2-methylpentanoic acid to the corresponding (*S*)-fluorinated ester product. At the end of the reaction, the biocatalyst is separated by filtration, and the recooling of the reaction mixture results in the retention of the fluorinated product into the FS, while the unreacted (*R*)-2-methylpentanoic acid remains in the hexane phase. The main problem of this strategy is the need to use substrate(s) that are miscible in the FS, as well as the long reaction time required (95–145 hours) to reach 49–53% conversion and 95% ee for the (*S*)-product.

Recently, Thomas and coworkers demonstrated that proteins such as cytochrome *c* (Cc) and CT can be solubilized in either fluorous solvents or supercritical CO_2 by hydrophobic ion pairing (HIP) with perfluorinated anionic.[105] By using perfluoro anionic surfactants, for example, perfluoropolyether carboxylate, the extraction of Cc, CT, and CRL from aqueous solutions into an FS, for example, perfluoromethylcyclohexane (PFMC), has been demonstrated: The anionic surfactant is able to interact by HIP with basic amino acid residues (e.g., Lys, His, Arg) placed on the surface of a protein dissolved in aqueous buffer at pH below its isoelectric point (Krytox, see Figure 4.10). Thus, the HIP–protein complexes containing the fluorophilicity surfactant can easily be extracted into the PFMC phase, resulting in a homogeneous and clear phase. Thus, the transesterification of *N*-acetyl-L-phenylalanine ethyl ester with n-butanol or *rac*-2-butanol catalyzed by the CT–perfluoropolyether carboxylate surfactants complex in a truly homogeneous fluorous biphasic system (hexane/perfluoromethylcyclohexane) can be performed (Figure 4.10).[105b] As expected, the homogenous system containing solubilized enzyme molecules showed higher activity than the same system with the suspended enzyme. Furthermore, the CT–Krytox complex, which is retained in the FS on cooling the solution, was successfully reused over four cycles with no loss of activity.

In another example, the KR of 1-phenylethanol was carried out by using the *Burkholderia cepacia* lipase–Krytox complex as catalyst in the PFMC/hexane biphasic system. The enzyme–Krytox complex showed a high stereospecificity (ca. ee 99%) with moderate catalytic efficiency (49% yield after 99 hours), and a high operational and storage stability. Temperature modulation of the fluorous biphase system (FBS) miscibility allowed the separation and recovery of the solubilized lipase for at least three operation cycles.[105c]

4.5.2 Bioprocesses in Water/scCO_2 Systems

The use of scCO_2 systems for biocatalytic transformation of hydrophobic substrates, mainly using lipases, has been widely demonstrated.[77] However, the low solubility of hydrophilic compounds (i.e., carbohydrates) hampers the wider application of biocatalytic processes in scCO_2. Although this limitation can be partially overcome by the use of a polar cosolvent, a more attractive strategy has emerged by the development of multiphasic strategies in water/scCO_2 systems. The first examples of water/scCO_2 biocatalytic processes were reported in water/scCO_2 microemulsion formed using specially

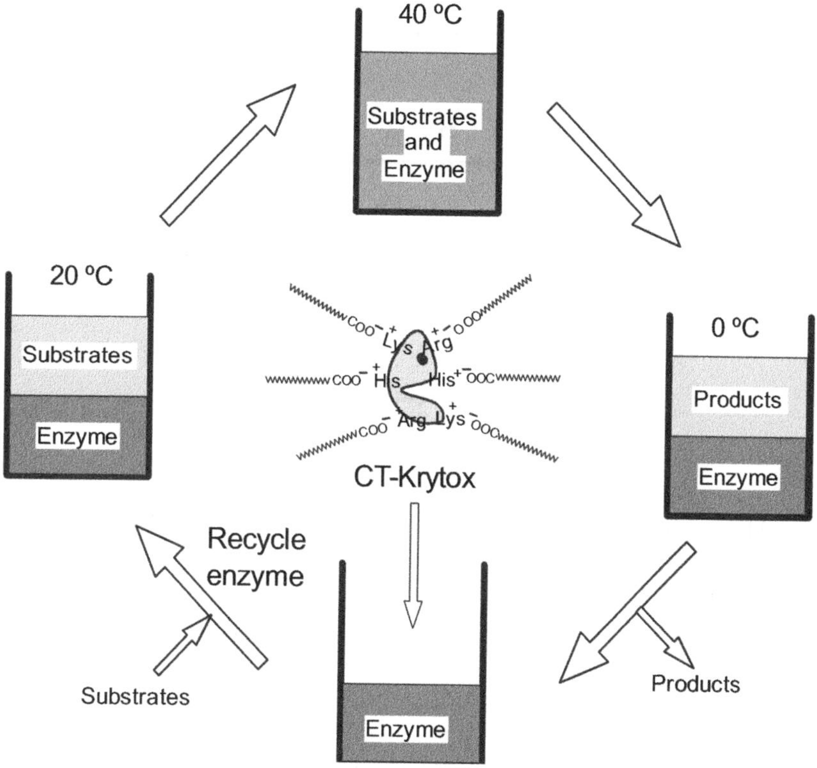

Figure 4.10. Truly homogeneous biphasic organic-flourous biocatalytic processes based on the hydrophobic ion pairing (HIP) technique with fluorophilicity surfactant to solubilize enzymes.[105]

designed surfactants, incorporating a"CO_2-philic"fluorocarbon moiety.[106] The authors showed the viability of enzyme-catalyzed reactions within a pH-controlled water/scCO_2 microemulsion and showed the turnover to be comparable with that obtained in water/organic solvent systems.[107]

The use of water/CO_2 biphasic systems as reaction media for biocatalytic processes required careful pH control in the aqueous phase. The combination of CO_2 with water leads to the formation and dissociation of carbonic acid, resulting in low pH values of about 3.[108] This low pH may contribute to the deactivation of the enzyme. However, pH control is possible up to a pH of approximately 6 by means of buffer salts.[109] Thus, by strict control of the pH by $NaHCO_3$ buffer, the asymmetric reduction of ketones by an alcohol dehydrogenase from *Geotrichum candidum* was reported.[110] The asymmetric reduction of *o*-fluoroacetophenone with the immobilized enzyme in water/scCO_2 biphasic systems proceeded with a 75% yield and >99% ee of the (*S*)-alcohol.

Another interesting example is the chemo-enzymatic cascade oxidation in scCO_2/water biphasic media to catalyze the enantioselective sulfoxidation of thioanisole proposed by Leitner's group (see Figure 4.11).[111] In this system, Pd(0) catalyzes the

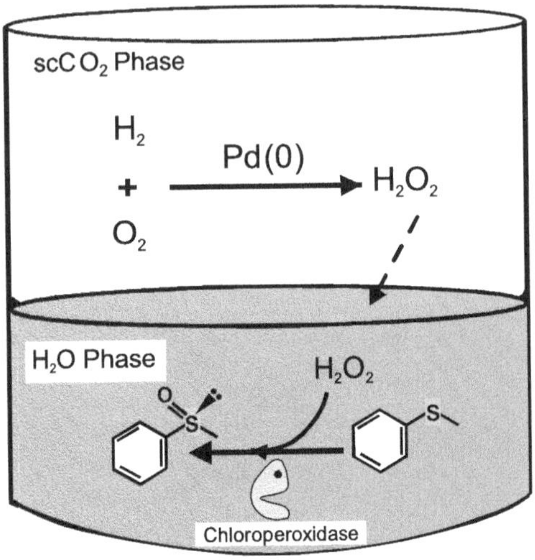

Figure 4.11. Chemo-enzymatic sulfoxidation of thioanisole in $scCO_2/H_2O$ biphasic media.[111]

formation of H_2O_2 from H_2 and O_2 in the supercritical phase; the peroxide is subsequently used by the chloroperoxidase as an oxidant for the asymmetric sulfoxidation in the aqueous phase. In spite of the moderate reaction yields (14–60%), and the important activity loss of the enzyme with time (up to 90% in 3 days at 40°C and 130 bar), this work exemplifies the potential of compartmentalization of the catalytic processes in multiphase systems.

Finally, an interesting methodology should be mentioned concerning the use of enzymes in organic-aqueous tunable solvent (OATS). This approach combines homogeneous enzymatic reactions with a built-in heterogeneous separation for pure products. The addition of moderate CO_2 pressure is used to easily separate enzyme from the monophasic reaction mixture after reaction cycle. The separate phases can then be sent for further processing and catalysts recycle. Greater than 99% ee is shown for catalyzed hydrolysis of *rac*-1-phenylethyl acetate with CALB both before and after CO_2-induced separation.[112]

4.5.3 Bioprocesses in Biphasic ILs System

Although enzymes show potential for recycling in ILs, their use in monophasic reaction media presents the drawback that after each cycle, the product and the unreacted substrates should be extracted with the same type of solvent. Since many molecular solvents (e.g., water, hexane, and toluene) are immiscible with ILs, biphasic systems based on IL/water or IL/organic solvent have been assayed for development of biocatalytic processes.

4.5.3.1 Biphasic ILs/Water System. In general, organic substrates are barely soluble in water and their use in purely aqueous systems is considerably limited. The addition of water-soluble organic solvents leads to monophasic aqueous/organic mixtures with increased reactant solubility but results in lower enzyme activities and stabilities due to the destabilizing effect of the added solvents. A biphasic system consisting of a water-immiscible organic solvent (e.g., diisopropylether [DIPE]) and an aqueous phase is currently widely used in enzymatic synthesis. Typically, the organic phase contains the dissolved reactants, while the enzyme is present in the aqueous phase. This biphasic reaction system has several advantages:

- The partition coefficients of substrates and products within the biphasic system result in an enrichment of the reactants in the organic phase, which facilitates product separation. Thus, the organic phase behaves as a substrate reservoir and *in situ* extractant for the product.[113]
- The low reactant concentration within the aqueous phase also prevents substrate- or product inhibition, thereby increasing the overall productivity.
- The relative ease of removing the organic phase enables the recycling of the aqueous phase and reuse of the enzyme.
- The low substrate contraction within the aqueous phase reduces reaction rates of possible side nonenzymatic reactions.

However, biphasic water/organic solvent presents some limitations related mainly with the negative effect of the organic solvent for the environment, as well as issues related to the stability of some enzymes and whole-cell biocatalysts in some organic solvents used as second phase. In order to maintain the above-mentioned advantages, water-immiscible ILs have emerged as ideal replacement of traditional organic solvents. Thus, for instance, the enantioselective reduction of 2-octanone to *R*-2-octanol, coupled with the oxidation of 2-propanol are developed in an aqueous/[Bmim][NTf_2] biphasic system. Both reactions are catalyzed by an $NADP^+$-dependent alcohol dehydrogenase, allowing continuous regeneration of the expensive cofactor $NADPH+H^+/NADP^+$ and providing excellent results (98% product yield and 99% ee). In this case, the hydrophobic IL phase is used as a reservoir of the hydrophobic substrate (2-octanone), and the favorable partition coefficients of the products shift the biotransformation toward the synthesis of *R*-2-octanol.[114] (see Figure 4.12).

Similarly, different biocatalytic processes catalyzed by the whole cell for the production of fine chemicals have been designed in biphasic water/hydrophobic IL systems.[115] Water/IL systems are superior to many organic solvents so far applied in whole-cell biotransformations. Impressive process intensification in simple biphasic batch processes was recently demonstrated on the examples of asymmetric synthesis with whole-cell biocatalysts.[116] Hence, 21 different ILs were screened as second liquid phase in whole-cell biotransformations combining a recombinant *Escherichia coli* co-expressing a *Lactobacillus brevis* alcohol dehydrogenase gene for the desired asymmetric reduction of prochiral ketones, and a *Candida boidinii* formate dehydrogenase for the regeneration of NAD^+ with formate.[117] The study identified up to 12 new ILs as being potentially suited for biphasic biotransformations. Furthermore, it was validated that ILs with PF_6 and NTf_2 anions lead to better performances. The use of

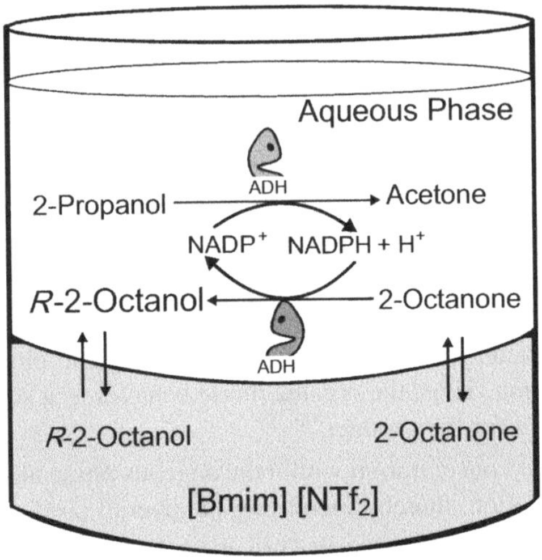

Figure 4.12. Alcohol dehydrogenase (ADH)-catalyzed asymmetric reduction of 2-octanone in IL–water system.[114]

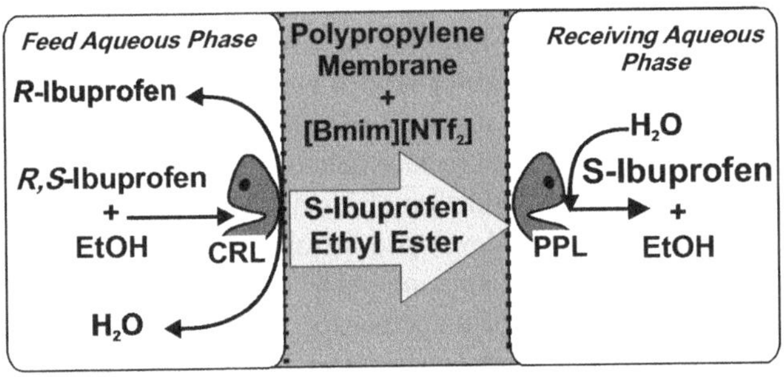

Figure 4.13. Facilitated enantioselective transport of *S*-ibuprofen through a supported IL membrane with immobilized lipase. CRL, *Candida rugosa* lipase; PPL, porcine pancreas lipase.[118]

1-hexyl-1-methylpyrrolidinium bis(trifluoromethylsulfonyl)imide ([HMPL][NTf_2]), for example, resulted in a high space–time yield (180 g (*R*)-2-octanol/L d), 95% chemical yield, and 99.7% ee.

An alternative approach involves the use of supported liquid membrane (SLM) to facilitate the selective separation of a target molecule either by exploiting the solubility differences between solutes in the liquid membrane phase or by the specific interaction between a carrier and the target molecule. Thus, SLM based in ILs can facilitate the development of processes in water-immiscible ILs as reaction solvents. Thus, the KR of *rac*-ibuprofen has been carried out in these concepts (Figure 4.13).[118] The system was operated by coupling two lipase reactions (esterification and hydrolysis, respec-

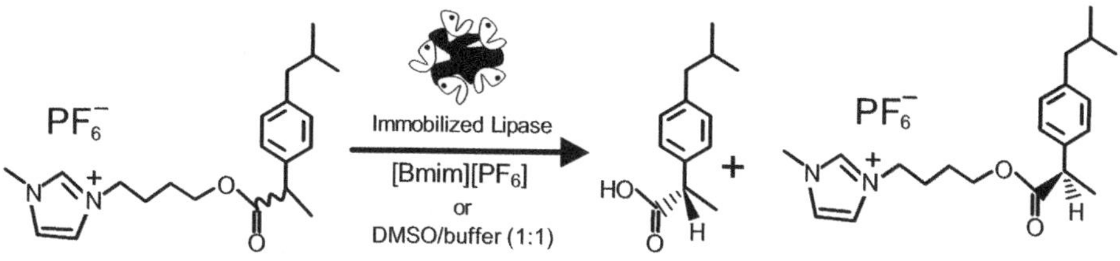

Figure 4.14. Lipase-catalyzed enantioselective hydrolysis of an imidazolium–ibuprofen ester cation combined with [PF_6] in either [Bmim][PF_6]: aqueous buffer biphasic system or DMSO: aqueous buffer mixture.[69b]

tively) with a membrane containing supported ILs. As the IL-based SLMs permit the selective transport of organic molecules, the system provides for the easy and selective permeation of the synthesized (*S*)-ibuprofen ester through the membrane. The ester is then hydrolyzed by another lipase that provides a successful resolution of the racemic mixture (see also Chapters 3 and 8).

In a similar example, an IL based on an imidazolium ibuprofen ester cation combined with [PF_6] was firstly synthesized (see Figure 4.14).[69b] Then, the lipase-catalyzed enantioselective hydrolysis of this ester in either [Bmim][PF_6]: aqueous buffer biphasic system, or DMSO: aqueous buffer mixture, was carried out, obtaining an 87% yield (86% ee) (*S*)-ibuprofen after 24 hours' reaction. Although both the long reaction times and the use of VOCs to recover products from the IL mixture are important drawbacks of this approach, these results show the possibility of designing biocatalytic processes with ILs playing an active role as co-substrates.

An alternative approach to medium engineering for biocatalytic processes in water/IL systems is to form microemulsions of water in ILs. In such systems, enzymes are entrapped in small droplets-water domains (nano/micrometer size), which are formed in a hydrophobic IL and stabilized by a layer of anionic surfactant, that is, sodium bis(2-ethyl-1-hexyl) sulfosuccinate (AOT) and the presence of 1-hexanol as a cosurfactant.[119] The IL microemulsions could have advantages over conventional microemulsions (prepared in organic solvents) as reaction media for carrying out biotranformations with polar or hydrophilic substrates such as amino acids, and carbohydrates, which are poorly soluble in most organic solvents (e.g., isooctane and hexane).[119a] Besides, the catalytic activity of one of the enzymes studied became higher than in microemulsions of AOT/isooctane. Thus, the oxidation of pyrogallol with hydrogen peroxide catalyzed by horseradish peroxidase (HRP) in IL microemulsuions was much more effective than in a conventional AOT/water/isooctane microemulsion, showing that HRP retained almost 70% of its initial activity after incubation at 28°C for 30 hours.[119b]

The origin of such stability was investigated by conformational studies via FT-IR and CD spectroscopy indicating that enzymes entrapped in water/IL microemulsions in most cases retain their native structure or adapt to a more rigid structure, compared with other microheterogeneous media, which correlated well with the stability results.[120]

Besides, a simple procedure suitable for ester separation and enzyme reuse was developed. Thus, *Thermomyces lanuginosa* lipase retained 90% of activity after 10 reaction cycles in water/IL microemulsions formulated with Tween 20.

4.5.3.2 Phase Behavior of IL/scCO$_2$ Biphasic Systems. The use of VOCs to recover solutes dissolved in ILs by extraction using liquid-liquid biphasic systems represents a clear breakdown in the greenness of any chemical processes. The pioneering work of Brennecke's group in 1999 showed that ILs (e.g., [Bmim][PF$_6$]) and scCO$_2$ form biphasic systems.[121] Additionally, although scCO$_2$ is highly soluble in the IL phase and is able to extract previously dissolved hydrophobic compounds (e.g., naphthalene), the same IL is not measurably soluble in the scCO$_2$ phase.[122] This discovery was crucial for further developments in multiphase green (bio)catalytic processes involving both chemical transformation and extraction steps.[123] Catalysis in multiphase operation offers promising opportunities for developing chemical processes (e.g., the catalyst operates in one phase and the product is extracted in the second phase).[124]

The use of enzymes in multiphase systems for biocatalysis based on ILs and scCO$_2$ was originally described in 2002 and represented the first operational approach for the development of fully green chemical processes in nonaqueous environments.[125] Using this approach, the scCO$_2$ flow can serve both to transport the substrate to the IL phase containing the biocatalyst, and to extract the product(s) from the IL. In this way, the product(s) obtained by decompression of the SCF are free from IL and from other organic solvent residues, whereas CO$_2$ can be recycled by recompression. Additionally, if the reaction product does not require any further purification, the approach enhances the economic benefit of the process because the system runs as a blackbox able to transform to pure substrate in pure products without waste generation (see Figure 4.15).

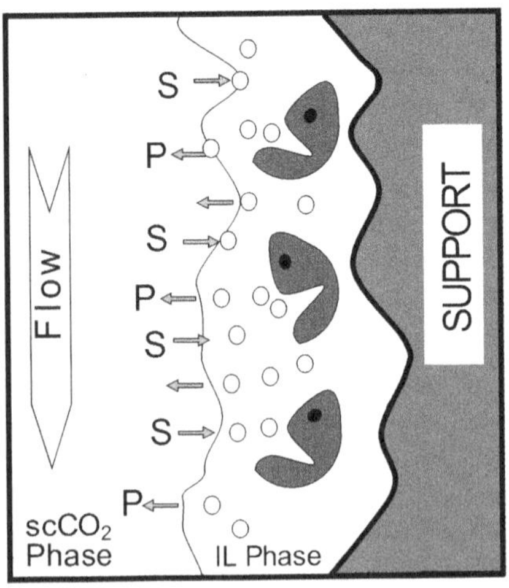

Figure 4.15. A continuous green biocatalytic transformation catalyzed by a supported enzyme in an ILs/scCO$_2$ biphasic medium.[125,136]

Knowledge of the phase behavior of IL/scCO_2 systems is essential for developing any process, because it determines the contact conditions between scCO_2 and solute. These acts include the partitioning behavior of organic compounds between both neoteric solvents, as well as the conditions for reducing the viscosity of the IL phase, thus enhancing the mass-transfer rate of any catalytic system.[126] Preliminary investigations into [Bmim][PF_6]/scCO_2 mixtures indicated that these systems behave as biphasic systems, where the solubility of CO_2 in the IL phase increases with pressure (up to 0.32 mole fraction at 93 bar and 40°C), while temperature has a low effect on CO_2 solubility.[127] At high pressures (up to 970 bar), the density of the CO_2 phase increases, but since the IL phase does not expand, the two distinct phases never become one phase.[128] In the same way, the water content of ILs has an important effect on these systems, as seen from the increased solubility of CO_2 when ILs are previously dried (e.g., the mole fraction of CO_2 in dry [Bmim][PF_6] is 0.54, whereas for the wet, that is, water saturated, IL sample, it is only 0.13). This behavior was also observed for other IL/scCO_2 systems (e.g., [Omim][PF_6], [Omim][BF_4], [Bmim][NO_3]), where the solubility of CO_2 in the IL-enriched phase increased proportionally with the increase in the alkyl chain length of the cation, being highest for the ILs with fluorinated anions (i.e., [PF_6]). In general, the solubility of CO_2 in IL increases with increasing pressures, but the exact amount of CO_2 dissolved in the IL phase varies significantly; for example, at 70 bar, the solubility of CO_2 in [Emim][EtSO_4] was 0.36 mole fraction, whereas it was 0.63 in [Omim][PF_6]).[128] Furthermore, it was observed how as CO_2 pressure is increased (up to 287 bar), the viscosity of several ILs, based on 1-alkyl-3-methyl-imidazolium with as NTf_2 counter ion, dramatically decreases. Also, while the ambient pressure viscosity of ILs increases significantly with chain length, the viscosity of all the CO_2-saturated ILs becomes very similar at high CO_2 pressures (2–3 mPa s).[129]Additionally, the presence of other compounds (e.g., acetone and ethanol) may enhance the solubility of ILs in the scCO_2 phase as a result of the strong interaction of these co-solvents with the IL due to their strong polarity, being in agreement with the increase in its dipole moment (i.e., acetonitrile > acetone > methanol > ethanol > hexane).[130]

The extraction of compounds dissolved in ILs with scCO_2 is probably the most attractive feature of these biphasic systems because both IL and CO_2 can be recycled and extraction does not involve cross-contamination. The extraction of naphthalene from [Bmim][PF_6] by scCO_2 was the first example of solute recovery in green nonaqueous conditions, the process providing a product extract containing no detectable liquid solvent.[121] Further studies determined the solubility of 20 organic solutes containing different substituent groups (e.g., halogen, alcohol, amide, ester, and ketone) in [Bmim][PF_6], and provided quantitative data on extraction recovery rates for these aromatic and aliphatic compounds from this IL with scCO_2. In this respect, it should be noted how compounds with a low solubility in the IL phase (e.g., benzene or chlorobenzene) required the least amount of CO_2 for recovery. Also, solutes with high dipole moment gave low distribution coefficient values (defined as the ratio between solute mole fractions in the scCO_2 and in the IL phases, respectively) because of their high affinity for IL and low affinity for CO_2, which makes extraction more difficult.[131] In the same context, the ability of CO_2 to manipulate the phase behavior in IL/organic or IL/water systems by increasing the pressure has been demonstrated. As example, the pressurization of a methanol solution in [Bmim][PF_6] with CO_2 induces the formation of three phases, two of them liquid: The densest liquid phase (L1) is rich in IL; the next liquid phase (L2) is rich in the organic compound; and the third vapor phase (V) is mostly CO_2 with some organic compounds. The pressure and temperature conditions in which the second liquid phase appears is called the lower critical end point, which is dependent on the initial amounts of methanol and IL. In these conditions, the L2 phase expands significantly with increased CO_2 pressure, while the IL-rich phase (L1) expands relatively little. As the pressure of CO_2 increases again and reaches another critical point, called K-point, the methanol-rich phase (L2) merges with the vapor phase (V), while the resulting scCO_2–organic compound phase contains no detectable IL. These *phase equilibrium switches* can be reversed by modifying the CO_2 pressure at a particular temperature and IL/methanol proportion, and they are very interesting for both reaction and separation processes using ILs.[132]

The growing number of possible applications of IL/scCO_2 biphasic systems in synthesis and extraction processes was recently increased by several phase equilibrium studies for compound/IL/scCO_2 systems with larger organic molecules (i.e., naphthalene, 1-phenylethanol, naproxen, etc.).[133] Further investigations into [Bmim][PF_6]-ethanol–water–CO_2 quaternary mixtures at 50°C found a varied phase behavior, ranging from total miscibility to partial miscibility to nearly complete phase separation, which can also be useful in reaction/separation cycles.[134] Another interesting feature of IL/scCO_2 phase concerns the change in the melting point of ILs. Prounounced melting point depressions, some even exceeding 100°C, induced by compressed CO_2 in some ILs based on ammonium or phosphonium cation, have been reported. In the case of tetrabutylammonium tetrafluoroborate ([Bu_4N][BF_4], m.p. 156°C), equilibration with high-pressure CO_2 at 150 bar resulted in a melting point depression of 120°C. This discovery may make available some new solvents for IL/scCO_2 biphasic catalysis, as well as facilitate IL recovery and reuse.[135] All these features underline the advantages of using multiphase (bio)catalytic systems based on IL and scCO_2, for the development of sustainable chemical processes in nonaqueous media.

4.5.3.3 Bioprocesses in IL/scCO_2 Biphasic Systems. Biotransformations based on ILs and scCO_2 are interesting alternatives to organic solvents for designing clean synthetic chemical processes that provide pure products directly. The classical advantages of scCO_2 to extract, dissolve, and transport chemicals are tarnished in the case of enzymatic processes because of its denaturative effect on enzymes, while ILs have shown themselves to be excellent stabilizing agents of enzymes. In this context, the use of IL/scCO_2 biphasic systems as reaction media for enzyme catalysis has opened up new opportunities for the development of integral green processes in nonaqueous environments.

Green biphasic biocatalytic systems in nonaqueous environments have been designed by immobilizing the free or supported enzyme molecules into an IL phase (catalytic phase), while substrates and products reside largely in the SCF phase (extractive phase), and directly pure products (see Figure 4.15). The system was firstly tested for two different reactions catalyzed by CALB: the synthesis of aliphatic esters by transesterification from 1-alkanols and vinyl esters (e.g., butyl butyrate from vinyl butyrate and 1-butanol), and the KR of *rac*-1-phenylethanol in a wide range of conditions (100–150 bar and 40–100°C). In these conditions, the enzyme showed an exceptional level of activity, enantioselectivity (ee > 99.9), and operational stability (e.g., the enzyme only lost 15% activity after 11 cycles of 4 hours).[125a] These excellent results obtained for biotransformations in $scCO_2$ using the enzyme coated with ILs were corroborated in extreme conditions, such as 100 bar pressure and 150°C temperature.[136] Further studies on these IL/$scCO_2$ biocatalytic systems attempted to understand the importance of mass-transport phenomena between both neoteric phases. By using two similar ILs based on the same ions, but with different degrees of hydrophobicity in the cation, [Btma][NTf_2] and [CNPrtma][NTf_2], the continuous synthesis of six different short chain alkyl esters (e.g., from butyl acetate to octyl propionate) catalyzed by CALB in $scCO_2$ was studied. Using Hansen's solubility parameter (δ) as criterion to compare the hydrophobicity of the main alkyl chain of cations in ILs with those of substrates and products, it was shown how the same values for this parameter in reagents and IL resulted in a clear improvement of productivity, as a consequence of favoring the mass-transfer phenomena between both the IL and the $scCO_2$ phases.[137]

A further step toward green biocatalysis in IL/$scCO_2$ biphasic systems was the appropriate selection of acyl donor in the CALB-catalyzed KR of *rac*-1-phenylethanol, because the selective separation of the synthetic product can be included as an integrated step in the full process. By using vinyl laurate as acyl donor, the stereoselective synthetic product ((*R*)-1-phenylethyl laurate) can be selectively separated from the unreacted (*S*)-1-phenylethanol with $scCO_2$ into two different cryo-traps (see Figure 4.16). This process takes advantage of the fact that the solubility of a compound in

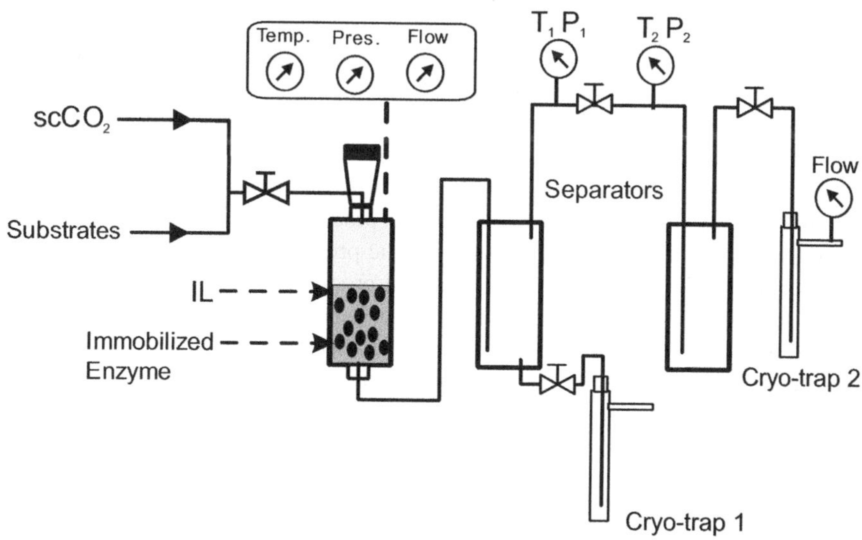

Figure 4.16. Setup of a reaction/separation system for continuous-flow combination of enzymatic kinetic resolution and enantiomer separation using an ionic liquid/$scCO_2$ medium.[138]

$scCO_2$ depends on both the polarity and vapour pressure. Thus, if the alkyl chain of an ester product is long enough, its low volatility should mean that it is less soluble in $scCO_2$ than the corresponding alcohol. Using this experimental approach, the introduction of two additional separation chambers connected with cryo-traps, and the selection of an appropriate pressure and temperature, resulted in the selective separation of the synthetic product and the unreacted alcohol from the reaction mixture (66% yield, ee > 99.9%).[138]

The selective extraction of products by $scCO_2$ after lipase-catalyzed transesterification IL media has been applied in other cases. For example, for CALB-catalyzed butanolysis of triolein, the biotransformation first occurred in 80% v/v [Toma][TFA] as a homogeneous liquid phase. Then, the butyl oleate product was extracted in a second step using $scCO_2$ at 85 bar and 35°C.[139] In another example, such as the KR of *rac*-2-octanol, by using succinic anhydride as acylating agent, the vapor–liquid equilibrium data for systems containing ILs, $scCO_2$, and reaction products were firstly studied, and the partition coefficients of the reaction products between the IL-enriched phase and the CO_2-enriched phase were calculated. Then, the postreaction mixture was placed in $scCO_2$ at 110 bar and 35°C, which allowed the unreacted enantiomer of 2-octanol to be completely recovered with a very high enantiomeric excess (98.42%).[140] In another example, the solubility of CALB in the IL 1-hydroxy-1-propyl-3-methylimidazolium nitrate was firstly studied, showing how pressurization of enzyme–IL solution with $scCO_2$ (35–70°C, 120 bar) did not produce precipitation of the enzyme. Also, at constant CO_2/IL ratios, the pressure of the bubble points remained almost unchanged at all the assayed enzyme concentrations, while the recovery of the pure IL was made possible by precipitating the enzyme using 2-propanol as an antisolvent.[141]

Enzymes other than lipases, for example, cutinase from *Fusarium solani pisi* immobilized onto zeolite NaY, were also tested in a [Bmim][PF_6]/$scCO_2$ system to carry out the KR of 2-phenyl-1-propanol. The protective effect of the IL against enzyme deactivation by $scCO_2$ was demonstrated, as well as the higher enzymatic activity than that observed for the cutinase/IL monophasic system. This enhanced activity was attributed to the CO_2 dissolved in the IL, which would decrease its viscosity and hence improve the mass transfer of substrates to the enzyme's active site.[142]

Two final approaches are worth mentioning to push toward the excellences of IL/$scCO_2$ biphasic systems, such as multicatalytic processes and reaction systems with reduced amounts of ILs. Integrated multicatalytic processes, whereby one initial substrate is catalytically transformed into one final product by two or more consecutive catalytic steps in the same reaction system, is of great interest for developing a new chemical industry.[143] On the other hand, some ILs have been described as being not fully green solvents because of their low biodegradability and high (eco)toxicological properties, so that reaction systems based on reduced amounts of ILs are preferred (see also Chapters 1 and 7).[144]

Enzyme-catalyzed KR is probably the most widely used method for separating the two enantiomers of a racemic mixture, the chemical yield of the process being limited to 50%. However, this drawback can be overcome by combining the enzymatic KR with *in situ* racemization of the undesired enantiomer, using so-called dynamic kinetic resolution (DKR), which theoretically can reach up to 100% of one enantiomeric

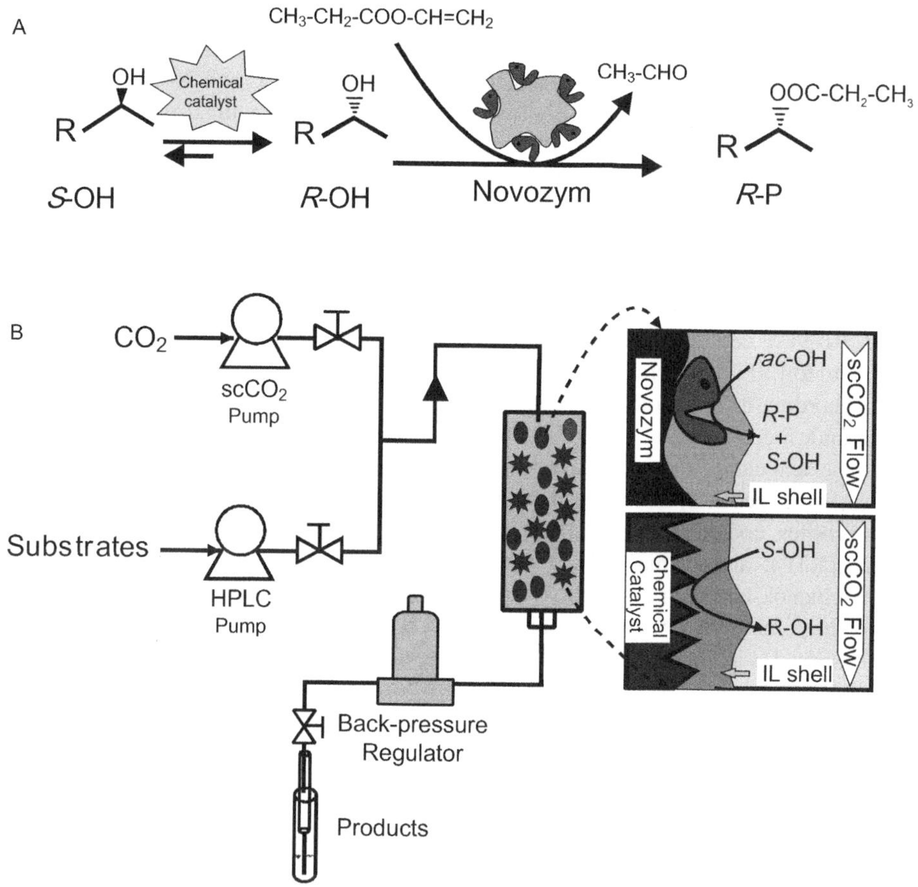

Figure 4.17. (A) DKR of *sec*-alcohols (*rac*-OH) catalyzed by the combined action of immobilized CALB (*Novozym 435*) and a chemical catalyst (e.g., zeolite). (B) Setup of a continuous packed bed reactor containing both *Novozym 435* and a chemical catalyst coated with ILs.[146b]

product (see Figure 4.17A). For example, the DKR of *rac*-1-phenylethanol was carried out by combining immobilized lipase (Amano PS CI) with a chemical catalyst (either the metal catalyst [Ru(p-cymene)Cl_2]$_2$, or the acid catalyst Nafion®) in a discontinuous way, and without the presence of ILs in $scCO_2$ at 100 bar and 40°C.[145] By this approach, the yield of the (*R*)-product (70% Ru-catalyst; 85% Nafion) was improved compared with the reaction carried out in hexane (30–35% yield), while the enantioselectivities of the products were slightly higher in $scCO_2$ (96% Ru-catalyst; 85% Nafion) than in hexane (91%; 81%). The moderately low enantioselectivities achieved for the acidic Nafion were attributed to the uncontrolled chemical esterification of the substrates catalyzed by this solid acid, so that a physical separation of the enzyme and the chemical catalyst, along with the use of a continuous flow system, is suggested as a way to prevent this undesirable side reaction.

In the same way, continuous DKR processes of *rac*-1-phenylethanol were carried out combining immobilized CALB with silica modified with benzenosulfonic acid groups as catalysts in a packed bed reactor under $scCO_2$ at 50°C and 100 bar. Both the chemical and the enzymatic catalysts were previously coated with ILs (e.g., [Emim][NTf_2], [Btma][NTf_2], or [Bmim][PF_6]) at a 1:1 (w : w) ratio, to prevent enzyme deactivation by $scCO_2$.[146a] The use of both catalysts as a simple mixture resulted in a complete loss of activity, probably due to the acid environment around the enzyme particles, which would lead to deactivation. However, the packaging of catalyst particles in three different layers (immobilized enzyme—acid catalyst—immobilized enzyme) physically separated by glass wool led to encouraging results for the (*R*)-ester product (76% yield, 91–98% ee).[146a] For this reactor configuration, the (*R*)-ester product yield may only tend to 100% if several enzymatic and acid catalyst layers are stacked in the packed bed, according to a dichotomist progression. It is also worth noting how the presence of the undesired ester (*S*)-ester and hydrolytic products in the $scCO_2$ flow was enhanced when the acid catalyst particles were assayed without IL coating. The use of weak solid acids, such as zeolites, as chemical catalyst clearly improved the results. Four different acid zeolites coated with ILs (e.g., [Bmim][PF_6], [Bdmim][PF_6], [Odmim][NTf_2], [Toma][NTf_2], and [Btma][NTf_2]) were able to catalyze the racemization of (*S*)-1-phenylethanol, and their suitability to perform the continuous DKR of *rac*-1-phenylethanol in combination with immobilized CALB under $scCO_2$ flow was successfully demonstrated (see Figure 4.17B).[146b] The best results (98% yield, 95% ee) were obtained for a heterogeneous mixture between fajausite type zeolite (CBV400) particles coated with [Btma][NTf_2] and *Novozym* particles coated with the same IL. Due to the low acidity of the assayed zeolites, the packaging of the heterogeneous mixture of catalyst particles coated with IL did not result in any activity loss of the immobilized CALB during 14 days of continuous operation in CO_2 under different supercritical conditions. This work clearly demonstrated the exciting potential of multicatalytic (enzymatic or chemo-enzymatic) systems in ILs/$scCO_2$ for synthesizing optically active pharmaceutical drugs by a sustainable approach.

A further step toward optimizing IL/$scCO_2$ biphasic systems arose from the immobilization of the IL species onto solid supports. The immobilization facilitates the separation processes and avoids a possible accidental spill to the environment. Besides, it reduces the cost of the process as lower amounts of ILs are employed in catalytic processes in IL/scCO2 biphasic systems.[147] The supported ionic species can be obtained either by adsorbing ILs onto solid supports (supported ionic liquid phases [SLIPs]) or by covalently bonding IL-like fragments on the surface of the solid support (supported ionic liquid-like phases [SILLPs]).[148] Covalently < SILLPs) are prepared by functionalization of the polystyrene-divinylbenzene (PS-DVB) surfaces with IL-like (imidazolium) moieties or by polymerization of the corresponding monomers. By this approach, ILs properties are transferred to the solid phase, leading to either particles or monolithic SILLP. Preliminary results proved quantitatively the increase in polarity of SILLPs compared with the original PS-DVB polymers by using pyrene as a fluorescent probe.[149] Thus, SILLPs provide microenvironments that present similar properties in terms of polarity to bulk molecular ILs. Hence, they might be regarded as "solid solvents" or as nanostructured materials with microenvironments of tunable polarity. These type of materials have been successfully used as supports for metallic catalysts[150] and as supported organocatalysts.[151]

There are only few examples of biocatalysis using covalently bonding supported IL. Thus, the CRL have been immobilized on magnetic nanoparticles coated with supported ILs. Materials based on imidazoliun cations with different chain lengths (C-1, C-4, and C-8) and anions ([Cl], [BF_4], and [PF_6]) were obtained by covalent bonding of IL–silane moieties on magnetic silica nanoparticles (55 nm diameter), which permits large amounts of lipase to be loaded (about 64 mg/100 mg carrier). Furthermore, the activity of bound lipase was 118.3% compared with that of the native lipase, when the esterification of oleic acid with butanol in free solvent media was used as activity test at 30°C.[152] Recently, HRP was encapsulated in microparticles composed of polymerized IL. The enzyme entrapped in this support exhibits higher activity than in conventional polyacrylamide microparticles and is easily recycled by centrifugation from reaction mixtures.[153]

Supported IL can also be used to develop enzymatic catalyzed processes in $scCO_2$ (see Figure 4.18).

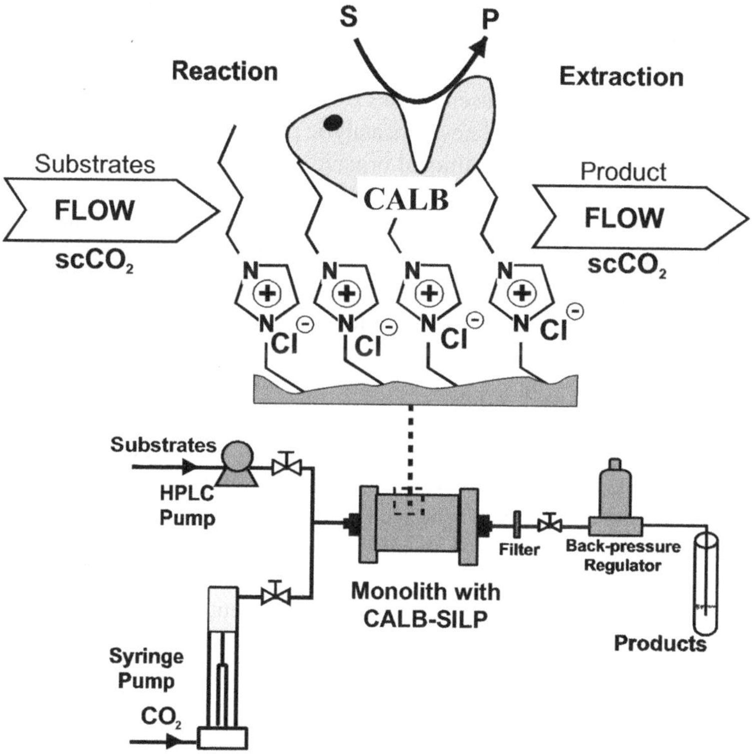

Figure 4.18. Setup of a reactor with immobilized CALB onto monolith-supported ionic liquid phase (M-SILP) for continuous operation under flow conditions in $scCO_2$.

Thus, bioreactors with covalently SILLP were prepared as polymeric monoliths based on styrene-divinylbenzene, containing imidazolium units in loadings ranging from ~55% to 40% wt IL per gram of polymer, which results in a liquid phase coating the surface of the solid support. These SILLPs were able to absorb CALB, leading to highly efficient and robust heterogeneous biocatalysts. The bioreactors were prepared as macroporous monolithic miniflow systems and tested for the continuous flow synthesis of citronellyl propionate by transesterification in $scCO_2$ at 100 bar and 40–100°C. The catalytic activity of these miniflow bioreactors remained practically unchanged for seven operational cycles of 5 hours each in different supercritical conditions.[154]

4.6 PROSPECTS

Biocatalytic approaches in green nonconventional reaction media hold much promise for the development of a sustainable chemical manufacturing industry. Biocatalytic processes in nonaqueous environments enhance the possible technological applications and expand the repertoire of enzyme-mediated transformations.

It has been demonstrated that some ILs have an exceptional ability to overstabilize enzymes. Besides, the unique properties of ILs, which can be tailored at the molecular level by an appropriate selection of the cation (e.g., aliphatic or heterocyclic), the length and nature of the alkyl chain attached, and the anion, has opened a new window of processing options not available using conventional organic solvents. For instance, the power of ILs to dissolve large concentrations of saccharides and carbohydrate biopolymers has resulted in an increase of new biocatalytic processes hampered by low solubility of this type of substrates in traditional organic media.

SCFs are being increasingly used to carry out enzymatic reactions. SCFs allow rapid reaction rates, simplify product recovery, and are ideal replacement for conventional organic solvents as they are more environmentally friendly. Furthermore, in the search of greener synthetic processes $scCO_2$ seems to be the perfect match with ILs. The unique phase behavior of IL/$scCO_2$ systems allows processes involving reaction and downstream isolation and purification steps, facilitating the easy reuse of the catalyst and IL phase.

The combination of multiphase neoteric systems with more complex enzymes (oxidoreductases, lyases, etc.) should be explored as a clear strategy for developing integral new green synthetic processes. The integration of the catalytic steps in multistep organic syntheses and downstream processing without the need to isolate intermediates is key to successful implementation of all these catalytic methodologies in chemical manufacture. Multienzymatic and/or multi-chemo-enzymatic green chemical processes in multiphase neoteric systems for synthesizing pharmaceutical drugs are only a beginning, but the door for the development of a sustainable chemical industry is open.

ACKNOWLEDGMENTS

We thank the CICYT-MICIN (CTQ2008-00877 and CTQ2011-28903), SENECA Foundation (08616/PI/08), CSIC (200980I092), and Bancaixa-UJI (P1 2009-58) for financial support.

REFERENCES

1 (a) P. T. Anastas, J. C. Warner, *Green Chemistry: Theory and Practice*, Oxford University Press, New York, **1998**; (b) I. Horvath, P. T. Anastas, *Chem. Rev.* **2007**, *107*, 2167; (c) P. Anastas, N. Eghbali, *Chem. Soc. Rev.* **2010**, *39*, 301.

2 (a) J. H. Clark, S. J. Tavener, *Org. Proc. Res. Dev.* **2007**, *11*, 149; (b) C. J. Li, B. M. Trost, *Proc. Natl. Acad. Sci. U.S.A.* **2009**, *105*, 13197.

3 (a) A. D. Curzons, D. J. C. Constable, D. N. Mortimerand, V. L. Cunningham, *Green Chem.* **2001**, *3*, 1; (b) D. J. C. Constable, A. D. Curzonsand, V. L. Cunningham, *Green Chem.* **2002**, *4*, 521.

4 E. Buncel, R. Stairs, H. Wilson, *The Role of the Solvent in Chemical Reactions*, Oxford University Press, Oxford, New York, **2003**.

5 (a) J. M. DeSimone, *Science* **2002**, *799*, 297; (b) P. T. Anastas, W. Leitner, P. G. Jessop, C. J. L. P. Wasserscheid, A. Stark, *Handbook of Green Chemistry—Green Solvents*, Wiley-VCH Verlag GmbH, New York, **2010**.

6 (a) P. Wasserscheid, T. Welton, *Ionic Liquids in Synthesis*, Wiley-VCH, Weinheim, **2008**; (b) N. V. Plechkova, K. R. Seddon, *Chem. Soc. Rev.* **2008**, *37*, 123; (c) H. Olivier-Bourbigou, L. Magna, D. Morvan, *Appl. Catal. A* **2010**, *373*, 1.

7 (a) P. J. Jessop, W. Leitner, *Chemical Synthesis Using Supercritical Fluids*, Wiley-VCH, Weinheim, **1999**; (b) E. J. Beckman, *J. Supercrit. Fluids* **2004**, *28*, 121.

8 (a) P. G. Jessop, B. Subramaniam, *Chem. Rev.* **2007**, *107*, 2666; (b) B. Subramaniam, *Coord. Chem. Rev.* **2010**, *254*, 1843–1853.

9 J. A. Gladysz, D. P. Curran, I. T. Horváth, *Handbook of Fluorous Chemistry*, Wiley-VCH Verlag GmbH & Co. KGaA, Weinheim, Germany, **2004**.

10 R. A. Sheldon, *Chem. Commun.* **2008**, 3352.

11 (a) P. T. Anastas, M. M. Kirchhoff, T. C. Williamson, *Appl. Catal. A* **2001**, *221*, 3; (b) R. A. Sheldon, I. Arends, U. Hanefeld, *Green Chemistry and Catalysis*, Wiley-VCH Verlag GmbH & Co. KGaA, Weinheim, **2007**; (c) G. Rothenberg, *Catalysis, Concepts and Green Chemistry*, Wiley-VCH Verlag GmbH, Weinheim, **2008**.

12 (a) P. T. Anastas, L. G. Heine, T. C. Williamson, *Green Engineering*, Oxford University Press, Oxford, **2000**; (b) P. T. Anastas, L. G. Heine, T. C. Williamson, *Green Chemical Syntheses and Processes*, American Chemical Society Publication, Washington, DC, **2000**; (c) J. H. Clark, D. Macquarrie, *Handbook of Green Chemistry & Technology*, Blackwell Science Ltd, Oxford, **2002**; (d) R. L. Lankey, P. T. Anastas, *Advancing Sustainability through Green Chemistry and Engineering*, ACS Publications, Washington, DC, **2002**.

13 (a) R. B. Silverman, *The Organic Chemistry of Enzyme-Catalyzed Reactions*, Academic Press, New York, **2002**; (b) A. S. Bommarius, B. R. Riebel, *Biocatalysis*, Wiley-VCH Verlag GmbH & Co. KGaA, Weinheim, **2004**; (c) P. Anastas, R. Crabtree, *Handbook of Green Chemistry-Green Catalysis: Biocatalysis*, Vol. 3, Wiley-VCH Verlag GmbH, New York, **2009**.

14 (a) V. Gotor, I. Alfonso, E. García-Urdiales, *Asymmetric Organic Synthesis with Enzymes*, Wiley-VCH Verlag GmbH & Co. KGaA, Weinheim, **2008**; (b) M. Sureshkumar, C. K. Lee, *J. Mol. Catal., B Enzym.* **2009**, *60*, 1; (c) T. Hudlicky, J. W. Reed, *Chem. Soc. Rev.* **2009**, *38*, 3117.

15 (a) S. M. Roberts, G. Casy, M.-B. Nielsen, S. Phythian, C. Todd, K. Wiggins, *Biocatalysts for Fine Chemicals Synthesis*, J. Wiley & Sons, New York, **1999**; (b) A. Liese, K. Seelbach,

C. Wandrey, *Industrial Biotransformations-A Comprehensive Handbook*, Willey-VCH, Weinheim, **2006**.

16 A. M. Klibanov, *Nature* **2001**, *409*, 241.

17 M. N. Gupta, I. Roy, *Eur. J. Biochem.* **2004**, *271*, 2575.

18 P. Lozano, *Green Chem.* **2010**, *12*, 555–569.

19 Y. L. Khmelnitsky, J. O. Rich, *Curr. Opin. Chem. Biol.* **1999**, *3*, 47.

20 H. E. Schoemaker, D. Mink, M. G. Wubbolts, *Science* **2003**, *299*, 1694.

21 N. J. Turner, *Nat. Chem. Biol.* **2009**, *5*, 567.

22 (a) M. T. Reetz, *Adv. Catal.* **2006**, *49*, 1; (b) R. A. Sheldon, *Adv. Synth. Catal.* **2007**, *349*, 1289; (c) R. J. Kazlauskas, U. T. Bornscheuer, *Nat. Biotechnol.* **2009**, *5*, 526; (d) J. M. Palomo, *Curr. Org. Synth.* **2009**, *6*, 1.

23 K. Hult, P. Berglund, *Trends Biotechnol.* **2007**, *25*, 231.

24 (a) E. P. Hudson, R. K. Eppler, D. S. Clark, *Curr. Opin. Biotechnol.* **2005**, *16*, 637; (b) R. J. Kazlauskas, *Curr. Opin. Chem. Biol.* **2005**, *9*, 195; (c) K. Hult, P. Berglund, *Curr. Opin. Biotechnol.* **2003**, *14*, 395.

25 (a) R. J. Kazlauskas, H. K. Weber, *Curr. Opin. Chem. Biol.* **1998**, *2*, 121; (b) A. Ghanem, *Tetrahedron* **2007**, *63*, 1721; (c) H. Akita, *Heterocycles* **2009**, *78*, 1667.

26 H. Frauenfelder, G. Chen, J. Berendzen, P. W. Fenimore, H. Jansson, B. H. McMahon, I. R. Stroe, J. Swenson, R. D. Young, *Proc. Natl. Acad. Sci. U.S.A.* **2009**, *106*, 5129.

27 (a) P. J. Halling, *Enzyme Microb. Technol.* **1994**, *16*, 178; (b) M. N. Gupta, I. Roy, *Eur. J. Biochem.* **2004**, *271*, 2575; (c) S. Torres, G. R. Castro, *Food Technol. Biotechnol.* **2004**, *42*, 271.

28 P. A. Fitzpatrick, A. C. U. Steinmetz, D. Ringe, A. M. Klibanov, *Proc. Natl. Acad. Sci. U.S.A.* **1993**, *90*, 8653.

29 (a) A. M. Azevedo, D. M. F. Prazeres, J. M. S. Cabral, L. P. Fonseca, *J. Mol. Catal., B Enzym.* **2001**, *5*, 147; (b) G. Pencreac'h, J. C. Baratti, *Enzyme Microb. Technol.* **2001**, *28*, 473; (c) S. Hazarika, P. Goswami, N. N. Dutta, *Chem. Eng. J.* **2002**, *85*, 61; (d) N. Krieger, T. Bhatnagar, J. C. Baratti, A. M. Baron, V. M. De Lima, D. Mitchell, *Food Technol. Biotechnol.* **2004**, *42*, 279.

30 Y. Mine, L. Zhang, K. Fukunaga, Y. Sugimura, *Biotechnol. Lett.* **2005**, *27*, 383.

31 T. Maruyama, S. Nagasawa, M. Goto, *Biotechnol. Lett.* **2002**, *24*, 1341.

32 Y. L. Khmelnitsky, S. H. Welch, D. S. Clark, J. S. Dordick, *J. Am. Chem. Soc.* **1994**, *116*, 2647.

33 A. M. O'Brien, A. T. Smith, C. O. O'Fagain, *Biotechnol. Bioeng.* **2003**, *81*, 233.

34 For immobilized enzymes in green solvent, see: (a) P. Lozano, T. de Diego, J. L. Iborra, in *Immobilization of Enzymes and Cells—Methods in Biotechnology Series*, Vol. 22 (Ed. J. M. Guisan), Humana Press Inc., Totowa, **2006**, Chapters 22 and 23; for review in immobilized enzymes, see: (b) R. A. Sheldon, *Adv. Synth. Catal.* **2007**, *349*, 1289–1307; (c) C. Mateo, J. M. Palomo, G. Fernandez-Lorente, J. M. Guisan, R. Fernandez-Lafuente, *Enzyme Microb. Technol.* **2007**, *40*, 1451; for nanostructures, see: (d) J. Kim, J. W. Grate, P. Wang, *Chem. Eng. Sci.* **2006**, *61*, 1017; M. Hartmann, D. Jung, *J. Mater. Chem.* **2010**, *20*, 844.

35 (a) E. Katchalski-Kazir, D. M. Kraemer, *J. Mol. Catal., B Enzym.* **2000**, *10*, 156; (b) Q. Wang, Z. Yang, L. Wang, M. Ma, B. Xu, *Chem. Commun.* **2007**, 1032; (c) P. Lozano,

A. B. Pérez-Marín, T. De Diego, D. Gómez, D. Paolucci-Jeanjean, M. P. Belleville, G. M. Rios, J. L. Iborra, *J. Memb. Sci.* **2002**, *201*, 55.

36 (a) P. Vidinha, V. Augusto, M. Almeida, I. Fonseca, A. Fidalgo, L. Ilharco, J. M. S. Cabral, S. Barreiros, *J. Biotechnol.* **2006**, *121*, 23; (b) S. H. Lee, T. T. N. Doan, S. H. Ha, Y. M. Koo, *J. Mol. Catal., B Enzym.* **2007**, *45*, 57.

37 Y. A. Shchipunov, T. Y. Karpenko, I. Y. Bakunina, Y. V. Burtseva, T. N. Zvyagintseva, *J. Biochem. Biophys. Methods* **2004**, *58*, 25; (b) C. M. F. Soares, O. A. Dos Santos, H. F. De Castro, F. F. De Moraes, G. M. Zanin, *J. Mol. Catal., B Enzym.* **2006**, *39*, 69.

38 (a) N. Burns, J. C. Tiller, *Nano Lett.* **2005**, *5*, 45; (b) Q. Wang, Z. Yang, L. Wang, M. Ma, B. Xu, *Chem. Commun.* **2007**, 1032; (c) J. Kim, J. W. Grate, P. Wang, *Trends Biotechnol.* **2008**, *26*, 639.

39 (a) P. Vidinha, V. Augusto, M. Almeida, I. Fonseca, A. Fidalgo, L. Ilharco, J. M. S. Cabral, S. Barreiros, *J. Biotechnol.* **2006**, *121*, 23; (b) H. R. Hobbs, B. Kondor, P. Stephenson, R. A. Sheldon, N. R. Thomas, M. Poliakoff, *Green Chem.* **2006**, *8*, 816; (c) Z. D. Dijkstra, R. Merchant, J. T. F. Keurentjes, *J. Supercrit. Fluids* **2007**, *41*, 102.

40 H. Olivier-Bourbigou, L. Magna, D. Morvan, *Appl. Catal. A* **2010**, *373*, 1.

41 Pioneering works using ILs in biotransformations: (a) R. Madeira Lau, F. van Rantwijk, K. R. Seddon, R. A. Sheldon, *Org. Lett.* **2000**, *2*, 4189; (b) M. Erbeldinger, A. J. Mesiano, A. J. Russell, *Biotechnol. Prog.* **2000**, *16*, 1129; (c) P. Lozano, T. De Diego, D. Carrié, M. Vaultier, J. L. Iborra, *Biotechnol. Lett.* **2001**, *23*, 1529; (d) T. Itoh, E. Akasaki, K. Kudo, S. Shikarami, *Chem. Lett.* **2001**, *30*, 262.

42 (a) L. Crowhurst, P. R. Mawdsley, J. M. Parez-Arlandis, P. A. Salter, T. Welton, *Phys. Chem. Chem. Phys.* **2003**, *5*, 2790; (b) C. Reichardt, *Green Chem.* **2005**, *7*, 339; (c) B. R. Mellein, S. N. V. K. Aki, R. L. Ladewski, J. F. Brenneke, *J. Phys. Chem. B* **2007**, *111*, 131; (d) J. M. Lee, S. Ruckes, J. M. Prausnitz, *J. Phys. Chem. B* **2008**, *112*, 1473; (e) C. Chiappe, M. Malvaldi, C. S. Pomelli, *Pure Appl. Chem.* **2009**, *81*, 767.

43 (a) C. Wakai, A. Oleinikova, M. Ott, H. Weingärtner, *J. Phys. Chem. B* **2005**, *109*, 17028; (b) H. Weingärtner, *Z. Phys. Chem.* **2006**, *220*, 1395.

44 M. Moniruzzaman, K. Nakashima, N. Kamiya, M. Goto, *Biochem. Eng. J.* **2010**, *48*, 295.

45 (a) Z. Yang, W. Pan, *Enzyme Microb. Technol.* **2005**, *37*, 19; (b) F. van Rantwijk, R. A. Sheldon, *Chem. Rev.* **2007**, *107*, 2757; (c) J. Durand, E. Teuma, M. Gómez, *C.R. Chim.* **2007**, *10*, 152; (d) J. R. Harjani, P. U. Naik, S. J. Nara, M. M. Salunkhe, *Curr. Org. Synth.* **2007**, *4*, 354; (e) C. Roosen, P. Muller, L. Greiner, *Appl. Microbiol. Biotechnol.* **2008**, *81*, 607; (f) M. Sureshkumar, C. K. Lee, *J. Mol. Catal., B Enzym.* **2009**, *60*, 1.

46 Z. Yang, *J. Biotechnol.* **2009**, *144*, 12.

47 (a) S. Wallert, J. I. Grayson, K. H. Drauz, H. Gröger, C. Bolm, P. Domínguez de María, F. Chamouleau, WO 2006/005409A1. **2006**; (b) S. Wallert, K. H. Drauz, J. I. Grayson, H. Gröger, P. Domínguez de María, C. Bolm, *Green Chem.* **2005**, *7*, 602.

48 (a) H. Zhao, S. V. Malhotra, *Biotechnol. Lett.* **2002**, *24*, 1257; (b) S. Shipovskov, H. Q. N. Gunaratne, K. R. Seddon, G. Stephens, *Green Chem.* **2008**, *10*, 806.

49 (a) Z. Guo, X. Xu, *Green Chem.* **2006**, *8*, 54; (b) T. De Diego, P. Lozano, M. A. Abad, K. Steffensky, M. Vaultier, J. L. Iborra, *J. Biotechnol.* **2009**, *140*, 234.

50 (a) C. F. Poole, *J. Chromatogr. A* **2004**, *1037*, 49; (b) H. Zhao, S. Q. Xia, P. S. Ma, *J. Chem. Technol. Biotechnol.* **2005**, *80*, 1089; (c) T. L. Greaves, C. J. Drummond, *Chem. Rev.* **2008**, *108*, 206.

51 (a) T. De Diego, P. Lozano, S. Gmouh, M. Vaultier, J. L. Iborra, *Biotechnol. Bioeng.* **2004**, *88*, 916; (b) T. De Diego, P. Lozano, S. Gmouh, M. Vaultier, J. L. Iborra, *Biomacromolecules* **2005**, *6*, 1457; (c) P. Lozano, T. De Diego, S. Gmouh, M. Vaultier, J. L. Iborra, *Biocatal. Biotransformation* **2005**, *23*, 169; (d) E. Feher, B. Major, K. Belafi-Bako, L. Gubicza, *Biochem. Soc. Trans.* **2007**, *35*, 1624; (e) H. X. Shan, Z. J. Li, M. Li, G. X. Ren, Y. J. Fang, *J. Chem. Technol. Biotechnol.* **2008**, *83*, 886; (f) N. M. Micaelo, C. M. Soares, *J. Phys. Chem. B* **2008**, *112*, 2566; (g) T. A. Page, N. D. Kraut, P. M. Page, G. A. Baker, F. V. Bright, *J. Phys. Chem. B* **2009**, *113*, 12825.

52 (a) J. K. Lee, M. J. Kim, *J. Org. Chem.* **2002**, *67*, 6845; (b) M. J. Kim, J. K. Lee, WO 03/057871A1, **2003**.

53 (a) S. H. Lee, T. T. N. Doan, S. H. Ha, W. J. Chang, Y. M. Koo, *J. Mol. Catal., B Enzym.* **2007**, *47*, 129; (b) S. H. Lee, T. T. Doan, S. H. Ha, Y. M. Koo, *J. Mol. Catal., B Enzym.* **2007**, *45*, 57.

54 (a) P. Lozano, T. De Diego, D. Carrié, M. Vaultier, J. L. Iborra, *J. Mol. Catal., B Enzym.* **2003**, *21*, 9; (b) J. Q. Tian, Q. Wang, Z. Y. Zhang, *Eur. Food Res. Technol.* **2009**, *229*, 357; (c) D. H. Zhang, S. Bai, M. Y. Ren, Y. Sun, *Food Chem.* **2008**, *109*, 72; (d) L. Banoth, M. Singh, A. Tekewe, U. C. Banerjee, *Biocatal. Biotransformation* **2009**, *27*, 263.

55 (a) J. A. Berberich, J. L. Kaar, A. J. Russell, *Biotechnol. Prog.* **2003**, *19*, 1029; (b) M. Adamczak, U. T. Bornscheuer, *Process Biochem.* **2009**, *44*, 257.

56 M. Y. Ren, S. Bai, D. H. Zhang, Y. Sun, *J. Agric. Food Chem.* **2008**, *56*, 2388.

57 P. Hara, U. Hanefeld, L. T. Kanerv, *Green Chem.* **2009**, *11*, 250.

58 Y. Abe, K. Kude, S. Hayase, M. Kawatsura, K. Tsunashima, T. Itoh, *J. Mol. Catal., B Enzym.* **2008**, *51*, 81.

59 (a) Z. Guo, D. Kahveci, B. Özçelik, X. B. Xu, *New Biotechnol.* **2009**, *26*, 37; (b) D. Kahveci, Z. Guo, B. Özçelik, X. Xu, *Food Chem.* **2010**, *119*, 880.

60 R. M. Lau, M. J. Sorgedrager, G. Carrea, F. van Rantwijk, F. Secundo, R. A. Sheldon, *Green Chem.* **2004**, *6*, 483.

61 (a) M. B. Turner, S. K. Spear, J. G. Huddleston, J. D. Holbrey, R. D. Rogers, *Green Chem.* **2003**, *5*, 443; (b) T. Maruyama, H. Yamamura, T. Kotani, N. Kamiya, M. Goto, *Org. Biomol. Chem.* **2004**, *2*, 1239; (c) K. Nakashima, T. Maruyama, N. Kamiya, M. Goto, *Chem. Commun.* **2005**, 4297; (d) K. Nakashima, T. Maruyama, N. Kamiya, M. Goto, *Org. Biomol. Chem.* **2006**, *4*, 3462.

62 (a) M. B. Turner, J. D. Holbrey, S. K. Spear, M. L. Pusey, R. D. Rogers, *ACS Symp. Ser.* **2005**, *902*, 233; (b) J. D. Holbrey, M. B. Turner, W. M. Reichert, R. D. Rogers, *Green Chem.* **2003**, *5*, 731.

63 (a) H. Zhao, G. A. Baker, Z. Song, O. Olubajo, T. Crittle, D. Peters, *Green Chem.* **2008**, *10*, 696; (b) H. Zhao, C. L. Jones, J. V. Cowins, *Green Chem.* **2009**, *11*, 1128; (c) S. H. Lee, T. V. Doherty, R. J. Linhardt, J. S. Dordick, *Appl. Biochem. Biotechnol.* **2009**, *102*, 1368.

64 F. Falcioni, H. R. Housden, Z. Ling, S. Shimizu, A. J. Walker, N. C. Bruce, *Chem. Commun.* **2010**, *46*, 749.

65 T. Itoh, E. Akasaki, Y. Nishimura, *Chem. Lett.* **2002**, *31*, 154.

66 B. Smitha, D. Suhanya, S. Sridhar, M. J. Ramakrishna, *J. Memb. Sci.* **2004**, *241*, 1.

67 (a) E. Feher, V. Illeova, I. Kelemen-Horvath, K. Belafi-Bako, M. Polakovic, L. Gubicza, *J. Mol. Catal., B Enzym.* **2008**, *50*, 28; (b) L. Gubicza, K. Belafi-Bako, E. Feher, T. Frater,

Green Chem. **2008**, *10*, 1284; (c) E. Feher, B. Major, K. Belafi-Bako, L. Gubicza, *Desalination* **2009**, *241*, 8.

68 P. Izak, S. Hovorka, A. Randova, L. Bartovska, C. A. M. Afonso, J. G. Crespo, *Planta Med.* **2009**, *75*, 1023.

69 (a) N. M. T. Lourenço, C. A. M. Afonso, *Angew. Chem. Int. Ed. Engl.* **2007**, *46*, 8178; (b) P. U. Naik, S. J. Nara, J. R. Harjani, M. M. Salunkhe, *J. Mol. Catal., B Enzym.* **2006**, *44*, 93.

70 (a) M. A. McHugh, V. J. Krukonis, *Supercritical Fluid Extraction Principles and Practice*, 2nd ed., Butterworth–Heinemann, London, **1994**; (b) M. Mukhopadhyay, *Natural Extracts Using Supercritical Carbon Dioxide*, CRC Press, London, **2000**.

71 (a) J. L. Kendall, D. A. Canelas, J. L. Young, J. M. De Simone, *Chem. Rev.* **1999**, *99*, 543; (b) F. Cansell, C. Aymonier, *J. Supercrit. Fluids* **2009**, *47*, 508.

72 (a) A. Baiker, *Chem. Rev.* **1999**, *99*, 453; (b) R. Ciriminna, M. L. Carraro, S. Campestrini, M. Pagliaro, *Adv. Synth. Catal.* **2008**, *350*, 221.

73 R. A. Bourne, J. G. Stevens, J. Ke, M. Poliakoff, *Chem. Commun.* **2007**, 4632.

74 (a) S. Pereda, E. A. Brignole, S. B. Bottini, *J. Supercrit. Fluids* **2009**, *47*, 336; (b) M. Nunes da Pontes, *J. Supercrit. Fluids* **2009**, *47*, 344.

75 (a) G. Brunner, ed., *Supercritical Fluids as Solvents and Reaction Media*, Elsevier BV, Amsterdam, **2004**; (b) J. M. DeSimone, W. Tumas, eds., *Green Chemistry Using Liquid and Supercritical Carbon Dioxide*, Oxford University Press, Oxford, **2003**; (c) P. G. Jessop, T. Ikariya, R. Noyori, *Chem. Rev.* **1999**, *99*, 475; (d) E. J. Beckman, *J. Supercrit. Fluids* **2004**, *28*, 121; (e) C. Rayner, *Org. Proc. Res. Dev.* **2007**, *11*, 121; (f) A. Kruse, H. Vogel, *Chem. Eng. Technol.* **2008**, *31*, 23.

76 (a) S. Damar, M. O. Balaban, *J. Food Sci.* **2006**, *71*, R1; (b) L. García-Gonzalez, A. H. Geeraerd, S. Spilimbergo, K. Elst, L. Van Ginneken, J. Debevere, J. F. van Impe, F. Devlieghere, *Int. J. Food Microbiol.* **2007**, *117*, 1.

77 For review on enzyme catalysis in SCFs, see: (a) A. J. Mesiano, E. J. Beckman, A. J. Russel, *Chem. Rev.* **1999**, *99*, 623; (b) H. R. Hobbs, N. R. Thomas, *Chem. Rev.* **2007**, *107*, 2786; (c) Z. Knez, *J. Supercrit. Fluids* **2009**, *47*, 357; (d) M. J. Eisenmenger, J. I. Reyes de Corcuera, *Enzyme Microb. Technol.* **2010**, *45*, 331.

78 S. K. Karmee, L. Casiraghi, L. Greiner, *Biotechnol. J.* **2008**, *3*, 104.

79 (a) S. Kamat, J. Barrera, E. J. Beckman, A. J. Russell, *Biotechnol. Bioeng.* **1992**, *40*, 158; (b) M. Habulin, Z. Knez, *J. Chem. Technol. Biotechnol.* **2001**, *76*, 1260.

80 S. Kamat, G. Critchley, E. J. Beckman, A. J. Russell, *Biotechnol. Bioeng.* **1992**, *46*, 606.

81 A. T. Fricks, E. G. Oestreicher, L. C. Filho, A. C. Feihrmann, Y. Cordeiro, C. Dariva, O. A. C. Antunes, *J. Supercrit. Fluids* **2009**, *50*, 162.

82 (a) A. G. Arreola, M. O. Balaban, M. R. Marshal, A. J. Peplow, C. I. Wei, J. A. Cornell, *J. Food Sci.* **1991**, *56*, 1030; (b) T. Yoshimura, M. Furutera, M. Shimoda, H. Ishikawa, M. Miyake, K. Matsumoto, Y. Osajima, I. Hayakawa, *J. Food Sci.* **2002**, *67*, 3227; (c) D. del Pozo, M. O. Balaban, S. T. Talcott, *Food Res. Intern.* **2007**, *40*, 894.

83 Some examples of lipase-catalyzed ester synthesis in $scCO_2$ are: (a) W. Chulalaksananukul, J. S. Condoret, D. Combes, *Enzyme Microb. Technol.* **1993**, *15*, 691; (b) A. Marty, D. Combes, J. S. Condoret, *Biotechnol. Bioeng.* **1994**, *43*, 497; (c) P. Lozano, G. Villora, D. Gomez, A. B. Gayo, J. A. Sanchez-Conesa, M. Rubio, J. L. Iborra, *J. Supercrit. Fluids* **2004**, *29*, 121; (d) M. D. Romero, L. Calvo, C. Alba, A. Daneshfar, H. S. Ghaziaskar, *J. Supercrit. Fluids* **2005**, *33*, 77; (e) I. H. Kim, S. N. Ko, S. M. Lee, S. H. Chung, H. Kim,

K. T. Lee, T. Y. Ha, *J. Am. Oil Chem. Soc.* **2004**, *81*, 537; (f) D. Oliveira, J. V. Oliveira, *J. Supercrit. Fluids* **2001**, *19*, 141; (g) Y. Ikushima, N. Saito, T. Yokoyama, K. Hatakeda, S. Ito, M. Arai, H. W. Blanch, *Chem. Lett.* **1993**, *22*, 109.

84 (a) A. Capewell, V. Wendel, U. Bornscheuer, H. H. Meyer, T. Scheper, *Enzyme Microb. Technol.* **1996**, *19*, 181; (b) T. Matsuda, K. Watanabe, T. Harada, K. Nakamura, Y. Arita, Y. Misumi, S. Ichikawa, T. Ikariya, *Chem. Commun.* **2004**, 2286; (c) E. Celia, E. Cernia, C. Palocci, S. Soro, T. Turchet, *J. Supercrit. Fluids* **2005**, *33*, 193; (d) J. F. Martins, T. Correa de Sampaio, I. Borges de Carvalho, S. Barreiros, *Biotechnol. Bioeng.* **1994**, *44*, 119; (e) M. Rantakyla, O. Aaltonen, *Biotechnol. Lett.* **1994**, *16*, 825.

85 (a) K. Rezaei, F. Temelli, E. Jenab, *Biotechnol. Adv.* **2007**, *25*, 272; (b) K. Rezaei, E. Jenab, F. Temelli, *Crit. Rev. Biotechnol.* **2008**, *27*, 183.

86 (a) P. Lozano, A. Avellaneda, R. Pascual, J. L. Iborra, *Biotechnol. Lett.* **1996**, *18*, 1345; (b) A. Giessauf, W. Magor, D. J. Steinberger, R. Marr, *Enzyme Microb. Technol.* **1999**, *24*, 577.

87 (a) T. Matsuda, T. Harada, K. Nakamura, *Curr. Org. Chem.* **2005**, *9*, 299; (b) T. Matsuda, T. Harada, K. Nakamura, T. Ikariya, *Tetrahedron Asymmetry* **2005**, *16*, 909.

88 (a) S. V. Kamat, E. J. Beckman, A. J. Russell, *Crit. Rev. Biotechnol.* **1995**, *15*, 41; (b) N. Fontes, M. C. Almeida, C. Peres, S. Garcia, J. Grave, R. M. Aires-Barros, C. M. Soares, J. M. S. Cabral, C. D. Maycock, S. Barreiros, *Ind. Eng. Chem. Res.* **1998**, *37*, 3189.

89 S. V. Luis, E. García-Verdugo, *Chemical Reactions and Processes under Flow Conditions*, RSC Green Chemistry Series, The Royal Society of Chemistry, Cambridge, **2010**.

90 P. Licence, J. Ke, M. Sokolova, S. K. Ross, M. Poliakoff, *Green Chem.* **2003**, *5*, 99.

91 N. N. Rao, S. Lütz, K. Würges, D. Minör, *Org. Process Res. Dev.* **2009**, *13*, 607.

92 W. Keim, *Green Chem.* **2003**, *5*, 105.

93 S. Jeong, B. Y. Hwang, J. Kim, B. G. Kim, *J. Mol. Catal., B Enzym.* **2000**, *10*, 597.

94 C. G. Laudani, M. Habulin, Z. Knez, G. D. Porta, E. Reverchon, *J. Supercrit. Fluids* **2007**, *41*, 74.

95 T. Matsuda, K. Watanabe, T. Kamitanaka, T. Harada, K. Nakamura, *Chem. Commun.* **2003**, 119.

96 (a) M. Habulin, M. Primozic, Z. Knez, *Ind. Eng. Chem. Res.* **2005**, *44*, 9619; (b) M. Primozic, M. Paljevac, M. Habulin, Z. Knez, *Desalination* **2009**, *241*, 14.

97 V. A. Soloshonok, K. Mikami, T. Yamazaki, J. T. Welch, J. Honek, *Current Fluoroorganic Chemistry. New Synthetic Directions, Technologies, Materials and Biological Applications*, Oxford University Press/American Chemical Society, Washington, DC, **2007**.

98 (a) I. T. Horváth, *Acc. Chem. Res.* **1998**, *31*, 641; (b) E. de Wolf, G. van Koten, B. J. Deelman, *Chem. Soc. Rev.* **1999**, *28*, 37; (c) L. P. Barthel Rosa, J. A. Gladysz, *Coord. Chem. Rev.* **1999**, *190–192*, 587; (d) G. Pozzi, I. Shepperson, *Coord. Chem. Rev.* **2003**, *242*, 115.

99 I. T. Horváth, in *Multiphase Homogeneous Catalysis*, Vol. 1, (Eds. B. Cornils, W. A. Herrmann, I. T. Horvath, W. Leitner, S. Mecking, H. Olivier-Bourbigou, D. Vogt), Wiley-VCH Verlag GmbH & Co. KGaA, Weinheim, Germany, **2005**, pp. 339–403.

100 (a) J. A. Gladysz, D. P. Curran, *Tetrahedron* **2002**, *58*, 3823; (b) W. Zhang, *QSAR Comb. Sci.* **2006**, *25*, 679.

101 (a) D. P. Curran, *Angew. Chem. Int. Ed. Engl.* **1998**, *37*, 1174; (b) W. Zhang, *Tetrahedron* **2003**, *59*, 4475; (c) W. Zhang, *Chem. Rev.* **2004**, *104*, 2531–2556; (d) D. P. Curran, *Aldrichmica Acta* **2006**, *39*, 3.

102 W. Zhang, C. Cai, *Chem. Commun.* **2008**, 5686.

103 (a) Z. Y. Luo, S. M. Swaleh, F. Theil, D. P. Curran, *Org. Lett.* **2002**, *4*, 2585; (b) B. Hungerhoff, H. Sonnenschein, F. Theil, *J. Org. Chem.* **2002**, *67*, 1781; (c) E. L. Teo, G. K. Chuah, A. R. J. Huguet, S. Jaenicke, G. Pande, Y. Z. Zhu, *Catal. Today* **2004**, *97*, 263.

104 P. Beier, D. O. Hagan, *Chem. Commun.* **2002**, 1680.

105 (a) H. R. Hobbs, H. M. Kirke, M. Poliakoff, N. R. Thomas, *Angew. Chem. Int. Ed. Engl.* **2007**, *46*, 7860; (b) K. Benaissi, M. Poliakoff, N. R. Thomas, *Green Chem.* **2010**, *12*, 54; (c) S. S. Adkins, H. R. Hobbs, K. Benaissi, K. P. Johnston, M. Poliakoff, N. R. Thomas, *J. Chem. Phys. B* **2008**, *12*, 4760; (d) S. Shipovskov, *Biotechnol. Prog.* **2008**, *24*, 1262.

106 (a) T. A. Hoefling, R. M. Enick, E. J. Beckman, *J. Phys. Chem.* **1991**, *95*, 7127; (b) G. J. McFann, K. P. Johnston, S. M. Howdle, *AlChEJ* **1994**, *40*, 54.

107 J. D. Holmes, D. C. Steytler, G. D. Rees, B. H. Robinson, *Langmuir* **1998**, *14*, 6371.

108 J. D. Holmes, K. J. Ziegler, M. Audriani, C. T. J. Lee, P. A. Bhargava, D. C. Steytler, K. P. Johnston, *J. Chem. Phys.* **1999**, *103*, 5703.

109 C. Roosen, M. Ansorge-Schumacher, T. Mang, W. Leitner, L. Greiner, *Green Chem.* **2007**, *9*, 455.

110 T. Harada, Y. Kubota, T. Kamitanaka, K. Nakamura, T. Matsuda, *Tetrahedron Lett.* **2009**, *50*, 4934.

111 S. K. Karmee, C. Roosen, C. Kohlmann, S. Lütz, L. Greiner, W. Leitner, *Green Chem.* **2009**, *11*, 1052.

112 (a) J. M. Broering, E. M. Hill, J. P. Hallett, C. L. Liotta, C. A. Eckert, A. S. Bommarius, *Angew. Chem. Int. Ed. Engl.* **2006**, *45*, 4670; (b) E. M. Hill, J. M. Broering, J. P. Hallett, A. S. Bommarius, C. L. Liotta, C. A. Eckert, *Green Chem.* **2007**, *9*, 888.

113 (a) G. J. Lye, J. M. Woodley, *Trends Biotechnol.* **1999**, *17*, 395; (b) D. Stark, U. Von Stockar, In situ product removal (ISPR) in whole cell biotechnology during the last twenty years, in *Advances in Biochemical Engineering/Biotechnology*, Vol. 80 (Ed. T. Scheper), Springer, Berlin, Heidelberg, **2003**, pp. 149–175.

114 M. Eckstein, M. Villela, A. Liese, U. Kragl, *Chem. Commun.* **2004**, 1084.

115 (a) H. Pfruender, R. Jones, D. Weuster-Botz, *J. Biotechnol.* **2006**, *124*, 182; (b) D. Weuster-Botz, *Chem. Rec.* **2007**, *7*, 334.

116 (a) H. Pfruender, M. Amidjojo, U. Kragl, D. Weuster-Botz, *Angew. Chem. Int. Ed. Engl.* **2004**, *43*, 4529; (b) W. Y. Lou, M. H. Zong, T. J. Smith, *Green Chem.* **2006**, *8*, 147; (c) R. J. Cornmell, C. L. Winder, S. Schuler, R. Goodacre, G. Stephens, *Green Chem.* **2008**, *10*, 685; (d) W. Y. Lou, L. Chen, B. B. Zhang, T. J. Smith, M. H. Zong, *BMC Biotechnol.* **2009**, *9*, 90.

117 S. Bräutigam, D. Dennewald, M. Schürmann, J. Lutje-Spelberg, W. R. Pitner, D. Weuster-Botz, *Enzyme Microb. Technol.* **2009**, *45*, 310.

118 E. Miyako, T. Maruyama, N. Kamiya, M. Goto, *Chem. Commun.* **2003**, 2926.

119 (a) M. Moniruzzaman, N. Kamiya, K. Nakashima, M. Goto, *Green Chem.* **2008**, *10*, 497; (b) M. Moniruzzaman, N. Kamiya, M. Goto, *Langmuir* **2009**, *25*, 977.

120 I. V. Pavlidis, D. Gournis, G. K. Papadopoulos, H. Stamatis, *J. Mol. Catal., B Enzym.* **2009**, *60*, 50.

121 L. A. Blanchard, D. Hancu, E. J. Beckman, J. F. Brennecke, *Nature* **1999**, *399*, 28.

122 C. Cadena, J. L. Anthony, J. K. Shah, T. I. Morrow, J. F. Brennecke, E. J. Maginn, *J. Am. Chem. Soc.* **2004**, *126*, 5300.

123 M. Arai, S. Fujita, M. Shirai, *J. Supercrit. Fluids* **2009**, *47*, 351.

124 W. Leitner, M. Hölscher, *Topics Organometallic Chem. vol. 23, Regulated Systems for Multiphase Catalysis*, Springer, Berlin, Heidelberg, **2008**.

125 (a) P. Lozano, T. De Diego, D. Carrié, M. Vaultier, J. L. Iborra, *Chem. Commun.* **2002**, 692; (b) M. T. Reetz, W. Wiesenhofer, G. Francio, W. Leitner, *Chem. Commun.* **2002**, 992.

126 (a) P. Lozano, T. De Diego, J. L. Iborra, *Chem. Today* **2007**, *25*, 76; (b) S. Keskin, D. Kayrak-Talay, U. Akman, O. Hortaçsu, *J. Supercrit. Fluids* **2007**, *43*, 150; (c) M. Roth, *J. Chromatogr. A* **2009**, *1216*, 1861.

127 L. A. Blanchard, Z. Y. Gu, J. F. Brennecke, *J. Phys. Chem. B* **2001**, *105*, 2437.

128 S. N. V. K. Aki, B. R. Mellein, E. M. Saurer, J. F. Brennecke, *J. Phys. Chem. B* **2004**, *108*, 20355.

129 A. Ahosseini, E. Ortega, B. Sensenich, A. M. Scurto, *Fluid Phase Equilib.* **2009**, *286*, 72.

130 W. Z. Wu, W. J. Li, B. X. Han, T. Jiang, D. Shen, Z. F. Zhang, D. H. Sun, B. Wang, *J. Chem. Eng. Data* **2004**, *49*, 1597.

131 L. A. Blanchard, J. F. Brennecke, *Ind. Eng. Chem. Res.* **2001**, *40*, 287.

132 (a) A. M. Scurto, S. N. V. K. Aki, J. F. Brennecke, *J. Am. Chem. Soc.* **2002**, *124*, 10276; (b) A. M. Scurto, S. N. V. K. Aki, J. F. Brennecke, *Chem. Commun.* **2003**, 572; (c) S. N. V. K. Aki, A. M. Scurto, J. F. Brennecke, *Ind. Eng. Chem. Res.* **2006**, *45*, 5574.

133 (a) D. Fu, X. Sun, Y. Qiu, X. Jiang, S. Zhao, *Fluid Phase Equilib.* **2007**, *251*, 114; (b) E. Kühne, S. Santarossa, E. Perez, G. J. Witkamp, C. J. Peters, *J. Supercrit. Fluids.* **2008**, *46*, 93; (c) E. Kühne, S. Santarossa, G. J. Witkamp, C. J. Peters, *Green Chem.* **2008**, *10*, 762.

134 V. Najdanovic-Visak, A. Serbanovic, J. M. S. S. Esperança, H. J. R. Guedes, L. P. N. Rebelo, M. Nunes da Ponte, *ChemPhysChem* **2003**, *4*, 520.

135 (a) A. M. Scurto, W. Leitner, *Chem. Commun.* **2006**, 3681; (b) A. M. Scurto, E. Newton, R. R. Weikel, L. Draucker, J. Hallett, C. L. Liotta, W. Leitner, C. A. Eckert, *Ind. Eng. Chem. Res.* **2008**, *47*, 493.

136 P. Lozano, T. De Diego, D. Carrié, M. Vaultier, J. L. Iborra, *Biotechnol. Prog.* **2003**, *19*, 380.

137 P. Lozano, T. De Diego, S. Gmouh, M. Vaultier, J. L. Iborra, *Biotechnol. Prog.* **2004**, *20*, 661.

138 M. T. Reetz, W. Wiesenhofer, G. Francio, W. Leitner, *Adv. Synth. Catal.* **2003**, *345*, 1221.

139 O. Miyawaki, M. Tatsuno, *J. Biosci. Bioeng.* **2008**, *105*, 61.

140 R. Bogel-Lukasik, V. Najdanovic-Visak, S. Barreiros, M. Nunes da Ponte, *Ind. Eng. Chem. Res.* **2008**, *47*, 4473.

141 M. D. Bermejo, A. J. Kotlewska, L. J. Florusse, M. J. Cocero, F. van Rantwijk, C. J. Peters, *Green Chem.* **2008**, *10*, 1049.

142 S. Garcia, N. M. T. Lourenço, D. Lousa, A. F. Sequeira, P. Mimoso, J. M. S. Cabral, C. A. M. Afonso, S. Barreiros, *Green Chem.* **2004**, *6*, 466–470.

143 E. Garcia-Junceda, *Multi-Steps Enzyme Catalysis: Biotransformation and Chemoenzymatic Synthesis*, Wiley-VCH, Weinheim, **2008**.

144 J. S. Torrecilla, J. Garcia, E. Rojo, F. Rodriguez, *J. Hazard. Mater.* **2009**, *164*, 182.

145 K. Benaissi, M. Poliakoff, N. R. Thomas, *Green Chem.* **2009**, *11*, 617.

146 (a) P. Lozano, T. De Diego, M. Larnicol, M. Vaultier, J. L. Iborra, *Biotechnol. Lett.* **2006**, *28*, 1559; (b) P. Lozano, T. De Diego, C. Mira, K. Montague, M. Vaultier, J. L. Iborra, *Green Chem.* **2009**, *11*, 538.

147 (a) M. H. Valkenberg, C. de Castro, W. H. Holderich, *Green Chem.* **2002**, *4*, 88; (b) C. P. Mehnert, *Chem. Eur. J.* **2004**, *11*, 50; (c) A. Riisager, R. Fehrmann, M. Haumann, P. Wasserscheid, *Top. Catal.* **2006**, *40*, 91; (d) Y. Gua, G. Li, *Adv. Synth. Catal.* **2009**, *351*, 817.

148 (a) D. W. Kim, D. Y. Chi, *Angew. Chem. Int. Ed. Engl.* **2004**, *43*, 483; (b) D. W. Kim, D. J. Hong, K. S. Jang, D. Y. Chi, *Adv. Synth. Catal.* **2006**, *348*, 1719; (c) D. W. Kim, H. J. Jeong, S. T. Lim, M. H. Sohn, D. Y. Chi, *Tetrahedron* **2008**, *64*, 4209.

149 M. I. Burguete, F. Galindo, E. Garcia-Verdugo, N. Karbass, S. V. Luis, *Chem. Commun.* **2007**, 3086.

150 (a) B. Altava, M. I. Burguete, E. Garcia-Verdugo, N. Karbass, S. V. Luis, A. Puzary, V. Sans, *Tetrahedron Lett.* **2006**, *47*, 2311; (b) N. Karbass, V. Sans, E. Garcia-Verdugo, M. I. Burguete, S. V. Luis, *Chem. Commun.* **2006**, 3095; (c) M. I. Burguete, E. García-Verdugo, I. Garcia-Villar, F. Gelat, P. Licence, S. V. Luis, V. Sans, *J. Catal.* **2010**, *269*, 150.

151 M. I. Burguete, H. Erythropel, E. Garcia-Verdugo, S. V. Luis, V. Sans, *Green Chem.* **2008**, *10*, 401.

152 Y. Y. Jiang, C. Guo, H. S. Xia, I. Mahmood, C. Z. Liu, H. Z. Liu, *J. Mol. Catal., B Enzym.* **2009**, *58*, 103.

153 K. Nakashima, N. Kamiya, D. Koda, T. Maruyamac, M. Goto, *Org. Biomol. Chem.* **2009**, *7*, 2353.

154 P. Lozano, E. Garcia-Verdugo, R. Piamtongkam, N. Karbass, T. De Diego, M. I. Burguete, S. V. Luis, J. L. Iborra, *Adv. Synth. Catal.* **2007**, *349*, 1077.

5

IONIC LIQUIDS AS (CO-)SOLVENTS FOR HYDROLYTIC ENZYMES

Hua Zhao

NOMENCLATURE OF ILs

Cations

$[AdMIM]^+$	1-allyl-2,3-dimethylimidazolium
$[aliq]^+$	trioctylmethylammonium (Aliquat 336®)
$[AMIM]^+$	1-allyl-3-methylimidazolium
$[BDMIM]^+$	1-butyl-2,3-dimethylimidazolium
$[BMIM]^+$	1-butyl-3-methylimidazolium
$[BMPy]^+$	3-methyl-*N*-butylpyridinium
$[BMPyrr]^+$	1-butyl-1-methylpyrrolidinium
$[btma]^+$	butyltrimethylammonium
$[C_2OHmim]^+$	1-(2-hydroxyethyl)-3-methylimidazolium
$[C_5O_2mim]^+$	1-(2-(2-methoxyethoxy)-ethyl)-3-methylimidazolium
$[C_5MIM]^+$	1-methyl-3-pentylimidazolium
$[C_7MIM]^+$	1-heptyl-3-methylimidazolium

Ionic Liquids in Biotransformations and Organocatalysis: Solvents and Beyond, First Edition. Edited by Pablo Domínguez de María.

$[CPMA]^+$	cocosalkyl pentaethoxy methylammonium
$[EMIM]^+$	1-ethyl-3-methylimidazolium
$[EMMIM]^+$	1-ethyl-2,3-dimethylimidazolium
$[EtNH_3]^+$	ethylammonium
$[(EtOH)NH_3]^+$	ethanolammonium
$[HMIM]^+$	1-hexyl-3-methylimidazolium
$[HMMIM]^+$	1-hexyl-2,3-dimethylimidazolium
$[HOPMIm]^+$	1-(1-hydroxypropyl)-3-methylimidazolium
$[MMEP]^+$	1-methyl-1-(2-methoxyethyl)pyrrolidinium
$[(MeOEt)NH_3]^+$	2-methoxyethylammoniun
$[MoeMIM]^+$	1-methoxyethyl-3-methylimidazolium
$[MomMIM]^+$	1-methoxymethyl-3-methylimidazolium
$[MTOA]^+$	methyl trioctylammonium
$[OMIM]^+$	1-methyl-3-octylimidazolium
$[ONIM]^+$	1-octyl-3-nonylimidazolium
$[PMMIM]^+$	2,3-dimethyl-1-propylimidazolium
$[PPMIM]^+$	1-(3′-phenylpropyl)-3-methylimidazolium
$[PrNH_3]^+$	propylammonium
$[TOMA]^+$	trioctylmethylammonium

Anions

$[beti]^-$	bis(perfluoroethanesulfonyl)imide
$[BF_4]^-$	tetrafluoroborate
$[dca]^-$	dicyanamide
$[EtSO_4]^-$	ethyl sulfate
$[HCOO]^-$	formate
$[MeSO_4]^-$	methyl sulfate
$[OAc]^-$	acetate
$[OctSO_4]^-$	octyl sulfate
$[OTf]^-$	triflate (or trifluoromethanesulfonate)
$[OTs]^-$	tosylate (or *p*-toluenesulfonate)
$[PF_6]^-$	hexafluorophosphate
$[Tf_2N]^-$	bis(trifluoromethane)sulfonimide

5.1 INTRODUCTION

5.1.1 Type of Hydrolases

In biochemistry, a hydrolase is an enzyme that promotes the cleavage of a chemical bond via hydrolysis. Hydrolases are classified as EC 3 in the EC number classification of enzymes. Furthermore, hydrolases are divided into several subclasses, based on the bonds they hydrolyze.

EC 3.1: ester bonds (such as **esterase**, nuclease, phosphodiesterase, **lipase**, phosphatase)

EC 3.2: sugars (such as DNA glycosylases, **lysozyme**, **β-glycosidase**, **cellulase**)

EC 3.3: ether bonds (such as **epoxide hydrolase**)

EC 3.4: peptide bonds (such as **proteases**/peptidases, **thermolysin**)

EC 3.5: carbon-nitrogen bonds, other than peptide bonds (such as **penicillin amidase**)

EC 3.6: acid anhydrides (such as acid anhydride hydrolases, including helicases and GTPase)

EC 3.7: carbon-carbon bonds

EC 3.8: halide bonds

EC 3.9: phosphorus–nitrogen bonds

EC 3.10: sulfur–nitrogen bonds

EC 3.11: carbon–phosphorus bonds

EC 3.12: sulfur–sulfur bonds

EC 3.13: carbon–sulfur bonds

Examples of main hydrolases that have been studied in ionic liquids (ILs) are highlighted above in **bold**. The functionalities of these hydrolases are briefly discussed below:

- Esterases (EC 3.1.1), such as carboxylesterase (EC 3.1.1.1) and cutinase (EC 3.1.1.74), hydrolyze esters bonds into acids and alcohols.
- Lipases (EC 3.1.1), such as triacylglycerol lipases (EC 3.1.1.3), may be considered the subclass of esterases, which catalyze the hydrolysis of ester bonds of lipid substrates.
- Lysozyme (EC 3.2.1.17), also known as muramidase or *N*-acetylmuramide glycanhydrolase, is an enzyme occurring naturally in egg white, human tears, saliva, and other body fluids. This enzyme damages bacterial cell walls by facilitating the hydrolysis of β(1→4) linkages between *N*-acetylmuramic acid and *N*-acetyl-D-glucosamine residues in a peptidoglycan, and between *N*-acetyl-D-glucosamine residues in chitodextrins.
- β-Glycosidases (EC 3.2.1) catalyze the hydrolysis of the glycosidic linkage to produce two smaller sugar units.
- Cellulase is an enzyme complex and includes several components such as endo-β-glucanase (EC 3.2.1.4), exo-β-glucanase (EC 3.2.1.91), and β-glucosidase (cellobiase) (EC 3.2.1.21), which are involved in different stages of cellulose hydrolysis.

- Epoxide hydrolases (EC 3.3.2) convert epoxides to *trans*-dihydrodiols, which is an important function of detoxication during the drug metabolism.
- Serine proteinases such as subtilisin (EC 3.4.21.14) and α-chymotrypsin (EC 3.4.21.1), and cysteine proteinases such as papain (EC 3.4.22.2), catalyze the hydrolysis of peptide bonds in proteins.
- Thermolysin (EC 3.4.24.27) is a thermophilic neutral metalloproteinase enzyme produced by the gram-positive bacteria *Bacillus thermoproteolyticus*, and specifically hydrolyzes protein bonds on the *N*-terminal side of hydrophobic amino acid residues.
- Penicillin amidase (also called penicillin G acylase [PGA], EC 3.5.1.11) is the key enzyme in the synthesis of β-lactam penicillin antibiotics. PGA is capable of hydrolyzing the side chain of penicillin G and related β-lactam antibiotics, and the hydrolysis product (6-aminopenicillanic acid) is the intermediate for producing semisynthetic penicillins. On the other hand, PGA can also catalyze the condensation of an acyl compound with 6-aminopenicillanic acid to synthesize semisynthetic β-lactam antibiotics.

By nature, hydrolases break down large molecules into small molecules via the action of hydrolysis in the presence of water. However, under low-water environments, such a reaction equilibrium can be reversed to synthesis, which is the principle behind the enzymatic synthesis in nonaqueous media.[1–3] For this reason, the manipulation of water content in hydrolase-facilitated reactions allows one to determine the fate of substrates: hydrolysis or synthesis. However, these two opposite reactions often compete with each other at low water contents. The organic synthesis catalyzed by hydrolases in conventional organic solvents has been well documented in a recent book by Bornscheuer and Kazlauskas.[4] This chapter will discuss the use of ionic liquids (ILs) and their aqueous solutions for enzymatic reactions catalyzed by hydrolases (mainly lipases and proteases).

5.1.2 Properties and Applications of ILs

Ionic liquids consist of ions and remain liquid at temperatures lower than 100°C. The so-called room-temperature ionic liquids (RTILs) are most desirable as solvents for chemical reactions and other applications. Being different from conventional organic solvents, ILs have many favorable properties such as low vapor pressure, a wide liquid range, low flammability, high ionic conductivity, high thermal conductivity, high dissolution capability toward many substrates, high thermal and chemical stability, and a wide electrochemical potential window.[5] Because of these unique properties, ILs have been widely recognized as solvents or (co-)catalysts in a variety of applications including organic catalysis,[5–13] inorganic synthesis,[14] biocatalysis,[12,15–20] polymerization,[21,22] and engineering fluids.[23–25] Typical IL cations are nitrogen-containing (such as alkylammonium, *N,N′*-dialkylimidazolium, *N*-alkylpyridinium, and pyrrolidinium), or phosphorous-containing (such as alkylphosphonium). The common choices of anions include halides, BF_4^-, PF_6^-, $CH_3CO_2^-$, $CF_3CO_2^-$, NO_3^-, Tf_2N^-, $[(CF_3SO_2)_2N^-]$, $[RSO_4]^-$, and $[R_2PO_4]^-$. Some representative cations and anions are illustrated in Figure 5.1.

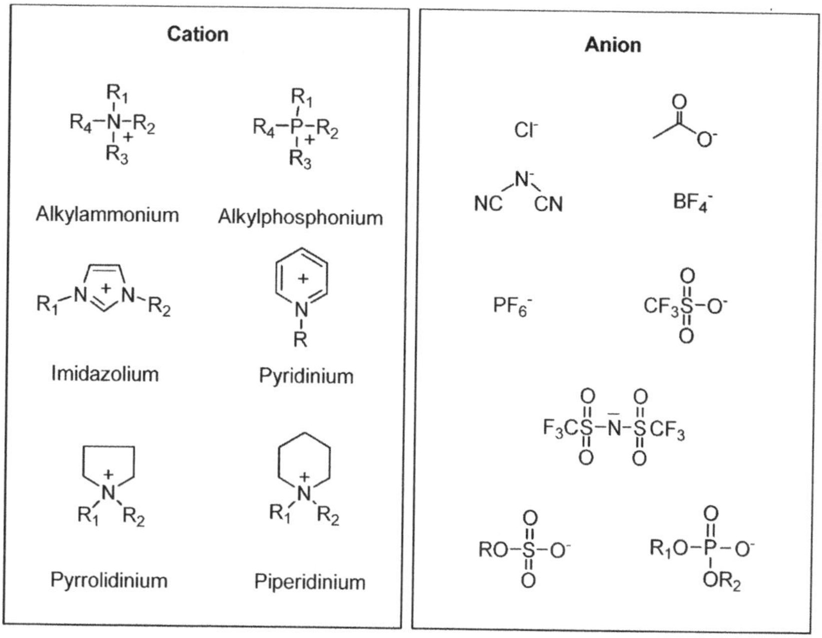

Figure 5.1. Structures of representative cations and anions in ILs.

More importantly, the physical properties of ILs (such as polarity, hydrophobicity, and hydrogen-bond basicity) can be finely tuned through the judicious selection of cations and anions. For example, ILs can be made to be water-miscible, partially miscible, or immiscible, can be made with different polarities, and can also be synthesized with a wide range of viscosities targeting individual applications.[5] All these adjustable properties are very important for enzyme stabilization and activation; therefore, numerous enzymatic reactions have been investigated in different types of ILs.[16,17,20,26]

This chapter focuses on the hydrolase-catalyzed synthetic and hydrolytic reactions in ILs or their aqueous solutions. The purpose of this chapter is to provide some empirical or theoretical guidelines for performing and improving the enzymatic reactions in IL solutions in the presence of hydrolases. The main contents of this chapter include factors influencing the hydrolase activity and stability in ILs and methods to improve the hydrolase performance in ILs (Section 5.2), (dynamic) kinetic resolutions in ILs (Section 5.3), esterifications of saccharides and cellulose derivatives in ILs (Section 5.4), glycosidase-catalyzed synthesis and hydrolysis (Section 5.5), and prospects (Section 5.6).

5.2 STATE-OF-THE-ART: LIPASES, ESTERASES, PROTEASES IN ILs AS (CO-)SOLVENTS

As hydrolase-catalyzed reactions in ILs have been reviewed by recent papers,[20,26,27] this section will not repeat each of these biocatalytic reactions. Instead, this section will focus on two major themes:* (1) How do physical properties of ILs (Section 5.2.1) and other factors (Section 5.2.2) influence the hydrolase activity and stability? (2) What methods (Section 5.2.3) are able to improve the hydrolase activity and stability in ILs?

5.2.1 Effect of Physical Properties of ILs on Hydrolase Activity and Stability

It is well known that the enzyme's performance in ILs is affected by common factors such as the water activity, pH, excipients, and impurities.[29] Since ILs have different physical properties from conventional organic solvents, many studies have realized that physical properties of ILs play critical roles in hydrolase's catalytic behaviors. The effects of these properties on the hydrolase's activity, stability and specificity are discussed below.

5.2.1.1 IL Polarity[19]**.** The common polarity parameters include dielectric constants (ε_r), Hildebrandt solubility (δ), dipole moments (μ), and the popular solvatochromic polarity scales (such as E_T^N and Kamlet-Taft scales). In particular, the E_T^N scale is a normalized polarity scale, which sets tetramethylsilane as 0.0 and water as 1.0.[30] The E_T^N values of many ILs have been measured and were recently evaluated (see Table 5.1).[31] Based on the solvatochromic studies, ILs were found to be moderately polar, being close to lower alcohols[32,33] or formamide.

TABLE 5.1. Polarity of ILs as Measured by the E_T^N Scale

Solvent	E_T^N (25°C)	Reference
Formamide	0.775	30
Methanol	0.762	30
Ethanol	0.654	30
1-propanol	0.617	30
[EMIM][Tf_2N]	0.676	Selected value by Reichardt[31]
[BMIM][Tf_2N]	0.642	41
[HMIM][Tf_2N]	0.654	Selected value by Reichardt[31]
[BMIM][PF_6]	0.667	41
[EMIM][BF_4]	0.710	35
[BMIM][BF_4]	0.673	41
[BMIM][OTf]	0.667	Selected value by Reichardt[31]
[BMIM][NO_3]	0.651	Selected value by Reichardt[31]
[BMIM][OAc]	0.571	Selected value by Reichardt[31]
[BMIM][CF_3COO]	0.630	Selected value by Reichardt[31]
[EMIM][dca]	0.648	Selected value by Reichardt[31]
[BMIM]Cl	0.614	Selected value by Reichardt[31]

* Parts of this section are also available in a review paper[28] published by John Wiley & Sons.

A common knowledge in enzymology suggests that high enzyme activities could be achieved in less polar, hydrophobic organic solvents, especially at high temperatures. On the other hand, Narayan and Klibanov[34] also indicated that the solvent polarity and water-miscibility could not be directly related to enzymatic activities of three lipases and one protease in organic solvents.

A number of studies have correlated the hydrolase activities in ILs with the IL polarity. Park and Kazlauskas[35] observed the lipase (from *Pseudomonas cepacia*) activity increasing with the IL polarity during the acetylation of racemic 1-phenylethanol with vinyl acetate (for example, the initial reaction rate in less polar [BMIM][PF_6] was three times slower than that in more polar [EMIM][BF_4]). In another study, lower synthetic activities of α-chymotrypsin were obtained in less polar ILs.[36] In the esterification of methyl-α-D-glucopyranoside with fatty acids catalyzed by Novozym® 435 (immobilized lipase B from *Candida antarctica*), Mutschler et al.[37] observed that the ester conversion increased with the IL polarity when ILs were employed as liquid film-coating (under solvent-free condition), but decreased with the IL polarity when ILs were used as solvents. However, the correlation between IL polarity and enzyme activity has not been established for other enzymatic reactions performed in ILs.[29,38–40] Based on the E_T^N polarity scale in Table 5.1, some enzyme-denaturing ILs (such as [EMIM][dca], [BMIM]Cl, and [BMIM][OAc]) have about the same polarities as those nondenaturing ILs based on Tf_2N^- and PF_6^-.

5.2.1.2 Hydrogen-bond (H-bond) Basicity and Nucleophilicity of Anions.

Hydrogen-bond basicity and nucleophilicity are two different concepts* but are often closely related. For molecules containing the same nucleophilic atoms of the same charge, the stronger base is usually the stronger nucleophile in aprotic solvents. Relying on the solvatochromic measurements, several studies have suggested the order of anion's basicity (nucleophilicity) as the following (in decreasing orders):

Basicity series #1:[41] OTf^- ($CF_3SO_3^-$) > Tf_2N^- > PF_6^-
Basicity series #2:[42] Cl^- > Br^- > SCN^- > OAc^- > I^- > NO_3^- > OTf^- > ClO_4^- > BF_4^-
Basicity series #3:[43] Cl^- > Br^- > OAc^- > OTf^- > ClO_4^- > BF_4^-
Basicity series #4:[44] Cl^- > Br^- > $CH_3OSO_3^-$ > SCN^- > BF_4^- ~ OTf^- > PF_6^-

Based on the above series and some discussions in the literature,[45,46] a summary of the basicity of selected anions is illustrated in Figure 5.2. These anions are divided into three categories (basic, neutral, and acidic), and some of them are ranked in the order of basicity. Basic anions include halides, acetate, dicyanamide (dca^-), lactate, and methyl sulfate; these anions are good H-bond donors and tend to form H-bonds with proteins, resulting in enzyme denaturation and/or inactivation. Neutral anions include those tending to form hydrophobic ILs (Tf_2N^- and PF_6^-) and others tending to form

* Basicity refers to the ability of a base to accept a proton (Brønsted–Lowry definition), and is a matter of equilibrium. Nucleophilicity of a Lewis base refers to the relative reaction rate of different nucleophilic reagents toward a common substrate, most usually involving the formation of a bond to carbon; nucleophilicity is a matter of kinetics (rate).

Basic anions	Neutral anions	Acidic anions
←	←	→
$Cl^- > Br^- > OAc^- >$	$OTf^- > Tf_2N^- > BF_4^- > PF_6^- >$	$H_2PO_4^-$, HSO_4^- (amphoteric)
dca⁻, lactate⁻, $MeSO_4^-$	SCN^-, NO_3^-, $CH_3SO_3^-$	

← **Increasing nucleophilicity**

Figure 5.2. Comparison of hydrogen-bond basicity of selected anions in ILs.

hydrophilic ILs (BF_4^-, OTf^-, SCN^-, NO_3^-, and $CH_3SO_3^-$). These anions have weak abilities in forming H-bonds; in another word, if enzymes are inactivated in ILs containing neutral anions, the H-bond basicity is unlikely the main reason. Acidic anions (such as amphoteric $H_2PO_4^-$ and HSO_4^-) are not common anions in ILs for biocatalysis. However, the Ohno group[47,48] found that choline dihydrogen phosphate (m.p. 119°C) containing 20% (wt) water could dissolve and stabilize cytochrome *c* (cyt *c*).*

Bernson and Lindgren[49] dissolved lithium salts LiX in poly(propylene glycol) (MW 3000) with hydroxyl end-groups. Using infrared (IR) spectroscopy, they observed that the shifts of –OH stretching band depended on the strength of H-bond formed between the –OH group and the anion, as well as the coordination of cations with the –OH group. The strength of anion coordination is further dependent on the H-bond basicity of the anion and is summarized from the IR band shifts as (in an increasing order),

$$PF_6^- < BF_4^- < ClO_4^- < OTf^- < I^- < Br^- < Cl^-$$

In general, this basicity series is consistent with the basicity order from solvatochromic measurements (Figure 5.2). From experimental data of IR and electrospray ionization mass spectrometry (ESI-MS). Dupont[50] suggested the strength of H-bond basicity in a similar increasing order of

$$BPh_4^- < PF_6^- < BF_4^- < CF_3COO^-$$

On the other hand, the *ionic association strength* of LiX salts has been investigated in a variety of aprotic solvents including glymes (see a short review in the Supporting Information of Henderson[51]). The approximate ionic association strength in aprotic solvents is listed below in an increasing order:[51,52]

$$beti^-, Tf_2N^- < PF_6^- < ClO_4^-, I < SCN^- < BF_4^- < CF_3SO_3^- < Br^- < NO_3^- < CF_3COO^- < Cl^-$$

* Although this small heme protein is not a hydrolase, the impact of ILs on this protein represents a general effect on other enzymes. Therefore, this chapter cited several studies on cyt *c*.

This order represents the strength of an anion in interacting with solvated cations through ionic attraction, or could be implied to represent the strength of interactions between the anions and charged regions of macromolecules (such as proteins). The exact mechanism of this ion–protein interaction is not well understood. However, this ionic association strength series resembles the anion's H-bond basicity order in Figure 5.2.

The following examples of enzymatic reactions in ILs demonstrate how the nucleophilicity and basicity of anions may be related to the hydrolase activity and stability.

The first group of examples focused on the discussion of the anion's nucleophilicity. Kaar et al.[38] observed that free *Candida rugosa* lipase (CRL) was only active in hydrophobic [BMIM][PF_6], but inactive in all hydrophilic ILs based on NO_3^-, OAc^-, and CF_3COO^- during the transesterification of methylmethacrylate with 2-ethyl-1-hexanol. They indicated that the latter three anions are more nucleophilic than PF_6^-, and thus could interact with the enzyme causing the protein conformation changes. In this example, the solvent hydrophobicity is another important factor in influencing the enzyme activity (see Section 5.2.1.6.). Hernández-Fernández et al.[53] reported that the stability of lipase B from *Candida antarctica* (CALB) in ILs was in the following order: [HMIM][PF_6] > [HMIM][Tf_2N] > [HMIM][BF_4], and [BMIM][PF_6] > [BMIM][dca], and the stability of PGA was in a similar order of [BMIM][Tf_2N] > [BMIM][PF_6] > [BMIM][BF_4]. They explained that the decreasing stability were in general consistent with the increasing order of nucleophilicity in Figure 5.2 (PF_6^- < BF_4^- < Tf_2N^- < dca^-), where the more nucleophilic anions tend to interact with the positively charged sites in enzymes and to modify the enzyme's conformation. On the other hand, they also pointed out that the enzyme stability was in agreement with the hydrophobicity of ILs: Both enzymes were more stable in hydrophobic ILs than in hydrophilic ones (see Section 5.2.1.6.). However, in another study, a contradictory result was reported. Irimescu and Kato[54] carried out the CALB-catalyzed enantioselective acylation of 1-phenylethylamine with 4-pentenoic acid and found that the reaction rates relied on the type of IL anions (reaction rates in a decreasing order of OTf^- > BF_4^- > PF_6^-, same cations). This example suggests higher anion nucleophilicity leading to higher enzymatic activity. In a second acylation reaction of 2-phenyl-1-propylamine with 4-pentenoic acid, however, Irimescu and Kato[54] observed that PF_6^- based ILs afforded fastest reaction rates, followed by OTf^- and BF_4^- based ILs. The rather confusing finding may be due to the reason that the enzymatic reaction is affected by multiple factors of ILs such as nucleophilicity, hydrophobicity, viscosity, and impurity. Lee et al.[55] measured the initial transesterification rates of three lipases (Novozym 435, *Rhizomucor miehei* lipase, and CRL) in different ILs under the same water activity (a_w), and observed the anion effect on the initial rates followed a decreasing order of Tf_2N^- > PF_6^- > OTf^- > SbF_6^- ~ BF_4^-. They explained that OTf^- and BF_4^- are more nucleophilic than PF_6^-. The second factor could be the IL hydrophobicity (see Section 5.2.1.6) because lipases seemed more active in hydrophobic ILs than in hydrophilic ones.

The second group of examples focused on the discussion of the anion's H-bond basicity. [BMIM][Cl] could effectively dissolve cellulose[56,57] because chloride ions (as H-acceptors) interact with the cellulose –OH group and break the H-bonding network of cellulose.[58] For the same H-bonding reason, this IL induced the inactivation of cellulase (from *Trichoderma reesei*).[59] Similarly, Lee et al.[60] observed a dramatic decrease of the lipase activity in [OMIM][Tf_2N] with the increasing addition of [OMIM][Cl]. Based on multiple salvation interactions, [BMIM][Cl] showed the largest H-bond basicity among ILs considered in a study by Anderson et al.,[61] and thus could dissolve complex polar molecules such as cyclodextrins and antibiotics.[62] Lou et al.[63] reported that Novozym 435 showed no ammonolysis activity toward (*R*,*S*)-*p*-hydroxyphenylglycine methyl ester in [BMIM][Br] and [BMIM][NO_3], implying the denaturing nature of these two ILs. Lau et al.[64] suggested that the low CALB activity in [BMIM][lactate] was caused by the secondary structure changes of the protein, which was further triggered by the H-bonding interaction between lactate anions and peptide chains. Dicyanamide (dca^-)-based ILs such as [BMIM][dca] are known able to dissolve carbohydrates[65,66]; however, [BMIM][dca] is an enzyme-denaturing IL[67–69] possibly due to the high H-bond basicity of the anion. Fujita et al.[48] detected low stabilities of cyt *c* in [BMIM][$MeSO_4$], [BMIM][lactate], and [BMIM][OAc] all containing 20 wt% water, implying the high H-bond basicity and enzyme-denaturing nature of $MeSO_4^-$, lactate, and OAc^-. Our group[70] also suggested both free and immobilized CALB in [EMIM][OTf] were about as active as in [BMIM][dca], which were less active than in hydrophobic ILs. Bermejo et al.[71] observed that free CALB lost 35% of its initial activity once being dissolved in [HOPMIm][NO_3], but maintained 80% of the remaining activity after 3 months of incubation in the IL. The CALB activity loss in [HOPMIm][NO_3] was primarily due to the denaturing effect of NO_3^- as discussed earlier. On the other hand, the less denaturing property of this IL (vs. [BMIM][NO_3]) may be explained by two reasons: (1) the $HOPMIm^+$ cation is larger than $BMIM^+$, and as a result the molar concentration of NO_3^- in [HOPMIm][NO_3] is lower than that in [BMIM][NO_3]; and (2) [HOPMIm][NO_3] contains a hydroxyl group, which may favorably interact with NO_3^- and thus reduce the interaction between NO_3^- and the lipase.

5.2.1.3 IL Network. ILs can form so-called organized nanostructures (hydrogen-bonded polymeric supramolecules, which are similar to water molecules) with polar and nonpolar regions in solid, liquid, and solution states, or even in the gas phase.[50,72] Dupont[50] suggested that the aqueous solution of free enzymes might be embedded in the IL network, which could protect the essential water of proteins and the solvophobic interactions that are critical for maintaining the native structure of proteins. In the case of imidazolium-based ILs, each cation coordinates with at least three anions, and in turn, each anion coordinates with three cations, forming a polymeric network. The inclusion of other molecules and macromolecules into this polymeric network induces the formation of polar and nonpolar regions.[50] When the enzyme-in-water droplets are dissolved (or dispersed) into the IL network (in polar regions), the enzyme's active conformation is maintained by the network (see Figure 5.3).[73] The embedding of enzyme molecules in such highly ordered supramolecular structures of ILs prevents the protein from thermal unfolding.[74]

However, since enzymes are not soluble in most common ILs, enzyme molecules (in particular, immobilized enzymes) are practically suspended in reaction media with low or little water (for example, CALB is still active in the absence of water[69,75]); as a

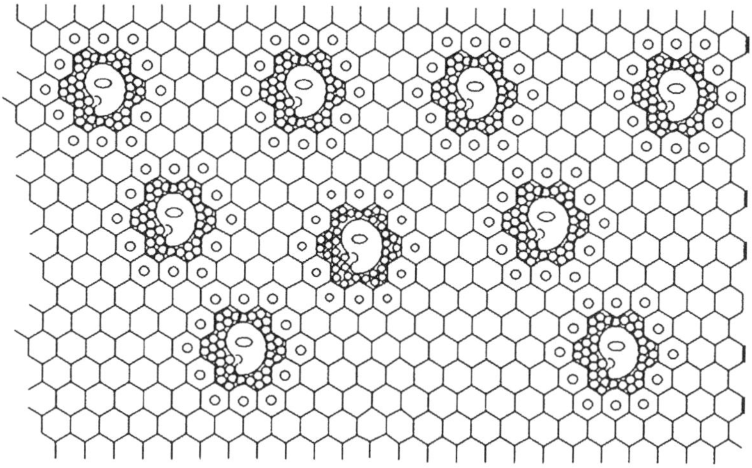

Figure 5.3. Enzymes with a small amount of water are firmly trapped in the network of ILs. Adapted from Fehér et al.,[73] reproduced with permission of Portland Press Limited.

result, the IL network theory is not always suitable for interpreting the enzyme activity and stability.

5.2.1.4 *Ion Kosmotropicity.* In aqueous solutions, hydrophilic ILs dissociate into individual ions. Therefore, the individual ion's effect on the enzyme behaviors in solutions takes a high priority. The realization of ion specificity began with Franz Hofmeister's observation of various ions displaying different abilities in precipitating the proteins (globulins from blood serum and hen's egg).[76,77] The order of these ions in salting out proteins is recognized as the "Hofmeister series" (Figure 5.4). To explain how and why ions influence the protein stability following the Hofmeister series, a number of theories have been proposed and are still under vigorous debates. Theses theories include salt-in and salt-out interactions,[78,79] water-structure changes (low/high density water) and protein preferential hydration,[80–86] hydrophobic interactions,[86–88] excluded volume,[89–91] preferential interactions,[92–94] electrostatic interactions,[95,96] and others. So far, there is no unified theory in interpreting the Hofmeister effect perhaps due to the complex nature of ion–protein interactions.

Nevertheless, the protein stability is often related to the hydration behaviors of ions.[78,97] Strongly hydrated ions (such as Mg^{2+}, Ca^{2+}, Li^+, CH_3COO^-, SO_4^{2-}, and HPO_4^{2-}) have strong interactions with water molecules and increase the structuring of water, resulting in a lower fluidity (or a higher viscosity) of the solution than that of pure water. Therefore, these ions are named "structure-makers" or "kosmotropes" (see Figure 5.4). On the other hand, some other ions are weakly hydrated in aqueous solutions, such as SCN^-, I^-, NO_3^-, BF_4^-, Cs^+, $(NH_2)_3C^+$ (guanidinium), and $(CH_3)_4N^+$ (tetramethylammonium). They have weak interactions with water and reduce the structuring of water, causing a higher fluidity of the solution. For this reason, this effect is known

← Protein stabilization Protein destabilization →

Anions (*kosmotropic*) PO_4^{3-} > SO_4^{2-} > $EtSO_4^-$ > OAc^- > $MeSO_4^-$ > Cl^- > Br^- > I^- > BF_4^- > PF_6^- (*chaotropic*)
B-coefficients: 0.495→0.206→0.265→0.246→0.188→-0.005→-0.033→-0.073→-0.093→-0.21

Cations: (*chaotropic*) $(CH_3)_4N^+$ > K^+ > Na^+ > Li^+ > Ca^{2+} > Mg^{2+} > Al^{3+} (*kosmotropic*)
B-coefficients: 0.123→0.009→0.085→0.146→0.284→0.385→0.744

Figure 5.4. The Hofmeister series as an order of the ion effect on protein stability.[82,393] (The viscosity *B*-coefficients in dm^3/mol at 25°C were taken from the Marcus collection,[103] except those of $EtSO_4^-$ and $MeSO_4^-$, which were from Tamaki et al.[125]; the positions of $EtSO_4^-$ and $MeSO_4^-$ are based on the consideration of *B*-coefficients, NMR *B′*-coefficients,[115] and enzyme stability studies[48,116,127,130]).

as "negative hydration,"[98,99] and these ions are often referred to as "structure breakers" or "chaotropes" (see Figure 5.4).

The capacity of an ion in strengthening the water structure, known as kosmotropicity (vs. chaotropicity), therefore, is directly related to the degree of ion hydration. As explained in our recent review,[100] the ion kosmotropicity can be quantified by different thermodynamic parameters including viscosity *B*-coefficients, structural entropies, structural volumes, structural heat capacities, nuclear magnetic resonance (NMR) *B′*-coefficients, ion mobility, and so on. These quantities allow us to understand the interactions involved in the ion hydration from various aspects, and possibly uncover the mechanism behind some phenomena and properties. Due to their wide availability, Jones-Dole viscosity *B*-coefficients are the most frequently used parameter for comparing the ion kosmotropicity. The *B*-coefficients are calculated from the Jones-Dole empirical equation (Eq. 5.1) of the relative viscosities of electrolyte solutions as functions of their concentrations,[101]

$$\eta / \eta_0 = 1 + Ac^{1/2} + Bc + Dc^2 \ldots \tag{5.1}$$

where η is the viscosity of the solution and η_0 is the viscosity of the solvent (both of them have the same unit, for example Pa s), while c is the molar concentration (mol/cm^3). The *A*-coefficients (also known as Falkenhagen coefficient[102]), representing the solute–solute or electrostatic interactions, can be calculated theoretically. However, *A*-values are usually small and negligible for nonelectrolytes[103]; therefore, they are often omitted in calculations. The *B*-coefficients represent the solute–solvent interactions (short-range dispersion forces), while *D*-coefficients reflect the solute–solute interactions as well as the solute–solvent interactions.[104] For most salts at low concentrations ([<0.5 M][103] or [<0.1 M for binary strong electrolytes][105]), the *D* or higher coefficients can be neglected, although they are necessary at higher concentrations.[103] Positive *B*-values typically suggest ions as kosmotropes since strongly hydrated ions exhibit a larger change in viscosity with concentration, while negative *B*-coefficients indicate chaotropes for weakly hydrated ions.[103] However, hydrophobic solutes tend to have unusually large *B*-coefficients due to the hydrophobic hydration.[100] For example, tetramethylammonium cation (Me_4N^+) has a positive *B*-value as high as 0.123,[103] but this ion is considered a structure breaker.[83,106–110] Some researchers[103,111–113] recommend the use of first derivatives of *B*-values over temperature because the sign of dB/dT might be more indicative for measuring the structure-making or breaking ability than the sign or quantity of *B*-coefficients. The negative sign of dB/dT means structure-making (kosmotropic), while the positive sign suggests structure-breaking (chaotropic).

In aqueous solutions of inorganic salts, many studies (see our recent review[114]) have suggested that the ion effect on the enzyme activity followed the ion kosmotropicity (Hofmeister series): Kosmotropic anions and chaotropic cations stabilize the enzyme, while chaotropic anions and kosmotropic cations destabilize it. A series of studies[40,100,115–121] by our laboratory have demonstrated that the same rule is loosely applicable to the enzyme activity in IL aqueous solutions. In our first study, the activities of Amano protease P6 (from *Aspergillus melleus*) in 0.7 M IL aqueous solutions were affected by anions in a decreased order of CH_3COO^-, CF_3COO^- > Cl^-, Br^- > OTs^- > BF_4^- (which is consistent with the decreasing order of anion's kosmotropicity), and influenced by cations in a decreasing order of $EMIM^+$, $BuPy^+$ > $BMIM^+$ > $EtPy^+$.[40] In a second study,[116] our group carried out the kinetic hydrolysis of enantiomeric phenylalanine methyl ester catalyzed by *Bacillus licheniformis* protease in aqueous solutions of several hydrophilic ILs (0.5 M). The protease enantioselectivity was in a decreasing order in the presence of these anions: PO_4^{3-} > citrate^{3-}, CH_3COO^-, $EtSO_4^-$, CF_3COO^- > Br^- > OTs^-, BF_4^- (decreasing kosmotropicity), and in the presence of these cations: $EMIM^+$ > $BMIM^+$ > $HMIM^+$ (decreasing chaotropicity). The overall IL kosmotropicity was quantified by the δ value (difference in viscosity *B*-coefficients of anion and cation). In general, a high enzyme enantioselectivity was observed in the solution of IL with a high δ value. After measuring the NMR B'-coefficients of a number of ions (see Figure 5.5, which is consistent with Figure 5.4), our group[115] further found a linear correlation between the enzyme enantioselectivity in aqueous solution and the δ' parameter (difference in NMR B'-coefficients of anion and cation) of ILs, concluding that high enzyme enantiomeric ratios (*E*) could be achieved in solutions of ILs with high δ' values. Other groups[15,122,123] reported low/no activities of β-glycosidase in aqueous solutions of [BMIM][BF_4], which may be explained by the chaotropic nature of BF_4^- in solutions[123] (Note: in pure ILs based on BF_4^-, the chaotropic property of

← ***Increasing kosmotropicity***

Anions Citrate^{3-} > CF_3COO^- > $EtSO_4^-$ > OAc^- > Cl^- > BF_4^-
B'-coefficients: 0.65 → 0.21 → 0.17 → 0.13 → -0.017 → -0.17

Cations Mg^{2+} > Na^+ > K^+ and
$HMIM^+$, $BuPy^+$ > $BMIM^+$ > $EMIM^+$ > Me_4N^+, $EtPy^+$
B'-coefficients: 0.60 → 0.050 → -0.017 and
0.40, 0.40 → 0.33 → 0.29 → 0.18, 0.11

Figure 5.5. The ion kosmotropicity based on NMR B'-coefficients.[115]

anion does not apply to the enzyme activity; therefore, many studies reported enzyme being active in BF_4^- ILs). Our group[117] also conducted the enzymatic hydrolysis of DL-phenylalanine methyl ester in aqueous solutions of ILs carrying anions of chiral- or ω-amino acids, and observed higher enantiomeric excesses and yields in ILs based on D-amino acids rather than in those derived from L-isomers. The likely reason was that amino acid anions are more kosmotropic than zwitterionic amino acids,[118] and D-amino acids are more kosmotropic than L-isomers.[119] The use of ILs with kosmotropic anions (OAc^- and CF_3COO^-) in activating hydrolases in aqueous solutions was further demonstrated in our two subsequent studies.[120,121]

Recently, Fujita et al.[47,48,124] correlated the stability of cyt *c* in ILs containing 20% (wt) water with the kosmotropicity of components ions; the cation's effect on the protein stability followed a decreasing order of choline$^+$ > BMPyrr$^+$ > BMIM$^+$, which is also a decreasing order of cation chaotropicity; the anion's effect on the protein stability followed a decreasing order of $H_2PO_4^-$ > $Bu_2PO_4^-$ > OAc^- > lactate$^-$ > $MeSO_4^-$, which is the decreasing order of anion kosmotropicity (*B*-coefficients at 25°C: $H_2PO_4^-$ = 0.340,[103] OAc^- = 0.246,[103] $MeSO_4^-$ = 0.188[125]; lactate might be considered a kosmotropic anion[126]).

Constantinescu et al.[127,128] confirmed that the thermal stability of ribonuclease A (RNase A) in aqueous solution of ILs (typically 0–2 M) followed the Hofmeister series. In that study, differential scanning calorimetry (DSC) was used to measure the effect of ILs on the thermal denaturation of RNase A near 60°C. In terms of decreasing protein stability, the cation series are

$$K^+ > Na^+ \sim Me_4N^+ > Li^+ > Et_4N^+ \sim EMIM^+ > BMPyrr^+$$
$$> BMIM^+ \sim Pr_4N^+ > HMIM^+ \sim Bu_4N^+$$

and

$$K^+ > Na^+ \sim Me_4N^+ > Cholinium^+ > EMIM^+ \sim Guanidinium^+ > BMIM^+$$

and the anion series follows

$$SO_4^{2-} > HPO_4^{2-} > Cl^- > EtSO_4^- > BF_4^- \sim Br^- > MeSO_4^- > OTf^- > SCN^- \sim dca^- > Tf_2N^-$$

The cation series suggests the higher the cation hydrophobicity (of organic cations), the higher the cation kosmotropicity, and the lower the protein stability in general. The anion series suggests the opposite: the higher the anion kosmotropicity, the higher the protein stability in general (with slight differences in the position of neighboring ions from earlier discussion).

An excellent review by Yang[129] systematically discussed the possible mechanisms of Hofmeister effects of ILs on the enzyme activity and stability. The above preliminary studies have shown that the kosmotropic effect of ILs on enzymes may be applicable to diluted aqueous solutions of ILs,[40,114,116] as well as some concentrated ILs (such as 20 wt% water[48]). However, it is not quite clear if such an effect exists in pure ILs or ILs with trace amount of water, and how the IL hydrophobicity may influence the kosmotropicity. For example, PF_6^- is a chaotropic anion[100] and denatures enzymes when dissolved in aqueous solutions as Na^+ or K^+ salt (more denaturing than BF_4^- and $MeSO_4^-$ for mushroom tyrosinase[130]). However, PF_6^- based ILs (such as [BMIM][PF_6]) are hydrophobic, and thus the solubility and degree of dissociation of ILs in water are limited. On the other hand, it is also known PF_6^- based ILs containing low water contents are typically enzyme stabilizing.[20] Therefore, the Hofmeister effect may not be suitable in explaining the enzyme's behaviors in these hydrophobic ILs or their mixtures with water. Without sufficient water to hydrate them, kosmotropic or borderline anions (such as acetate, lactate, and chloride) of ILs bearing high H-bond basicities tend to interact strongly with enzymes causing their inactivation (see Section 5.2.1.2). Consequently, the enzyme-stabilizing/activating kosmotropic anions (such as OAc^- and Cl^-) in aqueous solutions become enzyme-inactivating in pure ILs containing low water contents. For example, several papers[40,59,116,120] have mentioned the enzyme activation at low concentrations of chloride-based ILs in water, but inactivation at high concentrations. Chapter 2 supplied a complementary discussion on the Hofmeister series and protein–ion interactions.

5.2.1.5 *Viscosity.* ILs are rather more viscous fluids than conventional organic solvents (see viscosity data of ILs in Wasserscheid and Welton[5]); in addition, many enzymatic reactions in ILs are practically heterogeneous systems due to the low solubility of enzymes in most pure ILs. Therefore, the internal and external mass-transfer limitations should be considered.[16] Lozano et al.[36] indicated that besides the IL polarity, the activity of α-chymotrypsin also depended on the IL viscosity; a higher enzyme activity was observed in [EMIM][Tf_2N] than in [MTOA][Tf_2N] (MTOA = methyl trioctylammonium) because the former IL (34 mPa s) is less viscous that the latter one (574 mPa s). Eckstein et al.[131] explained the higher enantioselectivity of lipase in [BMIM][Tf_2N] at low water activities ($a_w < 0.53$) than in methyl *tert*-butyl ether (MTBE) from two aspects: (1) the higher viscosity of the IL (52 mPa s) than that of MTBE (0.34 mPa s) might lead to mass-transfer limitations and lower the reaction rate; (2) the lower solubility of substrates in the IL than in MTBE could cause a lower activation energy in the IL. Van Rantwijk and Sheldon[20] rationalized that the high viscosity of ILs slows down the conformation changes of proteins, allowing enzymes to maintain their native structures and activity. However, Basso et al.[132] observed that during the amide synthesis through immobilized penicillin G amidase, the viscosities of ILs (i.e., [BMIM][PF_6] and [BMIM][BF_4]) showed no effect on the initial reaction rates despite their much higher viscosities than toluene. Our recent study[69] of the CALB-catalyzed transesterification of ethyl butyrate and 1-butanol in more than 20 ILs also suggested that the IL viscosity might affect the reaction rates in some cases, but is not always the primary factor in controlling the enzyme activity.

5.2.1.6 *Hydrophobicity.* "Hydrophobicity" may be considered a narrower concept of "polarity." However, it is practically important to distinguish "hydrophobicity" from "polarity" because the former is often related to the miscibility with water.[19] The hydrophobicity of ILs is usually quantified by the $\log P$ scale, a concept derived from the partition coefficient of ILs between 1-octanol and water. The partition coefficient (K_{OW} or P) is a ratio of concentrations of un-ionized compound between the two phases. The $\log P$ is defined as the partition coefficient at the unlimited dilution concentration of solute,

$$\log P = \lim_{c \to 0} K_{OW} = \lim_{c \to 0} \frac{C^o}{C^w} \tag{5.2}$$

where C^o is the IL concentration in the octanol phase and C^w is the IL concentration in the aqueous phase. For the simplicity, it is common to use extremely low concentrations of IL in the experiment instead of extrapolating the IL concentration to zero (Eq. 5.2). However, since ILs dissociate into ions in water and current K_{OW} values were reported as the ratio of concentrations of both undissociated and dissociated ILs in two phases, most $\log P$ values of ILs (Table 5.2) should be strictly called $\log D$, where D is the *distribution coefficient*, the ratio of the total concentrations of all forms of IL (ionized and un-ionized) between two phases. Alternatively, the intrinsic partition coefficients of ILs should be calculated from the apparent partition coefficients (D).[133]

TABLE 5.2. $\log P$ (or $\log K_{OW}$ at Low Concentrations[a]) Values of ILs at 25°C

Solvent		$\log P/\log K_{OW}$	Reference
1	dichloromethane	1.25	Selected value by Sangster[394]
2	THF	0.46	Selected value by Sangster[394]
3	*t*-butanol	0.35	Selected value by Sangster[394]
4	acetone	−0.24	Selected value by Sangster[394]
5	acetonitrile	−0.34	Selected value by Sangster[394]
6	[EMIM][Tf_2N]	−1.18	[202]
		$\log K_{OW}$ (−1.05 to −0.96) (0.28–2.8 mM)	Calculated from Ropel et al.[134]
7	[BMIM][Tf_2N]	0.11	[69]
		$\log K_{OW}$ (−0.96 to −0.21) (0.15–2.2 mM)	Calculated from Ropel et al.[134]
		0.33	[395]
		−1.74	[133]
8	[HMIM][Tf_2N]	0.64	[69]
		$\log K_{OW}$ (0.15 to 0.22) (0.32–0.38 mM)	Calculated from Ropel et al.[134]
		0.65	[395]
9	[OMIM][Tf_2N]	0.79	[202]
		$\log K_{OW}$ (0.80–1.05) (0.099–0.21 mM)	Calculated from Ropel et al.[134]
10	[EMMIM][Tf_2N]	$\log K_{OW}$ (−1.15 to −0.92) (0.32–2.9 mM)	Calculated from Ropel et al.[134]
11	[PMMIM][Tf_2N]	$\log K_{OW}$ (−0.92 to −0.62) (1.4–2.8 mM)	Calculated from Ropel et al.[134]
12	[HMMIM][Tf_2N]	$\log K_{OW}$ (0.13 to 0.25) (0.36–0.49 mM)	Calculated from Ropel et al.[134]

TABLE 5.2 (*Continued*)

Solvent		log P/log K_{OW}	Reference
13	[BMIM][PF_6]	−1.66	Calculated from Ropel et al.[134]
		−2.39	38,395
		−2.38	133,396
		−2.06	202
		−2.35	142
14	[HMIM][PF_6]	−1.86	395
15	[OMIM][PF_6]	−0.35	202
		−1.33	395
16	[ONIM][PF_6]	−2.19	396
17	[BMIM]Cl	−2.40	Calculated from Ropel et al.[134]
18	[BMIM]Br	−2.48	Calculated from Ropel et al.[134]
19	[EMIM][OAc]	−2.53	69
20	[BMIM][OAc]	−2.77	38
21	[EMIM][CF_3COO]	−2.75	69
22	[HMIM][CF_3COO]	−2.30	69
23	[BMIM][NO_3]	−2.90	38
		−2.42	Calculated from Ropel et al.[134]
24	[BMIM][dca]	−2.32	69
25	[EMIM][BF_4]	−2.57	69
26	[BMIM][BF_4]	−2.51	69
		−2.44	142,396
		−2.52	Calculated from Ropel et al.[134]
27	[OMIM][BF_4]	−1.34	69
		−1.14	202
28	[EtPy][CF_3COO]	−2.57	69
29	[EtPy][Tf_2N]	−0.90	69
30	[BuPy][Tf_2N]	−0.26	69
31	[choline][Tf_2N]	log K_{OW} = −0.57 (calculated value)	397*b*

[a] log K_{OW} values calculated from Ropel et al.[134] were converted from initial values of K_{OW} measured at room temperature (24 ± 2°C), and the concentration range given for each log K_{OW} was the IL concentration range in water phase.

[b] This reference also provides K_{OW} values for a number of pyridinium and imidazolium ILs based on Tf_2N^- and $B(CN)_4^-$.

From a practical point of view, the log *P* values (or log K_{OW} at low concentrations) of ILs in Table 5.2 are valuable for comparing the hydrophobicity of ILs with conventional organic solvents. In general, ILs are very hydrophilic in nature based on the negative log *P* values (or log K_{OW}) for most ILs (including water-immiscible Tf_2N^- and PF_6^- ones). The discrepancy between different measurements of the same IL might be caused by different initial concentrations of ILs (as high concentrations leading to higher K_{OW} values[133,134]), and different experimental techniques.

The Russell group[38] measured the log *P* values (<−2.0) of several ILs and suggested that they are very hydrophilic in nature based on Laane's scale;[135–137] they also observed that free lipase (*Candida rugosa*) was only active in hydrophobic [BMIM][PF_6] ($\log P = -2.39$), but inactive in other hydrophilic ILs including [BMIM][CH_3COO] ($\log P = -2.77$), [BMIM][NO_3] ($\log P = -2.90$), and [BMIM][CF_3COO].[38] Similarly, Nara et al.[138] achieved higher transesterification activities of lipases in [BMIM][PF_6] than in [BMIM][BF_4]. The Goto group also reported higher activities of polyethylene glycol (PEG)-modified lipase[139] and subtilisin[140] in more hydrophobic ILs such as [EMIM][Tf_2N]. Zhang et al.[141] reported low penicillin acylase stabilities in [BMIM][BF_4] and [BMIM][dca]. Lou and Zong[142] studied the enantioselective acylation of (*R*,*S*)-1-trimethylsilylethanol with vinyl acetate catalyzed by lipases in several ILs, and indicated the activity, enantioselectivity, and thermostability of Novozym 435 increasing with the IL hydrophobicity ([BMIM][PF_6] > [OMIM][BF_4] > [C_7MIM][BF_4] > [HMIM][BF_4] > [C_5MIM][BF_4] > [BMIM][BF_4]). Paljevac et al.[143] reported that the cellulase activity decreased in the order of IL hydrophobicity: [BMIM][PF_6] > [BMIM][BF_4] > [BMIM][Cl]. The Víllora group[144] observed the lower stability of PGA in [BMIM][BF_4] than in hydrophobic ILs (Tf_2N^- and PF_6^-), particularly in the absence of substrate. A recent study[145] on the alcoholysis of vinyl butyrate and 1-butanol by free CALB suggested that the lipase activities were generally much lower in water-miscible ILs (such as BF_4^-, dca^-, NO_3^-, and OAc^-) than in water-immiscible ones (PF_6^- and Tf_2N^-), and the enzyme's activities increased with the cation's hydrophobicity ($EMIM^+ < BMIM^+ < HMIM^+ < OMIM^+$). Ha et al.[146] also found that Novozym 435 was less active and stable in hydrophilic ILs (BF_4^- and OTf^-) than in other hydrophobic ILs (Tf_2N^- and PF_6^-). Lee et al.[55] reported that Novozym 435 was more thermally stable in hydrophobic ILs than in hydrophilic ones, following the order of [BMIM][Tf_2N] > [BMIM][PF_6] > [BMIM][OTf] > [BMIM][BF_4] > [BMIM][SbF_6]. Shen et al.[147] noticed that during the kinetic resolution of racemic cyanohydrins, Amano lipase PS showed a high enantioselectivity (80% ee_p) in hydrophobic [OMIM][PF_6], but poor enantioselectivities (<5% ee_p) in hydrophilic [HMIM][BF_4] and [HMIM][Cl] (also see Section 5.2.2.1 about halide impurities). Hernández-Fernández et al.[53] concluded that both free CALB and PGA were more stable in hydrophobic ILs than in hydrophilic ones: In the case of CALB, the stability was in a decreasing order of [HMIM][PF_6] > [HMIM][Tf_2N] > [HMIM][BF_4], and [BMIM][PF_6] > [BMIM][dca], as well as [OMIM][PF_6] > [HMIM][PF_6] > [BMIM][PF_6]; in the case of PGA, the stability was in a decreasing order of [BMIM][Tf_2N] > [BMIM][PF_6] > [BMIM][BF_4]. However, the hydrophobic cations showed an adverse effect on the PGA stability: [EMIM][Tf_2N] > [BMIM][Tf_2N], and [BMIM][PF_6] > [OMIM][PF_6]. The effect of nucleophilicity of these anions is discussed in Section 5.2.1.2. These examples imply that the high hydrophobicity (large $\log P$) of ILs could be beneficial to the enzyme stabilization.

Through a systematic investigation of Novozym 435-catalyzed transesterification in over 20 ILs, our group[69] observed that the lipase activity increased with the $\log P$ value of ILs to a maximum and then declined with a further increase in $\log P$ (a bell shape). Our previous discussion implied that the enzyme is active in hydrophobic solvents (with high $\log P$). However, a higher $\log P$ of the solvent also means a more thermodynamic ground-state stabilization of substrates,[148] which reduces the conversion of substrates. This could explain the decreasing reaction rate in very hydrophobic ILs.

Similarly, Lou et al.[63] found the initial rates of Novozym 435-catalyzed ammonolysis of (*R*,*S*)-*p*-hydroxyphenylglycine methyl ester increased with the hydrophobicity of BF_4^- based ILs to a maximum and then decreased with a further increase in the IL hydrophobicity.

At present, we have not understood why the higher hydrophobicity of ILs may lead to a higher enzyme activity (up to the optimum activity). A common knowledge gained from biocatalysis in organic solvents is that polar solvents strip off the "essential" water from enzyme molecules, causing their inactivation.[137,149] This could be an important reason in many above reactions. However, relatively high enzyme activities could still be obtained in [BMIM][Tf_2N] when substrates, IL, and CALB were all intensively dried[69] (CALB is known to be active in organic solvents containing little or no water[75,150,151]). Therefore, the solvent-stripping capacity is unlikely the only mechanism of Novozym 435 inactivation by hydrophilic ILs. An alternative explanation may be derived from the correlation between $\log P$ and solvent hydrogen-bond basicity. A general regression equation has been established as the following,[152]

$$\log P = c + rR + s\pi^* + a\alpha + b\beta + vV_x \tag{5.3}$$

where R is the excess molar refraction obtained from refractive index measurements, V_x is the McGowan characteristic volume or simply the intrinsic volume of the solute, α reflects the H-bond acidity (H-bond donating ability, property of IL cation), β reflects the H-bond basicity (H-bond accepting ability, property of IL anion), and π^* reflects dipolarity/polarizability. The correlations of 613 organic solutes yielded the coefficient values as $c = 0.088$, $r = 0.562$, $s = -1.054$, $a = 0.034$, $b = -3.460$, and $v = 3.814$ (V_x in unit of cm^3/mol/100).[152,153] The near-zero a-coefficient suggests that the hydrogen-bond acidity does not contribute much to the partition. The s- and b-coefficients are negative because water is more dipolar and is a stronger hydrogen-bond acid than wet octanol.[154] The large and positive v-coefficient shows that larger solute molecules are more hydrophobic and tend to partition into the octanol layer. Other correlations using different solutes reached similar conclusions.[155–157] Although Equation (5.3) correlated from IL data is not available, those coefficients generated from 613 organic solutes offer some insight into solute–solvent interactions. Based on Equation (5.3), the higher is the hydrogen-bond basicity of an IL anion, the lower is the $\log P$ value, and thus the lower is the enzyme activity due to the H-bonding interactions between the anion and enzyme. This explains the increasing trend of the enzyme activity with the $\log P$ value up to a certain value. Therefore, the IL hydrophobicity is closely related to the H-bond basicity of IL anions. Why did the lipase activity decrease with a further increase in $\log P$? Since v-coefficient is quite positive in Equation (5.3), an increase in the cation's size results in the increase of $\log P$ value. As discussed previously, the stabilization of substrates could be one reason. But the possibility of hydrophobic interactions between large IL molecules and the enzyme cannot be completely excluded. For example, the Atkin group[158] investigated the stability and activity of hen's egg white lysozyme in aqueous solutions of four protic ILs (25–75 wt%); the protein denaturing–renaturing circular dichroism (CD) experiments and the activity measurements of lysozyme indicated that the highest catalytic activity and most complete refolding was achieved in solutions of

[(EtOH)NH_3][HCOO], followed by [PrNH_3][HCOO], and then [EtNH_3][HCOO] and [(MeOEt)NH_3][HCOO]. It is believed that the protein–IL interactions include the electrostatic interaction of IL cations with negatively charged residues in the protein, hydrogen bonds between amine protons and the protein, as well as the *hydrophobic interactions* between alkyl chains in ILs and hydrophobic regions of the protein. Since electrostatic interactions between $[(EtOH)NH_3]^+$ and lysozyme is about the same as for $[EtNH_3]^+$, the hydroxyl group in $[(EtOH)NH_3]^+$ probably reduces the strength of *hydrophobic interactions* with the protein. Another possibility is that the hydroxyl group interacts with the anion formate via hydrogen bonds, reducing the interaction of formate with the protein. The IL viscosity-induced mass transport was not a limiting factor in the study because [(EtOH)NH_3][HCOO] is several times more viscous that other three ILs. In summary, *the hydrophobicity factor of ILs is a combination effect of the anion's H-bond basicity and the cation's hydrophobic effect.*

Since hydrophobicity is not the only factor in controlling the hydrolase activity, complications arose in interpreting some biocatalytic reactions. De Diego et al.[159] conducted the transesterification of vinyl propionate and 1-butanol catalyzed by free and immobilized lipases from *Candida antarctica* (CALA and CALB), *Thermomyces lanuginosus* (TLL) and *Rhizomuncor miehei* (RML). Most of the enzyme preparations (except free CALA) showed higher activities in more hydrophobic [OMIM][PF_6] than in [BMIM][PF_6], but lower activities in other more hydrophobic based ILs ([OMIM][BF_4] < [HMIM][BF_4] < [BMIM][BF_4], and [BDMIM][PF_6] < [BDMIM][BF_4]). Another study by Irimescu and Kato[54] on the lipase-catalyzed acylation of primary amines indicated lower reaction rates in ILs with longer alkyl chains in cations, and the water miscibility of ILs was not a main factor in influencing the reaction rate. Some studies also obtained relatively high enzyme activities in hydrophilic ILs (such as [BMIM][BF_4], [EMIM][BF_4], [BMIM][OTf], and [MMIM][$MeSO_4$]).[35,64,160–163] Therefore, multiple factors must be considered in explaining the enzymatic systems like these.

5.2.1.7 Enzyme Dissolution. Hydrophobic ILs (typically consisting of PF_6^- and Tf_2N^-) do not dissolve appreciable amounts of enzymes. The aqueous enzyme in hydrophobic ILs (PF_6^- and Tf_2N^-) should be considered the enzyme inclusion by IL network (see 5.2.1.3) rather than the dissolution in the media.[164] On the other hand, hydrophilic ILs (such as those based on NO_3^-, lactate, $EtSO_4^-$, and CH_3COO^-) may dissolve some enzymes*; however, most of them tend to strongly interact with the protein (such as via H-bonds), resulting in the enzyme inactivation.[59,64,67–69,145] For example, cellulase was dissolved but inactivated in concentrated solutions of [BMIM][Cl].[59] Erbeldinger et al.[165] found that thermolysin became inactive when fully dissolved in [BMIM][PF_6] containing 5% (v/v) water (enzyme solubility up to 3.2 mg/mL), but the suspended thermolysin remained active in the IL. Currently, there are only a few pure (or concentrated) ILs that are known to dissolve considerable amounts of enzymes

* There are also some exceptions. For example, BF_4^- based ILs are hydrophilic but do not dissolve the enzyme.[64]

but do not inactivate them. For example, choline dihydrogen phosphate (m.p. 119°C) containing 20% (wt) water could solubilize and stabilize cyt *c*;[47,48] triethylmethylammonium methyl sulfate ([Et_3MeN][$MeSO_4$]) was reported able to dissolve >1.2 mg/mL CALB and retain its catalytic capability.[64,150] Recently, we synthesized a series of ether-functionalized ILs that are able to dissolve enzymes (>5 mg/mL CALB at 50°C) and a variety of substrates, but do not inactivate the lipase (more discussion in Sections 5.2.3.6 and 5.4).[68,70] Therefore, the enzyme dissolution in ILs is not a definite indication of enzyme denaturation.

5.2.2 Other Factors Influencing Hydrolase Activity and Stability

As many ILs are prepared from their halide precursors through metathesis, the residual halides in final IL products are often inevitable. These halide impurities have a strong influence on the enzyme's performance in ILs as discussed below in Section 5.2.2.1; therefore, it is always desirable to know the halide content in ionic media before using them in biocatalytic reactions. Since water also plays a critical role in enzymatic reactions in nonaqueous solvents,[166,167] it is imperative to control the water content (or "thermodynamic water activity") in biocatalysis carried out in ILs (Section 5.2.2.2). In particular, this becomes crucial for synthetic reactions involving hydrolysis side reactions. Another reason to control the water content is that ILs often absorb small amounts of water as another impurity (the hygroscopic nature of ILs is believed due to the preference of ion pair association[168]).

5.2.2.1 Halide Impurities in ILs. The purity of ILs is known to affect their physical properties, and further influence the reactions performed in them.[169] Recently, Lee et al.[60] reported that the activity of Novozym 435 in [OMIM][Tf_2N] decreased linearly with the higher chloride content, and a 1% (wt) increase in [OMIM][Cl] (~1540 ppm Cl^-) caused a 5% decrease in enzyme activity. However, another lipase (from *Rhizomucor miehei*) could tolerate much less Cl^-; its activity in [OMIM][Tf_2N] with 2% [OMIM][Cl] was only about 2% of the activity in pure [OMIM][Tf_2N]. Leet et al.'s model reaction was the transesterification of vinyl acetate with benzyl alcohol. In a second study by Lee et al.,[55] it was seen that the activity of CRL in the [OMIM][Tf_2N] containing 1% (w/w) [OMIM][Cl] was only about half of that in pure [OMIM][Tf_2N] (with <30 ppm Cl^-). Both groups of Kazlauskas[35] and Víllora[170] observed higher lipase activities when the impurities were removed from ILs through washing them with aqueous solutions of $NaHCO_3$ or Na_2CO_3. Shen et al.[147] reported that during the kinetic resolution of racemic cyanohydrins, Amano lipase PS showed a high enantioselectivity (80% ee_p) in [OMIM][PF_6], but a poor enantioselectivity (<5% ee_p) in [HMIM][Cl]. Our recent study[69] on the transesterification between ethyl butyrate and 1-butanol catalyzed by Novozym 435 suggested that when 3000 ppm Cl^- or Br^- anions were present in [BMIM][Tf_2N], the initial rates were about 60% of the activity in [BMIM][Tf_2N] with little halide (260 ppm Br^-). These data suggest that the degree of halide inhibition depends on specific substrates and enzymes. In our above model reaction, when the halide contents of ILs fell below 1000 ppm, there was no direct correlation between the lipase activity and the halide concentration.[69] However, it is always important to determine the halide contents of ILs before using them in enzymatic reactions. Halides and other impurities should be removed from ILs as much as possible.

5.2.2.2 Water Activity. It is a common strategy to add a fixed amount of water into ILs to provide the "essential" water needed by the enzyme. For example, 1–2% (v/v) water was added into α-chymotrypsin or free CALB-catalyzed transesterification reactions in Tf_2N^-, PF_6^-, and BF_4^- based ILs.[36,171,172] However, the water content does not represent the amount of water available for enzyme molecules because some water molecules may be tightened by the solvent molecules (such as through H-bonds). The actual water freely available for the enzyme is often quantified by a dimensionless property called thermodynamic water activity (a_w), whose definition is given in Equation (5.4).

$$a_w = \frac{p}{p_0} \tag{5.4}$$

where p is the vapor pressure of water in the substance, and p_0 is the vapor pressure of pure water at the same temperature. The water activity a_w is a function of temperature.

There are two common methods to maintain the water activity (a_w):

(1) *Use of salt hydrate pairs*. In this method, salt hydrate pairs (such as $NaOAc \cdot 3H_2O$ and NaOAc) are directly added into nonaqueous solvents. This method has been well developed for biocatalysis in conventional organic solvents.[166,167,173] This approach is also known for its accurate control of water activity in real time during the reaction, even if the reaction is producing water as the by-product (as long as a trace of each salt hydrate is present). Berberich et al.[174] controlled the water activity (a_w = 0.17–0.78) in [BMIM][PF_6] through different salt hydrate pairs and showed the initial rate of immobilized CALB-catalyzed transesterification reactions decreasing with the increase of water activity. However, one should be cautious before adding salt hydrates into ILs because some ionic solvents (especially hydrophilic ones) are able to dissolve salts.[5] The dissolved salt hydrates may affect the enzyme activity by binding to the enzyme surface.[174] Based on the relationship of water content in [BMIM][PF_6] and the water activity (a_w) reported by Berberich et al.,[174] Barahona et al.[175] investigated the influence of water activity on the esterification of geraniol with acetic acid in [BMIM][PF_6] promoted by Novozym 435, and observed the maximum initial reaction rate at a_w = 0.6. Recently, Yang et al.[176] studied the *Penicillium expansum* lipase-catalyzed hydrolysis of *p*-nitrophenyl palmitate in [BMIM][PF_6] with the presence of salt hydrates and observed two important roles of salt hydrates in influencing the lipase activity: (1) controlling the water activity and (2) acting as Hofmeister salts (see Section 5.2.1.4.) since salt hydrates might dissolve in microaqueous environments surrounding the enzyme molecules.

(2) *Pre-equilibrium over saturated salt solutions*. In this method, substrates and enzyme in separate containers are placed in a larger closed vessel containing saturated salt solution at a fixed temperature (typically is room temperature to prevent the thermal inactivation of enzyme). The equilibrium through the vapor phase can be reached from 24 hours to a few days. The commonly used salts and their water activities at saturation include LiCl (0.12), $MgCl_2$ (0.33), $Mg(NO_3)_2$ (0.54), NaCl (0.75), and KNO_3 (0.96).[173] The drawback of this method is that it only sets the initial water activity (at room temperature) for substrates and enzymes, and the actual water activity during the reaction may change because of two main reasons: (1) the reaction temperature may not be room temperature; (2) the reaction equilibrium has been shifted and water may be produced as the by-product. However, due to its simplicity, this method has gained more applications in biocatalysis in ILs. Eckstein et al.[177] used saturated salt solutions to control the water activity for α-chymotropsin-catalyzed transesterifications in [BMIM][Tf_2N] and [EMIM][Tf_2N], and observed the enzyme activity increasing with a_w but the selectivity (synthesis over hydrolysis) decreasing with a_w. The same group[131] also showed a higher enantioselectivity of *Pseudomonas* sp. lipase in [BMIM][Tf_2N] at low water activities ($a_w < 0.53$) than in MTBE. Garcia et al.[178] pre-equilibrated the reaction mixture components to a fixed a_w through saturated salt solutions (KOAc, $a_w = 0.22$; NaCl, $a_w = 0.75$) at 25°C for about 3 days; they observed the transesterification activity of immobilized cutinase in [BMIM][PF_6] increased with an increase in a_w; however, the transesterification and hydrolysis activities of Novozym 435 decreased with the increase in a_w. Noël et al.[179] equilibrated the substrate mixture and aqueous enzyme solution in [BMIM][Tf_2N] to fixed water activities (a_w) over saturated salt solutions in closed containers (10 mL total capacity) at 25°C for 4 days; during the kinetic resolution of *rac*-2-pentanol catalyzed by free CALB, the synthetic activity increased with an increase in water activity, but the selectivity decreased with an increase in water activity. Lee[55] studied the effect of a_w on lipase activity through the transesterification of benzyl alcohol with vinyl acetate in [OMIM][Tf_2N] and found the optimal initial rates of Novozym 435 at $a_w = 0.2$, *Rhizomucor miehei* lipase at $a_w = 0.5$, and CRL at $a_w = 0.4$. De Diego[159] observed [CPMA][$MeSO_4$] was able to retain water molecules, which induced a low water activity even at a high water content (for example, at water content of 8% (v/v), $a_w < 0.5$); they also reported the optimal synthetic activity of free CALB in this IL occurred at the water content of 2% (v/v). The humidity sensor used in that study was calibrated with saturated salt solutions.

Given the complexity of enzyme's catalytic properties in ILs, there is no universal method that can solve all the enzyme inactivation issues. Therefore, a number of methods have been adopted or developed to improve the enzyme's stability and to increase its activity and enantioselectivity. The following section is the discussion of some representative methods.

5.2.3 Methods to Improve Hydrolase Activity and Stability

Overall, methods in stabilizing and activating enzymes in ILs may be grouped into two general categories: modification of enzymes and modification of solvent environments. The first category includes the enzyme immobilization (on solid support, sol–gel, or cross-linked enzyme aggregate [CLEA]), physical or covalent attachment to PEG, rinsing with *n*-propanol methods (propanol-rinsed enzyme preparation [PREP] and enzyme precipitated and rinsed with *n*-propanol [EPRP]), three-phase partitioning (TPP), enzyme/protein-coated micro-crystals, and lyophilization with cyclodextrins, and so on. The second category includes water-in-IL microemulsions, IL coating, the use of additives in ILs, and the design of enzyme-compatible ionic media, and so on. Methods in the first category allow the enzyme to be more tolerant to those denaturing factors of ILs (discussed earlier), while methods in the second category minimize the denaturing properties of some ILs (such as reducing the anion's nucleophilicity and H-bonding basicity). The following sections will address some major advances in these methods with representative examples.

5.2.3.1 Enzyme Immobilization. The most common method for enzyme stabilization and activation in ILs is the use of immobilized entities instead of free forms. These immobilized methods generally fall into three categories: binding to a solid carrier, sol–gel encapsulation, and protein cross-linking (a carrier-free technique).[180–182]

IMMOBILIZATION ON A SOLID CARRIER. The binding to a solid support is a routine method for improving enzyme stability in conventional organic solvents as well as in ILs. One main reason is that the immobilization techniques (both physical adsorption and covalent attachment) are straightforward and easy to accomplish via regular laboratory procedures, and many immobilized enzymes are commercially available (such as Novozym 435). Two recent comprehensive reviews[20,26] have mentioned many of these examples; therefore, this report will not replicate the work. However, it is interesting to highlight some new carriers for enzyme immobilization, one of which is the carbon nanotubes.[183] The high surface area and unique nanoscopic dimensions of carbon nanotubes enable a high protein loading and low mass-transfer resistance. The noncovalent binding of proteinase K onto single-walled carbon nanotubes (SWNTs) resulted in a higher enzyme activity and higher thermal stability than its native form in ILs; the enzyme-SWNT conjugates were well dispersed in ILs.[184] Similarly, CRL adsorbed on multiwalled carbon nanotubes (MWNTs) displayed a higher transesterification activity and enantioselectivity than the pH-tuned free form in [BMIM][PF_6].[185] Du et al.[186] immobilized heme-containing proteins/enzymes (myoglobin, cyt *c*, and horseradish peroxidase) onto SWNTs coated with ILs (such as [BMIM][BF_4] or [BMIM][PF_6]), and observed the encapsulated enzymes retained their bioelectrocatalytic activities toward the reduction of oxygen and hydrogen peroxide.

Recently, another new enzyme support named "magnetic nanoparticles supported ILs" was constructed by a covalent bonding of ILs–silane on magnetic silica nanoparticles; CRL immobilized on this support showed a higher catalytic activity and thermal stability.[187] New nanoparticles formed from cross-linked fluoroalkyl end-capped acrylic acid co-oligomer containing poly(oxyethylene) units were shown effective in encapsulating cyt *c*; the immobilized enzyme exhibited greater catalytic activity toward the oxidation of guaiacol with hydrogen peroxide in 1:1 (v/v) mixture of water and 3-methylpyrazolium tetrafluoroborate than in water.[188]

Sol–Gel Encapsulation. The encapsulation is a technique for entrapping biomolecules in a polymer matrix via noncovalent interactions between the network and the biomolecule. The actively pursued encapsulation method for the enzyme stabilization in ILs is the sol–gel technology. It is a relatively simple method (see protocols in Reetz[189] and Campas and Marty[190]) and has been widely used for entrapping a variety of biological molecules such as proteins, enzymes, and antibodies.[191] This technique involves the acid- or base-catalyzed hydrolysis of tetraalkoxysilanes, such as tetraethyl orthosilicate (TEOS) or tetramethyl orthosilicate (TMOS), in aqueous solutions; the subsequent cross-linking condensation forms a SiO_2 matrix to encapsulate the biomolecules.[192,193] The structural rigidity of sol–gel matrixes protects the integrity of encapsulated enzyme molecules and prevents their leaching; the mesoporous structures and high pore volume of sol–gel polymer afford the free diffusion of small substrate molecules and their effective interactions with the enzyme.[194] However, the gel shrinkage and pore collapse have been major drawbacks of this method; in addition, there are other issues such as the alcohol release during the hydrolysis of silicon alkoxide.[195] To overcome these hurdles, different additives (such as sugars, amino acids, and *N*-methylglycine) have been added to reduce the gel shrinkage and adjust the protein hydration, and to further increase the activity and stability of enzymes.[195]

Recently, there is a growing interest in utilizing ILs as additives[196] for the protein/enzyme sol–gel immobilization. This is because room-temperature ILs are nonvolatile, thermally stable, and are tunable to be enzyme-compatible. Earlier studies[197,198] focused on the preparation of mesoporous silica through a high dispersion of ILs (such as [BMIM][BF_4]) in sol–gel, with potential application of inclusion of metal catalysts. More recently, silica xerogels were prepared with various morphologies via the sol–gel method in the presence of ether-functionalized ILs (such as 1-triethylene glycol monomethyl ether-3-methylimidazolium methanesulfonate); these ILs act as both morphology controller and acid precatalyst.[199,200] An independent study by the Deng group[201] further confirmed that anions of ILs had a strong impact on the pore structures of silica-gel materials. Several groups have clearly demonstrated the advantages of enzyme encapsulation in IL sol–gel hybrid carrier. The Koo group[202,203] examined a variety of ILs and their mixtures as additives during the sol–gel immobilization of CRL, and achieved greater hydrolytic and esterification activities of sol–gel encapsulated lipase coated with ILs than that without ILs. However, the solvents used in the above assay reaction catalyzed by IL sol–gel immobilized enzymes were either aqueous buffer or organic solvent (*n*-hexane); the use of ILs as solvents was not demonstrated. Sangeetha et al.[204] encapsulated subtilisin in sol–gel-derived silica glasses using alkoxysilane precursors carrying different chain lengths; the resulting immobilized enzyme showed an enhanced thermal stability and a high activity toward peptide synthesis in [BMIM][PF_6]. The Kanerva group[205] immobilized lipase PS from *Burkholderia cepacia* in a sol–gel, and obtained a higher enzyme stability in [EMIM][BF_4], [EMIM][Tf_2N], and [BMIM][PF_6] than the enzyme preparation via the CLEA method (see next section).

CLEAs®. As an early version of carrier-free immobilization, the cross-linked enzymes (CLEs) can be created by reacting glutaraldehyde with NH_2 groups on the protein surface.[206,207] However, intermolecular cross-linking of these biomolecules often results in low enzyme activity, poor reproducibility, and low mechanical stability.[180,208] At the same time, the cross-linking of a crystalline enzyme via glutaraldehyde was reported,[209] eventually leading to a commercial immobilization technology called cross-linked enzyme crystals (CLECs or CLCs).[210,211] This technique offers many advantages such as enhanced thermal, mechanical, and pH stability, designable particle sizes, and ease of recycling.[180,208] However, the preparation of CLECs requires a laborious purification and crystallization of enzymes. A more recent development in cross-linking enzymes has been pioneered by the Sheldon group,[180–182,208,212] a new technique called cross-linked enzyme aggregates (CLEAs®). This method involves the addition of salts, organic solvents, or nonionic polymers to precipitate the enzyme as physical aggregates from aqueous solutions; the enzyme aggregates are then cross-linked by glutaraldehyde. The use of CLEAs offers many advantages such as ease of preparation and recycling, improved enzyme activity, high stability and selectivity in organic solvents, and immobilization of biocatalysts containing more than one enzyme.[181,213–218]

Therefore, high enzyme stability is expected for CLEAs in ILs. Toral et al.[67] immobilized CALB through two methods: CLEA method without a solid support (resulting in CALB-CL), and adsorbed and cross-linked on a polypropylene carrier (resulting in CALB-PP). They carried out several reactions using these two enzyme preparations in denaturing ILs (such as [BMIM][dca], [BMIM][NO_3], [BMIM][OAc], and [BMIM][lactate]). During the transesterification between ethyl butyrate and 1-butanol, CALB-PP exhibited considerably higher catalytic activities in denaturing ILs than CALB-CL and free CALB. In addition, CALB-PP also showed high enantioselectivities in two resolution reactions (resolution of 1-phenylethanol in [BMIM][NO_3], and resolution of 1-phenylethylamine in [BMIM][NO_3] and [BMIM][dca]), although the cross-linked lipase was not active in some ILs (such as [BMIM][OAc] and [BMIM][lactate]). Shah and Gupta[219] observed a higher enzymatic activity of *Burkholderia cepacia* lipase in [BMIM][PF_6] after the CLEA preparation than the pH-tuned lyophilized free enzyme.

5.2.3.2 PEG-Modification. The modification of enzymes with polyethylene glycol (PEG) through either physical complexation or covalent interaction is a routine method for enzyme stabilization in denaturing environments. PEG has both hydrophilic and hydrophobic properties (*amphiphilic*); therefore, modified enzymes become soluble in some organic solvents (such as benzene, toluene, and chlorinated hydrocarbons)[220] and ILs,[140] and exhibit an increased stability in these solvents. By a strict definition, the PEG modification is one type of enzyme immobilization method. The most common PEG-modification is achieved through the physical adsorption because of its simple procedures, mild conditions, and the unchanged protein structures. The Goto group[221,222] applied PEG with different molecular weights (average M_n = 4000–35,000) as the enzyme-coating amphiphile for the preparation of PEG–lipase complexes. They investigated the PEG–lipases in several alcoholysis reactions in hydrophobic ILs (PF_6^- and Tf_2N^-), and observed higher enzyme activities (as high as 14-fold) of the PEG–lipase complex than that of free form,[221,222] as well as comparable or higher enantioselectivities of the enzyme complex than that of free lipase.[222] Turner et al.[59] reported higher enzyme activities of PEG–cellulase complex (PEG average M_n = 1000) in aqueous solution and [BMIM][Cl] solutions than that of free cellulase. On the other hand, the PEG-modified α-chymotrypsin (PEG average M_n = 1000) via physical adsorption showed no or marginal increase in the reaction rate in PF_6^- based ILs than its free form.[171] These differences could be due to a number of factors such as different preparation conditions, different PEG molecular weights, and different enzymes involved.

The second strategy for associating enzymes with PEG is through the covalent attachment. For example, PEG *p*-nitrophenyl carbonate is frequently used to connect PEG units with amino residues of proteins to form stable carbamates.[223,224] Alternatively, Kaar et al.[38] applied PEG–monoisocyanate to link PEG with lysine residues of the protein to form a PEG–lipase complex; however, this complex showed no improvement in lipase activity in [BMIM][PF_6], [BMIM][NO_3], and [MMEP][OAc] compared with the free form. However, PEG-modification of cyt*c* allowed the protein soluble in [EMIM][Tf_2N] without denaturation.[225] Meanwhile, the Goto group adopted an unusual comb-shaped PEG, PM_{13}, (SUNBRIGHT AM-1510K from NOF Co., Tokyo, Japan) for covalently conjugating PEG with subtilisin Carlsberg[140,226] and *Candida rugosa* lipase,[139] respectively. As shown in Scheme 5.1, PM_{13} is a copolymer of PEG derivatives with maleic anhydride (molecular weight ~15,000); the acid anhydrides react with amino groups of the enzyme.[227] The PM_{13}–lipase complex showed a higher activity and stability in benzene than its free form.[228] In Tf_2N^- based ILs, PM_{13}-subtilisin is soluble and exhibited much higher transesterification activity and stability than the free enzyme.[140,226] Similarly, PM_{13}-lipase is also soluble in Tf_2N^- based ILs and shows a high catalytic activity and storage stability in these ionic solvents.[139]

Meanwhile, there are also some disadvantages associated with the PEG-modification. For example, the preparation could be cumbersome in some cases, and the enzyme catalytic properties may vary with different immobilization batches.

5.2.3.3 *EPRP.* Propanol-rinsed enzyme preparation (PREP) is a method for stabilizing the immobilized enzymes, which is achieved by repeatedly rinsing the silica-immobilized enzyme with dry *n*-propanol.[229] The principle of this technique is that

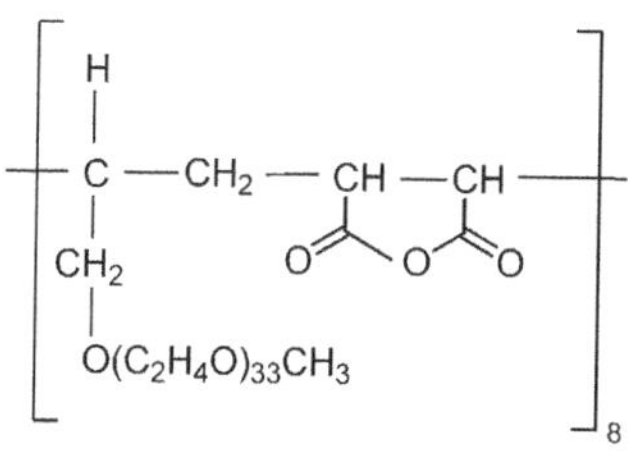

Scheme 5.1. Structure of comb-shaped polyethylene glycol PM_{13}.

n-propanol removes water from protein, which minimizes the protein denaturation and keeps the majority of enzyme molecules in an active conformation; only a small amount of water may be needed during the reaction to activate the catalysis.[229] The PREPs of subtilisin Carlsberg,[229] α-chymotrypsin,[229] and papain[230] have shown very high enzymatic activities in organic media (such as acetonitrile and *t*-butanol). A further evolvement of PREP becomes a preparation method called "enzyme precipitated and rinsed with *n*-propanol" (EPRP).[231,232] The EPRP technique is a combination of several methods: enzyme precipitation by alcohols, rinsing the precipitate with *n*-propanol, and the use of salt tuning during co-precipitation. The EPRP of α-chymotrypsin showed two orders of magnitude higher esterification activity than lyophilized powder in *n*-octane.[231] The EPRP-subtilisin exhibited >4000 times higher initial rate (for transesterification) than pH-tuned lyophilized powder in [BMIM][PF_6]; the pH-tuned lyophilized subtilisin showed no activity in [BMIM][BF_4], whereas the EPRP preparation gave a moderately high initial rate in this IL.[233] In another study, Shah and Gupta[219] conducted the enzymatic resolution of 1-phenylethanol in [BMIM][PF_6] catalyzed by various lipase preparations; they observed that the EPRP of CRL was more active than pH-tuned lyophilized CRL and CLEA preparation, and was more enantioselective than pH-tuned lyophilized CRL and PREP preparation. On the other hand, the EPRP of *Burkholderia cepacia* lipase (BCL) lost a substantial portion of hydrolytic activity, but two alternative formulations (i.e., enzyme precipitated and rinsed with acetone [EPRA] and acetone-rinsed enzyme preparation [AREP]) resulted in high enzyme activities and enantioselectivities.[219]

5.2.3.4 Water-in-IL Microemulsions. The above several methods focus on the modification of enzymes to improve their adaptability in ionic environments. The following several methods, on the other hand, intend to change the ionic media to improve their enzyme compatibility. The surfactant formation of micelles in ILs is being actively studied and has been reviewed.[234–236] Recently, the Goto group[237] reported the use of water-in-IL microemulsions as a new medium for dissolving various enzymes and proteins (such as lipase PS, CALB, α-chymotrypsin, horseradish peroxidase, and enhanced green fluorescent protein). The new medium was created by dissolving anionic surfactant sodium bis(2-ethyl-1-hexyl)sulfosuccinate (AOT) in hydrophobic [OMIM][Tf_2N] containing 10% (v/v) 1-hexanol (as a cosurfactant), followed by the addition of aqueous buffer to prepare a microemulsion. The lipase PS showed a higher hydrolytic activity in water-in-IL microemulsions than in water-saturated IL and in water-in-isooctane microemulsions.[237] The same group optimized the oxidation of pyrogallol in water-in-IL microemulsions catalyzed by horseradish peroxidase, and found that the reaction in IL microemulsions was much more effective than that in a conventional AOT/water/isooctane microemulsion.[238]

More recently, the Stamatis group[239] developed the water-in-IL microemulsions through using nonionic surfactants (Tween 20 or Triton X-100) in [BMIM][PF_6]. The catalytic properties of lipases from *Candida rugosa, Chromobacterium viscosum*, and *Thermomyces lanuginosa* in these novel microemulsions were investigated through the esterification of natural fatty acids with various aliphatic alcohols, and the hydrolysis of *p*-nitrophenyl butyrate. The operational stability of these lipases in water-in-IL

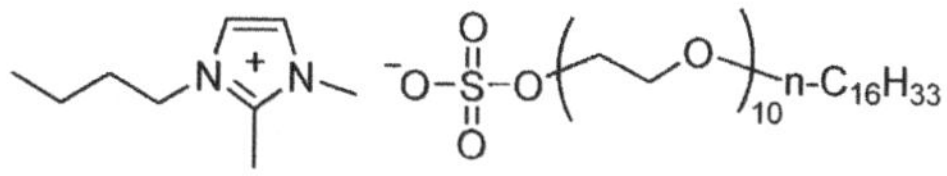

Scheme 5.2. Structure of 1-butyl-2,3-dimethylimidazolium cetyl-PEG10-sulfate.

microemulsions was much higher than that in other microheterogeneous media. Fourier transform infrared (FT-IR) and CD spectroscopy further confirmed that the lipase in water-in-IL microemulsions usually maintain its native conformation or adapt a more rigid structure compared with the incubation in other microheterogeneous media.

Recently, the Pandey group[240] developed a visual method for examining the formation of water-in-IL microemulsions: the Co (II) salt ($CoCl_2 \cdot H_2O$) was added into the microemulsions formed by using [BMIM][PF_6] as the oil phase and nonionic TX-100 as the surfactant; the Co(II) salt was chosen as the probe because of different colors of the hexa-coordinated and tetra-coordinated complexes of the cation, depending on the solvating environment; the color change can be detected by the UV-visible absorbance spectra.

5.2.3.5 *Coating Enzymes with ILs.* Lee and Kim[241] coated an IL, [PPMIM][PF_6] (PPMIM = 1-(3′-phenylpropyl)-3-methylimidazolium), onto lipase from *Pseudomonas cepacia*, and achieved a high enantioselectivity and activity from this enzyme preparation. The Itoh group[242,243] prepared an imidazolium IL carrying anions of cetyl-PEG10-sulfate (Scheme 5.2) and coated it onto lipases to stabilize the enzymes in organic solvents (such as diisopropyl ether).

The IL-coated lipases exhibited a high enantioselectivity and high reaction rates in several resolution reactions (up to 500- to 1000-fold acceleration for some substrates). The IL-coated lipase PS is also commercially available.[244] *Mucor javanicus* lipase coated with various ILs was found more active and more stable than the untreated lipase in aqueous solution, and the activation factors were 1.81, 1.66, 1.56, and 1.60 for [BMIM][PF_6], [EMIM][Tf_2N], [BMIM][BF_4], and [EMIM][BF_4] respectively.[245] Lozano et al.[246] carried out the enzymatic synthesis of citronellyl esters mediated by Novozym 435, and observed that the synthetic activity could be improved up to twofold by using the IL coating onto lipase; they also found the increase in activity followed the hydrophobicity of ILs ([OMIM][PF_6] > [HMIM][PF_6] > [BMIM][PF_6]). Mutschler et al.[37] coated various ILs onto Novozym 435 beads, which resulted in higher conversions in lipase-mediated esterification of methyl-α-D-glucopyranoside with fatty acids than the uncoated lipase.

5.2.3.6 *Designing Hydrolase-Compatible ILs.* As mentioned earlier, the structures of ILs (as defined by the types of cations and anions, and their combinations) greatly influence the IL physical properties that are crucial for IL–enzyme interactions and enzyme stabilization. These properties include the polarity, H-bond basicity, anion nucleophilicity, IL network, kosmotropicity, viscosity, hydrophobicity, and so on. Therefore, it is important to customize the structures of ILs for particular biocatalytic

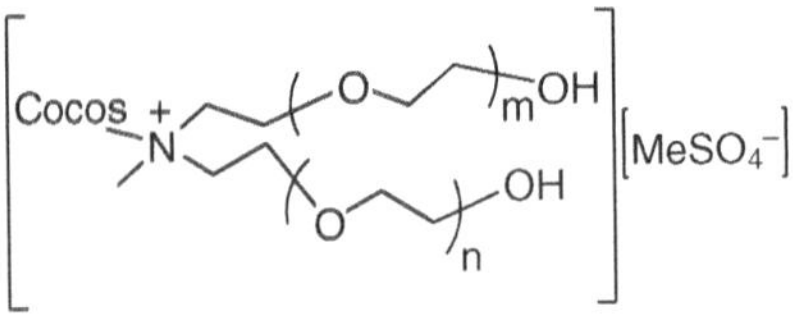

(a) AMMOENG 100
Cocos, C_{14} alkyl group; m + n = 4 – 14

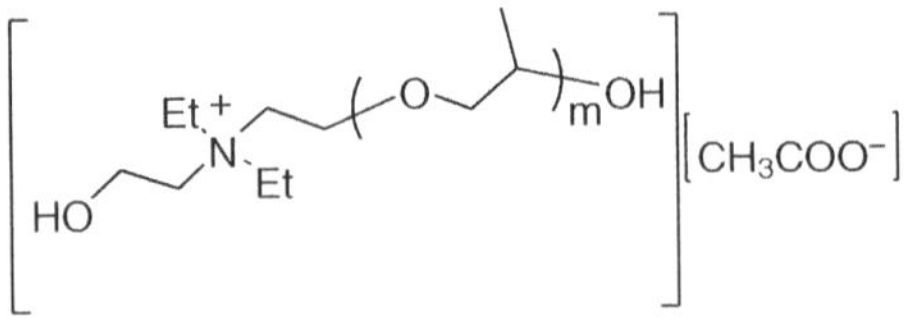

(b) AMMOENG 111
m = 50 – 60

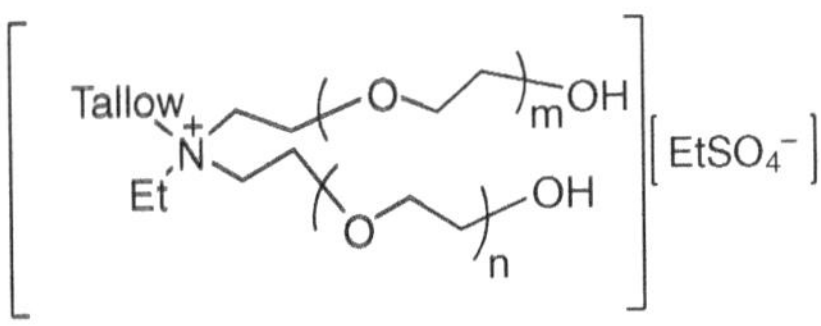

(c) AMMOENG 102
Tallow, C_{18} acyl group; m + n = 14 – 25

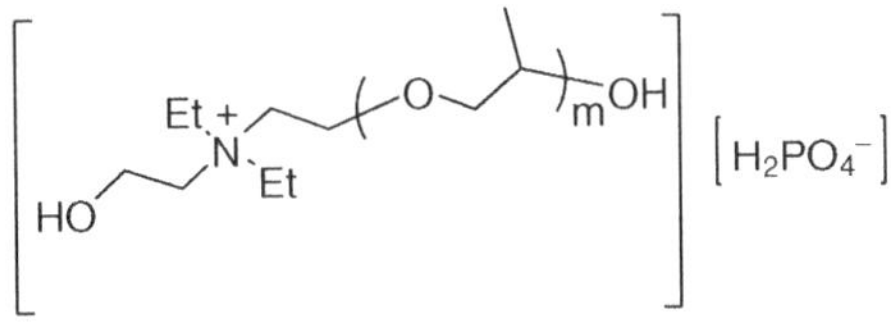

(d) AMMOENG 112
m = 50 – 60

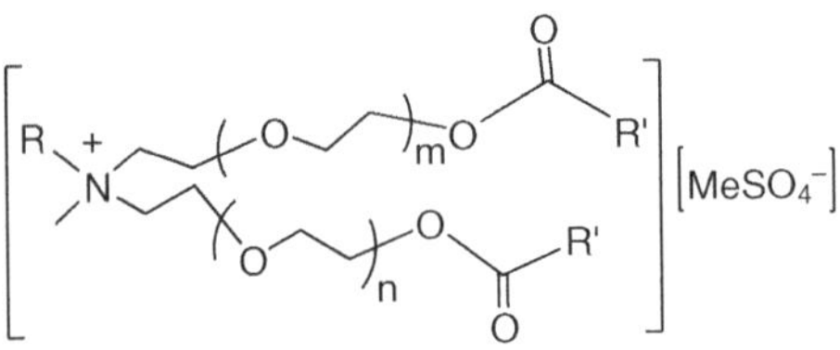

(e) AMMOENG 120
R, R', C_{18} acyl group; m, n, Unavailable

Scheme 5.3. Structures of tetraammonium-based ILs (AMMOENG™ series).

applications based on our current knowledge of IL structure and enzyme activity relationship. Several studies have tailored the IL structures to dictate the compatibility of enzymes. The Xu group[247–252] selectively studied a group of commercial tetraammonium-based ILs as reaction media for the enzymatic glycerolysis. As shown in Scheme 5.3, each of these ILs is an ionic mixture containing multiple alkyloxy groups, which have both hydrophilic and hydrophobic properties like PEG.

In particular, AMMOENG™ 100 (also known as [CPMA][$MeSO_4$]*) and 102 are capable of dissolving triglycerides and have shown to be lipase-compatible during the glycerolysis reaction;[248,249] trioctylmethylammonium bis(trifluoromethylsulfonyl)imide ([TOMA][Tf_2N]) and its mixture with AMMOENG 102 have also been demonstrated as suitable solvents for the enzymatic glycerolysis.[251–253] De Diego et al.[159] have further detected higher transesterification activities of both free and immobilized CALB in

* From the name of cocosalkyl pentaethoxy methylammonium methylsulfate.

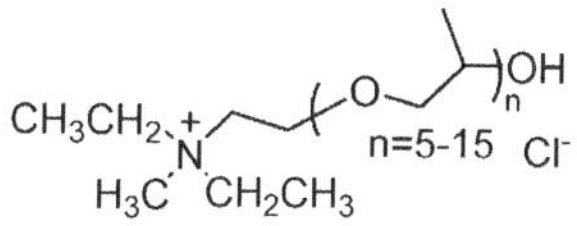

Scheme 5.4. Structure of AMMOENG™ 110.

[CPMA][$MeSO_4$] than in several PF_6^- and BF_4^- based ILs; however, the other two lipases (TLL and RML) seemed less active in [CPMA][$MeSO_4$] than in PF_6^- and BF_4^- based ILs. Xu and coworkers[248,250] utilized the conductor-like screening model for real solvents (COSMS-RS) to derive various parameters (such as misfit, H-bonding, and van der Waals interaction energy) to understand the multiple interactions in ILs; the model also provided some guidance in designing the structures of cations and anions.[254] The Kragl group[255] found that an IL in the AMMOENG family—AMMOENG 110 (Scheme 5.4)—was quite effective in forming aqueous two-phase (ATP) for the purification of active enzymes (two different alcohol dehydrogenases); the IL is capable of stabilizing the enzymes and enhancing the solubility of hydrophobic substrates. It is interesting to mention that oxygen-containing ILs (such as the AMMOENG series, and [C_2OHmim][Cl]) were used as additives in the enantioselective hydrolysis of diester malonates by pig liver esterase (PLE), and less than 1% of these ILs and 10% isopropanol/water were sufficient to improve the activity of PLE (up to four times) as well as the enantioselectivity.[256]

Abe et al.[244] synthesized an alkyloxy-containing hydrophobic IL named 2-methoxyethyl(tri-*n*-butyl)phosphonium bis(trifluoromethane)sulfonamide ([$MEBu_3P$] [Tf_2N]), and observed a faster reaction rate (lipase PS-catalyzed transesterification of secondary alcohols) in this IL than in diisopropyl ether. Lourenço et al.[257] measured little Novozym 435 activity in denaturing [BMIM][dca], but a high activity and enantioselectivity in [aliq][dca] (aliq = trioctylmethylammonium—from Aliquat 336®). The possible explanation is that the denaturing anion's molar concentration in [aliq][dca] is much lower than that in [BMIM][dca] due to the much higher molar mass of the former IL.[68,70,258] Vafiadi et al.[259] used two functionalized ILs, [C_2OHmim][PF_6] and [C_5O_2mim] [PF_6], as solvents for the feruloyl esterase-catalyzed esterification of glycerol with sinapic acid, and achieved high conversion yields (72.5% and 76.7%, respectively, in two ILs under optimal conditions). These two ILs are considered amphiphilic (hydrophilic cation and hydrophobic anion), and have relatively low viscosities.

Recently, the Abbott group[260–262] has demonstrated that a mixture of solid organic salt and a complexing agent can form a liquid at temperatures below 100°C, so-called *deep eutectic ILs*. The mechanism is that the complexing agent (typically an H-bond donor) interacts with the anion and increases its effective size, which in turn reduces the anion interaction with the cation and thus decreases the freezing point (T_f) of the mixture. A good example is the mixture of choline chloride (m.p. 302°C) and urea (m.p. 133°C) in a 1:2 molar ratio, resulting in a room-temperature IL (T_f = 12°C).[260] The major advantage of this approach is that inexpensive and nontoxic compounds can be used and the properties of the liquid can be finely tuned with different combinations of organic salts and complexing agents. Considering that many inexpensive quaternary ammonium salts are available and there is a wide variety of amides, amines, carboxylic acids, and alcohols that can be used as complexing agents, the possibility of forming new and inexpensive eutectic ILs is enormous. For example, choline (2-hydroxyethyl-trimethylammonium) chloride, so-called vitamin B_4, is produced on the scale of Mtonne (million metric tons) per year as an additive for chicken feed and many other applications. This ammonium salt is not only inexpensive and easy to make, but also it is nontoxic and biodegradable (as an essential micronutrient and human nutrient).[263] Therefore, eutectic ILs based on choline chloride have a promising future for large-scale applications. Eutectic ILs can dissolve many metal salts, aromatic acids, amino acids, glucose, citric acid, benzoic acid, and glycerol.[260,261,264–266] Gorke et al.[267] reported that several hydrolases (CALB, CALA, and epoxide hydrolase) maintained high activities in deep eutectic solvents (DESs) based on choline chloride or ethylammonium chloride (H-bond donors include acetamide, ethylene glycol, glycerol, urea, and malonic acid). Although some DESs contain potentially reactive alcohols (such as ethylene glycol or glycerol), ethylene glycol was ninefold less reactive than 1-butanol, and glycerol was >600-fold less reactive than 1-butanol in CALB-catalyzed transesterifications of butyl valerate. When used as co-solvents, DESs were capable of improving the reaction rate and/or conversion of hydrolases (esterases and epoxide hydrolase). The polarity study indicated that these DESs are more polar than common imidazolium-based ILs. The H-bond network in DESs is speculated to be responsible for reducing the chemical potential of the components of DESs and making them less reactive.

Recently, our group rationally synthesized a series of alkyloxyalkyl-containing ILs based on acetate (Scheme 5.5).[68] We found that these ILs can be designed to dissolve many substrates (such as cellulose, sugars, glucose, ascorbic acid, amino acids, betulinic acid, fatty acid, and triglycerides) that are not readily soluble in common organic solvents.[68,70,268–270] In addition, the IL structures can be optimized to be compatible with CALB, and thus become ideal solvents for producing the derivatives of many substrates via enzymatic reactions.[68,70,268] Therefore, it is possible to manipulate the structures of ILs to make them more compatible with enzymes, and to be able to dissolve a variety of compounds. This could significantly improve the catalytic efficiency of these reactions as some substrates may not be soluble in conventional organic media.

In summary, these enzyme-compatible ILs have some common structure features: (1) they usually have relatively large molecular structures so that H-bond basicity and nucleophilicity of anions are minimized; and (2) they often contain multiple ether and/

Scheme 5.5. Imidazolium- and ammonium-based ILs consisting of alkyloxyalkyl-substituted cation and acetate anion (abbreviated as [Me(OEt)$_n$-Et-Im][OAc] and [Me(OEt)$_n$-Et$_3$N][OAc] respectively) (n = 2, 3, . . .).

or hydroxyl groups so that the solvent properties (such as IL viscosity, H-bond basicity, and water affinity) are optimized for mild IL–enzyme interactions.

5.3 USE OF ILs FOR (DYNAMIC) KINETIC RESOLUTIONS ((D)KRs)

The enzymatic kinetic resolution can be accomplished through one of the two opposite reactions of the equilibrium system: hydrolysis or synthesis. Therefore, this section is divided into two general categories: resolution through enzymatic hydrolysis in aqueous solutions of ILs (5.3.1) and resolution through enzymatic synthesis in low-water ILs (Section 5.3.2).

5.3.1 Kinetic Resolutions via Hydrolysis in Aqueous Solutions of ILs

It is well known that organic solvents are routinely added into aqueous solutions to aid the dissolution of organic esters during the enzymatic resolution of these compounds. Because of the better substrate–enzyme interactions in homogeneous systems (vs. heterogeneous systems), high product yields and enantiomeric purities are often observed in aqueous solutions of organic solvents although some organic solvents are enzyme-denaturing.[271–274] As discussed in Section 5.2.1.4, the ion kosmotropicity is very important to the enzyme activity and stability in aqueous solutions of ILs. Therefore, it is possible to design the suitable IL structures (based on the empirical rule, kosmotropic anion + chaotropic cation) so that their aqueous solutions are enzyme-stabilizing and/or activating. The optimization of IL structures also depends on the specific enzymes, substrates, and IL concentrations.

5.3.1.1 Enantioselective Hydrolysis of Amino Acid Esters. A commercial form of *Bacillus licheniforms* protease, known as Alcalase, was found active in aqueous solutions of 15% (v/v) [EtPy][CF_3COO].[275] Consequently, the kinetic resolutions of several *N*-acetyl amino acids esters in this IL solution produced moderate to high product enantiomeric excess (86–97%). Similarly, porcine pancreas lipase (PPL) in the same IL solution yielded (*S*)-amino acids with optical purities between 73% and 98%.[276] The same enzyme Alcalase was also enantioselective in 15% (v/v) aqueous solutions of two BF_4^- based ILs ([EMIM][BF_4] and [EtPy][BF_4]) during the resolution of (*R*,*S*)-homophenylalanine ethyl ester, but it can be clearly seen that the enzyme's enantioselectivity dramatically decreased with the IL concentration above 15% (v/v).[277,278] As discussed in Section 5.2.1.4, CF_3COO^- is a kosmotropic while BF_4^- is chaotropic.

The Zong group[279–281] carried out the papain-catalyzed hydrolysis of (*R*,*S*)-*p*-hydroxyphenylglycine methyl ester in aqueous solutions of [BMIM][BF_4]. Papain exhibited a higher hydrolytic activity (~48% yield) and enantioselectivity (>96% ee) in phosphate buffer solution (pH 7.0) containing 12.5% (v/v) [BMIM][BF_4] than in solutions of organic solvents. A half-life time of 169 hours was obtained in 12.5% (v/v) [BMIM][BF_4] with the coexistence of the substrate, which was 9.2–16.8 times higher

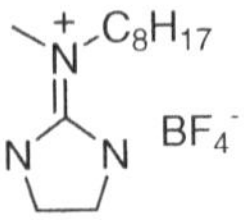

Scheme 5.6. Structure of alkylguanidinium-based IL [ETOMG][BF_4].

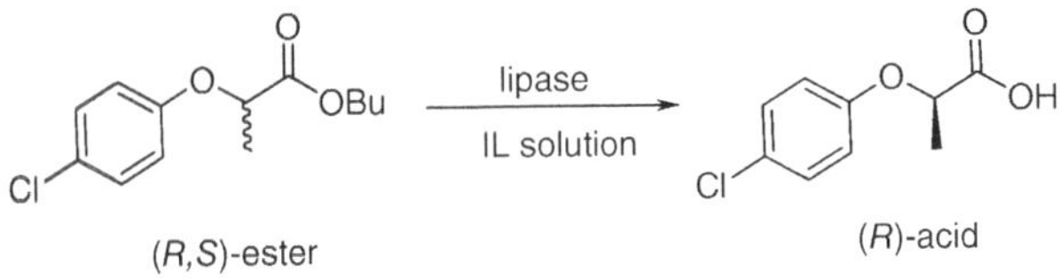

Scheme 5.7. Enzymatic hydrolysis of butyl 2-(4-chlorophenoxy)propionate in aqueous buffer of IL.

than those in aqueous solutions of organic solvents. The same group[282] also examined the Novozym 435-catalyzed enantioselective hydrolysis of (*R*,*S*)-phenylglycine methyl ester to produce enatiopure (*R*)-phenylglycine in aqueous solutions of [BMIM][BF_4]. A higher enantioselectivity was observed in 20% (v/v) [BMIM][BF_4] than in aqueous solutions of organic solvents. Under optimal conditions, a high enantioselectivity (ee_s = 93.8%, substrate conversion = 53.0%) was achieved in 20% (v/v) [BMIM][BF_4]. The Cao group[283] conducted the CALB-catalyzed kinetic resolution of *N*-(2-ethyl-6-methylphenyl)alanine methyl ester to yield (*S*)-amino acid in aqueous solutions of alkylguanidinium-based ILs (BF_4^- and PF_6^-). The best result was obtained in 50% aqueous solution of [ETOMG][BF_4] (Scheme 5.6): 38.4% conversion and 92.3% ee_p, while in the phosphate buffer the conversion was 45.3% but the ee_p was only 78.7%.

Recently, our group[116] studied the kinetic hydrolysis of phenylalanine methyl ester catalyzed by *Bacillus licheniformis* protease in aqueous solutions of several hydrophilic ILs (0.5 M). The protease enantioselectivity was in a decreasing order in the presence of the anions PO_4^{3-} > citrate^{3-}, CH_3COO^-, $EtSO_4^-$, CF_3COO^- > Br^- > OTs^-, BF_4^-, and in the presence of the cations $EMIM^+$ > $BMIM^+$ > $HMIM^+$. Our group[117] also carried out the enzymatic resolution of phenylalanine methyl ester in aqueous solutions of these ILs carrying anions of chiral- or ω-amino acids, and observed higher enantioselectivities and yields in ILs based on D-amino acids rather than in those derived from L-isomers. The enzymatic resolutions of several amino acids by *Bacillus licheniformis* protease were conducted in 2.0 M [EMIM][OAc], producing corresponding (*S*)-amino acids with high optical purities (82–98%).[121] The effect of ion kosmotropicity is discussed in Section 5.2.1.4.

5.3.1.2 *Enantioselective Hydrolysis of Other Esters.* Mohile et al.[284] investigated the CRL-catalyzed enantioselective hydrolysis of butyl 2-(4-chlorophenoxy) propionate in aqueous buffer of ILs (Scheme 5.7). When comparing with 47% ee in aqueous buffer, improved enantioselectivities (99% ees) were observed in 50% (v/v)

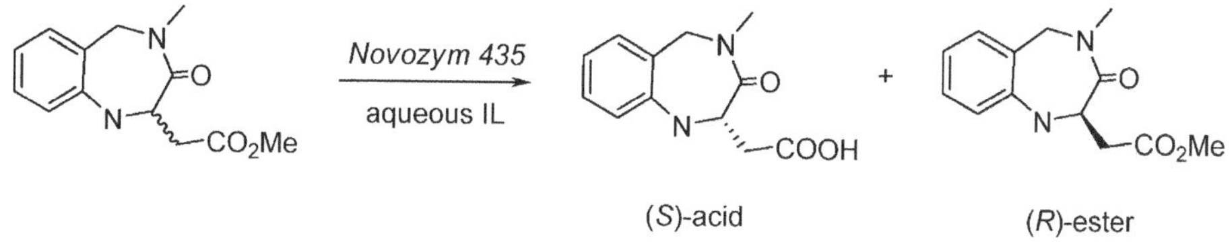

Scheme 5.8. Lipase-catalyzed enantioselective hydrolysis of 2,3,4,5-tetrahydro-4-methyl-3-oxo-1*H*-1,4-benzodiazepine-2-acetic acid methyl ester.

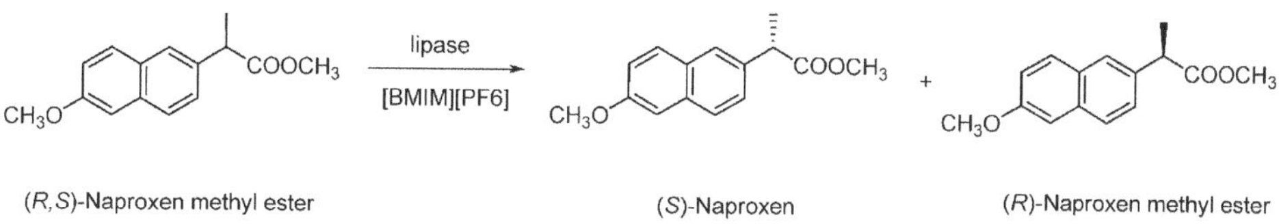

Scheme 5.9. Lipase-catalyzed kinetic resolution of (*R*,*S*)-Naproxen.

aqueous solutions of BF_4^- and PF_6^- based ILs, while the conversions increased with the IL hydrophobicity [BMIM][BF_4] < [HMIM][BF_4] < [BMIM][PF_6].

Roberts et al.[285] carried out the Novozym 435-catalyzed enzymatic resolution of 2,3,4,5-tetrahydro-4-methyl-3-oxo-1*H*-1,4-benzodiazepine-2-acetic acid methyl ester to (*S*)-acid (Scheme 5.8), which is a key intermediate in the synthesis of the glycoprotein antagonist Lotrafiban. The industrial process of this reaction is conducted in *t*-butanol (88% v/v). The use of ILs (>75 v% in water) increased the substrate solubility and the ability to operate at higher temperatures, which further improved the overall reaction rate by a factor of four while maintaining the same overall yield of 47% (99% product ee). The lipase activity and enantioselectivity in 88% (v/v) IL aqueous solutions decreased in the order of [BMIM][Tf_2N] > [BMIM][PF_6] > [BMIM][OTf] > [BMIM][BF_4]. The recyclability of the enzyme was demonstrated, but the authors also raised the concern about the solute mass transfer in ILs at large scale.

Xin et al.[286,287] used the water-saturated ILs as media for the kinetic resolution of (*R*,*S*)-Naproxen methyl ester catalyzed by CRL immobilized on silica (Scheme 5.9). Among several ILs based on PF_6^-, BF_4^- and heptylsulfate, [BMIM][PF_6] was selected as the best IL in terms of generating overall highest conversion and enantiomeric excess. The residual lipase activities in different ILs were in a decreasing order of [BMIM][PF_6] > [HMIM][PF_6] > [HMIM][heptylsulfate] > [HMIM][BF_4] > [BMIM][BF_4]. Under optimal conditions, (*S*)-Naproxen was obtained in 98.2% ee (28.3% conversion) after 72 hours of reaction. The immobilized lipase was reused for five times with a slight loss of activity.

Xanthakis et al.[288] performed the kinetic resolutions of isomers of 2-(4-methoxybenzyl) cyclohexyl acetates catalyzed by lipases (Novozym 435, Lipozyme IM [Novo Nordisk, Denmark], and free CALB) in aqueous buffer of <15% (v/v) [BMPy][Cl] or [MMIM]

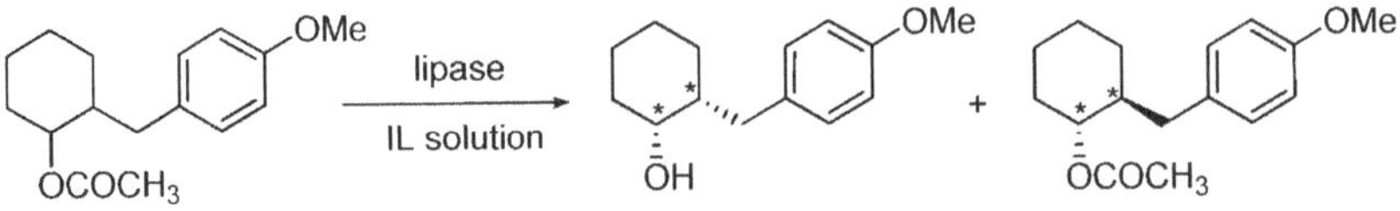

Scheme 5.10. Enzymatic resolution of 2-(4-methoxybenzyl)cyclohexyl acetates.

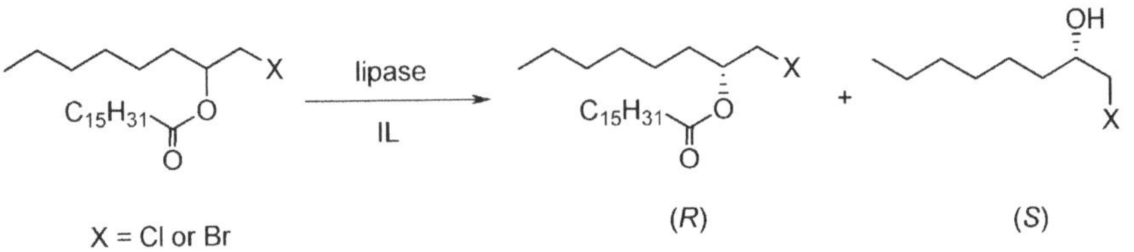

Scheme 5.11. Lipase-mediated kinetic resolution of 1-halo-2-octyl palmitates.

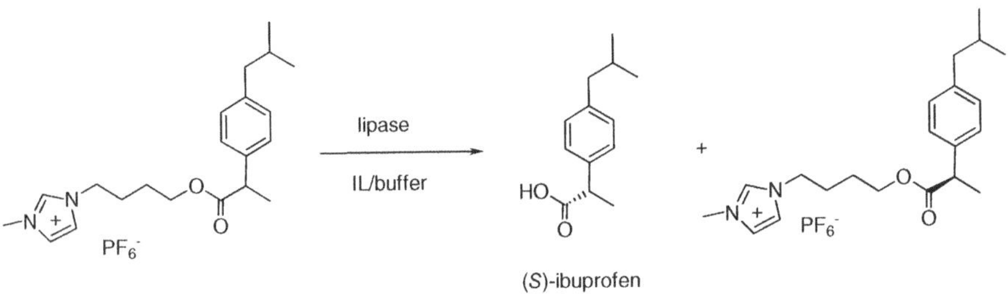

Scheme 5.12. Lipase-catalyzed kinetic resolution of IL-anchored ibuprofen ester.

[$MeSO_4$] (Scheme 5.10). The enantioselective hydrolysis led to stereoisomers of 2-(4-methoxybenzyl)cyclohexanol and in some cases stereoisomers of the deracemized substrate through controlling the reaction conditions (such as the type of IL, the type of lipase, and the addition of acetonitrile as auxiliary solvent).

Oromí-Farrús et al.[289] prepared (*S*)-1-chloro-2-octanol and (*S*)-1-bromo-2-octanol through the enantioselective hydrolysis of halohydrin palmitates (Scheme 5.11) catalyzed by lipases (Novozym 435, lipase from *Rhizomucor miehei*, and lipase from *Rhizopus oryzae*) in *t*-butanol or ILs ([BMIM][PF_6] and [OMIM][BF_4]) containing about 1% (v/v) water. When the reaction was catalyzed by Novozym 435 in [BMIM][PF_6], 98% ee_s and 51% conversion were obtained for 1-chloro-2-octanol palmitate, and >99% ee_s and 35% conversion were obtained for 1-bromo-2-octanol palmitate.

Interestingly, some studies incorporated the IL structure into the substrates for improved enzymatic resolutions. For example, Naik et al.[290] anchored ibuprofen with a hydroxyl group appended IL, and hydrolyzed the anchored ibuprofen by lipases in a 50/50 (v/v) mixture of organic solvent (or IL) and 0.1 M phosphate buffer (Scheme 5.12). When the reaction in DMSO/buffer was catalyzed by CALB, (*S*)-ibuprofen was obtained with 86% ee and the isolated yield of 87% of theory. The same reaction in

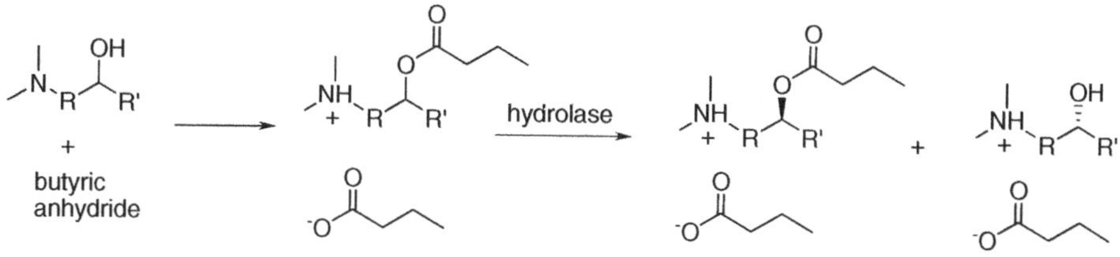

Scheme 5.13. One-pot enzymatic resolution of protic IL form of tertiary amino esters.

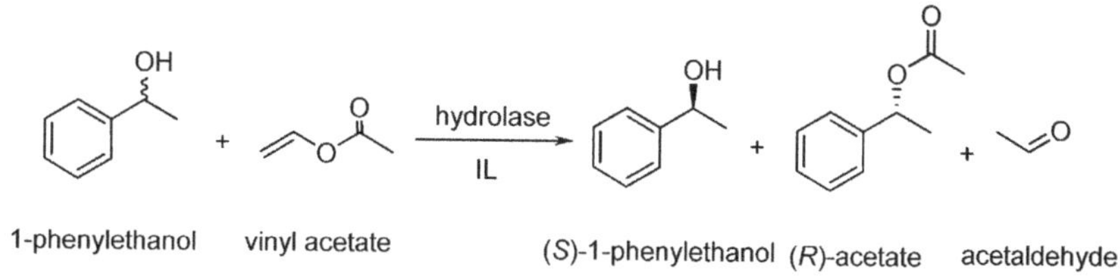

Scheme 5.14. Enantioselective transesterification of 1-phenylethanol and vinyl acetate.

[BMIM][PF_6]/buffer catalyzed by CALB produced 80% ee and the isolated yield of 80% of theory.

Brossat et al.[291] treated tertiary amino esters (such as quinuclidin-3-ol) with butyric anhydride to form the protic IL form of substrate (the first step in Scheme 5.13). The resulting IL forms of amino esters were hydrolyzed by the solution of subtilisin (protease from *Bacillus licheniformis*). (*S*)-Quinuclidin-3-ol was obtained in 90.4% purity (ee) and 52% conversion. Upon the completion of reaction, unreacted esters were easily separated from the product alcohols by extraction with hexane.

5.3.2 Kinetic Resolution via Synthesis in Nonaqueous Solutions of ILs

5.3.2.1 Evaluating Hydrolase's Enantioselectivity via the Kinetic Resolution of 1-phenylethanol. To evaluate the enantioselectivity of hydrolases in pure ILs (containing small amounts of water), it is common to conduct a model reaction such as the enantioselective transesterification of 1-phenylethanol with an acyl donor (such as vinyl acetate). As shown in Scheme 5.14, (*R*)-1-phenylethanol is usually selectively converted into corresponding ester while the (*S*)-enantiomer remains unchanged. Vinyl acetate is frequently used in this reaction because it is an activated acyl donor, and the by-product (i.e., acetaldehyde) is volatile (m.p. 21°C) and thus the equilibrium is irreversible; as a result, the reaction rate is faster than using regular esters. However, there have been some concerns that the accumulation of acetaldehyde oligomer may cause some degree of enzyme inactivation.[161,292–294]

The following section lists a few examples of using the model reaction (Scheme 5.14) in evaluating the enzyme enantioselectivity. Schöfer et al.[39] examined the model kinetic resolution (Scheme 5.14) catalyzed by several lipases in a variety of imidazolium and pyridinium ILs based on BF_4^-, OTf^-, $MeSO_4^-$, PF_6^-, and Tf_2N^-, and obtained comparable activity and enantioselectivity (in some cases, improved enantioselectivity) in these ILs than those in MTBE. Park and Kazlauskas[35] conducted the same reaction but catalyzed by *Pseudomonas cepacia* lipase (PCL), and observed that the reaction rate and enantioselectivity were similar to those in toluene, with a higher IL polarity leading to a higher lipase activity. Eckstein et al.[131] reported that *Pseudomonas* sp. lipase showed a higher enantioselectivity in [BMIM][Tf_2N] at low water activities ($a_w < 0.53$) than in MTBE, and the enzyme was also more thermally stable in [BMIM][Tf_2N] than in MTBE. Persson and Bornscheuer[295] studied the specific activity and enantioselectivity of celite-immobilized esterases from *Bacillus subtilis* and *Bacillus stearothermophilus* in organic solvents and ILs, and found the activities in ILs ([BMIM][Tf_2N] > [BMIM][PF_6] > [BMIM][BF_4]) were lower than that those in *n*-hexane, but the enantioselectivity was independent of the solvent; in addition, much higher stabilities of esterase from *Bacillus stearothermophilus* at 40°C was seen in [BMIM][BF_4] and [BMIM][PF_6] than those in *n*-hexane and MTBE. Maruyama et al.[222] examined the catalytic behavior of PEG–lipase complex using the model reaction in Scheme 5.14, and suggested the enantioselectivities obtained in ILs (Tf_2N^- and PF_6^-) were higher than or comparable to that in *n*-hexane.

Itoh et al.[243] prepared new imidazolium ILs carrying the anion of poly(oxyethylene) alkyl sulfate, and used them as additives or coating materials for lipase-catalyzed transesterification in diisopropyl ether; an increased enantioselectivity of *Burkholderia cepacia* lipase (lipase PS-C) was obtained when the IL was added at 3–10 mol% in the reaction; on the other hand, lipase PS coated with IL showed noticeably high activities and enantioselectivities in the kinetic resolution of a variety of secondary alcohols. Toral et al.[67] indicated that cross-linked CALB was able to maintain a reasonable reaction rate and high enantioselectivity in denaturing IL [BMIM][NO_3]. Shah et al.[185] immobilized the CRL on the MWNT; the immobilized enzyme showed a high enantioselectivity in the kinetic resolution of 1-phenylethanol in [BMIM][PF_6] (free CRL gave only 5% conversion after 36 hours, but the immobilized lipase yielded in 37% conversion and >99% ee). Ha et al.[146] observed that Novozym 435 was more active in hydrophobic ILs (Tf_2N^- and PF_6^-) than in hydrophilic ILs (BF_4^- and OTf^-), but the enantioselectivity was independent of the solvent. The kinetic resolution of 1-phenylethanol catalyzed by lipases was studied in [BMIM][Tf_2N], [BMIM][PF_6] and [BMIM][BF_4], and was further optimized with respect to several reaction parameters (such as enzyme concentration, temperature, and substrate concentration).[296,297]

Instead of using the most common vinyl acetate as the acyl donor, Lozano et al.[298] performed the kinetic resolution of 1-phenylethanol with vinyl propionate catalyzed by free CALB, and found a higher enzyme activity and stability in [btma][Tf_2N] than in hexane.

5.3.2.2 *Kinetic Resolutions of Other Alcohols.* Kim et al.[299] studied the kinetic resolutions of four secondary alcohols in [EMIM][BF_4] and [BMIM][PF_6] through the transesterification reactions with vinyl acetate (Scheme 5.15) catalyzed by immobilized CALB or free PCL. The results of this study demonstrated that the lipases

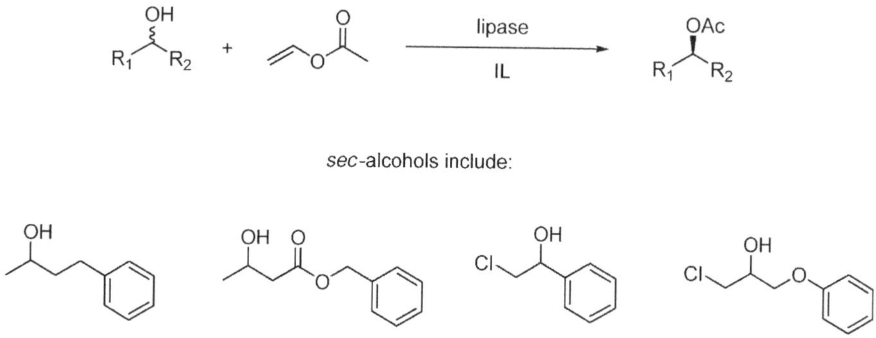

Scheme 5.15. Kinetic resolution of secondary alcohols via transesterification in ILs.

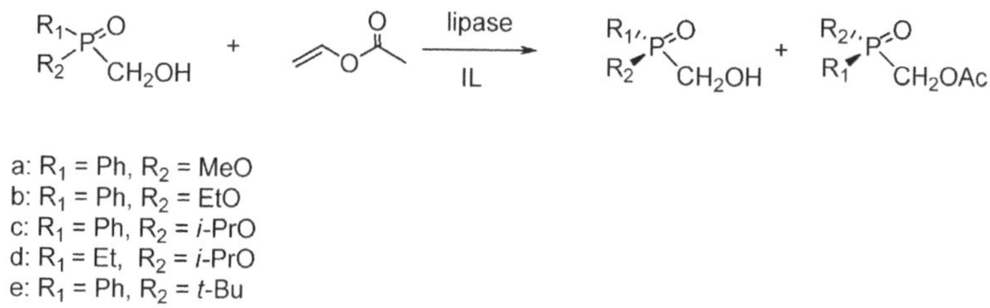

Scheme 5.16. Kinetic resolution of hydroxymethanephosphinates and hydroxymethylphosphine oxides.

were up to 25 times more enantioselective in ILs than in organic solvents (tetrahydrofuran [THF] and toluene).

Kielbasinski et al.[300] conducted the acetylation of racemic prochiral hydroxymethanephosphinates and hydroxymethylphosphine oxides catalyzed by lipase AK (Amano), or lipase from *Pseudomonas fluorescens* (Scheme 5.16). Both lipases were more enantioselective in [BMIM][PF_6] (up to six times) than in diisopropyl ether. However, none of the enzymes was enantioselective in [BMIM][BF_4] despite their comparable enzyme activities.

Itoh et al.[294] argued that oligomerization of acetaldehyde (by-product of vinyl acetate) was caused by the acidic proton from C2-position of the 1-butyl-3-methylimidazolium salts. As mentioned earlier, the accumulation of acetaldehyde oligomer may induce the enzyme inactivation.[161,292–294] Following this logic, Itoh et al.[294] synthesized a new IL without acidic proton at C2-position of the imidazolium ring, that is, 1-butyl-2,3-dimethylimidazolium tetrafluoroborate ([BDMIM][BF_4]). As expected, no accumulation of an acetaldehyde oligomer was observed in this IL as confirmed by the ^1H NMR analysis. The Novozym 435-catalyzed kinetic resolution of 5-phenyl-1-penten-3-ol in this IL (Scheme 5.17) exhibited a high reactivity and enantioselectivity, and the enzyme was reused for nine more times without losing any catalytic performance. However, the same enzyme gradually lost activities when being reused in

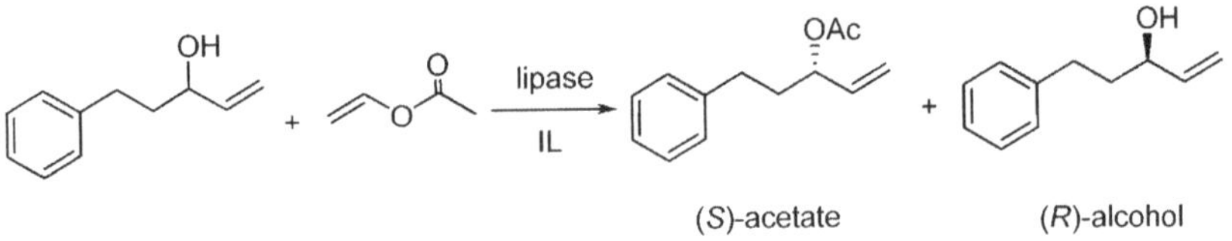

Scheme 5.17. Lipase-catalyzed enantioselective transesterification of 5-phenyl-1-penten-3-ol.

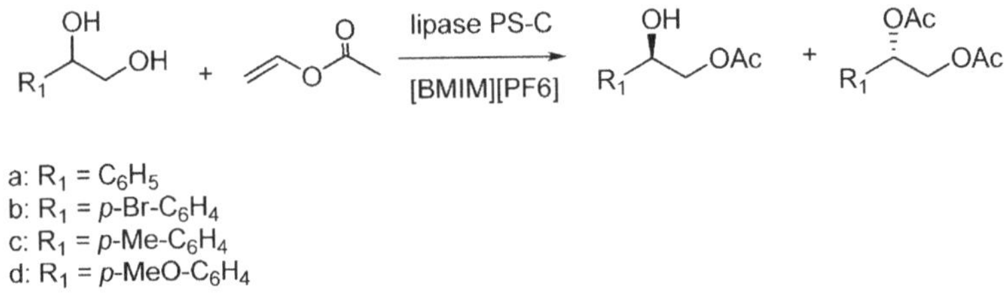

Scheme 5.18. Enzymatic resolution of 1,2-diols in [BMIM][PF_6].

[BMIM][BF_4] or [BMIM][PF_6]. Surprisingly, Novozym 435 showed no activity in [BDMIM][PF_6] for the same reaction.

In another study, Itoh et al.[161] investigated the same reaction as shown in Scheme 5.17 in different ILs, and observed higher Novozym 435 activities in ILs based on BF_4^- and PF_6^- but lower activities in CF_3COO^-, OTf^-, and SbF_6^- based ILs (although the enantioselectivity was almost independent of the solvent). To avoid the possible inhibitory action of acetaldehyde oligomer resulting from vinyl acetate, Itoh et al.[161] also switched the acyl donor to different methyl esters (such as methyl phenylthioacetate) in the enzymatic resolution of 5-phenyl-1-penten-3-ol; the by-product methanol was removed *in situ* via reduced pressure conditions (20–100 Torr), and a high reactivity and enantioselectivity of Novozym 435 were obtained in [BMIM][PF_6]. The Itoh group[301] also conducted the same transesterification (Scheme 5.17) catalyzed by Novozym 435 in different imidazolium alkyl sulfate; high enantioselectivities were achieved, but conversions were much lower than that in [BMIM][PF_6].

Enantiopure 1,2-diols are important precursors towards the synthesis of pharmaceutical molecules. Kamal et al.[302] carried out the kinetic resolutions of a variety of 1,2-diols (Scheme 5.18) in organic solvents and [BMIM][PF_6] catalyzed by lipase from *Pseudomonas cepacia* (PS-C, Amano). Higher enantioselectivities were obtained in [BMIM][PF_6] than in organic solvents (*n*-hexane/THF (4/l), diisopropyl ether and toluene).

Noël et al.[179] performed the enzymatic resolution of 2-pentanol catalyzed by free CALB using vinyl propionate at 2% (v/v) water content (Scheme 5.19). They observed the synthetic activity of CALB was higher in [BMIM][Tf_2N] than that in hexane (up to 2.5 times), and the enantioselectivity (ee) in [BMIM][Tf_2N] was as high as 99.99%. They also found the optimal temperature was 60°C and the optimal pH was 7. In addi-

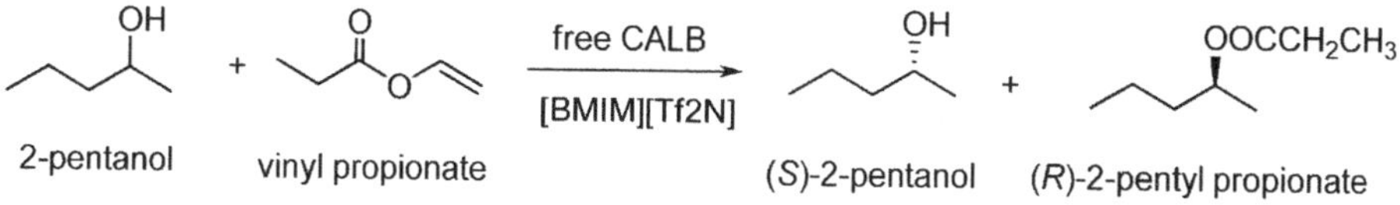

Scheme 5.19. Free CALB-catalyzed kinetic resolution of 2-pentanol.

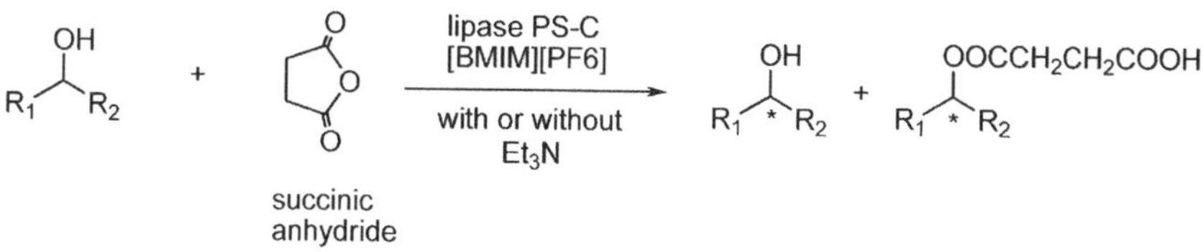

Scheme 5.20. Enzymatic acylation of secondary alcohols with succinic anhydride in [BMIM][PF_6].

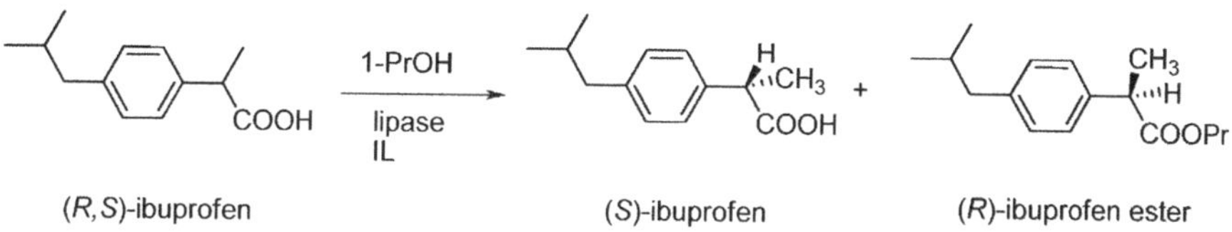

Scheme 5.21. Enantioselective esterification of (*R,S*)-ibuprofen with 1-propanol.

tion, they noticed that a higher water activity (a_w) induced a higher synthetic activity, but a lower selectivity.

Rasalkar et al.[303] resolved several secondary alcohols through enantioselective esterification with succinic anhydride catalyzed by *Pseudomonas cepacia* immobilized on celite (PS-C) in [BMIM][PF_6] (Scheme 5.20). The use of triethylamine (Et_3N) as an organic base was able to shorten the reaction times while producing comparable yields and enantioselectivities as those without any additive. The use of triethylamine as a nonreactive additive in enzymatic reactions has been known capable of improving the reaction rate and enantioselectivity for enzymes.[36,304–306]

Yu et al.[307] studied CRL-catalyzed esterification of ibuprofen with 1-propanol in seven ILs as well as in isooctane (Scheme 5.21). The lipase showed a higher enantioselectivity (ee_s = 38%) in [BMIM][PF_6] and a comparable enantioselectivity (32%) in [BMIM][BF_4] when comparing with that (33%) in isooctane, but very poor enantioselectivities (3–7%) in other ILs based on $MeSO_4^-$, $OctSO_4^-$, OTs^-, and N_3^-. In another report,[308] several lipases were screened for the same reaction (Scheme 5.21) in a biphasic system of isooctane and [BMIM][PF_6] (or [BMIM][BF_4]). The best enzyme with respect to the enantioselectivity was CRL (E = 8.5, ee_s = 60%), followed by the lipase

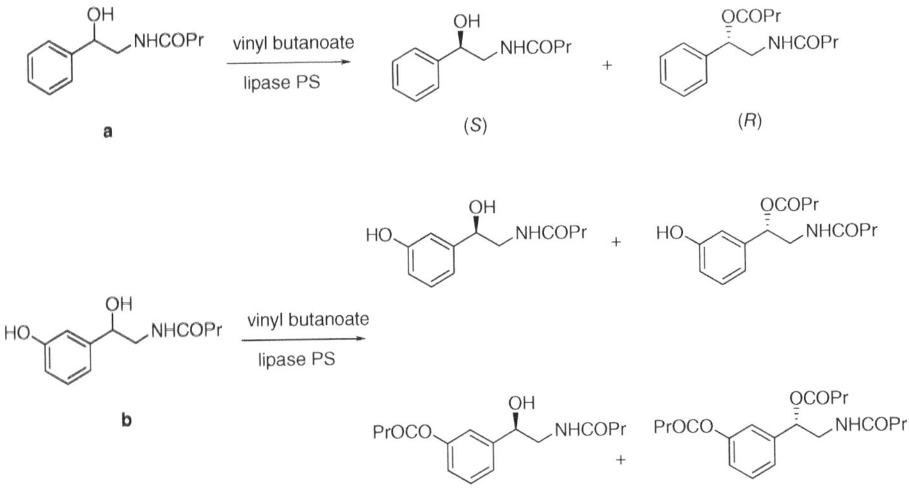

Scheme 5.22. Lipase PS-catalyzed acylation of *N*-acylated amino alcohols. (a) *N*-acylated 2-amino-1-phenylethanol, (b) *N*-acylated norphenylephrine.

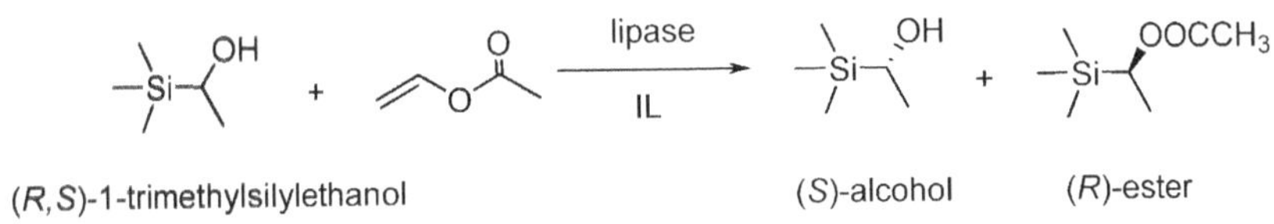

Scheme 5.23. Lipase-mediated kinetic resolution of (*R*,*S*)-1-trimethylsilylethanol.

from *Aspergillus niger* AC-54 ($E = 4.6$). [BMIM][PF_6] was shown more effective in improving the enantioselectivity than [BMIM][BF_4].

Lundell et al.[309] conducted the resolution of *N*-acylated 2-amino-1-phenylethanol (Scheme 5.22a) and *N*-acylated norphenylephrine (Scheme 5.22b) catalyzed by lipase PS-C II (*Burkholderia cepacia* lipase immobilized on ceramic particles) in imidazolium- and pyridinium-type ILs (Tf_2N^-, PF_6^-, BF_4^-, and OTf^-), MTBE, and their mixtures. Overall, a lower lipase reactivity was observed in ILs than in toluene/THF or in MTBE. The use of microwave irradiation showed no improvement in reactivity and selectivity versus the conventional heating when the acylation reactions (Scheme 5.22) were carried out in [EMIM][Tf_2N] or its mixture with MTBE.

As organosilicon compounds are important for asymmetry synthesis, Lou and Zong[142] investigated the enantioselective acylation of (*R*,*S*)-1-trimethylsilylethanol with vinyl acetate catalyzed by lipases in several ILs (Scheme 5.23). After examining a few lipases, Novozym 435 was identified as the best enzyme in delivering the highest enantioselectivity ($ee_s = 90.7\%$ and $E = 152$). An enhancement in reaction rates and enantioselectivities was observed in ILs (BF_4^- and PF_6^-) when comparing with those in organic solvents. The hydrophobic [BMIM][PF_6] produced the fastest initial rate and highest enantioselectivity (in contrast to more hydrophilic ILs based on BF_4^-). After

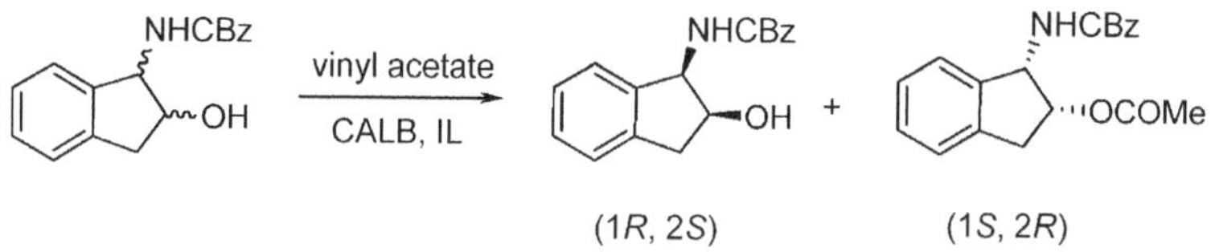

Scheme 5.24. Enzymatic resolution of (±)-*cis*-benzyl *N*-(1-hydroxyindan-2-yl)carbamate.

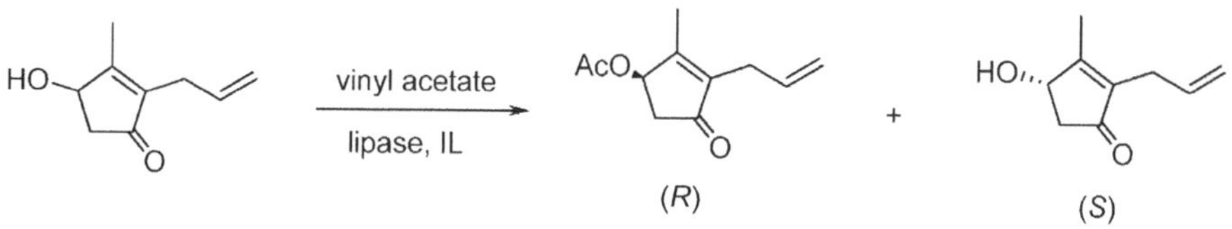

Scheme 5.25. Resolution of racemic allethrolone through lipase-catalyzed transesterification.

the reaction optimization, the enantiomeric excess (ee) of (*S*)-1-trimethylsilylethanol was 97.1% with the substrate conversion of 50.7% at 6 hours.

Wu et al.[306] carried out the kinetic resolution of secondary alcohols via the enantioselective transesterification with vinyl acetate catalyzed by lipases (reaction scheme is similar to Scheme 5.15 with different alcohols). Novozym 435 gave the highest enantioselectivities when comparing with other two lipases (porcine pancreatic lipase and CRL). In general, Novozym 435 showed a higher enantioselectivity in [BMIM][PF_6] than in [BMIM][BF_4] or hexane. Again, the addition of catalytic amounts of organic base (triethylamine or pyridine) considerably boosted the reaction rate and enantioselectivity in [BMIM][PF_6].

Lourenço et al.[257] conducted the enantioselective acylation of (±)-*cis*-benzyl *N*-(1-hydroxyindan-2-yl)carbamate catalyzed by Novozym 435 in ILs (Scheme 5.24). The substrate is poorly soluble in ILs based on BF_4^-, PF_6^-, and Tf_2N^-, resulting in low conversions; the enzyme was inactivated by the denaturing [BMIM][dca].[67–69] On the other hand, the substrate showed a high solubility in [aliq][dca], and Novozym 435 showed high activities and enantioselectivities in this IL.

Wang and Mei[310] lyophilized *Burkholderia cepacia* lipase (lipase PS) with cyclodextrins to formulate the enzyme preparation for use in the kinetic resolution of allethrolone in [BMIM][PF_6] (Scheme 5.25). They thoroughly evaluated the type and amount of cyclodextrins used in the lyophilization, and found that the activity of the lyophilized lipase increased with the increase in the amount of cyclodextrins. The lipase lyophilized with cyclodextrins showed higher initial rates but lower stability than the native enzyme.

Das et al.[311] developed an efficient process for resolving *cis*-(±)-4-*O*-TBS-2-cyclopentene-1-ol through porcine pancreatin lipase-catalyzed transesterification reaction in [OMIM][PF_6] (Scheme 5.26). This method afforded high enantiomeric purities

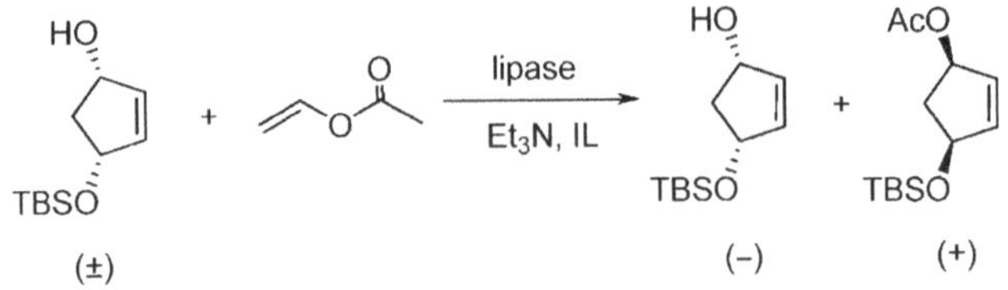

Scheme 5.26. Enantioselective transesterification of *cis*-(±)-4-*O*-TBS-2-cyclopentene-1-ol.

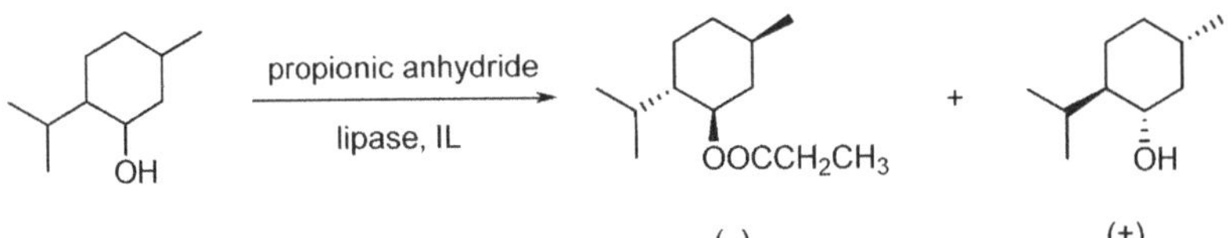

Scheme 5.27. Lipase-catalyzed enantioselective esterification of (±)-menthol.

and yields for both alcohol (ee > 99%, yield 49.9%) and the acetate (ee > 96%, yield 49.8%).

The Sun group[312] focused on the enantioselective esterification of (±)-menthol by CRL in [BMIM][PF_6], [BMIM][BF_4], and organic solvents (Scheme 5.27). Higher enantioselectivities were observed in [BMIM][PF_6] and hexane than that in [BMIM][BF_4]. They also optimized the reaction temperature and substrate molar ratio. The highest enantioselectivities were 88.5% (ee_p) and 43.4% (conversion) in hexane, 86.3% (ee_p) and 39.3% (conversion) in [BMIM][PF_6]. Meanwhile, the lipase showed a higher stability in [BMIM][PF_6] than in hexane. To further improve this resolution reaction, the same group[313] immobilized CRL on magnetic diethylaminoethyl chloride (DEAE-Cl)–modified poly(glycidyl methacrylate) (GMA)–ethylene glycol dimethacrylate (EDMA) microspheres, and found the pH value at the immobilization ("pH memory") was the key to the catalytic performance of lipase. When CRL was immobilized at the optimal pH = 5.0, (−)-menthyl propionate with ee > 90% was achieved by the esterification reaction in [BMIM][PF_6] ($a_w = 0.07$).

Shen et al.[147] reported an efficient one-pot synthesis of optically active *O*-acetyl cyanohydrins via lipase-catalyzed kinetic resolution of the *in situ* generated racemic cyanohydrins (Scheme 5.28a) or *O*-acetyl cyanohydrins (Scheme 5.28b) using [OMIM][PF_6]. Moderate to high enantioselectivities (73–98% ee) of *O*-acetyl cyanohydrins were obtained by both methods.

Hara et al.[205] studied the enantioselective acylations of three secondary alcohols (Scheme 5.29) catalyzed by different preparations of lipase PS from *Burkholderia cepacia* (free, immobilized in a sol–gel, and CLEA). The reactions were examined in dried organic solvents (toluene, diisopropyl ether, and MTBE), ILs (Tf_2N^-, PF_6^- and BF_4^-), or their mixtures. This study has shown that ILs reduced the initial reaction rates of the lipase preparations compared with the reactions in those organic solvents or in

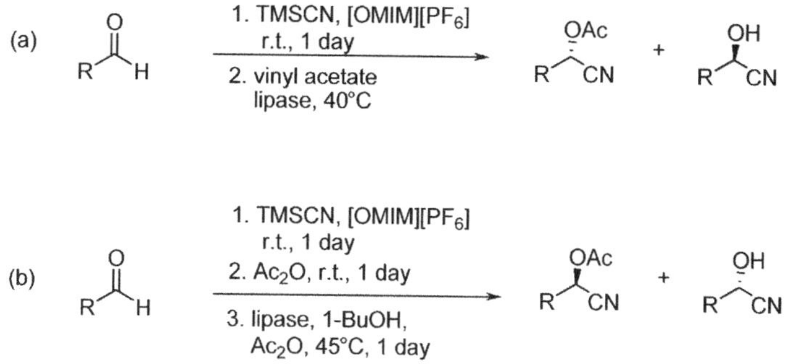

Scheme 5.28. One-pot kinetic resolution of (a) *in situ* generated racemic cyanohydrins and (b) *O*-acetyl cyanohydrins.

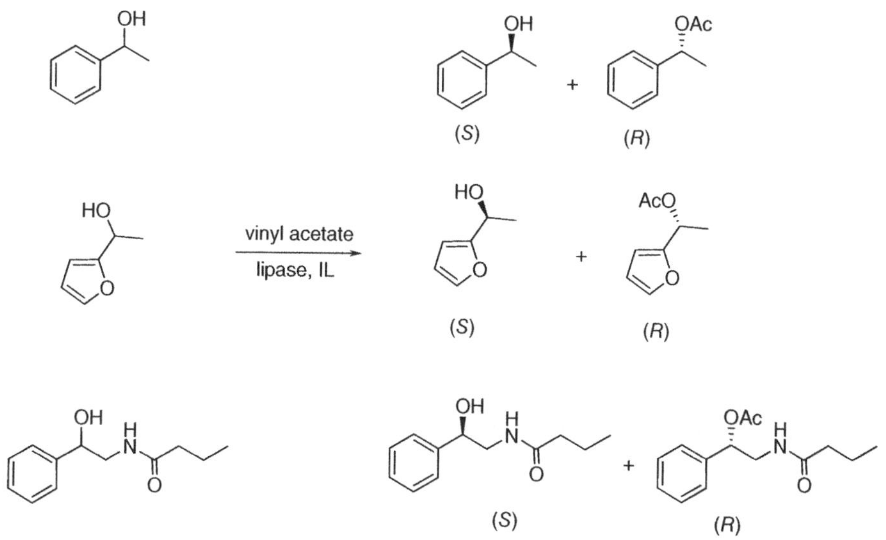

Scheme 5.29. Kinetic resolution of three secondary alcohols using *Burkholderia cepacia* lipase.

their mixtures with an IL. The authors also indicated that it is impossible to rank the performance of ILs without considering the nature of the substrate and the ester product.

As a key intermediate in the synthesis of chiral drug (*S*)-Lubeluzole, (*R*)-1-chloro-3-(3,4-difluorophenoxy)-2-propanol was resolved from its racemic mixture (Scheme 5.30) through the lipase-catalyzed transesterification with vinyl butyrate in a two-phase system consisting of hexane and an IL ([BIMI][PF_6] or [BMIM][BF_4]).[314] The highest conversion (>49%) and enantiomeric excess (ee > 99.9%) were obtained in 6 hours at 30°C with [BMIm][PF_6] as the co-solvent. The stability study on the lipase from *Pseudomonas aeruginosa* MTCC 5113 in ILs suggested that the enzyme stability and activity were in the order of [BMIM][PF_6] > [BMIM][BF_4] in two-phase systems.

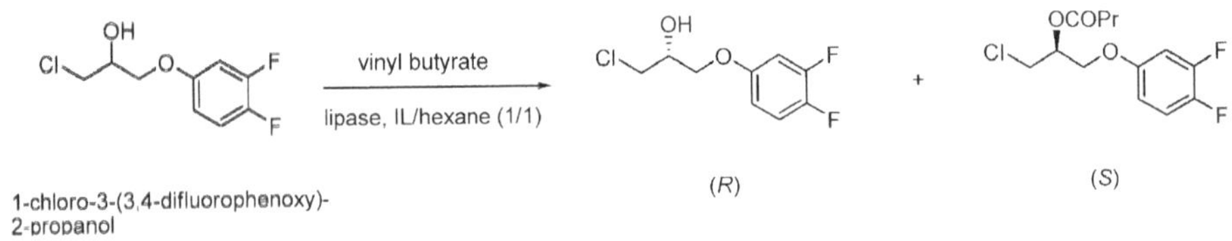

Scheme 5.30. Enzymatic transesterification of (*R*,*S*)-1-chloro-3-(3,4-difluorophenoxy)-2-propanol.

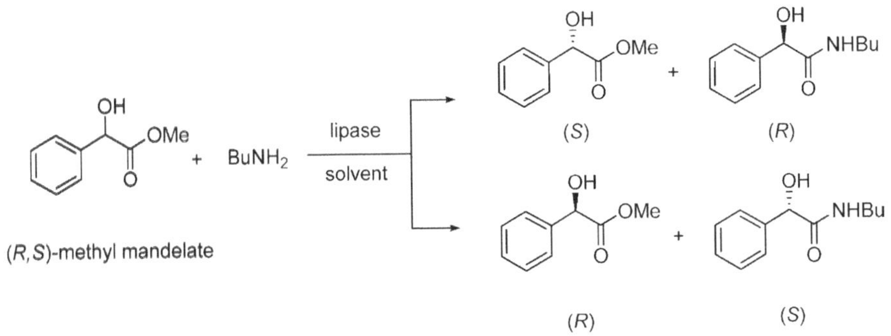

Scheme 5.31. Aminolysis of (*R*,*S*)-methyl mandelate with *n*-butylamine ($BuNH_2$).

As discussed above, the most common method in enzymatic resolutions of secondary alcohols is the (trans)esterification with acyl donors of (vinyl) esters, carboxylic acids, or anhydrides. However, as shown in the following example, it is possible to achieve the resolution through the enzymatic aminolysis reaction. Pilissão and Nascimento[315] carried out the kinetic resolution of (*R*,*S*)-methyl mandelate with *n*-butylamine by lipases in organic solvents (*n*-hexane, *t*-butanol, and chloroform) or in their mixtures with ILs ([BMIM][BF_4] and [BMIM][PF_6]) (organic solvent/IL, 10/1, v/v). As shown in Scheme 5.31, the amide configurations are dependent on the type of organic solvents. The high enantiomeric excess ($ee_p > 99\%$ and $E > 200$) of amides were obtained in mixtures of chloroform (or *t*-butanol)/ionic liquids (10:1 v/v) with Novozym 435 as the catalyst.

Lou et al.[63] compared the lipase-catalyzed asymmetric ammonolysis of (*R*,*S*)-*p*-hydroxyphenylglycine methyl ester with ammonium carbamate as the ammonia donor in different ILs and organic solvents (Scheme 5.32). The use of ILs as solvents improved the enantioselectivity of the ammonolysis when compared with organic solvents, but the reaction rates in ILs were much lower than that in *t*-butanol. However, a mixture of IL and *t*-butanol was able to enhance both the initial rate and enantioselectivity of the reaction ($ee_p > 90\%$ and yield $> 45\%$).

A recent trend of improving the resolution of racemic mixtures in ILs is the so-called dynamic kinetic resolution (DKR) technique. DKR is the coupling of the kinetic resolution of one enantiomer with the simultaneous racemization of its opposite enan-

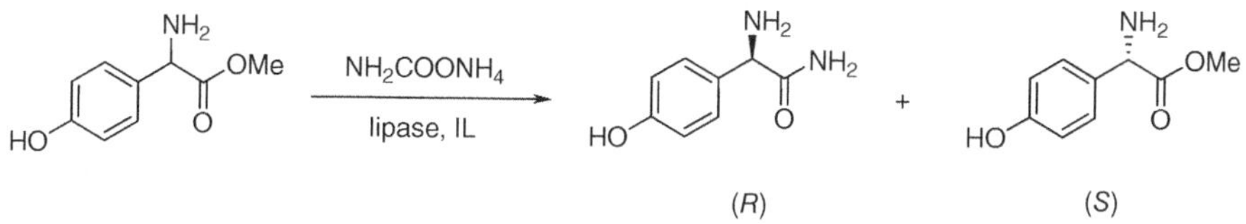

Scheme 5.32. Lipase-mediated ammonolysis of (*R*,*S*)-*p*-hydroxyphenylglycine methyl ester.

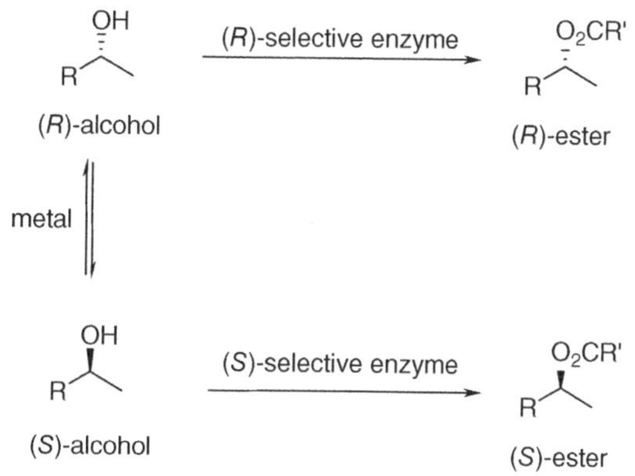

Scheme 5.33. DKR of secondary alcohols by enzyme–metal combo-catalysis.

tiomer in one step. The racemization process can be catalyzed by a variety of catalysts, such as metal complexes,[316] racemase,[317] silica modified with benzenosulfonic acid (SCX),[318] and acidic zeolite.[319] Kim et al.[316] studied the DKR reactions of secondary alcohols by lipase–ruthenium or subtilisin–ruthenium combo-catalysis in [BMIM][PF_6]. As shown in Scheme 5.33, the ruthenium complex catalysts racemized the secondary alcohols between (*R*)- and (*S*)- enantiomers, and the enzyme selectively acylated one of the enantiomers (in this case, (*R*)-selective is lipase-PS, and (*S*)-selective is subtilisin–CLEC). Using (*S*)-1-phenylethanol as the substrate, that study observed a much fast racemization catalyzed by the cymene–ruthenium complex in two ILs ([BMIM][PF_6] and [BMIM][BF_4]) than in organic solvents (THF, toluene, and dichloromethane). The (*R*)-selective DKR reactions of secondary alcohols by lipase–ruthenium combo-catalysis mostly completed in 2–3 days, achieving 85–92% yields, and 99% product ees. The (*S*)-selective DKR reactions of secondary alcohols by subtilisin–ruthenium combo-catalysis took a longer time (6 days), resulting in 78–92% yields and 82–97% product ees.

Another example of the DKR in ILs was demonstrated by Kaftzik et al.[317] in the resolution of mandelic acid by a lipase-mandelate racemase two-enzyme system (Scheme 5.34): mandelate racemase catalyzed the racemization of (*R*)-mandelic acid into (*S*)-enantiomer, while *Pseudomonas* sp. lipase PS selectively catalyzed the acylation of (*S*)-mandelic acid. Among a number of ILs (based on $MeSO_4^-$, $OctSO_4^-$, Tf_2N^-, PF_6^-, BF_4^-, and OTs^-) investigated, [MMIM][$MeSO_4$] and [BMIM][$OctSO_4$] at water

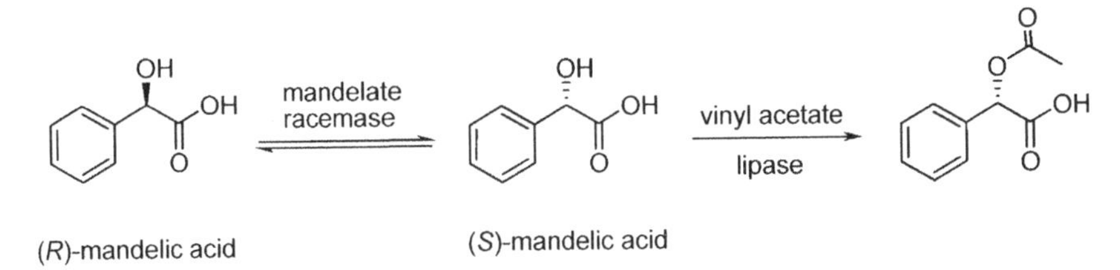

Scheme 5.34. DKR of mandelic acid by a lipase–mandelate racemase two-enzyme system.

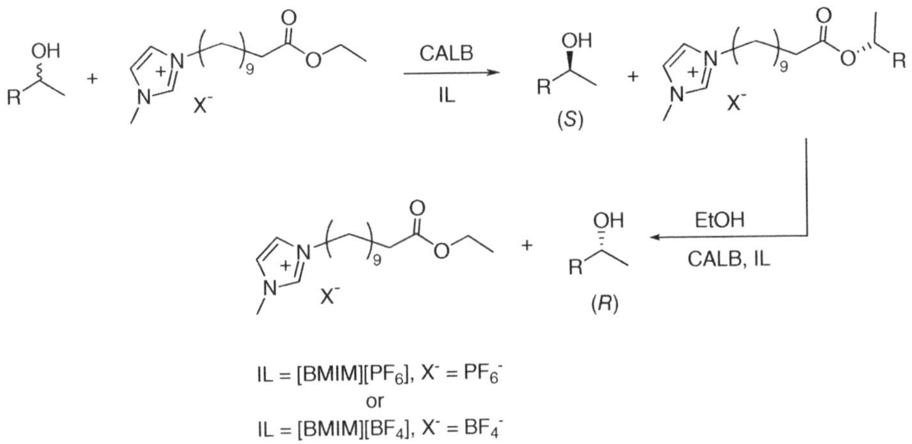

Scheme 5.35. One-pot enzymatic resolution and separation of *sec*-alcohols using ionic acylating agents in ILs.

activity $a_w > 0.74$, and a biphasic system of water and [OMIM][PF_6] (1/10, v/v) were found suitable for maintaining the activities of mandelate racemase. But the lipase exhibited the hydrolytic tendency (vs. synthesis) at $a_w > 0.74$. Therefore, the biphasic systems of water and [OMIM][PF_6] (1/9) was used in the DKR of mandelic acid, leading to the formation of (*S*)-acetyl mandelic acid. However, during the DKR operation, mandelate racemase seemed to be inactivated by vinyl acetate,[317] or acetaldehyde oligomer as discussed earlier.[161,292–294] More examples on DKR in IL/scCO_2 biphasic systems are discussed in Section 5.3.2.5.

Another interesting advance in the resolution of secondary alcohols is the use of ionic acylating agents for the enzymatic transesterification in ILs ([BMIM][PF_6] and [BMIM][BF_4]).[320] As shown in Scheme 5.35, in the first step, the (*R*)-alcohol in the racemic mixture reacts with the ionic acylation agent and the remaining (*S*)-enantiomer is removed via extraction; in the second step, the ionic acylation agent was regenerated by reacting with ethanol and the released (*R*)-alcohol is obtained via extraction. The main advantage of this method over other processes is the separation of both free enantiomers by enzymatic resolution in a one-pot reaction using one equivalent of acylating agent. The success of this approach relies on the ionic solvent being able to

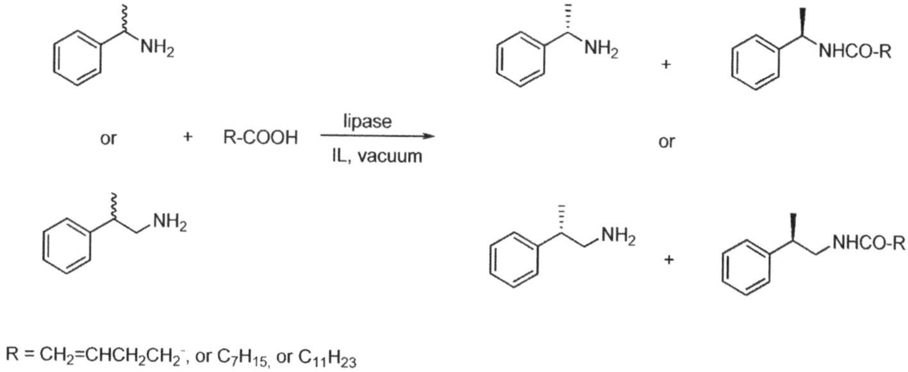

Scheme 5.36. Lipase-catalyzed acylation of 1-phenylethylamine and 2-phenyl-1-propylamine.

dissolve one of the enantiomers as an ionic ester moiety. Moderately high ees (50–81%) were obtained for (*S*)-alcohol, while very high ees (91–99%) were achieved for (*R*)-alcohol. The reusability of the reaction media was also demonstrated.

5.3.2.3 Kinetic Resolutions of Amines. Optically pure amines are valuable intermediates in asymmetric synthesis. As discussed earlier, during the enzymatic acylation of alcohols, vinyl esters are common acylating agents. However, for enantioselective acylation of amines, less reactive compounds (such as esters or carbonates) are sufficient for the much more nucleophilic amines to minimize spontaneous nonenzymatic reactions. Lipase-catalyzed enantioselective acylation of amines has been challenging due to several limiting factors.[54]

Irimescu and Kato[54,321] conducted the immobilized CALB-catalyzed enantioselective acylation of two primary amines (i.e., 1-phenylethylamine and 2-phenyl-1-propylamine) under reduced pressures in different ILs (Scheme 5.36). The structures of ILs strongly influence the lipase activity and enantioselectivity; the role of anion's nucleophilicity is discussed in Section 5.2.1.2; the effect of cation's hydrophobicity is discussed in Section 5.2.1.6. For 1-phenylethylamine, high enantioselectivities ($ee_p > 99\%$) were achieved in all ILs, but for 2-phenyl-1-propylamine, only moderate enantioselectivities ($ee_p < 60\%$) were obtained. The analytical scale kinetic resolution of 1-phenylethylamine produced amide (>99% purity) with 81% yield in 72 hours.

Toral et al.[67] carried out the cross-linked CALB-catalyzed acylation between 1-phenylethylamine and 2-methoxyacetate in denaturing [BMIM][NO_3], and observed a complete conversion (50%) of the (*S*)-enantiomer and a high enantiomeric purity of the remaining amine of 94% (corresponding to $E = 95$).

5.3.2.4 Kinetic Resolutions Integrated with Supported IL Membranes (SILMs) or Microfluidic Separation. The supported liquid membrane (SLM) is widely recognized as a selective separation technique. In the SLM, pores in porous supports are impregnated with a liquid. The SLM has numerous advantages including

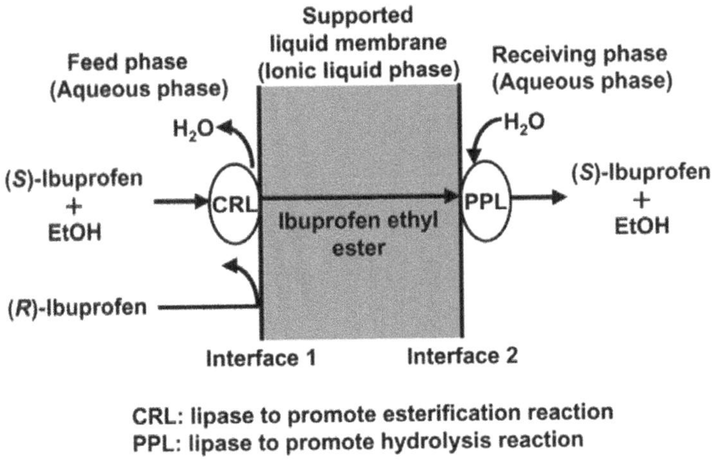

Figure 5.6. Enantioselective transport of (*S*)-ibuprofen through a lipase-facilitated supported IL membrane (SILM).[325] Reproduced with permission of The Royal Society of Chemistry.

the use of minimal amount of solvent, the combination of both extraction and stripping steps into one, and a high organic utilizing efficiency.[322] However, the industrial use of this technique is still limited and mainly restricted to the separation of metal ions and gases because of concerns of the SLM stability and long-term performance. SLMs with conventional liquids gradually deteriorate due to the liquid vaporization, dissolution into a contacting phase, and displacement from the porous structure under low-pressure gradient (<10 kPa).[323] On the other hand, ILs have been investigated as alternative solvents to SLM due to the negligible vapor pressure of most ILs. SILMs offer many advantages over conventional SLMs such as (1) the minimum loss of impregnated liquid in the membrane due to vaporization, (2) high carrier loadings, (3) tunable properties of the membrane solvent, (4) high stability and flux rates, and (5) improved selectivity of the liquid membrane.[324]

Miyako et al.[325] demonstrated the lipase-facilitated transport of (*S*)-ibuprofen through an SILM. As shown in Figure 5.6, CRL was used in the feed phase (Interface 1) to selectively convert (*S*)-ibuprofen to corresponding (*S*)-ester, and (*R*)-ibuprofen was collected from the feed phase; the (*S*)-ester then dissolved in the IL phase of SILM and diffused through the membrane; in the receiving phase (Interface 2), lipase from porcine pancreas (PPL) hydrolyzed the (*S*)-ester to produce the water-soluble (*S*)-ibuprofen. This study examined the effect of different ILs and organic solvents as the liquid membranes in SILMs. The SILMs based on ILs (PF_6^- and Tf_2N^-) enabled higher initial-permeate flux of ibuprofen than those based on organic solvents (C_5-C_{12} alkanes), with [BMIM][Tf_2N] being selected as the best IL. At 48 hours of operation, (*R*)-ibuprofen with 75.1% ee was obtained in the feed phase using CRL, and (*S*)-ibuprofen with 75.1% ee was collected in the receiving phase using lipase from porcine pancreas (PPL) (see Chapters 3 and 8).

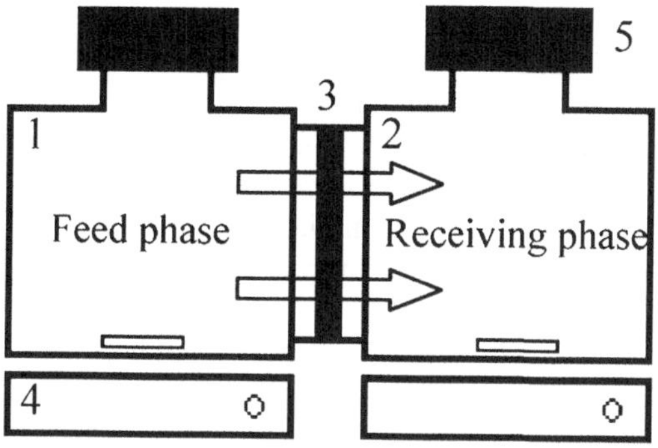

Figure 5.7. Schematic illustration of the glass diffusion cell set up with two independent compartments used for experiments: (1) feed solution containing solutes in *n*-hexane; (2) receiving solution containing fresh solvent of *n*-hexane; (3) supported liquid membrane; (4) magnetic stirrer; (5) septum.[326] Reproduced with permission of John Wiley & Sons.

The Víllora group[323] designed a membrane reactor containing an SILM (see Figure 5.7) and applied it to the kinetic resolution of 1-phenylethanol catalyzed by Novozym 435 (the reaction scheme is similar to Scheme 5.14 with various acyl donors). This group screened six ILs (based on BF_4^-, PF_6^-, and Tf_2N^-) as the liquid membrane phase supported in a nylon membrane, and observed the average permeoselectivity values increased for the same cation in the order of $Tf_2N^- < PF_6^- < BF_4^-$, and for the same anion, in the order of $OMIM^+ < BMIM^+$. [BMIM][BF_4] was identified as the best IL in this example for the selective separation of the target compounds in *n*-hexane. Since the solvent (*n*-hexane) for both feed and receiving phases is hydrophobic, the hydrophilic [BMIM][BF_4] might help the dissolution and transport of the solute molecules (alcohols, acids, and esters). At optimum conditions, (*S*)-phenylethanol was almost completely separated in the receiving phase after 24 hours of operation when vinyl laurate was used as the acyl donor. This group[326] further focused on the effect of water content in the medium on the synthetic activity, selectivity, and enantioselectivity of the lipase in the same reaction. They observed that for all three acyl donors (vinyl propionate, vinyl butyrate, and vinyl laurate), the synthetic activity of Novozym 435 versus the water content exhibited a bell-shape curve, with a maximum activity at ~100 ppm of water; on the other hand, the selectivity decreased with the increase in water content. At optimal conditions, (*S*)-1-phenylethanol was successfully separated in the receiving phase from the (*R*)-1-phenylethanol, which remained in the feed phase as the ester product ((*R*)-1-phenylethyl butyrate). The Víllora group[327] also applied this integrated reaction/separation process for the resolution of *rac*-2-pentanol using vinyl propionate and vinyl butyrate as acyl donors. The highest efficiency was found with the longest vinyl ester, vinyl butyrate. At optimal conditions established previously ([BMIM][BF_4] as liquid membrane phase; water content of 100 ppm and acyl-donor of vinyl butyrate), a separation factor of 72 at 450 minutes of operation was achieved.

In summary, the integration of the enzyme enantioselectivity with the selectivity of a highly stable SILM enables a promising development of new methodologies for the industrial enriching of chiral compounds.

Another interesting development is the coupling of SILM with microfluidic devices for the selective and rapid separation of racemic mixtures at the analytical scale.[328] Microfluidic technology has enabled the miniaturization of analytical devices for high-throughput applications including fast analysis of biomolecules.[329] The microfluid-based devices have many advantages such as smaller volumes of reagents and raw sample, speed of response, and cost-effectiveness.[330] Huh et al.[328] illustrated the configuration of the microchannel with a three-phase flow, and the transport of (*R*,*S*)-ibuprofen through the IL flow (ILF) in the microchannel. The involved reactions and selective diffusion of (*R*,*S*)-ibuprofen are similar to the discussion for Figure 5.6, where (*S*)-ibuprofen was selectively esterified and transported through the membrane via ILF. At an ILF rate of 0.30 mL/h, (*S*)-ibuprofen with 78% ee was detected in the receiving phase, and (*R*)-ibuprofen with 48% ee was obtained in the feed phase. Although the SILM system could transport more solutes from the feed phase to the receiving phase given sufficient reaction time (20–40 hours), the microfludic system affords a fast and selective separation of solutes (~30–60 seconds) at the analytical scale.

5.3.2.5 Kinetic Resolution Using IL/scCO$_2$ Biphasic Systems. The non-volatility of most ILs is the "green" feature of these new solvents, but also creates challenges for product separation and recovery. To overcome such a limitation, another type of "green" solvent, supercritical fluids (SCFs) have been explored for the product recovery from ILs (see also Chapter 4).[24] Among SCFs, supercritical CO_2 (scCO_2) is most commonly used in decaffeination, extraction of natural products, and dry cleaning.[331] Typically, the volatile and nonpolar scCO_2 forms two-phase systems with non-volatile and polar ILs. The principle of product recovery by these biphasic systems is based on the solubility of scCO_2 in the IL (controlled by pressure) to transfer organic products to scCO_2-rich phase, and the insolubility of the IL in scCO_2. The solubility of ILs in pure scCO_2 is normally very low.[332,333] However, when scCO_2 contained high contents (>10 mol%) of polar solutes (e.g., ethanol, acetone), the solubility of [BMIM][PF_6] in scCO_2 was considerably high.[332] On the other hand, since scCO_2 can be dissolved in ILs under high pressures, it is likely to dissolve some CO_2 in ILs to form a homogenous phase through the so-called miscibility switch.[334–336] For example, Bermejo et al.[71] formed a homogeneous system for enzymatic reaction by dissolving 3–12 wt% CALB, 5–20 mol% CO_2 in [HOPMIm][NO_3] in a temperature range of 30–75°C and pressures of up to 12 Mpa.

In addition to their applications in extractions, the biphasic IL/scCO_2 systems have been investigated in metal-catalyzed and enzyme-catalyzed reactions. This unique combination of reaction and separation offers unique advantages:[24] (1) organometallic and enzymatic catalysts are stable and may be soluble in ILs, but have low solubility in scCO_2; (2) many organic compounds are soluble in scCO_2, enabling the easy separation of products from ILs; (3) this process can be designed as batchwise or continuous operations. The biphasic IL/scCO_2 systems also offer an additional advantage for enzymatic reactions, that is, tunable solvent properties for enzyme activation and stabiliza-

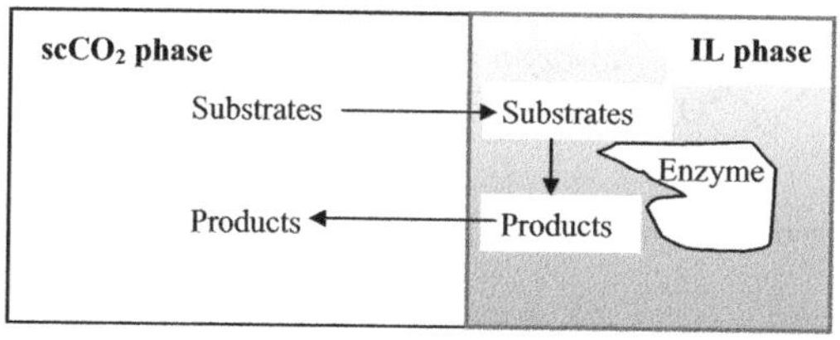

Figure 5.8. Illustration of an enzymatic reaction in the biphasic IL/$scCO_2$ system.

tion. ILs provide a rather polar environment for enzymes, but the polar environment can be adjusted by nonpolar $scCO_2$ to improve the enzyme activity.[171] It is also important to realize that $scCO_2$ alone has inactivation effects on enzymes, and the possible reason might be due to the local pH change induced by CO_2 molecules, or conformational changes during the pressurization/depressurization steps.[337,338]

Lozano et al.[163,339] developed a continuous biocatalytic process for enzymatic reactions in biphasic IL/$scCO_2$ systems. In this process (Figure 5.8), aqueous CALB was dissolved in the IL ([EMIM][Tf_2N] or [BMIM][Tf_2N]) phase ("working phase," where the enzymatic conversion takes place), and substrates and products mainly located in the supercritical phase ("extractive phase"). Using the IL/$scCO_2$ system, the kinetic resolution of 1-phenylethanol with vinyl propionate was shown to be very enantioselective and only (*R*)-1-phenylethyl propionate was produced. This study demonstrated the protective effect of ILs towards enzyme deactivation by temperature and/or CO_2. The protective effect of ILs was further confirmed by the same group[340] during the kinetic resolution of the same reaction in IL/$scCO_2$ systems catalyzed by free and immobilized CALB at denaturing temperatures (120 and 150°C). The explanation of such protective effects of ILs is give by the Iborra group[340,341]: As a liquid immobilization support, ILs interact with enzyme molecules through multipoint interactions (such as ionic, hydrogen bonds, and van der Waals), which form a supramolecular network to maintain the active protein conformation (see Section 5.2.1.3).

Similarly, Reetz et al.[342] conducted the batchwise CALB-catalyzed resolution of 1-phenylethanol with vinyl acetate in [BMIM][Tf_2N] using $scCO_2$ as the mobile phase. The products and the by-product (acetaldehyde) were collected in a cold trap from the gas stream upon venting. The substrate conversions were 43–51%, and (*R*)-phenylethyl acetate was produced in 99% ee_S. The same group[343] further demonstrated the batchwise and continuous flow enzymatic resolutions of several secondary alcohols with vinyl acetate (or vinyl laurate) using IL/$scCO_2$ media. High enantioselectivities were obtained when the alcohols were selectively converted to CALB to corresponding esters. The efficient product separation using SCF was also achieved when alcohols were extracted preferentially from their corresponding laurates.

Lozano et al.[344] immobilized free CALB by physical adsorption onto 12 silica supports modified with different side chains (such as alkyl, amino, carboxylic, and nitrile). The enzyme preparations were investigated in the kinetic resolution of 1-phenylethanol with vinyl propionate in IL/$scCO_2$ biphasic systems. Higher synthetic activities (up to six times) of immobilized enzymes coated with ILs were observed in supercritical

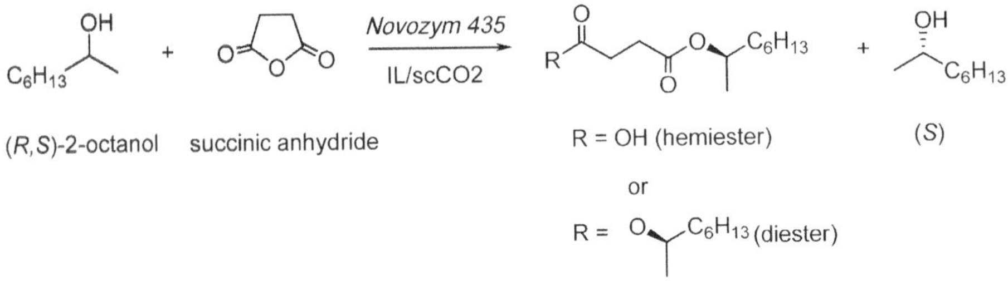

Scheme 5.37. Kinetic resolution of (*R*,*S*)-2-octanol with succinic anhydride in IL/$scCO_2$ media.

media than those in hexane. The resolution reaction catalyzed by CALB-C4-silica in [btma][Tf_2N]/$scCO_2$ medium produced 48% yield of (*R*)-1-phenylethyl propionate with high ee of >99.9% in continuous operation at 50°C and 10 MPa. They suggested that the "philicity" between the alkyl side chain of both IL and silica is a critical parameter for optimizing the immobilized enzyme/IL system for reactions in SCFs.

Bogel-Lukasik et al.[345] measured the vapor-liquid equilibrium (VLE) data for systems containing supercritical CO_2, [OMIM][PF_6] (or [OMIM][dca]), and products of the enzymatic resolution of (*R*,*S*)-2-octanol with succinic anhydride catalyzed by Novozym 435 (Scheme 5.37). The VLE data suggest the solubility of 2-octanol in CO_2 is over 1 order of magnitude higher than those of hemiester and diester, and the solubility of CO_2 in [OMIM][PF_6] is higher than that in [OMIM][dca]. It is also important to know that no IL was detected in the $scCO_2$ phase. Consequently, partition coefficients (the distribution of organic compounds between $scCO_2$ and IL phases) of 2-octanol are one order of magnitude higher than those of hemiester, and two orders of magnitude higher than those of diester under the same range of pressures and initial concentrations of the solutes. Therefore, an extraction experiment after the resolution reaction in [OMIM][PF_6] recovered >99.99 mol % of unreacted (*S*)-2-octanol with low coextraction of other products (3.69 mol % of hemiester and 0.73 mol % of diester), leading to a high ee_s of 98.42%.

In addition, several DKR reactions were also studied in IL/$scCO_2$ systems. The Iborra group[298,318] carried out the chemo-enzymatic DKR of 1-phenylethanol with vinyl propionate in IL/$scCO_2$ biphasic systems using both Novozym 435 and silica modified with benzenosulfonic acid (SCX) catalysts at 40°C and 10 MPa. The racemization rates of (*S*)-1-phenylethanol catalyzed by SCX in ILs (Tf_2N^- and PF_6^-) were 2–3 times faster than that in hexane. When both chemical and enzymatic catalysts were coated with ILs, good yields (78–82%) of (*R*)-1-phenylethyl propionate with high enantioselectivities (ee = 91–98%) were achieved in continuous chemo-enzymatic DKR processes. The same group[319,346] also investigated the same resolution reaction in IL/$scCO_2$ biphasic media catalyzed by both Novozym 435 and acidic zeolite at 50°C and 10 MPa. Among several zeolite catalysts studied, the H-Beta CP811E zeolite reduced the ee of (*S*)-1-phenylethanol from 100% to 53.6% in 3 hours. However, some acidic zeolites were also observed to catalyze the hydrolysis of vinyl propionate. The continuous DKR reactions in IL/$scCO_2$ biphasic media were performed in a (chemo)biocatalytic packed bed reactor containing Novozym 435 and zeolite particles both coated with ILs. Several ILs (Tf_2N^- and PF_6^-) were considered in combination with different acidic zeolite catalysts. Moderate to high yields (up to 98.0%) of (*R*)-phenylethyl propionate with high ees (up to 97.3%) were reported. The operational stability of the reactor was also evaluated (no activity loss during 14 days of operation).

Other examples of hydrolase-catalyzed kinetic resolutions include the enantioselective acylation of allylic alcohols by lipase in [BMIM][PF_6] and [BMIM][BF_4],[347] the kinetic resolution of 5-[4-(1-hydroxyethyl)phenyl]-10,15,20-tris(pentafluorophenyl) porphyrin by CALB in *i*-Pr_2O and in [BDMIM][PF_6],[348] the lipase-facilitated enantioselective transesterification of DL-3-phenyllactic acid with vinyl acetate in [EMIM][BF_4], [BMIM][BF_4] and [BMIM][PF_6] in combination with different organic solvents,[349] the enzymatic resolution of DL-phenylalanine in the toluene/water biphasic system containing IL catalyzed by α-chymotrypsin immobilized on super paramagnetic nanogels,[350] and so on. Due to the space limitation, these and other examples in the area are not further discussed.

5.4 HYDROLASE-CATALYZED ESTERIFICATIONS OF SACCHARIDES AND CELLULOSE DERIVATIVES IN ILs

It has been known that some ILs are capable of dissolving cellulose and other carbohydrates in a great extent. As summarized in Table 5.3, ILs based on chloride (Cl^-), dicyanamide (dca^-), acetate (OAc^-), and formate ($HCOO^-$) can dissolve up to 10–20% (wt) cellulose, and >100 g/L other carbohydrates such as β-D-glucose, sucrose, lactose, and β-cyclodextrin. More recently, ILs were used in dissolving portions of woods. For example, [BMIM][Cl] could dissolve considerable amounts of cellulosic materials and lignin from different wood samples over 24 hours at 100°C.[351] Another study[352] reported the dissolution of up to 8% (wt) wood in chloride-based ILs. A more comprehensive review of carbohydrate solubility in ILs was given by Zakrzewska et al.[353] The mechanism of dissolution is that the anions of ILs form strong hydrogen bonds with cellulose and other carbohydrates at elevated temperatures,[58,354–357] allowing these biomolecules to dissolve.

The solubilization of carbohydrates in ILs has enabled a number of chemical derivatizations of these natural products in homogeneous systems,[66,352,358–363] as well as the cellulose regeneration for a variety of applications (such as enzymatic hydrolysis,[352,364] blending with wool keratin,[365] and producing enzyme-encapsulated films[366]). The dissolution of saccharides and cellulose in ILs also makes the enzymatic enantioselective modification of these compounds possible, which has been very challenging in conventional organic solvents.[367]

The Kazlauskas group[35,368] conducted the Novozym SP435-catalyzed acetylation of β-D-glucose (0.5 M) in ILs and organic solvents (Scheme 5.38). In organic solvents, the regioselectivities were poor: 82% 6-*O*-acetyl D-glucose (~5:1 selectivity) with 42% conversion in acetone, and 85% 6-*O*-acetyl D-glucose (~6:1 selectivity) with 50% conversion in THF. The authors rationalized that the low selectivities were likely due to the poor solubility of D-glucose in organic solvents. On the other hand, in seven ILs

TABLE 5.3. Solubility of Carbohydrates in Some ILs

IL	Carbohydrate	Solubility
[BMIM][Cl]	Cellulose	10 wt% (100°C),[56] 25 wt% (microwave, 3–5 second pulses),[56] 10–18% (83°C)[358,359]
	Wool keratin fibers	11 wt% (130°C)[365]
	Eucalyptus pulp	≥13.6% (85°C)[398]
	Solucell® 1175 cellulose	16 wt% (100°C)[399]
[EMIM][Cl]	Eucalyptus pulp	≥15.8% (85°C)[398]
[AMIM][Cl]	Cellulose	8–14.5 wt% (80°C)[400,401]
	Solucell® 1175 cellulose	10 wt% (100°C)[399]
	KZO3 (1085) cellulose	12.5 wt% (100°C)[399]
[BDMIM][Cl]	Eucalyptus pulp	≥12.8% (85°C)[398]
[BMPy][Cl]	Cellulose	12-39% (105°C)[358]
[AdMIM][Br]	Cellulose	4–12% (80°C)[359]
[BMIM][dca]	β-D-glucose	145 g/L (25°C)[66]
	Sucrose	195 g/L (25°C),[66] 282 g/L (60°C)[66]
	Lactose	225 g/L (75°C)[66]
	β-cyclodextrin	750 g/L (75°C)[66]
	Amylose	4 g/L (25°C)[66]
[MoeMIM][dca]	β-D-glucose	91 g/L (25°C)[66]
	Sucrose	220 g/L (25°C)[66]
[MomMIM][dca]	Sucrose	249 g/L (25°C),[66] 352 g/L (60°C)[66]
[AMIM][HCOO]	Cellulose	10–20 wt% (60–85°C)[402]
[EMIM][OAc]	Eucalyptus pulp	≥13.5% (85°C)[398]
	Avicel® cellulose	15 wt% (110°C)[68]
	β-D-glucose	60 wt% (60°C)[68]
[BMIM][OAc]	Eucalyptus pulp	≥13.2% (85°C)[398]
[EMIM][(MeO)(R)PO_2]	Microcrystalline cellulose (DP 250)	10 wt% (45–65°C)[403]
[MoeMIM][BF_4]	β-D-glucose	5 mg/mL (55°C)[35]
[EMIM][$MeSO_4$]	β-D-glucose	89.6 g/L (25°C),[369] 133.2 g/L (60°C)[369]
	Sucrose	12.4 g/L (25°C)[369]
[EMIM][OTf]	β-D-glucose	6.1 g/L (25°C),[369] 27.8 g/L (60°C)[369]
	Fructose	32.8 g/L (25°C),[369] 123.9 g/L (60°C)[369]
	Sucrose	3.1 g/L (25°C),[369] 7.1 g/L (60°C)[369]
[BMIM][OTf]	β-D-glucose	4.8 g/L (25°C),[369] 18.1 g/L (60°C)[369]
	Fructose	27.0 g/L (25°C),[369] 87.5 g/L (60°C)[369]
	Sucrose	2.0 g/L (25°C),[369] 5.3 g/L (60°C)[369]
[BMIM][BF_4]	β-D-glucose	0.9 g/L (25°C),[369] 3.5 g/L (60°C)[369]
	Fructose	3.3 g/L (25°C),[369] 15.9 g/L (60°C)[369]
	Sucrose	0.5 g/L (25°C),[369] 0.6 g/L (60°C)[369]
[Me$(OEt)_3$-Et-Im][OAc]	β-D-glucose	80 wt% (60°C)[68]
	Avicel® cellulose	12 wt% (110°C)[68]
[Me$(Oet)_3$-Et_3N][OAc]	β-D-glucose	16 wt% (60°C)[68]
	Sucrose	16 wt% (60°C)[68]
	Avicel® cellulose	10 wt% (110°C)[68]

Note: $[(MeO)(R)PO_2]^-$ where R = H, Me or MeO.

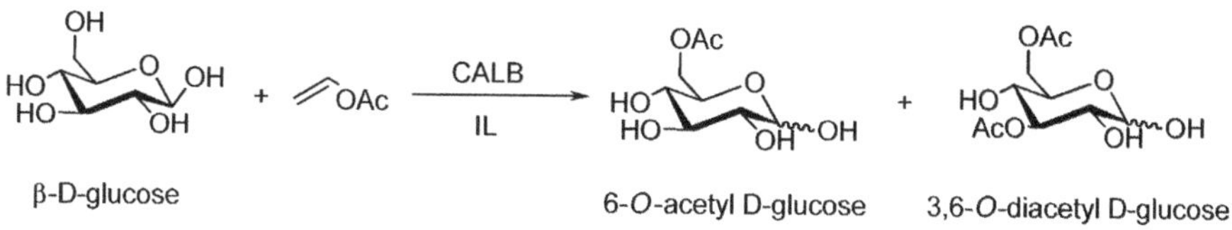

Scheme 5.38. Regioselective 6-*O*-acetylation of β-D-glucose in ILs.

based on BF_4^- and PF_6^-, the reaction resulted in 88–99% selectivities toward the 6-*O*-acetyl product, as well as 42–99% conversions. The best-performing IL was [MoeMIM][BF_4], which produced 93% 6-*O*-acetyl compound in 99% yield. The glucose solubility in this IL is about 5 mg/mL at 55°C, which is about 100 times higher than those in acetone and THF. The anomerization (α/β) of both reactant and products was observed, which is likely due to the high temperature (55°C) and the presence of traces of acetic acid (from vinyl acetate).

The Koo group[369] developed a method for preparing the supersaturated solutions of sugars (glucose, fructose, and sucrose) in OTf^- based ILs. The supersaturated glucose concentrations in [EMIM][OTf] and [BMIM][OTf] were respectively 19 and 10 times higher than the corresponding solubilities (6.1 and 4.8 g/L) of glucose in the same ILs at 25°C. In addition, Novozym 435 displayed activities in OTf^--based ILs to catalyze the (trans)esterifications between glucose with vinyl laurate (or lauric acid). The reaction using saturated glucose solution in [BMIM][OTf] only showed 8% conversion at 24 hours, but when using supersaturated solution (0.22 M) in the same IL (400% higher than its solubility), the conversion increased to 96% at 24 hours. However, it is also important to realize that hydrolases are not always active in OTf^--based ILs (at least not as active as in hydrophobic ILs) (see discussion and examples in Section 5.2.1.2.). During the investigation of the Novozym 435-catalyzed synthesis of 6-*O*-lauroyl-D-glucose, the same group[370] observed a high enzyme activity but low stability in [BMIM][OTf]. To circumvent the problem, a 1:1 (v/v) mixture of [BMIM][OTf] and [BMIM][Tf_2N] was used to afford the optimal lipase activity and stability. In particular, 86% of initial activity was detected in the IL mixture after five times reuse of the enzyme (vs. 36% residual activity in [BMIM][OTf]). In another study, this group[371] also optimized the use of mixtures of [BMIM][OTf] and [OMIM][Tf_2N] for the Novozym 435-catalyzed esterification of glucose and fatty acids. A mixture of 9:1 (v/v) ([BMIM][OTf] : [OMIM][Tf_2N]) enabled a highest yield of glucose fatty acid, while 1:1 volume ratio led to optimal stability and activity of CALB in the ionic solvent. A further attempt was made to improve the synthesis of glucose fatty acid esters in ILs by using the ultrasound irradiation.[372] The ultrasound irradiation increased the lipase activity during the acylation of glucose with vinyl laurate or lauric acid in both [BMIM][PF_6] and [BMIM][OTf], while the lipase stability was not affected.

Our group[68] has recently developed a new type of ether-functionalized ILs carrying the anion of acetate (see Scheme 5.5). These ILs are capable of dissolving up to 80 wt% of D-glucose at 60°C, and more than 10 wt% of cellulose. In addition, these ionic solvents could dissolve high concentrations of ascorbic acid, 3,4-dihydroxy-DL-phenylalanine, and betulinic acid, although common organic solvents are not able to dissolve much of these compounds.[70] More importantly, CALB in both free and immobilized forms are active and stable in these ILs, which allowed the enzymatic transformations of dissolved substances possible.[68,70,258,268] For these reasons, enzymatic transesterifications of D-glucose with methyl methacrylate[68] or vinyl laurate[70] by Novozym 435 were studied in ether-functionalized ILs, achieving moderate glucose conversions. However, the high sugar solubility may lead to low substrate ground-state energy, which may be responsible for the relatively low enzyme activities. This speculation was confirmed by our ^{13}C NMR experiments: In our ionic solvents, D-glucose solutions contained 80% β-anomers and 20% α-anomers; on the other hand, when dissolving in 2-methyl-2-butanol, D-glucose showed 40% β-anomer and 60% α-anomer.[373] It is known that strong solvation effects account for the high abundance of β-anomers in water (64%),[374] and β-glucose has a lower free energy and is thus more stable than α-anomer.[375] Therefore, D-glucose is highly solvated in some ILs, presumably due to the H-bond network between IL molecules and D-glucose, which results in low substrate ground-state energy and low enzymatic activity.

Our group[68] also demonstrated the lipase-catalyzed regioselective transesterification of cellulose dissolved in ether-functionalized ILs, resulting in 54–66% isolated yields of 6-*O*-cellulose ester. The FT-IR spectra of products confirmed the regioselectivity of this enzymatic reaction. As illustrated in Figure 5.9, the cellulose ester produced in [Me(OEt)$_7$-Et-Im][OAc] showed a characteristic peak at 1745/cm (Figure 5.9b), and no ester product was formed when 6-*O*-trityl-cellulose was used as the substrate (Figure

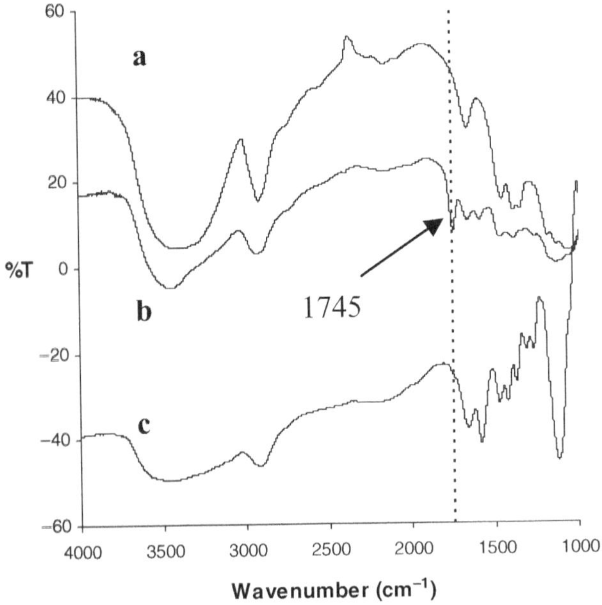

Figure 5.9. FT-IR KBr spectra of Avicel® PH-101 cellulose (a), and CALB-catalyzed transesterifications on (b) cellulose with methyl methacrylate and (c) 6-*O*-trityl-cellulose with methyl methacrylate (reactions were carried out in [Me(OEt)$_7$-Et-Im][OAc] at 60°C for 72 hours).[68] Reproduced by permission of The Royal Society of Chemistry.

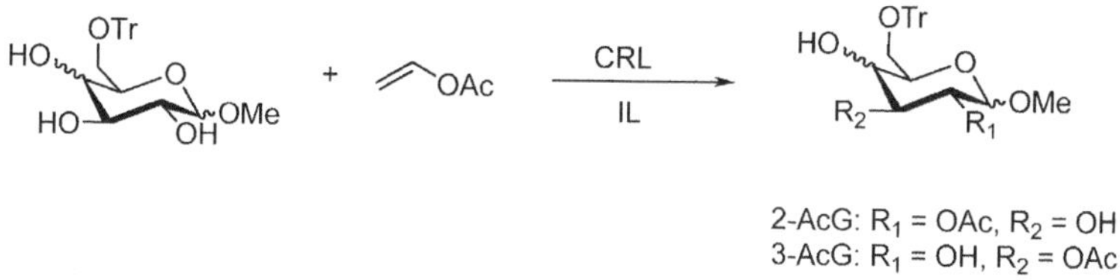

Scheme 5.39. CRL-catalyzed acylation of glycosides.

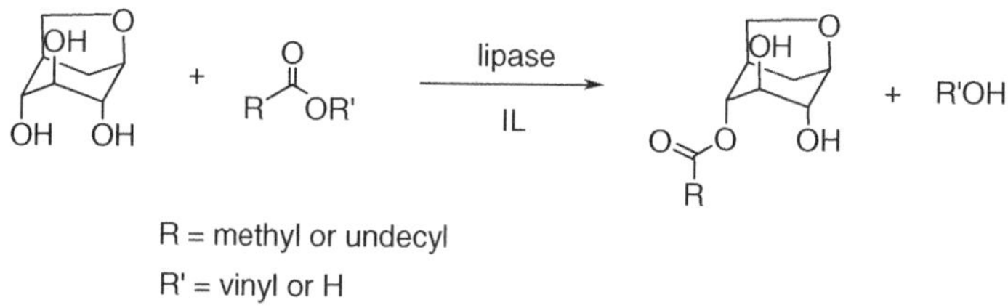

Scheme 5.40. Enzymatic acylation of levoglucosan in ILs.

5.9c). Therefore, the transesterification did not occur on the two secondary hydroxyls (2,3-OH) of cellulose, but on the primary hydroxyls (C-6 position). Such a regioselectivity has been seen in the subtilisin Carslberg-catalyzed transesterification of cellulose,[376] and enzyme-catalyzed transesterifications of sugars and amylose.[35,377–381]

Kim et al.[382] investigated the selective acylations of protected glycosides (methyl-6-*O*-trityl-glucosides and galactosides) (0.57 M) by CRL in organic solvents and ILs (Scheme 5.39). The reactions performed in [BMIM][PF_6] and [MoeMIM][PF_6] were faster and more selective (toward 2-*O*-acetyl-glycosides) than in THF and chloroform. The enhancement in reactivity in ILs was explained due to the increased solubility of substrates in ILs.

Katsoura et al.[383] performed the acylation of flavonoid glycosides (such as naringin and rutin) by lipases in two ILs ([BMIM][BF_4] and [BMIM][PF_6]). The regioselectivity in [BMIM][BF_4] was higher than [BMIM][PF_6] and organic solvents. Reaction rates in ILs were up to four times higher than those in organic solvents. The highest conversion yield (~65% after 96 hours) was achieved when a high molar ratio (10–15) of short chain acyl donors (up to four carbon atoms) was used in ILs. The acylated rutin, rutin-4‴-O-oleate, showed a considerable increase of antioxidant activity.

Galletti et al.[384] studied the regioselective acylation of levoglucosan (1,6-anhydroglucopyranose) (Scheme 5.40), an anhydro-sugar produced from cellulose by the pyrolytic treatment. Using different vinyl esters and carboxylic acids as acyl donors, acylated levoglucosans were produced with moderate to low yields after 5 days of reactions in ILs including [BMIM][BF_4], [MoeMIM][BF_4], and [MoeMIM][dca] when compared with higher yields in acetonitrile. Among three ILs considered, [MoeMIM][dca] performed the best in terms of high product yields. The possible reason could be the high solubility of sugar substrate in dca^- based ILs (see Table 5.3). However, as discussed earlier (Section 5.2.1.2.), dca^- based ILs could be enzyme-denaturing.

5.5 ILs FOR GLYCOSIDASES

Glycoside hydrolases (glycosidases) catalyze the hydrolysis of the glycosidic linkage to produce two smaller sugar units. The glycosidic bond is the linkage between monosaccharides in carbohydrate polymers. Glycoside hydrolases are usually named after the substrate that they act upon. For example, glucosidases catalyze the hydrolysis of glucosides. The hydrolysis reaction needs water as a co-reactant. Recent interests on the cellulase-catalyzed hydrolysis of cellulose in aqueous solutions of ILs for fuel ethanol production are discussed in Section 5.5.2. On the other hand, in low-water nonaqueous solvents, the glycosidase-catalyzed reaction is shifted away from the hydrolysis, leading to the enzymatic synthesis of novel glycoconjugates. This equilibrium shift can be achieved by using artificially high substrate concentrations, organic solvents, high salt buffers, or genetically modified glycosidases. An excellent review by Price[385] has described the fundamentals, mechanisms, and applications of glycosidase-catalyzed synthesis. In Section 5.5.1, the synthetic reactions in ILs facilitated by glycosidases are discussed.

5.5.1 Glycosidase-Catalyzed Synthesis in ILs

Husum et al.[15] noticed that the residual activity of β-galactosidase from *Escherichia coli* in 50% aqueous [BMIM][BF_4] was less than 6%. Kaftzik et al.[122] evaluated the enzyme activities of β-galactosidase from *Bacillus circulans* in 25–75% (v/v) aqueous solutions of ILs (based on $MeSO_4^-$, NO_3^-, $PhCOO^-$, $CF_3SO_3^-$, and $OctSO_4^-$), and observed the highest activity in 25% [MMIM][$MeSO_4$]. The residual activity in 25% [BMIM][BF_4] was 41% and no activity in 50% [BMIM][BF_4]. This group further conducted the transglycoslyation of lactose and *N*-acetylglucosamine catalyzed by β-galactosidase from *Bacillus circulans* (Scheme 5.41) and found that the use of 25% [MMIM][$MeSO_4$] suppressed the secondary hydrolysis of the product and thus increased the product yield to 60% (from 30% in buffer). The suppression effect of [MMIM] [$MeSO_4$] may have been caused by the strong interaction between $MeSO_4^-$ and water molecules resulting in a low water activity, or the interactions of ILs with charged

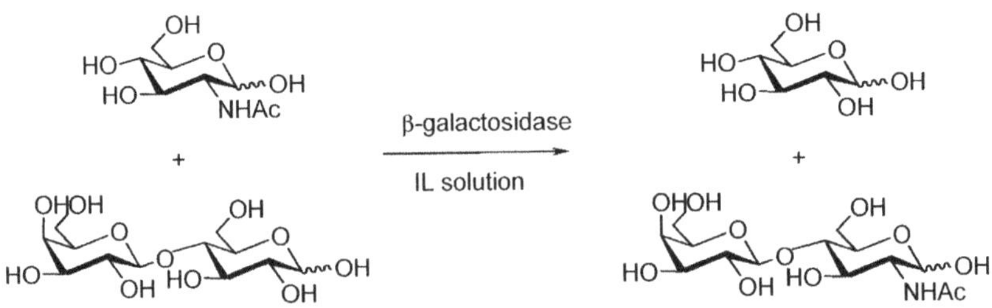

Scheme 5.41. β-Galactosidase-catalyzed transglycoslyation of lactose and *N*-acetylglucosamine in IL solutions.

groups in or near the active site of β-galactosidase. The high enzyme stability in pure [MMIM][$MeSO_4$] at 50°C was also reported.

A further study by Lang et al.[123] suggested that hyperthermostable β-glycosidase CelB from *Pyrococcus furiosus* retained a high enzyme activity in <50% (v/v) aqueous solutions of [MMIM][$MeSO_4$]. The CelB-catalyzed transglycoslyation of lactose with various galactosyl donors indicated that the presence of 45% [MMIM][$MeSO_4$] improved the enzyme selectivity and activity in most cases. However, the increase in yield by using [MMIM][$MeSO_4$] as the co-solvent was not impressive: no increase in the case of D-xylose or lactose as galactosyl donor, and moderate ~10% increase in the case of glycerol.

5.5.2 Cellulase-catalyzed Hydrolysis in ILs

The active research on the enzymatic production of cellulosic ethanol has led to recent studies of pretreatments of lignocelluloses by ILs.[269,270,364,386] These studies have further stimulated the investigation of hydrolytic activity of cellulase in aqueous solutions of ILs. Cellulase is an enzyme mixture and contains major components of endo-β-glucanase (EC 3.2.1.4), exo-β-glucanase (EC 3.2.1.91) and β-glucosidase (cellobiase) (EC 3.2.1.21). Endo-β-glucanase randomly attacks cellulose chains yielding glucose and cello-oligo saccharides; exo-β-glucanase breaks down 2–4 units from the ends of the exposed chains produced by endocellulase, producing the tetrasaccharides or disaccharide such as cellobiose; β-glucosidase further hydrolyzes cellobiose to glucose. It is well known the synergistic effect of these enzymes often exist.[387] The following examples will demonstrate recent evaluations of cellulase activity in aqueous solutions of ILs.

Turner et al.[59] conducted the enzymatic hydrolysis of cellulose azure catalyzed by cellulase from *T. reesei*, and found the enzyme was inactivated by high concentrations of [BMIM][Cl]. Paljevac et al.[143] carried out the hydrolysis of carboxymethyl cellulose (CMC) by cellulase from *Humicola insolens* (Celluzyme® 0.7T). In a low concentration (9% by volume) of ILs ([BMIM][PF_6], [BMIM][BF_4] and [BMIM][Cl]) in buffer, the cellulase activities were not considerably different from that in buffer solution. However, at a 50% (v/v) concentration of ILs, cellulase was greatly inactivated by [BMIM][Cl], and the cellulase activity decreased in the order of buffer ~ [BMIM][PF_6] > [BMIM][BF_4] > [BMIM][Cl].

Rayne and Mazza[388] determined the hydrolysis rate of cellulose azure catalyzed by *Trichoderma reesei* cellulase in *N,N*-dimethylethanolammonium akylcarboxylate ([MM(EtOH)NH][akylcarboxylate]), and measured the fluorescence intensities of cellulase in IL solutions. Their data suggested that the enzyme was active in all concentrations of [MM(EtOH)NH][OAc]; in particular, the cellulase activities in 20% and 40% (v/v) of ILs were identical to that in citrate buffer. However, in high concentrations of corresponding formate-* and octanoate-based ILs, lower cellulase activities and thus enzyme denaturation were observed.

* The formate anion is considered a kosmotrope[389] and has a viscosity *B*-coefficient of 0.052.[103]

Kamiya et al.[390] dissolved Avicel® PH-101 cellulose in [EMIM][Et_2PO_4] at 50°C. After the addition of citrate buffer, cellulose precipitated from the IL. Cellulase from *Trichoderma reesei* was then added into the mixture of cellulose suspension in aqueous IL solutions. The highest conversion (70%) was achieved at a low IL concentration (20%, v/v) after 24 hours; the cellulase was mostly inactivated at high IL concentrations (>60%, v/v). Another study[391] also investigated the effect of IL concentration on the enzymatic hydrolysis of cellulose regenerated from [EMIM][Et_2PO_4]. The CLEA preparation of *Trichoderma reesei* cellulase was optimized with regard to the precipitant solvent and glutaraldehyde concentration. Then, the immobilized cellulase was employed to catalyze the hydrolysis of regenerated cellulose in low concentrations of [EMIM][Et_2PO_4]. The addition of 2% (v/v) IL was able to increase the initial hydrolysis rate by 2.7 times of that in buffer. However, a higher IL concentration (such as 4%) caused a lower cellulase activity, which is contradictory with the earlier[390] finding about cellulase being active in 20% (v/v) of [EMIM][Et_2PO_4].

Recently, our group[269] also investigated the impact of IL type and IL concentration on the hydrolytic activity of *Trichoderma reesei* cellulase toward regenerated cellulose. The cellulase retained higher activities in low concentrations (<1.0 M) of hydrophilic ILs (chloride and acetate) than in high concentrations. In particular, cellulase maintained 82% initial activity in 1.0 M [BMIM][Cl] and 63% initial activity in 1.0 M [MM(EtOH)NH][OAc]. However, cellulase was much less active in 1.0 M other acetate-based ILs (such as [EMIM][OAc], [Me(OEt)$_n$-Et-Im][OAc] (n = 2–4) and [Me(OEt)$_3$-Et$_3$N][OAc]). The fluorescence study on the emission intensity of tryptophan (Trp) residues of cellulase at 346 nm in various IL solutions also confirmed the above kinetic study.

In seeking for new cellulase that are more tolerant to high concentrations of ILs, the Streit group[392] screened metagenomic libraries and identified 24 novel cellulase clones. Although 17 of cosmid clones showed measurable activities in the presence of 30% (v/v) ILs (based on Cl^-, OTf^-, and CF_3COO^-), most enzyme clones exhibited poor or no activities. Three enzyme clones (i.e., pCosJP10, pCosJP20, and pCosJP24) were moderately active and stable. The most active protein ($CelA_{10}$) was a GH5 family cellulase and maintained activities in 30% IL solutions, but became inactive at IL concentrations above 40%. Using SeSaM-technology, improved cellulase variants of $CelA_{10}$ were obtained in a directed evolution experiment. Examination of these variants suggested that the *N*-terminal cellulose binding domain is essential for the IL resistance.

5.6 PROSPECTS

Although some ILs (such as those based on PF_6^- and Tf_2N^-) are less denaturing than organic solvents and high catalytic activity and enantioselectivity have been reported for many reactions,[20,244] most hydrolases show the same magnitude of activities in ILs as in conventional organic solvents, which are considerably inferior to those in aqueous solutions. To further improve the enzyme activity and stability in ILs, future studies will likely focus on synthesizing new ILs that are more suitable for maintaining the active conformations of enzymes, and developing new methods for enzyme stabilization (such as immobilization and genetic engineering).

In the meantime, it remains a hard task to understand the mechanisms of IL–enzyme interactions. As discussed in Section 5.2, many factors could influence the enzyme's behaviors in ionic solvents, and it is important to know the determining factor(s) under controlled reaction systems. Future studies on the visualization of IL–enzyme interactions by spectroscopy, computer simulation, and other techniques will enable us to have a molecular-level understanding of biocatalysis in ILs.

On the other hand, the bottom line is to make ILs useful on industrial scales, the cost of ILs must be brought down, and the toxicity and biodegradability of ILs must be well understood. Inexpensive, biodegradable, and enzyme-compatible ILs are the dream solvents for the next generation of biocatalysis.

ACKNOWLEDGMENTS

The author is very grateful to Dr. Gary A. Baker (University of Missouri–Columbia) and Dr. Neil P.J. Price (National Center for Agricultural Utilization Research) for their kind assistance. Due to the overwhelming volume of publications on the subject covered by this chapter, the author regrets not being able to cover many other related papers. The chapter manuscript was completed in January 2010, and thus could cover only the literature up to then.

REFERENCES

1 Y. L. Khmelnitsky, A. V. Levashov, N. L. Klyachko, K. Martinek, *Enzyme Microb. Technol.* **1988**, *10*, 710–724.

2 Z. Yang, A. J. Russell, in *Enzymatic Reactions in Organic Media* (Eds. A. M. P. Koskinen, A. M. Kilbanov), Blackie Academic & Professional, New York, **1996**, pp. 43–69.

3 Y. L. Khmelnitsky, J. O. Rich, *Curr. Opin. Chem. Biol.* **1999**, *3*, 47–53.

4 U. T. Bornscheuer, R. J. Kazlauskas, *Hydrolases in Organic Synthesis*, 2nd ed., Wiley-VCH, Weinheim, **2006**.

5 P. Wasserscheid, T. Welton, *Ionic Liquids in Synthesis*, Vol. 1 & 2, 2nd ed., Wiley-VCH, Weinheim, **2008**.

6 C. M. Gordon, *Appl. Catal. A* **2001**, *222*, 101–117.

7 S. Houlton, *Chem. Week.* Feb. 25, **2004**, s10–s11.

8 K. R. Seddon, *J. Chem. Technol. Biotechnol.* **1997**, *68*, 351–356.

9 T. Welton, *Chem. Rev.* **1999**, *99*, 2071–2083.

10 H. Zhao, S. V. Malhotra, *Aldrichim. Acta* **2002**, *35*, 75–83.

11 M. Earle, A. Forestier, H. Olivier-Bourbigou, P. Wasserscheid, in *Ionic Liquids in Synthesis* (Eds. P. Wasserscheid, T. Welton), Wiley-VCH Verlag, Weinheim, **2003**, pp. 174–288.

12 N. Jain, A. Kumar, S. Chauhan, S. M. S. Chauhan, *Tetrahedron* **2005**, *61*, 1015–1060.

13 W. Wasserscheid, W. Keim, *Angew. Chem. Int. Ed. Engl.* **2000**, *39*, 3772–3789.

14 F. Endres, T. Welton, in *Ionic Liquids in Synthesis* (Eds. P. Wasserscheid, T. Welton), Wiley-VCH Verlag, Weinheim, **2003**, pp. 289–318.

15 T. L. Husum, C. T. Jorgensen, M. W. Christensen, O. Kirk, *Biocatal. Biotransformation* **2001**, *19*, 331–338.

16 U. Kragl, M. Eckstein, N. Kaftzik, *Curr. Opin. Biotechnol.* **2002**, *13*, 565–571.

17 S. Park, R. J. Kazlauskas, *Curr. Opin. Biotechnol.* **2003**, *14*, 432–437.

18 R. A. Sheldon, R. M. Lau, M. J. Sorgedrager, F. van Rantwijk, K. R. Seddon, *Green Chem.* **2002**, *4*, 147–151.

19 F. van Rantwijk, R. Madeira Lau, R. A. Sheldon, *Trends Biotechnol.* **2003**, *21*, 131–138.

20 F. van Rantwijk, R. A. Sheldon, *Chem. Rev.* **2007**, *107*, 2757–2785.

21 P. Kubisa, *Prog. Polym. Sci.* **2004**, *29*, 3–12.

22 A. J. Carmichael, D. M. Haddleton, in *Ionic Liquids in Synthesis* (Eds. P. Wasserscheid, T. Welton), Wiley-VCH Verlag, Weinheim, **2003**, pp. 319–335.

23 J. F. Brennecke, E. J. Maginn, US 6,579,343, **2003**.

24 H. Zhao, S. Xia, P. Ma, *J. Chem. Technol. Biotechnol.* **2005**, *80*, 1089–1096.

25 H. Zhao, *Chem. Eng. Commun.* **2006**, *193*, 1660–1677.

26 M. Moniruzzaman, K. Nakashima, N. Kamiya, M. Goto, *Biochem. Eng. J.* **2010**, *48*, 295–314.

27 M. Sureshkumara, C.-K. Lee, *J. Mol. Catal., B Enzym.* **2009**, *60*, 1–12.

28 H. Zhao, *J. Chem. Technol. Biotechnol.* **2010**, *85*, 891–907.

29 Z. Yang, W. Pan, *Enzyme Microb. Technol.* **2005**, *37*, 19–28.

30 C. Reichardt, *Chem. Rev.* **1994**, *94*, 2319–2358.

31 C. Reichardt, *Green Chem.* **2005**, *7*, 339–351.

32 A. J. Carmichael, K. R. Seddon, *J. Phys. Org. Chem.* **2000**, *13*, 591–595.

33 S. N. V. K. Aki, J. F. Brennecke, A. Samanta, *Chem. Commun.* **2001**, 413–414.

34 V. S. Narayan, A. M. Klibanov, *Biotechnol. Bioeng.* **1993**, *41*, 390–393.

35 S. Park, R. J. Kazlauskas, *J. Org. Chem.* **2001**, *66*, 8395–8401.

36 P. Lozano, T. de Diego, J.-P. Guegan, M. Vaultier, J. L. Iborra, *Biotechnol. Bioeng.* **2001**, *75*, 563–569.

37 J. Mutschler, T. Rausis, J.-M. Bourgeois, C. Bastian, D. Zufferey, I. V. Mohrenz, F. Fischer, *Green Chem.* **2009**, *11*, 1793–1800.

38 J. L. Kaar, A. M. Jesionowski, J. A. Berberich, R. Moulton, A. J. Russell, *J. Am. Chem. Soc.* **2003**, *125*, 4125–4131.

39 S. H. Schöfer, N. Kaftzik, P. Wasserscheid, U. Kragl, *Chem. Commun.* **2001**, 425–426.

40 H. Zhao, O. Olubajo, Z. Song, A. L. Sims, T. E. Person, R. A. Lawal, L. A. Holley, *Bioorg. Chem.* **2006**, *34*, 15–25.

41 M. J. Muldoon, G. M. Gordon, I. R. Dunkin, *J. Chem. Soc. Perkin Trans. II* **2001**, 433–435.

42 W. Linert, R. F. Jameson, A. Taha, *J. Chem. Soc., Dalton Trans.* **1993**, 3181–3186.

43 A. Camard, Y. Ihara, F. Murata, K. Mereiter, Y. Fukuda, W. Linert, *Inorg. Chim. Acta* **2005**, *358*, 409–414.

44 A. Oehlke, K. Hofmann, S. Spange, *New J. Chem.* **2006**, *30*, 533–536.

45 D. R. MacFarlane, J. M. Pringle, K. M. Johansson, S. A. Forsyth, M. Forsyth, *Chem. Commun.* **2006**, 1905–1917.

46 R. M. Pagni, in *Green Industrial Applications of Ionic Liquids* (Eds. R. D. Rogers, K. R. Seddon, S. Volkov), Kluwer Academic Publishers, Dordrecht, **2002**, pp. 105–127.

47 K. Fujita, D. R. MacFarlane, M. Forsyth, *Chem. Commun.* **2005**, 4804–4806.

48 K. Fujita, D. R. MacFarlane, M. Forsyth, M. Yoshizawa-Fujita, K. Murata, N. Nakamura, H. Ohno, *Biomacromolecules* **2007**, *8*, 2080–2086.

49 A. Bernson, J. Lindgren, *Polymer* **1994**, *35*, 4848–4851.

50 J. Dupont, *J. Braz. Chem. Soc.* **2004**, *15*, 341–350.

51 W. A. Henderson, *J. Phys. Chem. B* **2006**, *110*, 13177–13183.

52 W. A. Henderson, *Macromolecules* **2007**, *40*, 4963–4971.

53 F. J. Hernández-Fernández, A. P. de los Ríos, F. Tomás-Alonso, D. Gómez, G. Víllora, *Can. J. Chem. Eng.* **2009**, *87*, 910–914.

54 R. Irimescu, K. Kato, *J. Mol. Catal., B Enzym.* **2004**, *30*, 189–194.

55 S. H. Lee, Y.-M. Koo, S. H. Ha, *Korean J. Chem. Eng.* **2008**, *25*, 1456–1462.

56 R. P. Swatloski, S. K. Spear, J. D. Holbrey, R. D. Rogers, *J. Am. Chem. Soc.* **2002**, *124*, 4974–4975.

57 S. Zhu, Y. Wu, Q. Chen, Z. Yu, C. Wang, S. Jin, Y. Ding, G. Wu, *Green Chem.* **2006**, *8*, 325–327.

58 R. C. Remsing, R. P. Swatloski, R. D. Rogers, G. Moyna, *Chem. Commun.* **2006**, 1271–1273.

59 M. B. Turner, S. K. Spear, J. G. Huddleston, J. D. Holbrey, R. D. Rogers, *Green Chem.* **2003**, *5*, 443–447.

60 S. H. Lee, S. H. Ha, S. B. Lee, Y.-M. Koo, *Biotechnol. Lett.* **2006**, *28*, 1335–1339.

61 J. L. Anderson, J. Ding, T. Welton, D. W. Armstrong, *J. Am. Chem. Soc.* **2002**, *124*, 14247–14254.

62 D. W. Armstrong, L. He, Y. S. Liu, *Anal. Chem.* **1999**, *71*, 3873–3876.

63 W.-Y. Lou, M.-H. Zong, H. Wu, R. Xu, J.-F. Wang, *Green Chem.* **2005**, *7*, 500–506.

64 R. M. Lau, M. J. Sorgedrager, G. Carrea, F. van Rantwijk, F. Secundo, R. A. Sheldon, *Green Chem.* **2004**, *6*, 483–487.

65 S. A. Forsyth, D. R. MacFarlane, R. J. Thomson, M. von Itzstein, *Chem. Commun.* **2002**, 714–715.

66 Q. Liu, M. H. A. Janssen, F. van Rantwijk, R. A. Sheldon, *Green Chem.* **2005**, *7*, 39–42.

67 A. R. Toral, A. P. de los Ríos, F. J. Hernández, M. H. A. Janssen, R. Schoevaart, F. van Rantwijk, R. A. Sheldon, *Enzyme Microb. Technol.* **2007**, *40*, 1095–1099.

68 H. Zhao, G. A. Baker, Z. Song, O. Olubajo, T. Crittle, D. Peters, *Green Chem.* **2008**, *10*, 696–705.

69 H. Zhao, G. A. Baker, Z. Song, O. Olubajo, L. Zanders, S. M. Campbell, *J. Mol. Catal., B Enzym.* **2009**, *57*, 149–157.

70 H. Zhao, C. L. Jones, J. V. Cowins, *Green Chem.* **2009**, *11*, 1128–1138.

71 M. D. Bermejo, A. J. Kotlewska, L. J. Florusse, M. J. Cocero, F. van Rantwijk, C. J. Peters, *Green Chem.* **2008**, *10*, 1049–1054.

72 U. Schröder, J. D. Wadhawan, R. G. Compton, F. Marken, P. A. Z. Suarez, C. S. Consorti, R. F. de Souza, J. Dupont, *New J. Chem.* **2000**, *24*, 1009–1015.

73 E. Fehér, B. Major, K. Bélafi-Bakó, L. Gubicza, *Biochem. Soc. Trans.* **2007**, *35*, 1624–1627.

74 P. Lozano, T. De Diego, S. Gmouh, M. Vaultier, J. L. Iborra, *Biocatal. Biotransformation* **2005**, *23*, 169–176.

75 B. Réjasse, T. Besson, M.-D. Legoy, S. Lamare, *Org. Biomol. Chem.* **2006**, *4*, 3703–3707.

76 F. Hofmeister, *Arch. Exp. Pathol. Pharmakol.* **1888**, *24*, 247–260.

77 W. Kunz, J. Henle, B. W. Ninham, *Curr. Opin. Colloid Interface Sci.* **2004**, *9*, 19–37.

78 R. L. Baldwin, *Biophys. J.* **1996**, *71*, 2056–2063.

79 T. Arakawa, S. N. Timasheff, *Biochemistry* **1984**, *23*, 5912–5923.

80 K. D. Collins, *Methods* **2004**, *34*, 300–311.

81 P. Bauduin, A. Renoncourt, D. Touraud, W. Kunz, B. W. Ninham, *Curr. Opin. Colloid Interface Sci.* **2004**, *9*, 43–47.

82 K. D. Collins, M. W. Washabaugh, *Q. Rev. Biophys.* **1985**, *18*, 323–422.

83 H. S. Frank, W.-Y. Wen, *Discuss. Faraday Soc.* **1957**, *24*, 133–140.

84 F. Franks, *Water, a Comprehensive Treatise. Volume 3. Aqueous Solutions of Simple Electrolytes*, Plenum Press, New York, **1973**.

85 P. H. von Hippel, K.-Y. Wong, *J. Biol. Chem.* **1965**, *240*, 3909–3923.

86 P. M. Wiggins, *Physica A.* **1997**, *238*, 113–128.

87 C. Tanford, *Science* **1978**, *200*, 1012–1018.

88 G. Némethy, *Angew. Chem. Int. Ed. Engl.* **1967**, *6*, 195–206.

89 A. P. Minton, *Biopolymers* **1981**, *20*, 2093–2120.

90 P. R. Davis-Searles, A. J. Saunders, D. A. Erie, D. J. Winzor, G. J. Pielak, *Annu. Rev. Biophys. Biomol. Struct.* **2001**, *30*, 271–306.

91 J. A. Schellman, *Biophys. J.* **2003**, *85*, 108–125.

92 D. R. Robinson, W. P. Jencks, *J. Am. Chem. Soc.* **1965**, *87*, 2462–2470.

93 E. S. Courtenay, M. W. Capp, M. T. J. Record, *Protein Sci.* **2001**, *10*, 2485–2497.

94 J. K. Myers, C. N. Pace, J. M. Scholtz, *Protein Sci.* **1995**, *4*, 2138–2148.

95 B. Honig, A. Nicholls, *Science* **1995**, *268*, 1144–1149.

96 M. F. Perutz, *Science* **1978**, *201*, 1187–1191.

97 M. Boström, D. R. M. Williams, B. W. Ninham, *Curr. Opin. Colloid Interface Sci.* **2004**, *9*, 48–52.

98 O. Y. Samoilov, *Discuss. Faraday Soc.* **1957**, *24*, 141–146.

99 O. Y. Samoilov, *Bull. Acad. Sci. USSR, Div. Chem. Sci.* **1953**, 2, 219–225.

100 H. Zhao, *J. Chem. Technol. Biotechnol.* **2006**, *81*, 877–891.

101 G. Jones, M. Dole, *J. Am. Chem. Soc.* **1929**, *51*, 2950–2964.

102 A. Ali, S. Hyder, Y. Akhtar, *Indian J. Phys.* **2005**, *79*, 157–160.

103 H. D. B. Jenkins, Y. Marcus, *Chem. Rev.* **1995**, *95*, 2695–2724.

104 K. B. Belibagli, E. Ayranci, *J. Solution Chem.* **1990**, *19*, 867–882.

105 K. D. Collins, *Biophys. Chem.* **1997**, *72*, 65–76.

106 Y. Marcus, *J. Solution Chem.* **1994**, *23*, 831–848.

107 R. L. Kay, T. Vituccio, C. Zawoyski, D. F. Evans, *J. Phys. Chem.* **1966**, *70*, 2336–2341.

108 H. S. Frank, M. W. Evans, *J. Chem. Phys.* **1945**, *13*, 507–532.

109 E. R. Nightingale, *J. Phys. Chem.* **1962**, *66*, 894–897.

110 H. S. Frank, *J. Phys. Chem.* **1963**, *67*, 1554–1558.

111 T. S. Sarma, J. C. Ahluwalia, *Chem. Soc. Rev.* **1973**, *2*, 203–232.

112 J. M. Tsangaris, R. B. Martin, *Arch. Biochem. Biophys.* **1965**, *112*, 267–272.

113 W. Devine, B. M. Lowe, *J. Chem. Soc. A, Inorg. Phys. Theor.* **1971**, 2113–2116.

114 H. Zhao, *J. Mol. Catal., B Enzym.* **2005**, *37*, 16–25.

115 H. Zhao, Z. Song, *J. Chem. Technol. Biotechnol.* **2007**, *82*, 304–312.

116 H. Zhao, S. Campbell, L. Jackson, Z. Song, O. Olubajo, *Tetrahedron Asymmetry.* **2006**, *17*, 377–383.

117 H. Zhao, L. Jackson, Z. Song, O. Olubajo, *Tetrahedron Asymmetry* **2006**, *17*, 1549–1553.

118 H. Zhao, *Biophys. Chem.* **2006**, *122*, 157–183.

119 P. M. Wiggins, US Patent 6638360, **2003**.

120 H. Zhao, S. Campbell, J. Solomon, Z. Song, O. Olubajo, *Chin. J. Chem.* **2006**, *24*, 580–584.

121 H. Zhao, L. Jackson, Z. Song, O. Olubajo, *Tetrahedron Asymmetry* **2006**, *17*, 2491–2498.

122 N. Kaftzik, P. Wasserscheid, U. Kragl, *Org. Proc. Res. Dev.* **2002**, *6*, 553–557.

123 M. Lang, T. Kamrat, B. Nidetzky, *Biotechnol. Bioeng.* **2006**, *95*, 1093–1100.

124 K. Fujita, M. Forsyth, D. R. MacFarlane, R. W. Reid, G. D. Elliott, *Biotechnol. Bioeng.* **2006**, *94*, 1209–1213.

125 K. Tamaki, Y. Ohara, Y. Isomura, *Bull. Chem. Soc. Jpn* **1973**, *46*, 1551–1552.

126 H. Tanaka, Y. Oka, *Biochim. Biophys. Acta, Gen. Subj.* **2005**, *1724*, 173–180.

127 D. Constantinescu, H. Weingärtner, C. Herrmann, *Angew. Chem. Int. Ed. Engl.* **2007**, *46*, 8887–8889.

128 D. Constatinescu, C. Herrmann, H. Weingärtner, *Phys. Chem. Chem. Phys.* **2010**, *12*, 1756–1763.

129 Z. Yang, *J. Biotechnol.* **2009**, *144*, 12–22.

130 Z. Yang, Y.-J. Yue, W.-C. Huang, X.-M. Zhuang, Z.-T. Chen, M. Xing, *J. Biochem.* **2009**, *145*, 355–364.

131 M. Eckstein, P. Wasserscheid, U. Kragl, *Biotechnol. Lett.* **2002**, *24*, 763–767.

132 A. Basso, S. Cantone, P. Linda, C. Ebert, *Green Chem.* **2005**, *7*, 671–676.

133 S. H. Lee, S. B. Lee, *J. Chem. Technol. Biotechnol.* **2009**, *84*, 202–207.

134 L. Ropel, L. S. Belvèze, S. N. V. K. Aki, M. A. Stadtherr, J. F. Brennecke, *Green Chem.* **2005**, *7*, 83–90.

135 R. Hilhorst, R. Spruijt, C. Laane, C. Veeger, *Eur. J. Biochem.* **1984**, *144*, 459–466.

136 C. Laane, S. Boeren, K. Vos, *Trends Biotechnol.* **1985**, *3*, 251–252.

137 C. Laane, S. Boeren, K. Vos, C. Veeger, *Biotechnol. Bioeng.* **1987**, *30*, 81–87.

138 S. J. Nara, J. R. Harjani, M. M. Salunkhe, *Tetrahedron Lett.* **2002**, *43*, 2979–2982.

139 K. Nakashima, J. Okada, T. Maruyama, N. Kamiya, M. Goto, *Sci. Technol. Adv. Mater.* **2006**, *7*, 692–698.

140 K. Nakashima, T. Maruyama, N. Kamiya, M. Goto, *Org. Biomol. Chem.* **2006**, *4*, 3462–3467.

141 W.-G. Zhang, D.-Z. Wei, X.-P. Yang, Q.-X. Song, *Bioprocess Biosyst. Eng.* **2006**, *29*, 379–383.

142 W.-Y. Lou, M.-H. Zong, *Chirality* **2006**, *18*, 814–821.

143 M. Paljevac, M. Habulin, Z. Knez, *Chem. Ind. Chem. Eng. Q.* **2006**, *12*, 181–186.

144 A. P. de los Ríos, F. J. Hernández-Fernández, M. Rubio, D. Gómez, G. Víllora, *J. Chem. Technol. Biotechnol.* **2007**, *82*, 190–195.

145 A. P. de los Ríos, F. J. Hernández-Fernández, F. A. Martínez, M. Rubio, G. Víllora, *Biocatal. Biotransformation* **2007**, *25*, 151–156.

146 S. H. Ha, S. H. Lee, D. T. Dang, M. S. Kwon, W.-J. Chang, Y. J. Yu, I. S. Byun, Y.-M. Koo, *Korean J. Chem. Eng.* **2008**, *25*, 291–294.

147 Z.-L. Shen, W.-J. Zhou, Y.-T. Liu, S.-J. Ji, T.-P. Loh, *Green Chem.* **2008**, *10*, 283–286.

148 K. Ryu, J. S. Dordick, *Biochemistry* **1992**, *31*, 2588–2598.

149 A. Zaks, A. M. Klibanov, *J. Biol. Chem.* **1988**, *263*, 3194–3201.

150 F. van Rantwijk, F. Secundo, R. A. Sheldon, *Green Chem.* **2006**, *8*, 282–286.

151 B. Réjasse, S. Lamare, M.-D. Legoy, T. Besson, *Org. Biomol. Chem.* **2004**, *2*, 1086–1089.

152 M. H. Abraham, H. S. Chadha, G. S. Whiting, R. C. Mitchell, *J. Pharm. Sci.* **1994**, *83*, 1085–1100.

153 M. H. Abraham, H. S. Chadha, J. P. Dixon, A. J. Leo, *J. Phys. Org. Chem.* **1994**, *7*, 712–716.

154 M. H. Abraham, H. S. Chadha, F. Martins, R. C. Mitchell, M. W. Bradbury, J. A. Gratton, *Pestic. Sci.* **1999**, *55*, 78–88.

155 M. H. Abraham, *Chem. Soc. Rev.* **1993**, *22*, 73–83.

156 M. J. Kamlet, R. M. Doherty, M. H. Abraham, Y. Marcus, R. W. Taft, *J. Phys. Chem.* **1988**, *92*, 5244–5255.

157 N. El Tayar, R.-S. Tsai, B. Testa, P.-A. Carrupt, A. Leo, *J. Pharm. Sci.* **1991**, *80*, 590–598.

158 J. P. Mann, A. McCluskey, R. Atkin, *Green Chem.* **2009**, *11*, 785–792.

159 T. De Diego, P. Lozano, M. A. Abad, K. Steffensky, M. Vaultier, J. L. Iborra, *J. Biotechnol.* **2009**, *140*, 234–241.

160 R. M. Lau, F. van Rantwijk, K. R. Seddon, R. A. Sheldon, *Org. Lett.* **2000**, *2*, 4189–4191.

161 T. Itoh, Y. Nishimura, M. Kashiwagi, M. Onaka, in *Ionic Liquids as Green Solvents: Progress and Prospects* (Eds. R. D. Rogers, K. R. Seddon), American Chemical Society, Washington DC, **2003**, pp. 251–261.

162 N. Kaftzik, S. Neumann, M.-R. Kula, U. Kragl, in *Ionic Liquids as Green Solvents: Progress and Prospects* (Eds. R. D. Rogers, K. R. Seddon), American Chemical Society, Washington DC, **2003**, pp. 206–211.

163 P. Lozano, T. de Diego, D. Carrié, M. Vaultier, J. L. Iborra, in *Ionic Liquids as Green Solvents: Progress and Prospects* (Eds. R. D. Rogers, K. R. Seddon), American Chemical Society, Washington DC, **2003**, pp. 239–250.

164 P. Lozano, T. De Diego, D. Carrié, M. Vaultier, J. L. Iborra, *Biotechnol. Lett.* **2001**, *23*, 1529–1533.

165 M. Erbeldinger, A. J. Mesiano, A. J. Russell, *Biotechnol. Prog.* **2000**, *16*, 1129–1131.

166 P. J. Halling, *Biotechnol. Tech.* **1992**, *6*, 271–276.

167 G. Bell, P. J. Halling, B. D. Moore, J. Partridge, D. G. Rees, *Trends Biotechnol.* **1995**, *13*, 468–473.

168 C. Spickermann, J. Thar, S. B. C. Lehmann, S. Zahn, J. Hunger, R. Buchner, P. A. Hunt, T. Welton, B. Kirchner, *J. Chem. Phys.* **2008**, *129*, 104505.

169 K. R. Seddon, A. Stark, M.-J. Torres, *Pure Appl. Chem.* **2000**, *72*, 2275–2287.

170 F. J. Hernández-Fernández, A. P. de los Ríos, M. Rubio, D. Gómez, G. Víllora, *J. Chem. Technol. Biotechnol.* **2007**, *82*, 882–887.

171 J. A. Laszlo, D. L. Compton, *Biotechnol. Bioeng.* **2001**, *75*, 181–186.

172 P. Lozano, T. de Diego, D. Carrie, M. Vaultier, J. L. Iborra, *J. Mol. Catal., B Enzym.* **2003**, *21*, 9–13.

173 G. V. Chowdary, S. G. Prapulla, *Process Biochem.* **2002**, *38*, 393–397.

174 J. A. Berberich, J. L. Kaar, A. J. Russell, *Biotechnol. Prog.* **2003**, *19*, 1029–1032.

175 D. Barahona, P. H. Pfromm, M. E. Rezac, *Biotechnol. Bioeng.* **2006**, *93*, 318–324.

176 Z. Yang, K.-P. Zhang, Y. Huang, Z. Wang, *J. Mol. Catal., B Enzym.* **2010**, *63*, 23–30.

177 M. Eckstein, M. Sesing, U. Kragl, P. Adlercreutz, *Biotechnol. Lett.* **2002**, *24*, 867–872.

178 S. Garcia, N. M. T. Lourenco, D. Lousa, A. F. Sequeira, P. Mimoso, J. M. S. Cabral, C. A. M. Afonso, S. Barreiros, *Green Chem.* **2004**, *6*, 466–470.

179 M. Noël, P. Lozano, M. Vaultier, J. L. Iborra, *Biotechnol. Lett.* **2004**, *26*, 301–306.

180 L. Cao, L. van Langen, R. A. Sheldon, *Curr. Opin. Biotechnol.* **2003**, *14*, 387–394.

181 R. A. Sheldon, R. Schoevaart, L. M. van Langen, in *Immobilization of Enzymes and Cells*, 2nd ed. (Ed. J. M. Guisan), Humana Press, Totowa, NJ, **2006**, pp. 31–45.

182 R. A. Sheldon, R. Schoevaart, L. M. Van Langen, *Biocatal. Biotransformation* **2005**, *23*, 141–147.

183 S. S. Karajanagi, A. A. Vertegel, R. S. Kane, J. S. Dordick, *Langmuir* **2004**, *20*, 11594–11599.

184 B. Eker, P. Asuri, S. Murugesan, R. J. Linhardt, J. S. Dordick, *Appl. Biochem. Biotechnol.* **2007**, *143*, 153–163.

185 S. Shah, K. Solanki, M. N. Gupta, *Chem. Cent. J.* **2007**, *1*, 30.

186 P. Du, S. Liu, P. Wu, C. Cai, *Electrochim. Acta* **2007**, *52*, 6534–6547.

187 Y. Jiang, C. Guo, H. Xia, I. Mahmood, C. Liu, H. Liu, *J. Mol. Catal., B Enzym.* **2009**, *58*, 103–109.

188 Y. Okada, H. Sawada, *Colloid Polym. Sci.* **2009**, *287*, 1359–1363.

189 M. T. Reetz, in *Immobilization of Enzymes and Cells*, 2nd ed. (Ed. J. M. Guisan), Humana Press, Totowa, NJ, **2006**, pp. 65–76.

190 M. Campas, J.-L. Marty, in *Immobilization of Enzymes and Cells*, 2nd ed. (Ed. J. M. Guisan), Humana Press, Totowa, NJ, **2006**, pp. 77–85.

191 C. J. Brinker, G. W. Scherer, *Sol-Gel Science: The Physics and Chemistry of Sol-Gel Processing*, Academic Press, Boston, **1990**.

192 L. M. Ellerby, C. R. Nishida, F. Nishida, S. A. Yamanaka, B. Dunn, J. S. Valentine, J. I. Zink, *Science* **1992**, *255*, 1113–1115.

193 G. A. Baker, *Aust. J. Chem.* **2005**, *58*, 721.

194 I. Gill, *Chem. Mater.* **2001**, *13*, 3404–3421.

195 D. Avnir, T. Coradin, O. Lev, J. Livage, *J. Mater. Chem.* **2006**, *16*, 1013–1030.

196 A. Vioux, L. Viau, S. Volland, J. Le Bideau, *C.R. Chim.* **2010**, *13*, 242–255.

197 Y. Zhou, J. H. Schattka, M. Antonietti, *Nano Lett.* **2004**, *4*, 477–481.

198 F. Shi, Q. Zhang, D. Li, Y. Deng, *Chem. Eur. J.* **2005**, *11*, 5279–5288.

199 M. V. Migliorini, R. K. Donato, M. A. Benvegnu, R. S. Goncalves, H. S. Schrekker, *J. Sol-Gel Sci. Technol.* **2008**, *48*, 272–276.

200 R. K. Donato, M. V. Migliorini, M. A. Benvegnu, M. P. Stracke, M. A. Gelesky, F. A. Pavan, C. M. L. Schrekker, E. V. Benvenutti, J. Dupont, H. S. Schrekker, *J. Sol-Gel Sci. Technol.* **2009**, *49*, 71–77.

201 J. Zhang, Y. Ma, F. Shi, L. Liu, Y. Deng, *Microporous Mesoporous Mater.* **2009**, *119*, 97–103.

202 S. H. Lee, T. T. N. Doan, S. H. Ha, Y.-M. Koo, *J. Mol. Catal., B Enzym.* **2007**, *45*, 57–61.

203 S. H. Lee, T. T. N. Doan, S. H. Ha, W.-J. Chang, Y.-M. Koo, *J. Mol. Catal., B Enzym.* **2007**, *47*, 129–134.

204 K. Sangeetha, V. B. Morris, T. E. Abraham, *Appl. Catal. A* **2008**, *341*, 168–173.

205 P. Hara, U. Hanefeld, L. T. Kanerva, *Green Chem.* **2009**, *11*, 250–256.

206 A. F. S. A. Habeeb, *Arch. Biochem. Biophys.* **1967**, *119*, 264–268.

207 E. F. Jansen, A. C. Olson, *Arch. Biochem. Biophys.* **1969**, *129*, 221–227.

208 R. A. Sheldon, M. Sorgedrager, M. H. A. Janssen, *Chim. Oggi.* **2007**, *25*, 62–67.

209 F. A. Quiocho, F. M. Richards, *Proc. Natl. Acad. Sci. U.S.A.* **1964**, *52*, 833–839.

210 N. L. St. Clair, M. A. Navia, *J. Am. Chem. Soc.* **1992**, *114*, 7314–7316.

211 A. L. Margolin, M. A. Navia, *Angew. Chem. Int. Ed. Engl.* **2001**, *40*, 2204–2222.

212 R. A. Sheldon, *Biochem. Soc. Trans.* **2007**, *35*, 1583–1587.

213 L. Cao, F. van Rantwijk, R. A. Sheldon, *Org. Lett.* **2000**, *2*, 1361–1364.

214 L. Cao, L. M. van Langen, F. van Rantwijk, R. A. Sheldon, *J. Mol. Catal., B Enzym.* **2001**, *11*, 665–670.

215 P. Lopez-Serrano, L. Cao, F. van Rantwijk, R. A. Sheldon, *Biotechnol. Lett.* **2002**, *24*, 1379–1383.

216 R. Schoevaart, M. W. Wolbers, M. Golubovic, M. Ottens, A. P. G. Kieboom, F. van Rantwijk, L. A. M. van der Wielen, R. A. Sheldon, *Biotechnol. Bioeng.* **2004**, *87*, 754–762.

217 C. Mateo, J. M. Palomo, L. M. van Langen, F. van Rantwijk, R. A. Sheldon, *Biotechnol. Bioeng.* **2004**, *86*, 273–276.

218 H. W. Yu, H. Chen, X. Wang, Y. Y. Yang, C. B. Ching, *J. Mol. Catal., B Enzym.* **2006**, *43*, 124–127.

219 S. Shah, M. N. Gupta, *Bioorg. Med. Chem. Lett.* **2007**, *17*, 921–924.

220 Y. Inada, K. Takahashi, T. Yoshimoto, A. Ajima, A. Matsushima, Y. Saito, *Trends Biotechnol.* **1986**, *4*, 190–194.

221 T. Maruyama, S. Nagasawa, M. Goto, *Biotechnol. Lett.* **2002**, *24*, 1341–1345.

222 T. Maruyama, H. Yamamura, T. Kotani, N. Kamiya, M. Goto, *Org. Biomol. Chem.* **2004**, *2*, 1239–1244.

223 F. M. Veronese, R. Largajolli, E. Boccú, C. A. Benassi, O. Schiavon, *Appl. Biochem. Biotechnol.* **1985**, *11*, 141–152.

224 C. A. Woodward, E. N. Kaufman, *Biotechnol. Bioeng.* **1996**, *52*, 423–428.

225 H. Ohno, C. Suzuki, K. Fukumoto, M. Yoshizawa, K. Fujita, *Chem. Lett.* **2003**, *32*, 450–451.

226 K. Nakashima, T. Maruyama, N. Kamiya, M. Goto, *Chem. Commun.* **2005**, 4297–4299.

227 Y. Kodera, H. Tanaka, A. Matsushima, Y. Inada, *Biochem. Biophys. Res. Commun.* **1992**, *184*, 144–148.

228 M. Hiroto, A. Matsushima, Y. Kodera, Y. Shibata, Y. Inada, *Biotechnol. Lett.* **1992**, *14*, 559–564.

229 J. Partridge, P. J. Halling, B. D. Moore, *Chem. Commun.* **1998**, 841–842.

230 T. Theppakorn, P. Kanasawud, P. J. Halling, *Biotechnol. Lett.* **2004**, *26*, 133–136.

231 I. Roy, M. N. Gupta, *Bioorg. Med. Chem. Lett.* **2004**, *14*, 2191–2193.

232 K. Solanki, M. N. Gupta, *Chem. Cent. J.* **2008**, *2*, 2.

233 S. Shah, M. N. Gupta, *Biochim. Biophys. Acta, Gen. Subj.* **2007**, *1770*, 94–98.

234 Z. Qiu, J. Texter, *Curr. Opin. Colloid Interface Sci.* **2008**, *13*, 252–262.

235 O. Zech, S. Thomaier, A. Kolodziejski, D. Touraud, I. Grillo, W. Kunz, *Chem. Eur. J.* **2010**, *16*, 783–786.

236 A. Safavi, N. Maleki, F. Farjami, *Colloids Surf. A* **2010**, *355*, 61–66.

237 M. Moniruzzaman, N. Kamiya, K. Nakashima, M. Goto, *Green Chem.* **2008**, *10*, 497–500.

238 M. Moniruzzaman, N. Kamiya, M. Goto, *Langmuir* **2009**, *25*, 977–982.

239 I. V. Pavlidis, D. Gournis, G. K. Papadopoulos, H. Stamatis, *J. Mol. Catal., B Enzym.* **2009**, *60*, 50–56.

240 K. Behera, N. I. Malek, S. Pandey, *ChemPhysChem* **2009**, *10*, 3204–3208.

241 J. K. Lee, M.-J. Kim, *J. Org. Chem.* **2002**, *67*, 6845–6847.

242 T. Itoh, S. Han, Y. Matsushita, S. Hayase, *Green Chem.* **2004**, *6*, 437–439.

243 T. Itoh, Y. Matsushita, Y. Abe, S. Han, S. Wada, S. Hayase, M. Kawatsura, S. Takai, M. Morimoto, Y. Hirose, *Chem. Eur. J.* **2006**, *12*, 9228–9237.

244 Y. Abe, K. Kude, S. Hayase, M. Kawatsura, K. Tsunashima, T. Itoh, *J. Mol. Catal., B Enzym.* **2008**, *51*, 81–85.

245 D. T. Dang, S. H. Ha, S.-M. Lee, W.-J. Chang, Y.-M. Koo, *J. Mol. Catal., B Enzym.* **2007**, *45*, 118–121.

246 P. Lozano, R. Piamtongkam, K. Kohns, T. De Diego, M. Vaultier, J. L. Iborra, *Green Chem.* **2007**, *9*, 780–784.

247 Z. Guo, X. Xu, *Org. Biomol. Chem.* **2005**, *3*, 2615–2619.

248 Z. Guo, B. Chen, R. L. Murillo, T. Tan, X. Xu, *Org. Biomol. Chem.* **2006**, *4*, 2772–2776.

249 Z. Guo, X. Xu, *Green Chem.* **2006**, *8*, 54–62.

250 B. Chen, Z. Guo, T. Tan, X. Xu, *Biotechnol. Bioeng.* **2008**, *99*, 18–29.

251 D. Kahveci, Z. Guo, B. Özçelik, X. Xu, *Process Biochem.* **2009**, *44*, 1358–1365.

252 Z. Guo, D. Kahveci, B. Özçelik, X. Xu, *New Biotechnol.* **2009**, *26*, 37–43.

253 D. Kahveci, Z. Guo, B. Özçelik, X. Xu, *Food Chem.* **2010**, *119*, 880–885.

254 Z. Guo, B.-M. Lue, K. Thomasen, A. S. Meyer, X. Xu, *Green Chem.* **2007**, *9*, 1362–1373.

255 S. Dreyer, U. Kragl, *Biotechnol. Bioeng.* **2008**, *99*, 1416–1424.

256 S. Wallert, K. Drauz, I. Grayson, H. Gröger, P. Domínguez de María, C. Bolm, *Green Chem.* **2005**, *7*, 602–605.

257 N. M. T. Lourenço, S. Barreiros, C. A. M. Afonso, *Green Chem.* **2007**, *9*, 734–736.

258 H. Zhao, Z. Song, *Biochem. Eng. J.* **2010**, *49*, 113–118.

259 C. Vafiadi, E. Topakas, V. R. Nahmias, C. B. Faulds, P. Christakopoulos, *J. Biotechnol.* **2009**, *139*, 124–129.

260 A. P. Abbott, G. Capper, D. L. Davies, R. K. Rasheed, V. Tambyrajah, *Chem. Commun.* **2003**, 70–71.

261 A. P. Abbott, D. Boothby, G. Capper, D. L. Davies, R. K. Rasheed, *J. Am. Chem. Soc.* **2004**, *126*, 9142–9147.

262 A. P. Abbott, G. Capper, S. Gray, *ChemPhysChem* **2006**, *7*, 803–806.

263 J. K. Blusztajn, *Science* **1998**, *281*, 794–795.

264 A. P. Abbott, G. Capper, B. G. Swain, D. A. Wheeler, *Trans. Inst. Met. Finishing* **2005**, *83*, 51–53.

265 Y. Hou, Y. Gu, S. Zhang, F. Yang, H. Ding, Y. Shan, *J. Mol. Liq.* **2008**, *143*, 154–159.

266 A. P. Abbott, P. M. Cullis, M. J. Gibson, R. C. Harris, E. Raven, *Green Chem.* **2007**, *9*, 868–872.

267 J. T. Gorke, F. Srienc, R. J. Kazlauskas, *Chem. Commun.* **2008**, 1235–1237.

268 H. Zhao, Z. Song, O. Olubajo, J. V. Cowins, *Appl. Biochem. Biotechnol.* **2010**, *162*, 13–23.

269 H. Zhao, C. L. Jones, G. A. Baker, S. Xia, Z. Song, O. Olubajo, V. N. Person, *J. Biotechnol.* **2009**, *139*, 47–54.

270 H. Zhao, G. A. Baker, J. V. Cowins, *Biotechnol. Prog.* **2010**, *26*, 127–133.

271 S.-T. Chen, K.-T. Wang, C.-H. Wong, *Chem. Commun.* **1986**, 1514–1516.

272 S.-T. Chen, W.-H. Huang, K.-T. Wang, *Chirality* **1994**, *6*, 572–576.

273 J.-Y. Houng, M.-L. Wu, S.-T. Chen, *Chirality* **1998**, *8*, 418–422.

274 T. Kijima, K. Ohshima, H. Kise, *J. Chem. Technol. Biotechnol.* **1994**, *59*, 61–65.

275 H. Zhao, S. V. Malhotra, *Biotechnol. Lett.* **2002**, *24*, 1257–1260.

276 S. V. Malhotra, H. Zhao, *Chirality* **2005**, *17*, S240–S242.

277 H. Zhao, R. G. Luo, S. V. Malhotra, *Biotechnol. Prog.* **2003**, *19*, 1016–1018.

278 S. V. Malhotra, H. Zhao, in *Ionic Liquids IIIB: Fundamentals, Progress, Challenges, and Opportunities (Transformations and Processes)* (Eds. R. D. Rogers, K. R. Seddon), American Chemical Society, Washington DC, **2005**, pp. 111–123.

279 W.-Y. Lou, M.-H. Zong, H. Wu, *Biocatal. Biotransformation* **2004**, *22*, 171–176.

280 W. Lou, M. Zong, H. Wu, *Biotechnol. Appl. Biochem.* **2005**, *41*, 151–156.

281 Y.-Y. Liu, W.-Y. Lou, M.-H. Zong, R. Xu, X. Hong, H. Wu, *Biocatal. Biotransformation* **2005**, *23*, 89–95.

282 W.-Y. Lou, M.-H. Zong, Y.-Y. Liu, J.-F. Wang, *J. Biotechnol.* **2006**, *125*, 64–74.

283 L. Zheng, S. Zhang, X. Yu, L. Zhao, G. Gao, X. Yang, H. Duan, S. Cao, *J. Mol. Catal., B Enzym.* **2006**, *38*, 17–23.

284 S. S. Mohile, M. K. Potdar, J. R. Harjani, S. J. Nara, M. M. Salunkhe, *J. Mol. Catal., B Enzym.* **2004**, *30*, 185–188.

285 N. J. Roberts, A. Seago, J. S. Carey, R. Freer, C. Preston, G. J. Lye, *Green Chem.* **2004**, *6*, 475–482.

286 J.-Y. Xin, Y.-J. Zhao, G.-L. Zhao, Y. Zheng, X.-S. Ma, C.-G. Xia, S.-B. Li, *Biocatal. Biotransformation* **2005**, *23*, 353–361.

287 J.-Y. Xin, Y.-J. Zhao, Y.-G. Shi, C.-G. Xia, S.-B. Li, *World J. Microbiol. Biotechnol.* **2005**, *21*, 193–199.

288 E. Xanthakis, M. Zarevúcka, D. Šaman, M. Wimmerová, F. N. Kolisis, Z. Wimmer, *Tetrahedron Asymmetry* **2006**, *17*, 2987–2992.

289 M. Oromí-Farrús, J. Eras, N. Sala, M. Torres, R. Canela, *Molecules* **2009**, *14*, 4275–4283.

290 P. U. Naik, S. J. Nara, J. R. Harjani, M. M. Salunkhe, *J. Mol. Catal., B Enzym.* **2007**, *44*, 93–98.

291 M. Brossat, T. S. Moody, S. J. C. Taylor, J. W. Wiffen, *Tetrahedron Asymmetry* **2009**, *20*, 2112–2116.

292 B. Berger, K. Faber, *J. Chem. Soc. Chem. Commun.* **1991**, 1198–1200.

293 T. Itoh, E. Akasaki, Y. Nishimura, *Chem. Lett.* **2002**, *31*, 154–155.

294 T. Itoh, Y. Nishimura, N. Ouchi, S. Hayase, *J. Mol. Catal., B Enzym.* **2003**, *26*, 41–45.

295 M. Persson, U. T. Bornscheuer, *J. Mol. Catal., B Enzym.* **2003**, *22*, 21–27.

296 M. Paljevac, Z. Knez, M. Habulin, *Acta Chim. Slov.* **2009**, *56*, 399–409.

297 M. Habulin, Z. Knez, *J. Mol. Catal., B Enzym.* **2009**, *58*, 24–28.

298 P. Lozano, T. De Diego, S. Gmouh, M. Vaultier, J. L. Iborra, *Int. J. Chem. Reactor Eng.* **2007**, *5*, A53.

299 K.-W. Kim, B. Song, M.-Y. Choi, M.-J. Kim, *Org. Lett.* **2001**, *3*, 1507–1509.

300 P. Kielbasinski, M. Albrycht, J. Luczak, M. Mikolajczyk, *Tetrahedron Asymmetry* **2002**, *13*, 735–738.

301 T. Itoh, N. Ouchi, S. Hayase, Y. Nishimura, *Chem. Lett.* **2003**, *32*, 654–655.

302 A. Kamal, G. Chouhan, *Tetrahedron Lett.* **2004**, *45*, 8801–8805.

303 M. S. Rasalkar, M. K. Potdar, M. M. Salunkhe, *J. Mol. Catal., B Enzym.* **2004**, *27*, 267–270.

304 F. Theil, H. Sonnenschein, T. Kreher, *Tetrahedron Asymmetry* **1996**, *7*, 3365–3370.

305 M.-C. Parker, S. A. Brown, L. Robertson, N. J. Turner, *Chem. Commun.* **1998**, 2247–2248.

306 X.-M. Wu, J.-Y. Xin, W. Sun, C.-G. Xia, *Chem. Biodivers.* **2007**, *4*, 183–188.

307 H. Yu, J. Wu, C. B. Ching, *Chirality* **2005**, *17*, 16–21.

308 F. J. Contesini, P. de Oliveira Carvalho, *Tetrahedron Asymmetry* **2006**, *17*, 2069–2073.

309 K. Lundell, T. Kurki, M. Lindroos, L. T. Kanerva, *Adv. Synth. Catal.* **2005**, *347*, 1110–1118.

310 Y. Wang, L. Mei, *J. Biosci. Bioeng.* **2007**, *103*, 345–349.

311 S. Das, S. Chandrasekhar, J. S. Yadav, A. V. R. Rao, R. Grée, *Tetrahedron Asymmetry* **2008**, *19*, 2543–2545.

312 Y. Yuan, S. Bai, Y. Sun, *Food Chem.* **2006**, *97*, 324–330.

313 M.-Y. Ren, S. Bai, D.-H. Zhang, Y. Sun, *J. Agric. Food Chem.* **2008**, *56*, 2388–2391.

314 M. Singh, R. S. Singh, U. C. Banerjee, *J. Mol. Catal., B Enzym.* **2009**, *56*, 294–299.

315 C. Pilissão, M. G. Nascimento, *Tetrahedron Asymmetry* **2006**, *17*, 428–433.

316 M.-J. Kim, H. M. Kim, D. Kim, Y. Ahn, J. Park, *Green Chem.* **2004**, *6*, 471–474.

317 N. Kaftzik, W. Kroutil, K. Faber, U. Kragl, *J. Mol. Catal., A Chem.* **2004**, *214*, 107–112.

318 P. Lozano, T. De Diego, M. Larnicol, M. Vaultier, J. L. Iborra, *Biotechnol. Lett.* **2006**, *28*, 1559–1565.

319 P. Lozano, T. De Diego, C. Mira, K. Montague, M. Vaultier, J. L. Iborra, *Green Chem.* **2009**, *11*, 538–542.

320 N. M. T. Lourenço, C. A. M. Afonso, *Angew. Chem. Int. Ed. Engl.* **2007**, *46*, 8178–8181.

321 R. Irimescu, K. Kato, *Tetrahedron Lett.* **2004**, *45*, 523–525.

322 X. J. Yang, A. G. Fane, K. Soldenhoff, *Ind. Eng. Chem. Res.* **2003**, *42*, 392–403.

323 F. J. Hernández-Fernández, A. P. de los Ríos, F. Tomás-Alonso, D. Gómez, M. Rubio, G. Víllora, *Chem. Eng. Process* **2007**, *46*, 818–824.

324 P. Scovazzo, A. E. Visser, J. H. J. Davis, R. D. Rogers, C. A. Koval, D. L. DuBois, R. D. Noble, in *Ionic Liquids: Industrial Applications for Green Chemistry* (Eds. R. D. Rogers, K. R. Seddon), Oxford University Press, Washing, DC, **2002**, pp. 69–87.

325 E. Miyako, T. Maruyama, N. Kamiya, M. Goto, *Chem. Commun.* **2003**, 2926–2927.

326 F. J. Hernández-Fernández, A. P. de los Ríos, F. Tomás-Alonso, D. Gómez, G. Víllora, *J. Chem. Technol. Biotechnol.* **2009**, *84*, 337–342.

327 F. J. Hernández-Fernández, A. P. de los Ríos, F. Tomás-Alonso, D. Gómez, G. Víllora, *J. Memb. Sci.* **2008**, *314*, 238–246.

328 Y. S. Huh, Y.-S. Jun, Y.-K. Hong, W.-H. Hong, D. H. Kim, *J. Mol. Catal., B Enzym.* **2006**, *43*, 96–101.

329 K. Sato, A. Hibara, M. Tokeshi, H. Hisamoto, T. Kitamori, *Adv. Drug Deliv. Rev.* **2003**, *55*, 379–391.

330 Y. Huang, E. L. Mather, J. L. Bell, M. Madou, *Anal. Bioanal. Chem.* **2002**, *372*, 49–65.

331 R. Marr, T. Gamse, *Chem. Eng. Process* **2000**, *39*, 19–28.

332 W. Wu, J. Zhang, B. Han, J. Chen, Z. Liu, T. Jiang, J. He, W. Li, *Chem. Commun.* **2003**, 1412–1413.

333 S. V. Dzyuba, R. A. Bartsch, *Angew. Chem. Int. Ed. Engl.* **2003**, *42*, 148–150.

334 A. M. Scurto, S. N. V. K. Aki, J. F. Brennecke, *J. Am. Chem. Soc.* **2002**, *124*, 10276–10277.

335 A. M. Scurto, S. N. V. K. Aki, J. F. Brennecke, *Chem. Commun.* **2003**, 572–573.

336 V. Najdanovic-Visak, A. Serbanovic, J. M. S. S. Esperança, H. J. R. Guedes, L. P. N. Rebelo, M. Nunes da Ponte, *ChemPhysChem* **2003**, *4*, 520–522.

337 S. V. Kamat, E. J. Beckman, A. J. Russell, *Crit. Rev. Biotechnol.* **1995**, *15*, 41–71.

338 P. Lozano, A. Avellaneda, R. Pascual, J. L. Iborra, *Biotechnol. Lett.* **1996**, *18*, 1345–1350.

339 P. Lozano, T. de Diego, D. Carrie, M. Vaultier, J. L. Iborra, *Chem. Commun.* **2002**, 692–693.

340 P. Lozano, T. De Diego, D. Carrié, M. Vaultier, J. L. Iborra, *Biotechnol. Prog.* **2003**, *19*, 380–382.

341 P. Lozano, T. de Diego, S. Gmouh, M. Vaultier, J. L. Iborra, *Biotechnol. Prog.* **2004**, *20*, 661–669.

342 M. T. Reetz, W. Wiesenhofer, G. Francio, W. Leitner, *Chem. Commun.* **2002**, 992–993.

343 M. T. Reetz, W. Wiesenhöfer, G. Franciò, W. Leitner, *Adv. Synth. Catal.* **2003**, *345*, 1221–1228.

344 P. Lozano, T. De Diego, T. Sauer, M. Vaultier, S. Gmouh, J. L. Iborra, *J. Supercrit. Fluids* **2007**, *40*, 93–100.

345 R. Bogel-Lukasik, V. Najdanovic-Visak, S. Barreiros, M. Nunes da Ponte, *Ind. Eng. Chem. Res.* **2008**, *47*, 4473–4480.

346 P. Lozano, T. De Diego, M. Vaultier, J. L. Iborra, *Int. J. Chem. Reactor Eng.* **2009**, *7*, A79.

347 T. Itoh, E. Akasaki, K. Kudo, S. Shirakami, *Chem. Lett.* **2001**, *30*, 262–263.

348 T. Ema, T. Doi, T. Sakai, *Chem. Lett.* **2008**, *37*, 90–91.

349 L. Banoth, M. Singh, A. Tekewe, U. C. Banerjee, *Biocatal. Biotransformation* **2009**, *27*, 263–270.

350 J. Yu, C. Wang, J. Hong, J. Huang, *J. Macromol. Sci., Part A Pure Appl. Chem.* **2009**, *46*, 943–948.

351 D. A. Fort, R. C. Remsing, R. P. Swatloski, P. Moyna, G. Moyna, R. D. Rogers, *Green Chem.* **2007**, *9*, 63–69.

352 I. Kilpeläinen, H. Xie, A. King, M. Granstrom, S. Heikkinen, D. S. Argyropoulos, *J. Agric. Food Chem.* **2007**, *55*, 9142–9148.

353 M. E. Zakrzewska, E. Bogel-Lukasik, R. Bogel-Lukasik, *Energy Fuels* **2010**, *24*, 737–745.

354 T. G. A. Youngs, C. Hardacre, J. D. Holbrey, *J. Phys. Chem. B* **2007**, *111*, 13765–13774.

355 T. G. A. Youngs, J. D. Holbrey, M. Deetlefs, M. Nieuwenhuyzen, M. F. C. Gomes, C. Hardacre, *ChemPhysChem* **2006**, *7*, 2279–2281.

356 N. P. Novoselov, E. S. Sashina, V. E. Petrenko, M. Zaborsky, *Fibre Chem.* **2007**, *39*, 153–158.

357 R. C. Remsing, G. Hernandez, R. P. Swatloski, W. W. Massefski, R. D. Rogers, G. Moyna, *J. Phys. Chem. B* **2008**, *112*, 11071–11078.

358 T. Heinze, K. Schwikal, S. Barthel, *Macromol. Biosci.* **2005**, *5*, 520–525.

359 S. Barthel, T. Heinze, *Green Chem.* **2006**, *8*, 301–306.

360 C. F. Liu, R. C. Sun, A. P. Zhang, J. L. Ren, *Carbohydr. Polym.* **2007**, *68*, 17–25.

361 S. Köhler, T. Heinze, *Cellulose* **2007**, *14*, 489–495.

362 K. Schlufter, H.-P. Schmauder, S. Dorn, T. Heinze, *Macromol. Rapid Commun.* **2006**, *27*, 1670–1676.

363 A. Biswas, R. L. Shogren, D. G. Stevenson, J. L. Willett, P. K. Bhowmik, *Carbohydr. Polym.* **2006**, *66*, 546–550.

364 A. P. Dadi, S. Varanasi, C. A. Schall, *Biotechnol. Bioeng.* **2006**, *95*, 904–910.

365 H. Xie, S. Li, S. Zhang, *Green Chem.* **2005**, *7*, 606–608.

366 M. B. Turner, S. K. Spear, J. D. Holbrey, R. D. Rogers, *Biomacromolecules* **2004**, *5*, 1379–1384.

367 J. F. Kennedy, H. Kumar, P. S. Panesar, S. S. Marwaha, R. Goyal, A. Parmar, S. Kaur, *J. Chem. Technol. Biotechnol.* **2006**, *81*, 866–876.

368 S. Park, F. Viklund, K. Hult, R. J. Kazlauskas, in *ACS Symposium Series 856: Ionic Liquids as Green Solvents* (Eds. R. D. Rogers, K. R. Seddon), American Chemical Society, Washington, DC, **2003**, pp. 225–238.

369 S. H. Lee, D. T. Dang, S. H. Ha, W.-J. Chang, Y.-M. Koo, *Biotechnol. Bioeng.* **2008**, *99*, 1–8.

370 S. H. Lee, S. H. Ha, N. M. Hiep, W.-J. Chang, Y.-M. Koo, *J. Biotechnol.* **2008**, *133*, 486–489.

371 S. H. Ha, N. M. Hiep, S. H. Lee, Y.-M. Koo, *Bioprocess Biosyst. Eng.* **2010**, *33*, 63–70.

372 S. H. Lee, H. M. Nguyen, Y.-M. Koo, S. H. Ha, *Process Biochem.* **2008**, *43*, 1009–1012.

373 M. V. Flores, K. Naraghi, J.-M. Engasser, P. J. Halling, *Biotechnol. Bioeng.* **2002**, *78*, 815–821.

374 S. Ha, J. Gao, B. Tidor, J. W. Brady, M. Karplus, *J. Am. Chem. Soc.* **1991**, *113*, 1553–1557.

375 S. J. Angyal, *Angew. Chem. Int. Ed. Engl.* **1969**, *8*, 157–166.

376 J. Xie, Y.-L. Hsieh, *J. Polym. Sci. Part A Polym. Chem.* **2001**, *39*, 1931–1939.

377 T. Maruyama, S.-I. Nagasawa, M. Goto, *J. Biosci. Bioeng.* **2002**, *94*, 357–361.

378 M. Therisod, A. M. Klibanov, *J. Am. Chem. Soc.* **1986**, *108*, 5638–5640.

379 F. F. Bruno, J. A. Akkara, M. Ayyagari, D. L. Kaplan, R. Gross, G. Swift, J. S. Dordick, *Macromolecules* **1995**, *28*, 8881–8883.

380 S. Riva, J. Chopineau, A. P. G. Kieboom, A. M. Klibanov, *J. Am. Chem. Soc.* **1988**, *110*, 584–589.

381 J. O. Rich, B. A. Bedell, J. S. Dordick, *Biotechnol. Bioeng.* **1995**, *45*, 426–434.

382 M.-J. Kim, M. Y. Choi, J. K. Lee, Y. Ahn, *J. Mol. Catal., B Enzym.* **2003**, *26*, 115–118.

383 M. H. Katsoura, A. C. Polydera, L. Tsironis, A. D. Tselepis, H. Stamatis, *J. Biotechnol.* **2006**, *123*, 491–503.

384 P. Galletti, F. Moretti, C. Samorì, E. Tagliavini, *Green Chem.* **2007**, *9*, 987–991.

385 N. P. Price, in *Biocatalysis and Biotechnology for Functional Foods and Industrial Products*, 1st ed. (Eds. C. T. Hou, J.-F. Shaw), CRC Press, Taylor and Francis Group, New York, **2006**, pp. 451–466.

386 L. Liu, H. Chen, *Chin. Sci. Bull.* **2006**, *51*, 2432–2436.

387 Y.-H. P. Zhang, L. R. Lynd, *Biotechnol. Bioeng.* **2004**, *88*, 797–824.

388 S. Rayne, G. Mazza, *Nature Precedings* **2007**, hdl:10101/npre.2007.632.1.

389 P. Linsdell, S.-X. Zheng, J. W. Hanrahan, *J. Physiol.* **1998**, *512*, 1–16.

390 N. Kamiya, Y. Matsushita, M. Hanaki, K. Nakashima, M. Narita, M. Goto, H. Takahashi, *Biotechnol. Lett.* **2008**, *30*, 1037–1040.

391 P. O. Jones, P. T. Vasudevan, *Biotechnol. Lett.* **2010**, *32*, 103–106.

392 J. Pottkämper, P. Barthen, N. Ilmberger, U. Schwaneberg, A. Schenk, M. Schulte, N. Ignatiev, W. R. Streit, *Green Chem.* **2009**, *11*, 957–965.

393 P. H. von Hippel, T. Schleich, *Acc. Chem. Res.* **1969**, *2*, 257–265.

394 J. Sangster, *J. Phys. Chem. Ref. Data* **1989**, *18*, 1111–1229.

395 S.-M. Lee, W.-J. Chang, A.-R. Choi, Y.-M. Koo, *Korean J. Chem. Eng.* **2005**, *22*, 687–690.

396 O. Ulbert, T. Frater, K. Belafi-Bako, L. Gubicza, *J. Mol. Catal., B Enzym.* **2004**, *31*, 39–45.

397 A. Chapeaux, L. D. Simoni, M. A. Stadtherr, J. F. Brennecke, *J. Chem. Eng. Data* **2007**, *52*, 2462–2467.

398 B. Kosan, C. Michels, F. Meister, *Cellulose* **2008**, *15*, 59–66.

399 G. Laus, G. Bentivoglio, H. Schottenberger, V. Kahlenberg, H. Kopacka, T. Röder, H. Sixta, *Lenzinger Berichte* **2005**, *84*, 71–85.

400 H. Zhang, J. Wu, J. Zhang, J. He, *Macromolecules* **2005**, *38*, 8272–8277.

401 J. Wu, J. Zhang, H. Zhang, J. He, Q. Ren, M. Guo, *Biomacromolecules* **2004**, *5*, 266–268.

402 Y. Fukaya, A. Sugimoto, H. Ohno, *Biomacromolecules* **2006**, *7*, 3295–3297.

403 Y. Fukaya, K. Hayashi, M. Wada, H. Ohno, *Green Chem.* **2008**, *10*, 44–46.

6

IONIC LIQUIDS AS (CO-) SOLVENTS FOR NONHYDROLYTIC ENZYMES

Daniela Gamenara, Patricia Saenz Méndez, Gustavo Seoane, and Pablo Domínguez de María

NOMENCLATURE OF ILs

[BDMIM][PF_6]	1-butyl-2,3-dimethylimidazolium hexafluorophosphate
[BMIM][BF_4]	1-butyl-3-methylimidazolium tetrafluoroborate
[BMIM][Br]	1-butyl-3-methylimidazolium bromide
[BMIM][Cl]	1-butyl-3-methylimidazolium chloride
[BMIM][$(CF_3SO_2)_2N$]	1-butyl-3-methylimidazolium bis((trifluoromethyl)sulphonyl)amide
[BMIM][CF_3SO_3]	1-butyl-3-methylimidazolium trifluoromethanesulfonate
[BMIM][F_3CSO_3]	1-butyl-3-methylimidazolium trifluoromethanesulfonate
[BMIM][glycolate]	1-butyl-3-methylimidazolium glycolate
[BMIM][lactate]	1-butyl-3-methylimidazolium (L)-lactate
[BMIM][$MeSO_4$]	1-butyl-3-methylimidazolium methylsulfate
[BMIM][$OctSO_4$]	1-butyl-3-methylimidazolium octylsulfate

Ionic Liquids in Biotransformations and Organocatalysis: Solvents and Beyond, First Edition. Edited by Pablo Domínguez de María.

[BMIM][OTf]	1-butyl-3-methylimidazolium triflate
[BMIM][PF_6]	1-butyl-3-methylimidazolium hexafluorophosphate
[BMIM][Tf_2N]	1-butyl-3-methylimidazolium bis(trifluoromethylsulfonyl)amide
[BMP][NTf_2]	butylmethylpyrrolidinium bis(trifluoromethylsulfonyl) imide
[BMPy][BF_4]	*N*-butyl-3-methylpyridinium tetrafluoroborate
[Choline][DHP]	choline dihydrogen phosphate
[C_8MIM][Tf_2N]	1-octyl-3-methylimidazolium bis(trifluoromethylsulfonyl)amide
[EA][NO_3]	ethylammonium nitrate
[EMIM][BF_4]	1-ethyl-3-methylimidazolium tetrafluoroborate
[EMIM][Cl]	1-ethyl-3-methylimidazolium chloride
[EMIM][CF_3SO_3]	1-ethyl-3-methylimidazolium trifluoromethanesulfonate
[EMIM][$PhCO_2$]	1-ethyl-3-methylimidazolium benzoate
[EMIM][TOS]	1-Ethyl-3-methylimidazolium tosylate
[EtPy][BF_4]	*N*-ethylpyridinium tetrafluoroborate
[EtPy][PF_6]	*N*-ethylpyridinium hexafluorophosphate
[EtPy][TFA]	*N*-ethylpyridinium trifluoroacetate
[Et_3NH][$MeSO_4$]	triethylammonium methylsulfate
[Et_3NMe][$MeSO_4$]	triethylmethylammonium methylsulfate
[HMIM][BF_4]	1-hexyl-3-methylimidazolium tetrafluoroborate
[HPMIM][Cl]	1-(3-hydroxypropyl)-3-methylimidazolium chloride
[HPMIM][glycolate]	1-(3-hydroxypropyl)-3-methylimidazolium glycolate
[HPMIM][PF_6]	1-(3-hydroxypropyl)-3-methylimidazolium hexafluorophosphate
[4-MBP][BF_4]	4-methyl-*N*-butylpyridinium tetrafluoroborate
[MMIM][$MMPO_4$]	1,3-dimethylimidazolium dimethylphosphate
[MMIM][Me_2PO_4]	1,3-dimethylimidazolium dimethylphosphate
[MMIM][$MeSO_4$]	1,3-dimethylimidazolium methylsulfate
[Mor_{11}][$MeSO_4$]	*N,N*-dimethylmorpholinium methylsulfate
[MTEOA][$MeSO_4$]	tris(2-hydroxyethyl)methylammonium methylsulfate
[PMIM][BF_4]	1-propyl-3-methylimidazolium tetrafluoroborate
[PeMIM][BF_4]	1-pentyl-3-methylimidazolium tetrafluoroborate
[N-BPy][Br]	*N*-butylpyridinium bromide
[$N_{1112}OH$][H_2PO_4]	cholinium phosphate
[$N_{1112}OH$][OAc]	cholinium acetate
[$N_{1112}OH$][Citr]	cholinium citrate

6.1 IONIC LIQUIDS AND NONHYDROLYTIC ENZYMES

Previous chapters of this book have shown how the use of room-temperature ionic liquids (RTILs) as pure solvents, co-solvents, or being part of biphasic systems has become a promising strategy to overcome usual drawbacks associated with biocatalytic processes.[1] In this respect, challenges such as substrate solubility, catalyst stability and selectivity, or cofactor recycling procedures can be tackled (or at least improved), by the use of these alternative reaction systems.[2]

With regard to enhancing substrate solubility in aqueous media, first attempts involved the use of organic solvents or biphasic systems.[3] Enzymes proved to be active in nonpolar solvents ($\log P > 2$) not miscible with water. However, enzyme activities often decrease in polar or protic solvents ($\log P < 2$) miscible with water, due to the stripping off of the crucial water from the enzyme surface. As it was previously reported, many ILs show polarity values comparable to polar solvents like ethanol, methanol, or DMF (see Table 6.2, Chapter 5), but surprisingly they do not inactivate enzymes. Moreover, their nonvolatility, nonflammability, and thermal stability (with some remarks, see Chapter 1) have converted ILs into a promising environmentally engaged alternative to conventional organic solvents. Therefore, they are suitable solvents for performing enzymatic reactions in polar media, exploiting the advantages that they could provide, thus being a promising alternative to bridge the gap between water and conventional solvents.[4] Although many enzymes display reduced catalytic activity (or no activity at all) when dissolved in pure ILs (attributed to changes in their conformational state), ILs as co-solvents or as second phase are relevant alternatives,[5] since some enzymes often exhibit higher selectivities, faster rates, and enhanced enzyme stabilities.[6] In this respect, the role of ILs in biocatalysis is currently focused not only in replacing organic solvents, but also in the development of reaction protocols involving water-miscible ILs in aqueous systems, or even biphasic systems with the concomitant possibility of enzyme reutilization.[7–10] Even the use of three-phase systems was investigated as well in which the IL, together with organic solvents and water, modulates the solubility of the substrates and the enzyme activity.[8]

Despite the promising features of RTILs, their use in biotransformations involving nonhydrolytic enzymes has been scarce so far.[8–10] Most of the conducted research has focused on hydrolases, presumably due to the enhanced stability of those biocatalysts in nonconventional media and also because cofactors are not needed for the catalytic performances. Therefore, while in hydrolase-catalyzed processes involving ILs there is already a basis for a fundamental understanding of how anions and cations can interact with the enzymes (see Chapters 2 and 5), in other enzymatic performances, a systematic and rational treatment of reported data is not yet feasible. In any case, it can be assumed that information already acquired for hydrolases may serve as useful basis for further studies in other nonhydrolytic systems (chaotropicity, kosmotropicity, etc.). As an example of this, very recently an improvement of the stability and activity of formate dehydrogenase (FDH) from *Candida boidinii* in buffer–IL mixtures was reported.[11] Native FDH was found to be inactivated in the presence of high concentrations (>50 v/v%) of dimethylimidazolium dimethyl phosphate [MMIm][$MMPO_4$], a water-miscible IL. To enhance the FDH performance in that IL, a chemical modification of the enzyme was conducted by covalently grafting different cations to lysines present in the enzyme (via carbamate formation). As cations, cholinium, hydroxyethyl-methylimidazolium, and hydroxypropyl-methylimidazolium were used (Scheme 6.1).

All cations used in the grafting are considered chaotropic. It was therefore assumed that the usual Hofmeister series (see Chapters 2 and 5) might explain the enhanced stability and activity as well, since at molecular level the grafted cations obviously make an impact on the water distribution along the enzyme. Although the stabilizing effects on FDH were still moderate, this study is an interesting example on how the current state-of-the-art in understanding interactions between ILs and enzymes can be used for nonhydrolytic performances as well. Moreover, it represents a further option to use ILs not only as *mere solvents* for biocatalytic reactions, but also in "nonsolvent" applications (see also Chapter 8 of this book, entirely dedicated to those nonsolvent applications).

6.2 USE OF ILs IN OXIDOREDUCTASE-CATALYZED ENZYMATIC REACTIONS

Oxidoreductases are a large class of enzymes that catalyze *in vivo* oxidation and reduction reactions that are engaged in essential roles in living cells metabolism, and account for up to 25% of known enzymes.[13,14] The development of biocatalytic applications of oxidoreductases has been one of the most important research fields since the beginning of biotechnology. These biotechnological applications range from environmental goals, such as biodegradation and bioremediation, to synthetic performances involving the formation of chiral centers, or C–O (or other heteroatoms such as S or halogens) bond formation in organic substrates.[15] In fact, enzyme redox-catalyzed processes have passed from being a mere *lab curiosity* to becoming the first option in pharmaceutical and fine chemistry industries when asymmetric reductions or oxidations are intended. Key to this is the large availability of enzymes at accessible prices, which has been due to the impressive development of molecular biology techniques.[16]

ILs may provide several added values in the field of oxidoreductases. One of them is the (already mentioned) enhanced solubility of organic substrates in aqueous solutions when water-miscible ILs are used. As a relevant example for redox-catalyzed enzymatic performances, the addition of 40% v/v of 1,3-dimethylimidazolium methyl-sulfate [MMIM][$MeSO_4$] to water increases the solubility of acetophenone from 20 to 200 mmol/L.[2] In the following sections, the state-of-the-art on the use of ILs in redox processes catalyzed by enzymes will be provided.

6.2.1 Dehydrogenases

Among oxidoreductases,[13] the most extensively used enzymes in preparative-scale biocatalysis are dehydrogenases or reductases:

- EC 1.1.1. Acting upon hydroxyl groups as hydrogen donors.
- EC 1.2.1. Acting upon carbonylic or carboxylic groups as donors.
- EC 1.4.1. Acting upon amino groups as hydrogen donors.

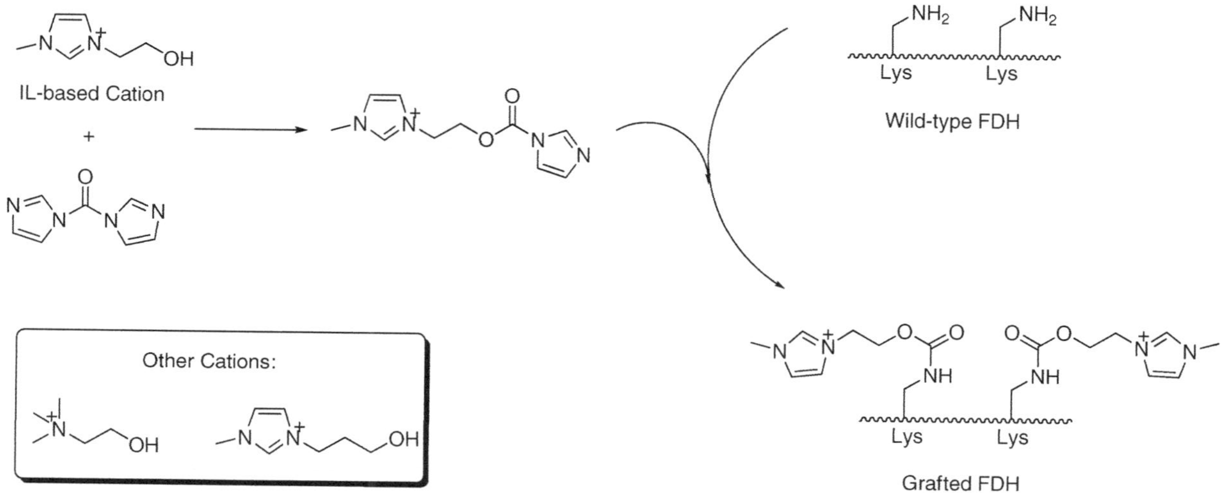

Scheme 6.1. Stabilization of FDH by grafting different cations commonly used in ILs.[11,12]

All these enzymes use NAD^+ or $NADP^+$ as cofactors, and to cope with economic costs (avoiding stoichiometric additions of them), they must be *in situ* regenerated.[17] Through dehydrogenase-catalyzed reduction processes, highly enantiopure alcohols, hydroxy-acids, hydroxy-esters, amines, and amino acids can be obtained. These enzymes catalyze a reversible reaction, in which one hydrogen atom and two electrons from the donor are transferred to the cofactor (NAD^+ or $NADP^+$). The inverse reaction involves the transfer of a hydrogen atom and two electrons from the reduced form of the cofactor (NADH or NADPH) to the carbonyl group, resulting in the reduction of the substrate. This reaction is often highly stereospecific since the transfer of the hydrogen atom takes place either to the *Si* face or to the *Re* face of the carbonyl group, thus resulting in the corresponding (*S*)- or (*R*)-alcohols. As already stated, dehydrogenase-catalyzed redox reactions require stoichiometric oxidation or reduction of costly nicotinamide cofactors. Therefore, efficient cofactor regeneration must be accomplished for practical applications. To this end, enzymatic, chemical, electrochemical, or photochemical approaches can be applied in isolated dehydrogenase-catalyzed processes, representing a crucial point of consideration when designing the enzymatic process.[16]

Following the successful applications of ILs in hydrolases by focusing on enzymatic activity or stability (see Chapter 5),[8–10,18] the interest in using ILs in other enzymatic nonhydrolytic performances has started to grow in the past few years.[5,19] In this respect, the association of oxidoreductases with ILs is very promising due to the vast field of application of these enzymes.[4,5,20] However, research involving ILs in combination with isolated oxidoreductases—and in particular cofactor-dependent dehydrogenases—is still incipient.[21–29]

The first attempt to use ILs as co-solvents in isolated oxidoreductase-catalyzed processes was reported by Kragl et al.,[21] who studied the effect of different ILs on the activity of several oxidoreductases such as formate dehydrogenase (FDH) from *Candida boidinii*, alcohol dehydrogenase from yeast (YADH), and *Candida parapsilosis* carbonyl-reductase (CPCR). The activity of these enzymes was assessed, depending on the IL content in the aqueous buffer (that is, ILs were used as co-solvents in aqueous solutions), and compared with the corresponding activity in pure buffer solution. As a model reaction, FDH activity was tested on NAD^+ reduction with simultaneous oxidation of formate. For YADH activity, the oxidation of ethanol with concomitant NAD^+ reduction was used, and CPCR activity was evaluated using the reduction of acetophenone under oxidation of NADH. An ample number of different ILs as co-solvents, such as [MMIM][$MeSO_4$], triethylammonium methylsulfate [Et_3NH][$MeSO_4$], triethylmethylammonium methylsulfate [Et_3NMe][$MeSO_4$], propylammonium nitrate [$PrNH_3$][NO_3], 1-butyl-3-methylimidazolium tetrafluoroborate [BMIM][BF_4], 1-ethyl-3-methylimidazolium benzoate [EMIM][$PhCO_2$], 1-butyl-3-methylimidazolium trifluoromethanesulfonate [BMIM][F_3CSO_3], and 1-butyl-3-methylimidazolium octylsulfate [BMIM][$OctSO_4$], were evaluated in different proportions (25%, 50%, and 75%v/v). The best results were obtained for FDH in mixtures of the IL [MMIM][$MeSO_4$] with buffer (Table 6.1). High residual activity was achieved for the enzyme, compared with pure buffer solution as solvent, even using 75% of [MMIM][$MeSO_4$]. Also using [Et_3NMe][$MeSO_4$] as a co-solvent, 82% of enzyme activity was retained when 25% (v/v) of IL was used in buffer, but it decreased to 55% when the proportion of IL was

TABLE 6.1. FDH Activities in the Presence of Variable IL Proportions. Residual Activity (%) Compared with the Activity in Pure Buffer Solution

Ionic liquid (% v/v)	25	50	75
[MMIM][$MeSO_4$]	65	73	98
[Et_3NH][$MeSO_4$]	–	–	–
[Et_3NMe][$MeSO_4$]	82	55	–
[$PrNH_3$][NO_3]	–	–	–
[BMIM][BF_4]	–	–	–
[EMIM][$PhCO_2$]	–	–	–
[BMIM][F_3CSO_3]	38	3	–
[BMIM][$OctSO_4$]	–	–	–

raised to 50%. This result was found to be highly promising since FDH can be used as a coupled enzyme for cofactor regeneration in many enzymatic redox-catalyzed reactions, and systems involving a high content of IL could thus be used without reduction of enzymatic activity. Interestingly, as stated in the previous section, the stability of FDH in 1,3-dimethylimidazolium dimethylphosphate [MMIM][$MMPO_4$] was recently enhanced by covalently grafting different chaotropic cations to the enzyme.[11]

Conversely, oxidation of ethanol using YADH was only possible in the presence of 25% (v/v) of [BMIM][F_3CSO_3], with a residual activity of less than 5%. In CPCR-catalyzed reduction of acetophenone, it was shown that the solubility of substrate could be significantly increased in the presence of 25% (v/v) [MMIM][$MeSO_4$]. Yet no catalytic activity was observed in the presence of any of the investigated ILs. Regarding YADH, a more in-depth study of the stability and activity of this enzyme in [BMIM][PF_6] was conducted by other researchers.[30] Interestingly, it was shown that when a surfactant (e.g., Triton X-100) was added to the IL–buffer system, a microemulsion was formed. YADH remained active in this emulsion, thus leading to another promising application of ILs in biotransformations, as discussed broadly in other sections of this book (see below in this chapter, and also Chapter 2, Section 5.3, Chapter 4, Section 5.2.3.4, and Chapter 8).

In another study, the activity of alcohol dehydrogenase from *Lactobacillus brevis* (LB-ADH) in the enantioselective reduction of 2-octanone to (*R*)-2-octanol was improved when the organic solvent methyl *tert*-butyl ether (MTBE) was substituted by 1-butyl-3-methylimidazolium bis(trifluoromethane sulfonyl)imide [BMIM][$(CF_3SO_2)_2N$] in biphasic reaction systems.[22] Kragl and coworkers focused their attention on the exploitation of the differences in the partition coefficients of 2-propanol and acetone in buffer/organic solvents and buffer/ILs systems, increasing the reaction rate when using the oxidation of 2-propanol to acetone as a substrate-coupled approach for NADPH regeneration (Scheme 6.2).

While partition coefficients of 2-propanol and acetone were approximately equal ($P \sim 1$) in a two-phase system buffer–MTBE, the partition behavior of the co-substrate and the co-product when the organic solvent was replaced by [BMIM][$(CF_3SO_2)_2N$]

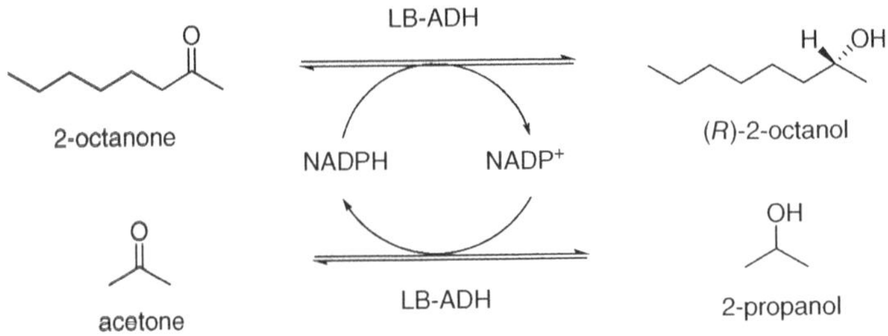

Scheme 6.2. Reduction of 2-octanone catalyzed by alcohol dehydrogenase from *Lactobacillus brevis* (LB-ADH). Substrate-coupled NADPH regeneration with 2-propanol.

led to a positive shift in the thermodynamics for a more efficient cofactor regeneration. Thus, the conversion of 2-octanone into (*R*)-2-octanol in the LB-ADH-catalyzed reaction was faster in the biphasic system containing the IL than in the one containing MTBE, leading to a conversion of 88% in the first 180 minutes, while the reaction reached a conversion of 61% in the presence of MTBE. Herein the cofactor regeneration was the rate-limiting step of the investigated reduction. Therefore, the more effective the regeneration step, the faster the main reaction. Regeneration equilibrium can be thus shifted using a higher concentration of 2-propanol, and/or by removing acetone. In the stated biphasic approach (Scheme 6.2), acetone was extracted by the IL due to its favorable partition coefficient, whereas 2-propanol was permanently available in the aqueous phase, thus leading to fast cofactor regeneration. Furthermore, acetone inhibits the studied alcohol dehydrogenase, and thus its *in situ* extraction from the buffer phase to the IL reduces such inhibitory effect. Clearly in this example, the expected advantages of using a biphasic system combining buffer and IL (compared with a buffer–MTBE approach) were then confirmed.

Since ILs can be finely tuned, it is important to assess how different ILs could interact with enzyme-catalyzed redox processes. In this area, the effect of the IL functionalization (either in cation or in anion) in cofactor-dependent enzyme-catalyzed oxidations was studied by Walker et al. as well.[23] As different ILs, 1-butyl-3-methylimidazolium hexafluorophosphate [BMIM][PF_6], 1-butyl-3-methylimidazolium glycolate [BMIM][glycolate], 1-butyl-2,3-dimethylimidazolium hexafluorophosphate [BDMIM][PF_6], 1-(3-hydroxypropyl)-3-methylimidazolium hexafluorophosphate [HPMIM][PF_6], 1-(3-hydroxypropyl)-3-methylimidazolium glycolate [HPMIM][glycolate], and 1-(3-hydroxypropyl)-3-methylimidazolium chloride [HPMIM][Cl], were synthesized and evaluated as co-solvents for $NADP^+$-dependent-catalyzed reactions. As model reaction morphine dehydrogenase (MDH) from *Pseudomonas putida* M10 acting upon codeine was selected. To regenerate the cofactor either alcohol dehydrogenase (ADH) from *Thermoanaerobium brockii*, or glucose dehydrogenase (GDH) from *Cryptococcus uniguttulatus* were chosen (Figure 6.1).

In this case, the use of hydrophobic ILs prevented the codeinone hydrolysis to the corresponding hydrated side product, which would take place in aqueous media (Figure 6.1). In a subsequent article, the reaction performance was slightly improved, affording

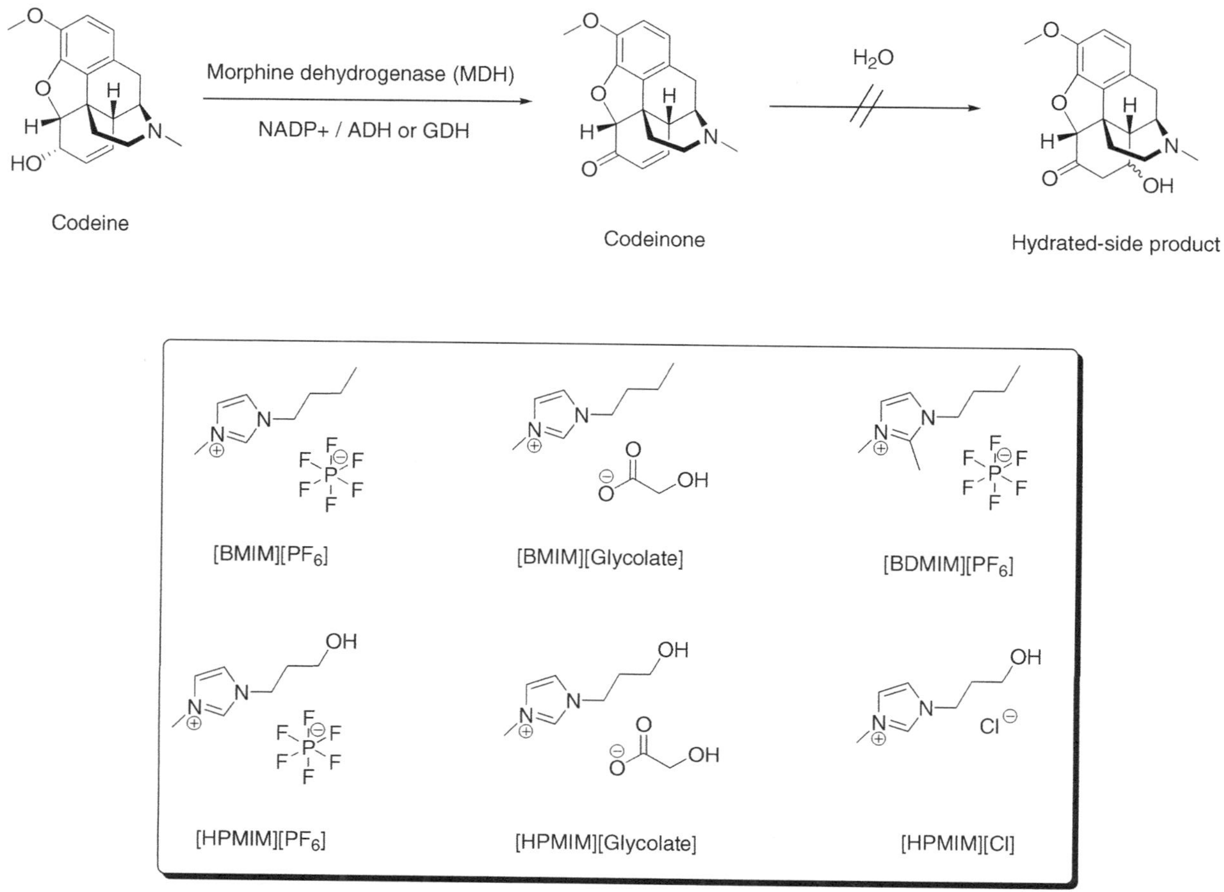

Figure 6.1. Ionic liquids synthesized and evaluated as co-solvents for oxidation of codeine to codeinone, catalyzed by morphine dehydrogenase (MDH) from *Pseudomonas putida* M10, with $NADP^+$ recycling using either alcohol dehydrogenase (ADH) from *Thermoanaerobium brockii* or glucose dehydrogenase (GDH) from *Cryptococcus uniguttulatus*. Use of a water-free system (IL) prevented product degradation.

up to 20% of codeinone, by using GDH and gluconolactone as cofactor $NADP^+$ regenerating system.[31] Enzyme activities in different ILs were compared, as well as with different molecular solvents, including pure water. Catalytic activity in water was higher than in other solvents, although product yield was depleted through hydrolysis (Figure 6.1). Furthermore, no conversion was observed in organic solvents. In reactions performed in ionic media, low to moderate enzyme activities were found even at very low levels of water, and product solvolysis was thus effectively suppressed. Enzymatic activities in hydrophobic RTILs strongly correlated with water content, and thus low activity was observed when the reaction was performed in [BMIM][PF_6] with less than 100 ppm of water, and no activity was found in [BDMIM][PF_6] in the same conditions, although both solvents differ only in the presence or absence of the acidic proton at the 2-position. Major differences were reported between these solvents, and the more strongly hydrogen-bonding ones, such as hydrophilic ILs [BMIM][glycolate], [HPMIM][PF_6], [HPMIM][glycolate], and [HPMIM][Cl]. The presence of the hydroxyl group in the cation in [HPMIM][PF_6] altered the water dependence of the catalytic activity, allowing enzymatic activities at lower water levels than those present in hydrophobic ILs, or even in organic solvents. On the other hand, the activity in hydrogen-bonding anion ILs was lower, and that could be seen by comparing results obtained with [BMIM][glycolate] and [BMIM][PF_6]. When ILs with hydroxylated cations were used, replacing [PF_6] with more hydrophilic anions, practically no activity was observed (e.g., with [HPMIM][Cl]). With regard to the variation of water levels, in most cases a progressive decrease of activity was observed with decreasing water content, except in the case of hydrophilic cations combined with hydrophobic anions in [HPMIM][PF_6], in which the enzyme activity increased with decreasing water levels. Authors concluded then that the combination of solvent properties was crucial for enzyme activity since simple variations in composition hardly affected the enzyme behavior. The physical properties of the anion seemed to be determinant—although not exclusively—for enzyme activity, and the influence of impurities, particularly water, needs to be further studied. Overall, this example shows the advantages that ILs may bring and how their properties can be finely tuned for a specific biocatalytic application.

Horse liver alcohol dehydrogenase (HLADH) was evaluated by Shi et al. in terms of its catalytic activity in biphasic systems involving ILs and buffer.[25] Enzyme activity in systems involving 1-butyl-3-methylimidazolium chloride [BMIM][Cl] has been studied previously by the same group.[24] The oxidation of ethanol was used as model reaction for enzyme activity test, and conformational changes of HLADH caused by ILs were evaluated by UV techniques. The ILs [BMIM][Cl], 1-butyl-3-methylimidazolium bromide [BMIM][Br], [BMIM][PF_6], [BMIM][BF_4], 1-butyl-3-methylimidazolium triflate [BMIM][OTf], and 1-ethyl-3-methylimidazolium chloride [EMIM][Cl] were used in concentrations varying from 0 to 0.4 g/mL. HLADH showed higher catalytic activities in the oxidation of ethanol in systems containing [BMIM][Cl], [BMIM][Br], [EMIM][Cl], or [BMIM][PF_6], with the proportion of the IL being <0.10 g/mL for [BMIM][Cl] and [BMIM][Br], and <0.075 g/mL for [EMIM][Cl] and [BMIM][PF_6], when compared with pure buffer (pH 8.8). Highest activity was obtained in the system containing 0.05 g/mL of [BMIM][Cl] or [BMIM][Br], and 0.025 g/mL of [EMIM][Cl] or [BMIM][PF_6]. These results suggest that the presence of these ILs in proper concentrations could activate the enzyme. However, an increasing addition of the same ILs resulted in a drastic decrease in enzymatic activity, presumably because higher IL concentrations might cause not only a higher ionic strength, which could inactivate the enzyme, but also an enhanced viscosity of the reaction system, leading to mass-transfer limitations. Experiments have shown that the structure of the IL has a significant impact on the catalytic performances of HLADH, and the activity and stability of the enzyme also seems to be mainly anion-dependent.

Likewise, the reductive enzymatic production of androsterone from androstandione was investigated in biphasic buffer–organic solvent systems containing different ILs as co-solvents in the aqueous phase, such as 1-ethyl-3-methylimidazolium trifluoromethanesulfonate [EMIM][CF_3SO_3], 1-butyl-3-methylimidazolium trifluoromethanesulfonate [BMIM][CF_3SO_3], [BMIM][BF_4], and 1-butyl-3-methylimidazolium (L)-lactate [BMIM][lactate]. A commercial 3α-hydrosteroid dehydrogenase (HSDH) from *Pseudomonas testosterone* was used as catalyst, and FDH from *Candida boidinii* as coupled enzyme for NADH regeneration (Scheme 6.3).[27]

The HSDH activity decreased slightly after incubation with 10% [EMIM][CF_3SO_3], and drastically when [BMIM][CF_3SO_3] and [BMIM][BF_4] were used as co-solvents. In contrast, when [BMIM][lactate] was added, the enzymatic activity increased 1.6-fold compared with the activity obtained without the addition of IL. Furthermore, activity of FDH significantly decreased by the addition of all ILs tested, except when using [BMIM][lactate], in which enzymatic activity was retained. Therefore, optimal conditions were achieved by the addition of 5% of [BMIM][lactate], since activity of FDH was retained in 70% for at least 24 hours, and HSDH activity was retained as well.

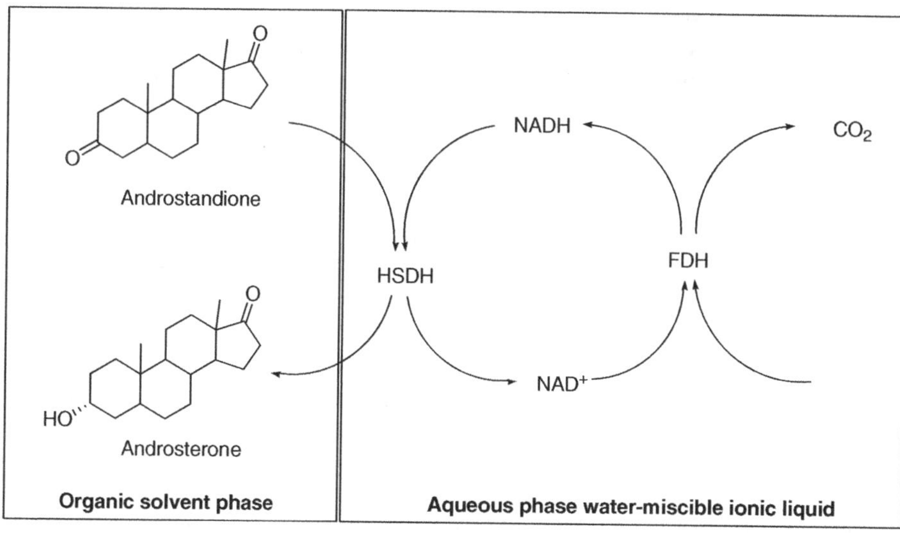

Scheme 6.3. Enzymatic production of androsterone using HSDH and FDH as a coupled-enzyme for NADH regeneration in a biphasic system using ionic liquids as co-solvents in the aqueous phase.

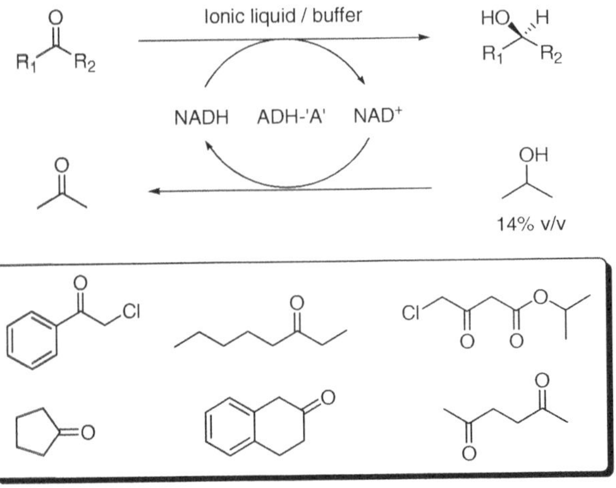

Scheme 6.4. Ionic liquid/buffer as reaction medium for the coupled-substrate approach in the alcohol dehydrogenase from *Rhodococcus ruber* ADH-′A′-catalyzed reduction of ketones.[26,32]

Kroutil and coworkers used bi- and monophasic IL–buffer systems for ketone reductions using a partially purified alcohol dehydrogenase ADH-′A′ from *Rhodococcus ruber* (Scheme 6.4).[26] That enzyme was a suitable catalyst for the reduction of ketones, as well as the recycling of the cofactor, working according to a coupled-substrate approach.[32]

For biphasic systems, conversions were acceptable with the addition of up to 20% v/v of ILs. Conversely, the use of water-miscible second-generation hydroxyl-functionalized "Tris-like" ILs resulted (in some cases) in successful enzymatic outcomes in monophasic systems, even with additions of ILs of up to 90% v/v. The studied ILs were tris(2-hydroxyethyl)methylammonium methylsulfate [MTEOA][$MeSO_4$], AMMOENG™ 100, AMMOENG™ 101, and AMMOENG™ 102 (Figure 6.2).

Reactions were performed using aromatic and aliphatic ketones as substrates, and 2-propanol as sacrificial co-substrate for the NADH recycling. At RTIL concentrations of up to 50% v/v, almost no effects were observed in conversions in the cases of [MTEOA][$MeSO_4$] and AMMOENG 101, whereas moderately diminished conversions were obtained when AMMOENG 100 and AMMOENG 102 were used. When higher concentrations of the ILs were added, AMMOENG 102 led to a considerable decrease in enzyme activity. Interestingly [MTEOA][$MeSO_4$] behaved as a suitable co-solvent with retained enzyme activities with IL additions of up to 70% v/v. At 80% v/v, still 80% of apparent residual activity could be measured, whereas for concentrations close to 90% v/v, only 40% of enzyme activity was retained. As another important asset, combinations of water and AMMOENG resulted totally immiscible with ethyl acetate. Therefore, clean and green downstream processing could be realized with that setup.[23] Likewise, it is worth mentioning that some ILs of the AMMOENG series have also been used for protein extraction in two-phase buffer–IL systems. In this particular case,

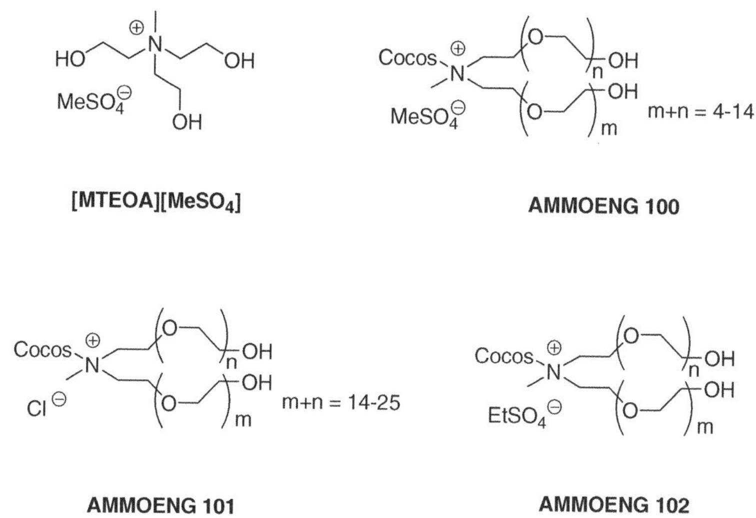

Figure 6.2. Water-miscible second-generation hydroxy-functionalized ionic liquids used in the biocatalytic reduction of ketones catalyzed by partially purified alcohol dehydrogenase from *Rhodococcus ruber* ADH-"A."[26,32]

water-immiscible AMMOENG-based ILs were used, representing a further example of the broad spectra of possibilities that tailor-made ILs can bring (for further discussion of ILs and protein extraction, see also Chapter 2, Section 5).[33]

Moreover, ADH-′A′ from *Rhodococcus ruber* in biphasic buffer–IL combinations has also been employed in cascade reactions involving a metal-catalyzed Suzuki coupling in the IL, together with the simultaneous enzymatic asymmetric reduction of the ketone, to afford chiral alcohols.[34] This is another relevant example of how ILs may provide novel approaches in synthetic setups. Herein, 1-butyl-3-methylimidazolium bis(trifluoromethylsulfonyl)amide, [BMIM][Tf_2N], was used since Suzuki coupling proceeded efficiently in that IL, whereas ADH-′A′ remained active in the buffer phase. [BMIM][Tf_2N] was totally immiscible with water, and interestingly diethylether was insoluble in both phases (IL and buffer) as well. Therefore, a straightforward work-up of the asymmetric alcohols could be conducted. Both phases, IL and buffer, together with metal- and biocatalysts, were reused up to four times without apparent loss of activity. The enantioselectivities shown by ADH-′A′ were always optimal (>99%), and even in some cases the simultaneous enantioselective reduction of two keto groups, to afford important biaryl alcohols, was also successfully conducted (Scheme 6.5).

Another example described the asymmetric reductions of 4′-bromo-2,2,2-trifluoroacetophenone to (*R*)-4′-bromo-2,2,2-trifluoroacetophenyl alcohol, and 6-Br-β-tetralone to its corresponding alcohol, (*S*)-6-Br-β-tetraol, using isolated alcohol dehydrogenase from *Rhodococcus erythropolis* (ADH RE).[29] Both reactions were performed using GDH 103 as second enzyme to recycle the cofactor NADH through the oxidation of glucose to gluconic acid (Scheme 6.6).

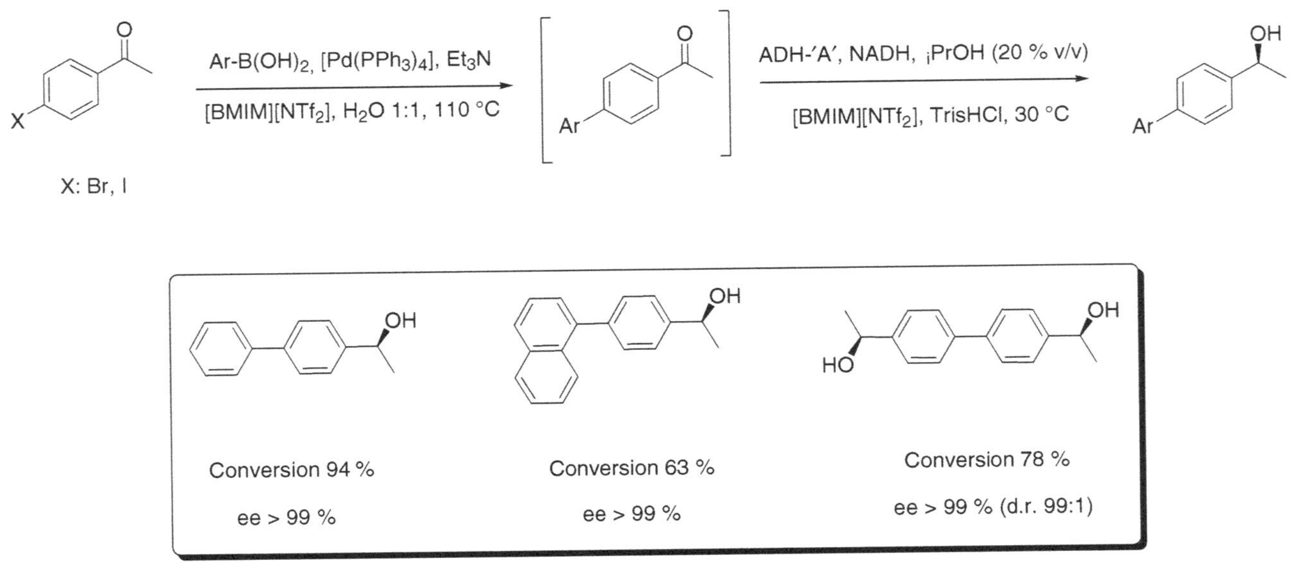

Scheme 6.5. Chemo-enzymatic cascade reaction, Suzuki coupling, and enzymatic enantioselective reduction performed in a two-phase system IL–buffer. ADH-'A' was used as lyophilized whole cells (for a broad description of whole-cell biotransformations and ILs, see Chapter 7).[34]

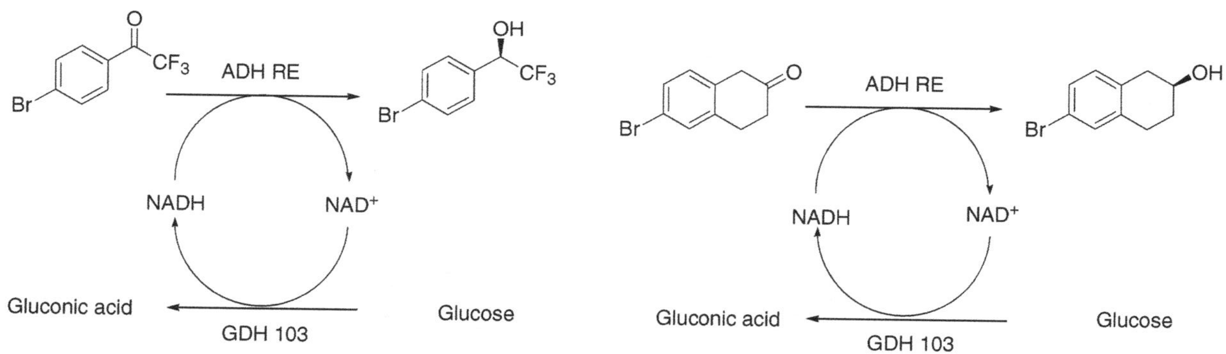

Scheme 6.6. Asymmetric reduction of 4′-Br-2,2,2-trifluoroacetophenone and 6-Br-β-tetralone to the corresponding (*R*)- and (*S*)- alcohols, respectively, by alcohol dehydrogenase from *Rhodococcus erythropolis* (ADH RE), with cofactor recycling by the glucose dehydrogenase 103 (GDH 102)-mediated oxidation of glucose.

TABLE 6.2. Initial Reaction Rates, Conversions, Residual ADH-RE Activities, and Substrate Solubilities in the IL Bioconversion Screened for the Reduction of 50 g/L of 4′-Br-2,2,2-trifluoroacetophenone to (R)-4′-Br-2,2,2-trifluoroacetophenyl Alcohol[29]

Co-solvent	Initial rate g prod/L h	Conversion (%)	Residual ADH activity (%)	Substrate solubility g/L
None	3.1	5.6	0	6.5
[BMP][NTf_2]	12	100	40	5.1
[BMIM][PF_6]	11	99.6	47	3.8
AMMOENG™ 102	5.5	75.8	4	14
[EMIM][TOS]	2.5	46.1	46	8.2
Toluene	6.1	20.8	1	8.2

The motivation for this study with ILs differed from one case to the other. For the enzymatic reduction of 4′-Br-2,2,2-trifluoroacetophenone to (*R*)-4′-Br-2,2,2-trifluoroacetophenyl alcohol, a biphasic system in which the IL acts as a ketone reservoir would minimize the exposure of the biocatalyst to the deactivating effect of the substrate. In the case of the reduction of 6-Br-β-tetralone, the intention was either to improve enantioselectivities achieved by chemical means[35] or to substitute whole-cell routes in aqueous media that, while leading to high selectivities in (*S*)-enantiomer,[36] are limited to low substrate concentrations (<2 g/L).[37] To assess the potential applications of ILs, 11 commercial ILs and organic solvents were screened as co-solvents for both reactions, and compared with buffer. Four of the ILs tested for the reduction of 4′-Br-2,2,2-trifluoroacetophenone led to improved conversions over the best organic co-solvent (toluene). Furthermore, three of these ILs significantly protected the enzyme from the deactivating effect of the substrate (Table 6.2).

As observed (Table 6.2), the water-immiscible IL butylmethylpyrrolidinium bis(trifluoromethylsulfonyl)imide [BMP][NTf_2] was the most suitable co-solvent for performing desired reductions. All bioconversions were carried out using 50 g/L as initial substrate concentration, and in [BMP][NTf_2] rapid reaction rates were achieved, together with full conversion of substrates in less than 10 hours. The use of AMMOENG 102, [BMIM][PF_6], and 1-ethyl-3-methylimidazolium tosylate, [EMIM][TOS], also improved enzymatic activities, compared with reactions conducted in a pure buffer solution or using organic co-solvents. Yet AMMOENG 102 drastically diminished ADH RE residual activity (Table 6.2). Furthermore, regarding enzyme stability, several of the screened ILs offered advantages toward the stability of both ADH RE and GDH 103 compared with organic solvents. In particular, the stability of ADH RE was improved in the presence of [BMP][NTf_2] and [BMIM][PF_6], where ~80% of the original activity was maintained respectively over 64 hours of exposure, compared to ~57% activity retention in buffer. Nevertheless, in AMMOENG 102, ~95% enzyme activity was lost after 15 hours. It was considered that the exceptional enzyme stability denoted in [BMP][NTf_2] is not sufficient to account for the significant increases in initial reaction rates observed in the biocatalytic processes, suggesting that further effect of

TABLE 6.3. Initial Reaction Rates, Conversions, Residual ADH-RE Activities, and Substrate Solubilities in the Ionic Liquid Bioconversion Screened for the Reduction of 50 g/L of 6-Br-β-tetralone to the Corresponding (S)-alcohol[29]

Co-solvent	Initial rate g prod/L h	Conversion (%)	Residual ADH activity (%)	Substrate solubility g/L
None	3.1	100	74	0.1
[BMP][NTf$_2$]	14	100	43	0.2
[BMIM][PF$_6$]	14	98.5	28	0.2
AMMOENG™ 102	12	100	9.4	6.0
[EMIM][TOS]	6.5	89.0	5.1	3.1
[BMIM][BF$_4$]	9	99.5	38	1.2
Toluene	5.8	100	41	1.0
DMSO	2.3	58.4	42	0.5
THF	2	24.9	0.8	0.3
DMF	3.6	95.8	7.6	0.6

improved mass transfer of the substrate from the IL phase to the aqueous phase should be also taken into account.

With regard to the reduction of 6-Br-β-tetralone, immiscible ILs [BMP][NTf$_2$] and [BMIM][PF$_6$] constituted the best of all co-solvent systems analyzed, achieving moderate enzyme stabilities (43% and 28%, respectively) after 24 hours (Table 6.3). The high conversions and initial rates achieved when using these co-solvents suggested enhanced mass transfer rates of the substrate from the IL. Likewise, conversion and residual ADH RE activities were similar to values observed when toluene (10%v/v) was used as co-solvent. Interestingly, toluene provided a higher substrate solubility (1 g/L) than that observed for [BMP][NTf$_2$] (0.2 g/L), suggesting a higher reaction rate in toluene. However, the rate is almost significantly lower in toluene than in [BMP][NTf$_2$], indicating that solute mass transfer is higher from the ionic phase. It is noteworthy that when no co-solvent was added, the reaction started as a two-phase system (aqueous buffer and solid substrate), due to the low solubility of tetralone. As product was formed, an immiscible oil phase was formed too, and the dispersed oil and the solid phase enhance the mass-transfer rate of the substrate into the aqueous phase, yielding full conversion in 24 hours. However, the initial rate of reaction in [BMP][NTf$_2$] was still higher than in buffer, suggesting that the IL provided a more efficient mass transfer than the substrate–product–buffer oil approach.

Recently, cellobiose dehydrogenase (CDH) from *Phanerochaete chrysosporium* was studied, in particular its capability to oxidize cellobiose in hydrated choline dihydrophosphate [choline][DHP], transferring electrons to a suitable acceptor, cytochrome *c* (cyt *c*) (Scheme 6.7).[28] The reaction was performed by dissolving CDH in [choline][DHP] containing 35% water. In this system, the electron transfer from cellobiose to the two domains of CDH (FAD and heme domains), and then to cyt *c*, was successful. Yet reaction rates were slower than in aqueous systems.

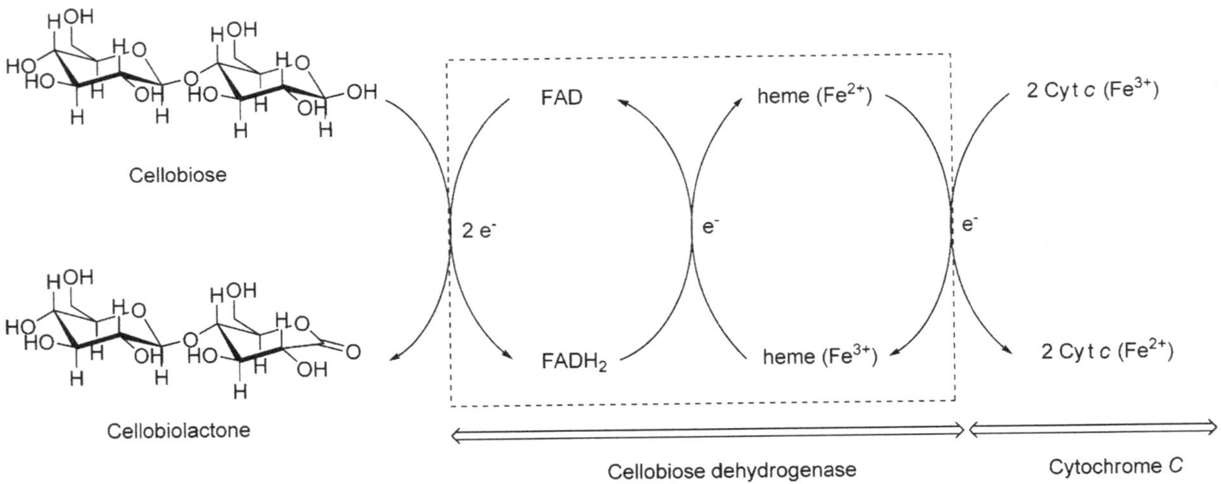

Scheme 6.7. Reaction catalyzed by cellobiose dehydrogenase to form cellobiolactone.[28]

In this section, several examples of isolated oxidoreductases coupled with proper cofactor regeneration, using monophasic or biphasic media containing ILs have been discussed. Combinations of ILs and this type of enzyme clearly seem to be a promising approach to developing novel enzymatic redox processes. The amount of data reported is, however, still scarce, and thus based on the current state-of-the-art, a rational protocol for IL selection is not yet possible. In principle, important factors in IL selection when used in biphasic systems appear to be the partition coefficient between them and the aqueous phase, mass-transfer rates of substrates from immiscible ILs into the aqueous phase, and the stability of biocatalysts in the aqueous/co-solvent mixtures. On the other hand, in monophasic systems the hydrophylicity of the ionic solvent must be considered, as well as the viscosity of the IL/water mixtures. There is clearly some unexplored potential in the topic, and probably more research will be carried out in the coming years leading to novel tailored applications of ILs.

6.2.2 Laccases, Peroxidases, Oxidases, and Oxygenases

Oxygenases are enzymes that enable the introduction of one (monooxygenases) or two (dioxygenases) oxygen atoms into substrates with high efficiency and selectivity. These enzymes generally utilize NADH or NADPH to provide reducing potential for the supply of electrons to the substrate. Therefore, herein an adequate cofactor regeneration is needed as well, and this requirement has been a major reason to perform oxygenase-catalyzed reactions using whole microbial cells.[38] Within this group, peroxidases have been widely studied due to their ubiquitous distribution in nature and their ability to catalyze the oxidation of broad substrate spectra. Peroxidase-catalyzed oxidation proceeds by using hydrogen peroxide or an organic hydroperoxide (such as *tert*-butyl hydroperoxide) as oxidant. Most peroxidases are heme proteins, whereas other have magnesium, vanadium, or selenium at their active sites.[39] Laccases are multi-copper-containing oxidases accessible from numerous sources, especially from fungi. Substrates of laccases include alkenes, phenols, aminophenols, aryl-amines, and polyphenols even as complex as lignin.[40] However, unlike peroxidases, laccases have low redox potentials, permitting only the oxidation of low redox potential substrates. The scope of laccase-catalyzed reactions can be broadened in the presence of synthetic redox mediators (e.g., 1-hydroxybenzotriazole).[41] Finally, oxidases use oxygen as the only oxidant agent, without the need of cofactor, making them a very interesting class of enzymes. One of the most important enzymes of this class is glucose oxidase. It is a highly specific enzyme that catalyzes the oxidation of glucose to glucuronic acid and hydrogen peroxide and has found several applications in food, chemical, and pharmaceutical areas.[5]

As stated as well with oxidoreductases, biocatalytic applications involving ILs and oxygenases, peroxidases, oxidases, and laccases, are not extensive and only some reactions have been explored so far. In most reported cases, results are promising since enzymes retain or even improve their activity in ILs. Reported applications of ILs in these areas range from *simple* oxidation reactions to design and development of biosensors, as well as some examples of polymerization reactions.[42] RTILs exhibit many properties that make them appealing media for oxidative biocatalysis. In addition to features already mentioned in previous chapters, the solubility of gases such as molecular oxygen in ILs is generally high, making them satisfactory solvents for aerobic oxidation reactions. ILs are often composed of weakly coordinating anions (BF_4^- and PF_6^-) and therefore have the potential to be highly polar yet noncoordinating solvents. This is particularly important in reactions catalyzed by metal-containing enzymes.[43]

Peroxidases and to a less extent, laccases, have been the most studied biocatalysts within this group of enzymes. With the aim of establishing the enzyme stability and activity in ILs, several studies have been conducted.[44,45] Hinckley and coworkers investigated the use of ILs as reaction medium for oxidative enzymes. They showed for the first time that laccase C from *Trametes* sp., horseradish peroxidase (HRP), and soybean peroxidase (SBP) were catalytically active in reaction systems containing 4-methyl-*N*-butylpyridinium tetrafluoroborate [4-MBP][BF_4] and [BMIM][PF_6].[44] The activity was assayed using syringaldazine, an effective reagent for detecting laccase and peroxidase, as substrate in a mixture of sodium citrate–phosphate buffer and IL (10–75%). Enzymes retained catalytic activity in moderate concentrations of ILs (less than 50%).

Similar studies were conducted for HRP in mixtures of phosphate buffer with water-miscible ILs, [BMIM][BF_4] and [BMIM][Cl]. The catalytic activity of HRP was retained and a remarkable recovery of peroxidase activity after thermal deactivation was also observed.[46] Another study on the stability and activity of HRP in mixtures of water and water-miscible [BMIM][BF_4] emphasized the high stability of this enzyme in such IL/water mixtures, displaying high activities even at IL proportions of up to 90% v/v.[47] Since then, several reports have been published discussing the activity of peroxidases and laccases in common and designed ILs.[48]

As stated in previous chapters, one of the captivating aspects of ILs is the concept of "tailor-made" solvents, which could be in the future specifically designed for a given reaction.[49] Related to this, an important task in biocatalysis is the possibility of reusing catalysts to cope with the associated costs in industrial applications, which has led to considerable research in the field of enzyme immobilization. As "tailor-made" solvents, ILs may be employed for enzyme encapsulation allowing the reuse of biocatalysts. In these kinds of systems, enzymes are incorporated in ILs supported on host matrices such as cellulose,[50] chitosan,[51] or silica gel.[52] Likewise, ILs bearing a vinyl group in the cationic component can be directly polymerized by free radical polymerization. Polymerized ILs can be used as supporting materials for enzyme immobilization, providing an IL-like microenvironment for immobilized enzymes. Recently, Nakashima and coworkers reported the encapsulation of HRP in a modified polyethylene glycol.[53] Encapsulated HRP exhibits higher activity than in conventional polyacrylamide microparticles, and is easily recycled by centrifugation from reaction mixtures. Once again it has been shown that the field of ILs exceeds the scope of *mere solvents* for reactions, since many other approaches and useful applications can be envisaged.

HRP was also immobilized by simple dissolution in [BMIM][PF_6] for the synthesis of polyaniline (PANI),[54] one of the most widely used conducting polymers, due to its stability and electrical and optical properties. The HRP/IL mixture was added to an aqueous solution of aniline, dodecyl-benzene sulfonic acid (DBSA) and H_2O_2 as oxidant, at pH 4.3. The IL phase was easily extracted from the aqueous phase by liquid–liquid (IL/water) phase separation, and the enzyme was then recovered. The resulting

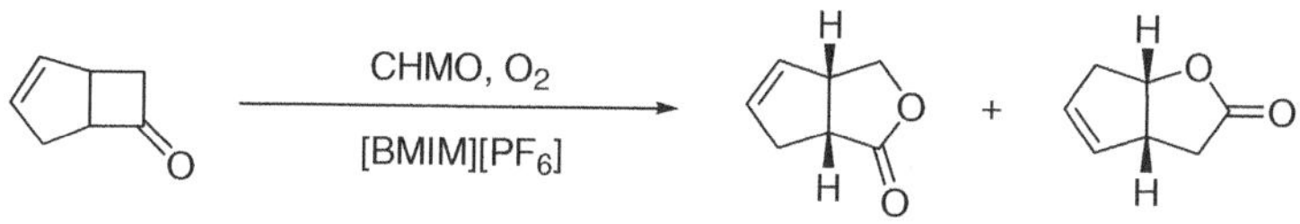

Scheme 6.8. Microbial Baeyer–Villiger oxidation of bicyclo[3.2.0]-hept-2-en-6-one.[10,58] (For a detailed discussion of whole-cell biotransformations and ILs, see Chapter 7).

polyaniline had conductivities comparable to polyaniline obtained with HRP in organic solvents.[55]

Another strategy to enhance enzyme activity in ILs is to form nano- or micrometer-sized water domains in an IL continuous phase, called w/IL microemulsions, stabilized by a suitable surfactant. The formation of aqueous droplets in a hydrophobic IL, 1-octyl-3-methylimidazolium bis(trifluoromethylsulfonyl)amide [C_8MIM][Tf_2N], stabilized by the anionic surfactant AOT (sodium bis(2-ethyl-1-hexyl)sulfosuccinate), has been reported.[56] Authors investigated the activity of HRP entrapped in w/IL microemulsions and results demonstrated that the rate of HRP-catalyzed reactions was increased. Analogous microemulsion systems have also been put forward for lignin peroxidase and laccases.[57] The potential of w/IL microemulsions is also discussed in other sections of this book (Chapter 2, Section 2.5.3; Chapter 4, Section 4.5.3.1; and Chapter 8).

As mentioned previously, monooxygenase-catalyzed oxidations are often carried out with whole-cell biocatalysts, due to the complex and expensive cofactor recycling in such enzymatic systems. A recombinant *Escherichia coli* expressing a Baeyer–Villiger monooxygenase, cyclohexanone monooxygenase (CHMO), catalyzed the regioselective oxidation of bicyclo[3.2.0]-hept-2-en-6-one into lactones (Scheme 6.8). The reaction takes place in [BMIM][PF_6] at a similar rate to that in aqueous solutions.[10,58] Both lactones achieved by regio-divergent biotransformation are valuable building blocks for the synthesis of prostaglandins, brown algae pheromones, and other relevant compounds.[59]

Data on free monooxygenases are rather scarce. A study on the influence of different ILs in the activity and stability of monooxygenase P450 BM-3 has been reported.[60] From several 1-alkyl-3-methylimidazolium-based ILs bearing different chain lengths, it was observed that higher inihibitory effects in the enzyme were observed when larger chains (octyl or hexyl) were present in the imidazolium cation of the IL. Thus, [EMIM][Cl] was reported as the best for that enzyme from the tested group of ILs.

Lutz-Wahl and coworkers described the use of ILs for D-amino acid oxidase (DAAO).[61] DAAO catalyzes the oxidative deamination of various D-amino acids to the corresponding imino acids, which undergo a nonenzymatic hydrolysis to the α-ketoacid. DAAO is used in the industrial production of 7-aminocephalosporanic acid from cephalosporin-C. In this study, both the activity and the stability of DAAO in different ILs were examined. The ILs used were [BMIM][BF_4], [BMIM][PF_6], [MMIM][$MMPO_4$], and [BMIM][Tf_2N]. DAAO was active in ILs, although enzymatic activities decreased with high IL proportions.

Some oxidations have been reported using HRP as biocatalyst. For instance, veratryl alcohol is a secondary metabolite in delignification mediated by *Phanerochaete*

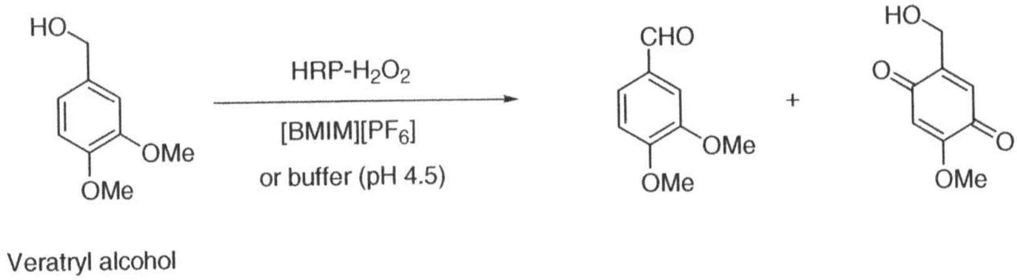

Scheme 6.9. Biomimetic oxidation of veratryl alcohol with H_2O_2 catalyzed by HRP in ionic liquids.

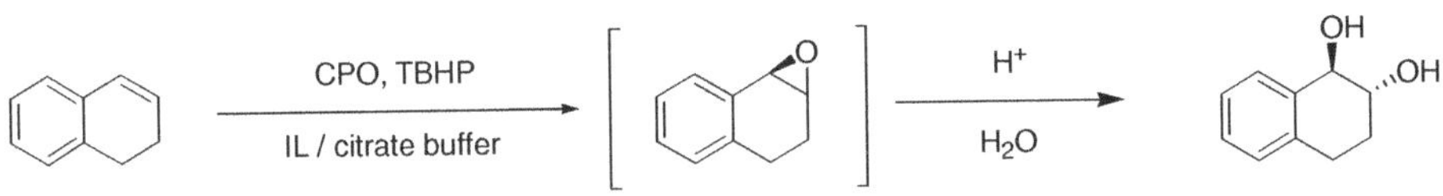

Scheme 6.10. CPO-catalyzed epoxidation of 1,2-dihydronaphtalene.

chrysosporium, and it can be considered a model compound for lignin substructures.[62] Biomimetic oxidation of veratryl alcohol with hydrogen peroxide and HRP in ILs is shown in Scheme 6.9.[63]

When the reaction was carried out in [BMIM][PF_6] instead of buffer, higher yields were achieved. The reaction of hydrogen peroxide with HRP forms high-valent oxo-iron(IV) π-cation radical as reactive intermediate. The high yield of the oxidized products in the reaction of veratryl alcohol in ILs may be explained for the noncoordinating nature of ILs, which may accelerate the reaction by stabilizing the highly charged intermediate.

Enantioselective epoxidations of olefins and oxidations of sulfides to sulfoxides are valuable tools in organic synthesis due to the high versatility of these functional groups. Considerable research has been performed for the development of catalytic methods for asymmetric epoxidation and synthesis of chiral sulfoxides.[64,65] Also, in the recent years, enantioselective epoxidation of olefins in ILs has been investigated. Sanfilippo and coworkers studied the catalytic efficiency of chloroperoxidase (CPO) in the presence of ILs. The epoxidation of 1,2-dihydronaphtalene using *tert*-butyl hydroperoxide (TBHP) as oxygen source was used as model reaction (Scheme 6.10).[66]

Considering the substrate solubility, several ILs were evaluated: [MMIM][$MeSO_4$], [BMIM][$MeSO_4$], [BMIM][Cl], [BMIM][BF_4], and [BMIM][PF_6]. Enzymatic activities were observed; epoxide was rapidly hydrolyzed and therefore product was recovered as (1*R*,2*R*)-dihydroxytetrahydronaphtalene. These promising results showed that CPO is able to catalyze the stereoselective epoxidation of the substrate in ILs. It is important to note that CPO from *Caldariomyces fumago* is frequently the peroxidase of first choice for sulfoxidation reactions, allowing the formation of desired products with high enantiopurity.[64] In the field of ILs, sulfoxidation of thioanisole by hydrogen peroxide

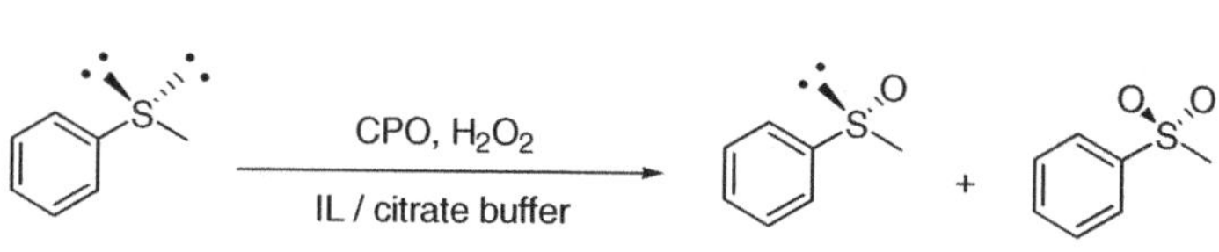

Scheme 6.11. CPO-catalyzed sulfoxidation of thioanisole.

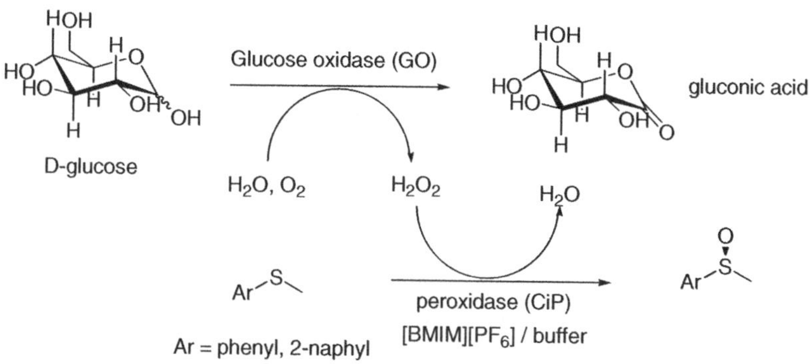

Scheme 6.12. Enantioselective sulfoxidation catalyzed by CPO in ILs.

in the presence of CPO was investigated in pure citrate buffer solution and in the presence of increasing amounts of ILs (Scheme 6.11).[67]

In some ILs such as [MMIM][$MeSO_4$], *N,N*-dimethylmorpholinium methylsulfate [Mor_{11}][$MeSO_4$] and cholinium phosphate [$N_{1112}OH$][H_2PO_4], CPO lost completely its activity. However, enantiomerically pure sulfoxide was obtained in the presence of 1,3-dimethylimidazolium dimethylphosphate [MMIM][Me_2PO_4], cholinium acetate [$N_{1112}OH$][OAc], and cholinium citrate [$N_{1112}OH$][Citr]. When active, as compared with the behavior observed in conventional organic solvents, CPO in ILs presented enhanced activity, stability, and selectivity.

In situ generation of hydrogen peroxide may be an attractive alternative in peroxidase-catalyzed reactions. Hydrogen peroxide can be generated from the autooxidation of glucose in the presence of glucose oxidase (GO). Subsequently, the peroxidase from *Coprinus cinereus* (CiP) oxidizes sulfides into their respective sulfoxides (Scheme 6.12).[68]

Reactions were carried out in several [BMIM][PF_6]/buffer mixtures. The optimum mixture reported was 90:10 with modest conversions (<30%), and moderate to high enantioselectivities of the sulfoxides (70–90%). Other authors have also addressed the activity of glucose oxidase in [BMIM][BF_4], with decreasing enzymatic activities at higher IL concentrations.[69]

Apart from the asymmetric organic reactions reported previously, some other articles describe the use of ILs in enzymatic polymerizations. The production of polymeric materials by means of biocatalysis represents an important interface between the fields of biotechnology and materials engineering. In this area, peroxidase-mediated oxidative coupling of phenols is one of the most studied redox reactions in biochemistry. The major drawback that interferes with the use of peroxidases in chemical synthesis is the difficulty of controlling free radical-mediated reactions. Peroxidase-catalyzed reaction of phenolic compounds in organic solvents frequently yields homogeneous polymeric products with the same reaction performed in aqueous buffers.[70]

Recently, investigations on the activity of HRP[71] and SBP[72] in ILs have been conducted, aiming to establish the mechanism of action and properties of obtained products. With regard to HRP, three substituted phenols were analyzed: 1-naphtol, 4-phenylphenol, and 1,5-di-*tert*-butylphenol. The peroxidase-mediated oxidative coupling in water buffers conducted to heterogeneous polymeric adducts. When using DMF or DMSO as co-solvents, a wide distribution of colored polymeric products was obtained. In the presence of [BMIM][BF_4], remarkably, polymeric species were never obtained, and high yields of dimeric or hydroxylated products were afforded. Likewise, SBP-catalyzed oxidative polymerization of phenols was performed in several ILs starting from *p*-cresol. Interestingly, polymers of higher molecular weight than those formed in aqueous buffer were obtained in ILs. This is probably due to the precipitation of the polymers formed in water before they reach a high molecular weight, while the high dissolution ability of ILs allows the growing polymer to propagate into larger chains. Although this field is clearly in its infancy, and only some proof-of-principles have been reported, it is clear that there is significant potential for developing novel and useful applications in combination with ILs.

ILs have found many applications not only in organic synthesis, but also in chemical analysis. A relevant example is the bioanalytic assay to determine the presence of organophosphate pesticides, based on inhibition of the enzyme acetylcholinesterase (AChE). However, thio-forms of organophosphate pesticides (methylparathion) show low inhibition potential toward AChE. Therefore, such compounds have to be previously oxidized to their oxo-forms, and enzymes like CPO in citrate buffer can be used to this end.[73] It has been shown that an enhancement in the sensitivity of the assay can be achieved by introducing ILs. Therefore, the reaction of CPO-catalyzed oxidation of methylparathion in the presence of various ILs as co-solvents was studied (Scheme 6.13).[74]

ILs tested were *N*-ethylpyridinium tetrafluoroborate [EtPy][BF_4], *N*-ethylpyridinium trifluoroacetate [EtPy][TFA], *N*-ethylpyridinium hexafluorophosphate [EtPy][PF_6], *N*-butylpyridinium bromide [N-BPy][Br], [BMIM][PF_6], [BMIM][Br], 1-butyl-3-methylimidazolium methylsulfate [BMIM][$MeSO_4$], ethylammonium nitrate [EA][NO_3], and *N*-butyl-3-methylpyridinium tetrafluoroborate [BMPy][BF_4]. Oxidation effi-

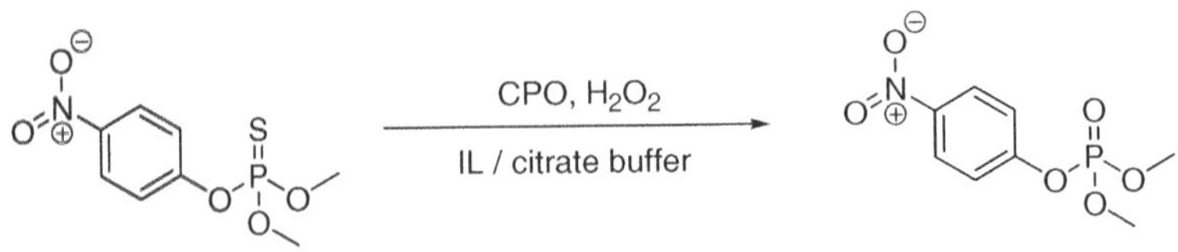

Scheme 6.13. CPO-catalyzed oxidation of methylparathion.

ciency was complete in the presence of ILs such as [N-BPy][Br], [EtPy][BF_4], [BMPy][BF_4], [BMIM][Br], or [BMIM][$MeSO_4$]. However, some ILs affected the ability of CPO to oxidize methylparathion, revealing that the effect of ILs depends on their chemical structure.

Bioenzymatic sensors for the determination of different targets are of great interest from several perspectives such as health or environment. There has been a continuous development of a variety of enzymatic electrodes based on oxidoreductase biocatalysts.[75]

ILs, in particular imidazolium-based, possess physicochemical properties such as excellent ionic conductivity, high chemical and thermal stability, and wide electrochemical windows. Such properties make ILs ideal electrolytes in electrochemical biosensors based, for instance, on laccase[76] or HRP.[77] Electrochemical biosensors combine the analytical potential of electrochemical techniques with the specificity of biological recognition processes. The aim is to biologically produce an electrical signal that relates to the concentration of the analyte. The success of this approach depends on the immobilization of the enzyme. Recently, the development of biosensors using chitosane as support for peroxidase immobilization has been reported.[78] In that study, pine nut peroxidase was immobilized on chitosan, cross-linked with citrate. Biosensor containing this enzyme and 1-butyl-3-methylimidazolium bis(trifluoromethylsulfonyl)amide [BMIM][Tf_2N] was constructed, evaluated, and used for rosmarinic acid determination, with good results (for more information on enzymes and biosensors combined with ILs, see Chapter 8).

6.3 ILs IN LYASE-CATALYZED REACTIONS

Lyases are enzymes that catalyze the formation/cleavage of chemical bonds by other means than by hydrolysis or oxidation. They differ from other enzymes in that two substrates are involved in one reaction direction, but only one in the other direction. Due to the central role of the C–C bond in the chemical processes, enzymes involved in its cleavage/formation are relevant from a synthetic viewpoint. Yet some drawbacks associated with enzyme availability, cost, and reaction conditions have hampered a more widespread use of lyases such as aldolases, oxynitrilases, transketolases, and related enzymes. Despite the increasing number of industrial and academic applications employing these enzymes, they are seldom used in IL-containing media.

6.3.1 Aldolases

The use of aldolases in IL-containing media has not been reported so far. However, an aldolase mimetic, the commercially available aldolase antibody 38C2, has been used for the aldol addition of aromatic aldehydes to 1-hydroxypropanone in neat [BMIM][PF_6], with moderate yields, and recovering the catalyst in the IL (Scheme 6.14).[79]

Herein, the field of biotransformations partly overlaps with proline-like organocatalytic reactions. A broad discussion of these biomimetic approaches in the field of ILs is widely reported in Chapters 9 and 10.

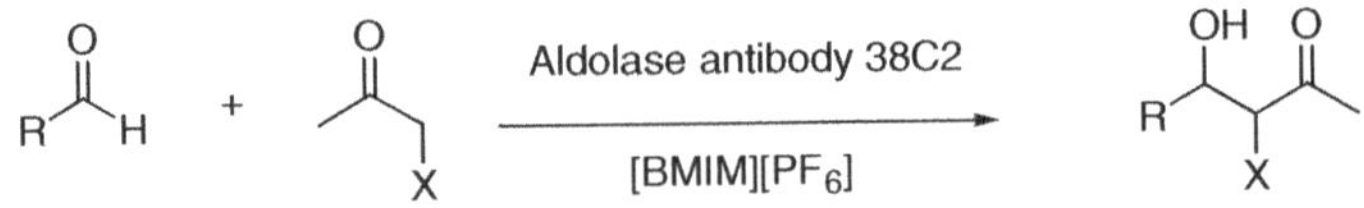

Scheme 6.14. Antibody 38C2-promoted aldol reaction in ionic liquid.

6.3.2 Oxynitrilases

Hydroxynitrile lyases (HNL, also referred to as oxynitrilases) catalyze the reversible decomposition of cyanohydrins to give carbonyl compounds (aldehydes or ketones) and HCN. They have been isolated from more than 20 plant species and some of them are commercially available, namely (*R*)-HNL from *Prunus amygdalus* (*Pa*HNL, EC 4.1.2.10), (*S*)-HNL from *Sorghum bicolor* (*Sb*HNL, EC 4.1.2.1), (*R*)-HNL from *Linum usitatissimum* (*Lu*HNL, EC 4.1.2.37), (*S*)-HNL from *Manihot esculenta* (*Me*HNL, EC 4.1.2.37), and (*S*)-HNL from *Hevea brasiliensis* (*Hb*HNL, EC 4.1.2.37).[80–82] Structurally, HNL can be classified in four types, according to the enzymatic family to which they are related (FAD containing enzymes using mandelonitrile as substrate [*Pa*HNL], α/β-hydrolases [*Hb*HNL and *Me*HNL in one type and *Sb*HNL in the other], Zn-dependent alcohol dehydrogenases [*Lu*HNL]).[81] In the last two decades their synthetic applications increased considerably with the observation that the use of water-immiscible solvents dramatically diminished the undesired chemical addition of HCN to the carbonyl group, while keeping the enzymatic activity.[83] Currently, the nonselective chemical addition is controlled mostly by performing the reaction at low pH and temperature in a biphasic system (buffer/organic medium), using immobilized enzymes in organic solvents, or else by keeping a low concentration of HCN throughout the reaction by means of a transcyanation using acetone cyanohydrin as the HCN carrier.[80,84] These strategies to retard the undesired chemical hydrocyanation are based on the reduction of the reactant concentrations in the aqueous phase since this reaction requires a polar medium and does not take place readily in ethers or toluene.

In this context, the use of ILs in this type of enzymatic reactions is scarce. Griengl and coworkers reported HNL-catalyzed hydrocyanations, which when carried out using both neat ILs and aqueous IL mixtures, displayed different results.[85] When performed in the neat ILs, 1-ethyl-3-methylimidazolium tetrafluoroborate [EMIM][BF_4], 1-propyl-3-methylimidazolium tetrafluoroborate [PMIM][BF_4], and [BMIM][BF_4], containing 1% water (necessary for enzyme activity), the hydrocyanation of benzaldehyde in the presence of *Pa*HNL or *Hb*HNL was fast, but only racemic products were formed. Clearly, the contribution of the enzymatic reaction is negligible under these conditions (room temperature, pH 7). The use of other ILs as solvents for enzymatic hydrocyanations has not been reported so far.

On the other hand, interesting results were obtained by the same group when using 50% aqueous mixtures of ILs (Scheme 6.15). In these 1:1 aqueous mixtures, the reactions were much faster than those in biphasic aqueous–MTBE medium, for both enzymes. The control of the undesired chemical hydrocyanation achieved by lowering the pH was successful with *Pa*HNL but could not be effectively used with *Hb*HNL,

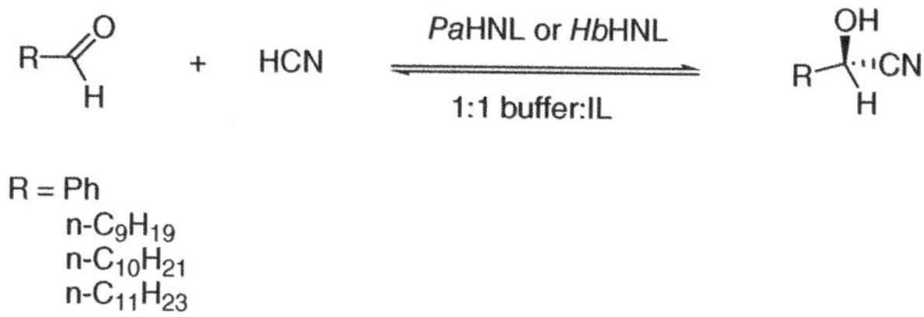

Scheme 6.15. Hydroxynitrile lyase-catalyzed hydrocyanation of carbonyl compounds.

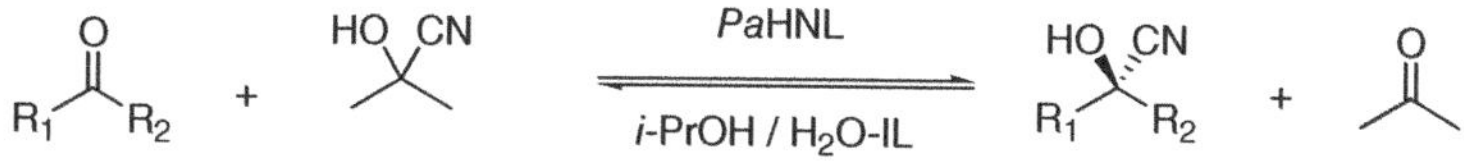

	% IL	
	0%	Optimum (2–6%)
a: R_1 = TMS, R_2 = Me	86% ee	94% ee
b: R_1 = Me_3C, R_2 = Me	75% ee	82% ee
c: R_1 = Ph, R_2 = H	73% ee	86% ee

Scheme 6.16. *Pa*HNL-catalyzed transcyanation of carbonyl substrates.

given its reduced stability at lower pH. Consequently, the latter reaction suffered from lower enantioselectivities because of the concurrent nonenzymatic reaction. With *Pa*HnL, in contrast, rapid and efficient hydrocyanation of benzaldehyde and the less reactive long chain aliphatic aldehydes (Scheme 6.15) was accomplished at pH 3.7 in 1/1 buffer/[PMIM][BF_4] with a satisfactory enantioselectivity. In those conditions hardly any conversion was detected using TBME/aqueous buffer.[85]

The control of undesired chemical hydrocyanation by using acetone cyanohydrin as the HCN carrier has also been used in IL-containing systems with *Pa*HNL and *Me*HNL as the biocatalyst.[86] A few percent (from 2% to 6% v/v) of [PMIM][BF_4], [BMIM][BF_4], 1-pentyl-3-methylimidazolium tetrafluoroborate [PeMIM][BF_4], and 1-hexyl-3-methylimidazolium tetrafluoroborate [HMIM][BF_4] in water activated the *Pa*HNL-catalyzed transcyanation of benzaldehyde and two aliphatic ketones (Scheme 6.16). The activation represented higher enantiomeric excesses both in aldehydes and in ketones, and 15% increase in the initial rate for benzaldehyde, and 40% increase in the case of the ketones. Conversely, IL concentrations higher than 10% (data for [BMIM][BF_4] only) caused a loss of activity down to 50% residual activity at 50% concentration of the IL and poor product enantiomeric excess.

In contrast, the HNL from cassava (*Me*HNL, belonging to a different structural type of oxynitrilases) was strongly inhibited even by small amounts of these ILs in the aqueous phase. For example, an increase in [PeMIM][BF_4] from 0% to 10% (v/v) resulted in a 28% drop in the initial rate and a 27% decrease in the enantiomeric excess of the product.[86] This is in agreement with the ample dependence of the enzyme used on the effect of ILs on enzymatic reactions.[8,87] It may be worth noting that both HNLs showed increased thermal stability in the IL in comparison to organic solvents. Thus, the incubation of these enzymes at 80°C during 24 hours in [BMIM][BF_4], followed by rehydration, resulted in the recovery of approximately 70% of the initial activity, compared with the 4–22% obtained when using acetonitrile or THF.[86]

6.4 PROSPECTS

ILs have been widely used in biotransformations, as extensively reported in several chapters of this book. Starting from proof-of-principles of applications, mostly in hydrolases, their use has started to be a matter of interest also when other nonhydrolytic enzymes are used. Although in this latter case there is still a lack of data that may provide the basis for a fundamental understanding of the influence of ILs in enzymatic systems, surely part of the already existing know-how in hydrolases and ILs may be of value for other enzymes as well. Much more research is needed with ILs and their application in several areas, including nonhydrolytic enzymes, which will surely provide more efficient and better applications in the future. Aspects such as enzyme stability and activity, substrate and product solubility, as well as added value that ILs may bring, represent important factors in IL research. Clearly, ILs exceed their initial scope as mere alternative solvents, and can presently be used in other areas, or to pursue other aims, as reported in this chapter, as well as in this whole book.

REFERENCES

1 C. M. Gordon, *Appl. Catal. A* **2001**, *222*, 101–117.

2 S. Klembt, S. Dreyer, M. Eckstein, U. Kragl, in *Ionic Liquids in Synthesis*, Vol. 1, 2nd ed. (Eds. P. Wasserscheid, T. Welton), Wiley-VCH GmbH & Co., Weinheim, **2008**, pp. 641–662.

3 (a) A. M. Klibanov, *Chemtech* **1986**, *1*, 354–359.; (b) A. M. Klibanov, *Nature* **2001**, *409*, 241–246.

4 S. Cantone, U. Hanefeld, A. Basso, *Green Chem.* **2007**, *9*, 954–971.

5 P. C. A. G. Pinto, L. M. F. S. Saraiva, J. L. F. C. Lima, *Anal. Sci.* **2008**, *4*, 1231–1238.

6 Z. Yang, W. Pan, *Enzyme Microb. Technol.* **2005**, *37*, 19–28.

7 (a) P. Dominguez de Maria, *Angew. Chem. Int. Ed. Engl.* **2008**, *47*, 6960–6968.; (b) E. Fehér, B. Major, K. Bélafi-Bakó, L. Gubicza, *Biochem. Soc. Trans.* **2007**, *35*, 1624–1627.; (c) E. S. Song, *Chem. Commun.* **2004**, *9*, 1033–1043.; (d) H. Zhao, *J. Mol. Catal., B Enzym.* **2005**, *37*, 16–25.

8 U. Kragl, M. Eckstein, N. Kaftzik, *Curr. Opin. Biotechnol.* **2002**, *13*, 565–571.

9 R. A. Sheldon, R. M. Lau, M. J. Sorgedrager, F. van Rantwijk, K. R. Seddon, *Green Chem.* **2002**, *4*, 147–151.

10 F. van Rantwijk, R. A. Sheldon, *Chem. Rev.* **2007**, *107*, 2757–2785.

11 M. Bekhouche, B. Doumeche, L. J. Blum, *J. Mol. Catal., B Enzym.* **2010**, *65*, 73–78.

12 C. O. Beauchamp, S. L. Gonias, D. P. Menapace, S. V. Pizzo, *Anal. Biochem.* **1983**, *131*, 25–33.

13 *Enzyme Nomenclature*, Academic Press, San Diego, **1992**.

14 K. Buchholz, V. Kasche, U. Bornscheuer, *Biocatalysts and Enzyme Technology*. Wiley-VCH Verlag GmbH & Co., Weinheim, **2005**.

15 S. W. May, *Curr. Opin. Biotechnol.* **1999**, *10*, 370–375.

16 S. M. A. De Wildeman, T. Sonke, H. E. Schoemaker, O. May, *Acc. Chem. Res.* **2007**, *40*, 1260–1266.

17 M. R. Kula, U. Kragl, *Stereoselective Biocatalysis* (Ed. R. N. Patel), Marcel Dekker, Inc, New York, **2000**, pp. 839–866.

18 F. J. Hernández-Fernández, A. P. De los Ríos, F. Tomás-Alonso, D. Gómez, G. Víllora, *Can. J. Chem. Eng.* **2009**, *87*, 910–914.

19 J. Gorke, F. Srienc, R. J. Kazlauskas, *Biotechnol. Bioprocess Eng.* **2010**, *15*, 40–53.

20 S. Park, R. J. Kazlauskas, *Curr. Opin. Biotechnol.* **2003**, *14*, 432–437.

21 N. Kaftzik, P. Wasserscheid, U. Kragl, *Org. Process Res. Dev.* **2002**, *6*, 553–557.

22 M. Eckstein, M. VIllela Filho, A. Liese, U. Kragl, *Chem. Commun.* **2004**, *9*, 1084–1085.

23 A. J. Walker, N. C. Bruce, *Chem. Commun.* **2004**, *22*, 2570–2571.

24 X. A. Shi, M. H. Zong, C. Meng, Y. H. Guo, *Chin. J. Catal.* **2005**, *26*, 10512.

25 X. A. Shi, M. H. Zong, W. Y. Lou, *Chin. J. Chem.* **2006**, *24*, 1643–1647.

26 G. de Gonzalo, I. Lavandera, K. Durchschein, D. Wurm, K. Faber, W. Kroutil, *Tetrahedron Asymmetry* **2007**, *18*, 2541–2546.

27 M. N. Okochi, I. Nakagawa, T. Kobayashi, S. Hayashi, S. Furusaki, H. Honda, *J. Biotechnol.* **2007**, *128*, 376–382.

28 K. Fujita, N. Nakamura, K. Igarashi, M. Samejima, H. Ohno, *Green Chem.* **2009**, *11*, 351–354.

29 W. Hussain, D. J. Pollard, M. Truppo, G. J. Lye, *J. Mol. Catal., B Enzym.* **2008**, *55*, 19–29.

30 Y. Zhang, X. Huang, Y. Li, *J. Chem. Technol. Biotechnol.* **2008**, *83*, 230–1235.

31 A. J. Walker, N. C. Bruce, *Tetrahedron* **2004**, *60*, 561–568.

32 (a) B. Kosjek, W. Stampfer, M. Pogorevc, C. Goessler, K. Faber, W. Kroutil, *Biotechnol. Bioeng.* **2004**, *86*, 55–62.; (b) W. Stampfer, B. Kosjek, C. Moitzi, W. Kroutil, K. Faber, *Angew. Chem. Int. Ed. Engl.* **2002**, *41*, 1014–1017.

33 (a) S. Dreyer, U. Kragl, *Biotechnol. Bioeng.* **2008**, *99*, 1416–1424.; (b) S. Dreyer, P. Salim, U. Kragl, *Biochem. Eng. J.* **2009**, *46*, 176–185.

34 V. Gauchot, W. Kroutil, A. R. Schmitzer, *Chem. Eur. J.* **2010**, *16*, 6748–6751.

35 D. M. Tschaen, L. Abramson, D. W. Cai, R. Desmond, U. H. Dolling, L. Frey, S. Karady, Y. J. Shi, T. R. Verhoeven, *J. Org. Chem.* **1995**, *60*, 4324–4330.

36 (a) J. Reddy, D. M. Tschaen, S. Yao Jun, V. Pecore, L. Katz, R. Greasham, M. Chartrain, *J. Ferment. Bioeng.* **1996**, *81*, 304–309.; (b) S. Stahl, R. Greasham, M. Chartrain, *J. Biosci. Bioeng.* **1997**, *89*, 367–371.

37 W. Hussain, D. J. Pollard, G. J. Lye, *Biocatal. Biotransformation* **2007**, *25*, 443–452.

38 H. L. Holland, in *Biotransformations*, Vol. 8a (Ed. D. R. Kelly), Wiley-VCH, Weinheim, **1998**, pp. 475–533.

39 (a) W. Adam, M. Lazarus, C. R. Saha-Möller, O. Weichold, U. Hoch, D. Häring, P. Schreier, in *Biotransformations*, Vol. 63 (Ed. K. Faber), Springer, Berlin/Heildelberg, **1999**, pp. 73–108.; (b) S. G. Burton, *Trends Biotechnol.* **2003**, *21*, 543–549.

40 S. G. Burton, *Curr. Org. Chem.* **2003**, *7*, 1317–1331.

41 (a) V. K. Gochev, A. I. Krastanov, *Bulg. J. Agric. Sci.* **2007**, *13*, 75–83.; (b) A. T. Martínez, in *Industrial Enzymes* (Eds. J. Polaina, A. P. MacCabe), Springer, Dordrecht, **2007**.; (c) F. Xu, J. J. Kulys, K. Duke, K. Li, K. Krikstopaitis, H.-J. W. Deussen, E. Abbate, V. Galinyte, P. Schneider, *Appl. Environ. Microbiol.* **2000**, *66*, 2052–2056.

42 M. Sureshkumar, C. K. Lee, *J. Mol. Catal., B Enzym.* **2009**, *60*, 1–12.

43 R. A. Sheldon, *Chem. Commun.* **2001**, 2399–2407.

44 G. Hinckley, V. V. Mozhaev, C. Budde, Y. L. Khmelnitsky, *Biotechnol. Lett.* **2002**, *24*, 2083–2087.

45 J. A. Laszlo, D. L. Compton, *J. Mol. Catal., B Enzym.* **2002**, *18*, 109–120.

46 M. F. Machado, J. M. Saraiva, *Biotechnol. Lett.* **2005**, *27*, 1233–1239.

47 Y. M. Lee, O. Y. Kwon, I. K. Yoo, K. G. Ryu, *J. Microbiol. Biotechnol.* **2007**, *17*, 600–603.

48 (a) D. Das, A. Dasgupta, P. Kumar Das, *Tetrahedron Lett.* **2007**, *48*, 5635–5639.; (b) E. S. Hong, J. H. Park, I. K. Yoo, K. G. Ryu, *J. Microbiol. Biotechnol.* **2009**, *19*, 713–717.; (c) S. Shipovskov, H. Q. N. Gunarathe, K. R. Seddon, G. Stephens, *Green Chem.* **2008**, *10*, 806–810.; (d) Q. Cao, L. Quan, C. He, N. Li, K. Li, F. Liu, *Talanta* **2008**, *77*, 160–165.

49 C. Baudequin, J. Baudoux, J. Levillain, D. Cahard, A. C. Gaumont, J. C. Plaquevent, *Tetrahedron Asymmetry* **2003**, *14*, 3081–3093.

50 M. B. Turner, S. K. Spear, J. D. Holbrey, R. D. Rogers, *Biomacromolecules* **2004**, *5*, 1379–1384.

51 X. Lu, J. Hu, X. Yao, Z. Wang, J. Li, *Biomacromolecules* **2006**, *7*, 975–980.

52 Y. Liu, M. Wang, J. Li, Z. Li, P. He, H. Liu, J. Li, *Chem. Commun.* **2005**, *13*, 1778–1780.

53 K. Nakashima, N. Kamiya, D. Koda, T. Maruyama, M. Goto, *Org. Biomol. Chem.* **2009**, *7*, 2353–2358.

54 V. Rumbau, R. Marcilla, E. Ochoteco, J. A. Pomposo, D. Mecerreyes, *Macromolecules* **2006**, *39*, 8547–8549.

55 S. Alvarez, S. Manolache, F. Denes, *J. Appl. Polym. Sci.* **2003**, *88*, 369–379.

56 M. Moniruzzaman, N. Kamiya, M. Goto, *Langmuir* **2009**, *125*, 977–982.

57 G. P. Zhou, Y. Zhang, X. R. Huang, C. H. Shi, W. F. Liu, Y. Z. Li, Y. B. Qu, P. J. Gao, *Colloids Surf. B. Biointerfaces* **2008**, *66*, 146–149.

58 N. J. Roberts, A. Seago, G. J. Lye, *International Congress on Biocatalysis*, Biocat2002, Hamburg, Germany, **2002**.

59 (a) V. Alphand, A. Archelas, R. Furtoss, *Tetrahedron Lett.* **1989**, *30*, 3663–3664.; (b) J. Lebreton, V. Alphand, R. Furtoss, *Tetrahedron Lett.* **1996**, *37*, 1011–1014.; (c) M. D. Mihovilovic, B. Müller, P. Stanetty, *Eur. J. Org. Chem.* **2002**, *22*, 3711–3730.

60 K. L. Tee, D. Roccatano, S. Stolte, J. Aming, B. Jastorff, U. Schwaneberg, *Green Chem.* **2008**, *10*, 117–123.

61 S. Lutz-Wahl, E. M. Trost, B. Wagner, A. Manns, L. Fischer, *J. Biotechnol.* **2006**, *124*, 163–171.

62 K. Chattopadhyay, S. Mozumdar, *Biochemistry* **2000**, *39*, 263–270.

63 A. Kumar, N. Jain, S. M. S. Chauhan, *Synlett* **2007**, *3*, 411–414.

64 I. Fernández, N. Khiar, *Chem. Rev.* **2003**, *103*, 3651–3705.

65 T. Katsuki, K. B. Sharpless, *J. Am. Chem. Soc.* **1980**, *102*, 5974–5976.

66 C. Sanfilippo, N. D'Antona, G. Nicolosi, *Biotechnol. Lett.* **2004**, *26*, 1815–1819.

67 C. Chiappe, L. Neri, D. Pieraccini, *Tetrahedron Lett.* **2006**, *47*, 5089–5093.

68 K. Okrasa, E. Guibé-Jampel, M. Therisod, *Tetrahedron Asymmetry* **2003**, *14*, 2487–2490.

69 B. Takács, L. Nagy, L. Kollár, G. Nagy, *Anal. Lett.* **2010**, *43*, 1734–1745.

70 J. S. Dordick, M. A. Marletta, A. M. Klibanov, *Biotechnol. Bioeng.* **1987**, *30*, 31–36.

71 S. Sgalla, G. Fabrizi, S. Cacchi, A. Macone, A. Bonamore, A. Boffi, *J. Mol. Catal., B Enzym.* **2007**, *44*, 144–148.

72 B. Eker, D. Zogorevski, G. Zhua, R. J. Linhardta, J. S. Dordick, *J. Mol. Catal., B Enzym.* **2009**, *59*, 177–184.

73 J. Hernandez, N. R. Robledo, L. Velasco, R. Quintero, M. A. Pickard, R. Vazquez-Duhalt, *Pestic. Biochem. Physiol.* **1998**, *61*, 87–94.

74 A. Boskin, C. D. Tran, M. Franko, *Environ. Chem. Lett.* **2009**, *7*, 267–270.

75 (a) C. Ding, M. Zhang, F. Zhao, S. Zhang, *Anal. Biochem.* **2008**, *378*, 32–37.; (b) Y. Liu, L. Shi, M. Wang, Z. Li, H. Liu, J. Li, *Green Chem.* **2005**, *7*, 655–658.; (c) J. Wan, J. Bi, P. Du, S. Zhang, *Anal. Biochem.* **2009**, *386*, 256–261.

76 A. C. Franzoi, J. Dupont, A. Spinelli, I. C. Vieira, *Talanta* **2009**, *77*, 1322–1327.

77 H. Chen, Y. Wang, Y. Liu, Y. Wanf, L. Qi, S. Dong, *Electrochem. Commun.* **2007**, *9*, 469–474.

78 K. dos Santos Maguerroski, S. C. Fernandes, A. C. Franzoi, I. C. Vieira, *Enzyme Microb. Technol.* **2009**, *44*, 400–405.

79 T. Kitazume, Z. Jiang, K. Kasai, Y. Mihara, M. Suzuki, *J. Fluorine Chem.* **2003**, *121*, 205–212.

80 T. Purkarthofer, W. Skranc, C. Schuster, H. Griengl, *Appl. Microbiol. Biotechnol.* **2007**, *76*, 309–320.

81 K. Gruber, C. Kratky, *J. Polym. Sci. [A1]* **2004**, *42*, 479–486.

82 (a) M. H. Fechter, H. Griengl, *Food Technol. Biotechnol.* **2004**, *42*, 287–294.; (b) G. Gartler, C. Kratky, K. Gruber, *J. Biotechnol.* **2007**, *129*, 87–97.; (c) M. Sharma, N. N. Sharma, T. C. Bhalla, *Enzyme Microb. Technol.* **2005**, *37*, 279–294.; (d) J. Holt, U. Hanefeld, *Curr. Org. Synth.* **2009**, *6*, 15–37.

83 F. Effenberger, T. Ziegler, S. Forster, *Angew. Chem. Int. Ed. Engl.* **1987**, *26*, 458–460.

84 (a) M. Avi, H. Griengl, in *Organic Synthesis with Enzymes in Non-Aqueous Media* (Eds. G. Carrea, S. Riva), Wiley-VCH, Weinheim, **2008**, pp. 211–225.; (b) S. Han, G. Lin, Z. Li, *Tetrahedron Asymmetry* **1998**, *9*, 1835–1838.; (c) G. Lin, S. Han, Z. Li, *Tetrahedron* **1999**, *55*, 3531–3540.; (d) V. I. Ognyanov, V. K. Datcheva, K. S. Kyler, *J. Am. Chem. Soc.* **1991**, *113*, 6992–6996.

85 R. P. Gaisberger, M. H. Fechter, H. Griengl, *Tetrahedron Asymmetry* **2004**, *15*, 2959–2963.

86 W.-Y. Lou, R. Xu, M.-H. Zong, *Biotechnol. Lett.* **2005**, *27*, 1387–1390.

87 Z. Yang, *J. Biotechnol.* **2009**, *144*, 12–22.

7

IONIC LIQUIDS AND WHOLE-CELL–CATALYZED PROCESSES

Danielle Dennewald and Dirk Weuster-Botz

ABBREVIATIONS

Abbreviations of Ionic Liquid Cations

[(E2OH)MIM]	1-(2-hydroxyethyl)-3-methylimidazolium
[(EO2E)MPL]	1-(ethoxycarbonyl)methyl-1-methylpyrrolidinium
[(EOE)MMO]	4-(2-ethoxyethyl)-4-methylmorpholinium
[(MOE)MPL]	1-(2-methoxyethyl)-1-methylpyrrolidinium
[(MOP)MPI]	1-(3-methyoxypropyl)-1-methylpiperidinium
[(NEMM)EO2E]	ethyl-dimethyl-(ethoxycarbonyl)methylammonium
[(NEMM)MOE]	ethyl-dimethyl-2-methoxyethylammonium
[(P3OH)PYR]	N-(3-hydroxypropyl)pyridinium
[BMIM]	1-butyl-3-methylimidazolium
[BMMIM]	1-butyl-2,3-dimethylimidazolium
[BMPL]	1-butyl-1-methylpyrrolidinium
[C_2OHMIM]	1-(2′-hydroxy)ethyl-3-methylimidazolium

Ionic Liquids in Biotransformations and Organocatalysis: Solvents and Beyond, First Edition. Edited by Pablo Domínguez de María.

[CABHEM]	PEG-5 cocomonium
[EMIM]	1-ethyl-3-methylimidazolium
[EtOHNMe$_3$]	2-hydroxy ethyl trimethylammonium
[EWTMG]	N,N,N,N-tetramethyl-N-ethylguanidinium
[HMIM]	1-hexyl-3-methylimidazolium
[HMPL]	1-hexyl-1-methylpyrrolidinium
[HMPyr]	1-hexyl-3-methylpyridinium
[NDecPy]	N-decylpyridinium
[Oc$_3$MeN]	methyltrioctylammonium
[OMIM]	1-octyl-3-methylimidazolium
[OMPyr]	1-octyl-3-methylpyridinium
[OPy]	N-octylpyridinium
[P$_{6,6,6,14}$]	trihexyltetradecylphosphonium

Abbreviations of Ionic Liquid Anions

[BF$_4$]	tetrafluoroborate
[Br]	bromide
[CF$_3$SO$_3$]	trifluoromethanesulfonate
[Cl]	chloride
[E$_3$FAP]	tris(pentafluoroethyl)trifluorophosphate
[EtSO$_4$]	ethylsulfate
[MDEGSO$_4$]	ethylenglycolmonomethyl ethersulfate
[Me$_2$PO$_4$]	dimethylphosphate
[MeSO$_4$]	methylsulfate
[NO$_3$]	nitrate
[NTF]	bis(trifluoromethylsulfonyl)imide
[OcSO$_4$]	octylsulfate
[PF$_6$]	hexafluorophosphate
[SbF$_6$]	hexafluoroantimonate
[TOS]	tosylate

Abbreviation of Ionic Liquid

DMEAA	N,N-dimethylethanolammonium acetate

List of Abbreviations

BSP	biomass support particles
CFU	colony forming units

D	partition coefficient ionic liquid–water phase
DEPC	1,2-dielaidoylphosphocholine
DES	deep eutectic solvent
DFA	discriminant function analysis
DMPG	1,2-dimyristoylphosphoglycerol
ee	enantiomeric excess
EC_{50}	half maximal effective concentration
LUV	large unilamellar vesicles
MBC	minimal bactericidal concentration
MIC	minimal inhibitory concentration
MTBE	methyltertbutylether
P	partition coefficient octanol–water
PBS	phosphate buffer saline
PCA	principal component analysis
SPB	supported phospholipid bilayers
TGA	thermogravimetric analysis

Biocatalysis has become a method of choice for producing fine chemicals at high yields and excellent selectivities under mild reaction conditions. Whole-cell biotransformations present a number of advantages as compared with the enzymatic approach, which includes, for example, the fact that no more purification steps are needed once the biocatalyst is produced and cell-intern cofactor regeneration. As in many other domains, increased interest has been turned toward applications of ionic liquids in this area of catalysis.

This chapter will discuss the use of this new kind of solvents in whole-cell biocatalysis. To begin with, a general overview of what properties an ionic liquid should present to constitute a suitable solvent for applications in whole-cell biocatalysis will be given in the first part of this chapter. The second part will provide more precise information about the tolerance of microorganisms used in whole-cell biotransformations toward this relatively new class of materials, and the mechanism of their interaction will be discussed. In the third part, the state-of-the-art of whole-cell biocatalysis in ionic liquids will be presented. Finally, future trends will be described in the fourth and last part of this chapter.

7.1 IONIC LIQUIDS COMPATIBLE WITH WHOLE-CELL BIOCATALYSIS: FUNDAMENTALS AND DESIGN

Several factors should be taken into account when choosing an ionic liquid for an application in whole-cell biocatalysis. In the following, the most important features will be discussed.

7.1.1 Biocompatibility

The first condition a solvent must satisfy to be considered for whole-cell biocatalysis is biocompatibility. This notion includes the impact of the solvent not only on cell growth and on cell division, but also on the activity of the organism in the presence or after exposure to the solvent. If the solvent is toxic to the biocatalyst, the biotransformation can hardly be successful. A minimal tolerance of the microorganism toward the solvent is therefore essential. In contrast to organic solvents, which have been studied since a few decades ago, not very much is known on the toxicity of ionic liquids toward microorganisms. The little data available for the large number of possible solvents and microorganisms make the establishment of general guidelines difficult.

The biocompatibility of the ionic liquid largely depends on the nature of its composing anion and cation. A significant effect of the length of the alkyl substituents on the cation has been repeatedly observed: The longer these chains are, the lower the tolerance of the organisms toward the solvent.[1–4] The explanation to this observation is the surfactant-like structure of cations with long alkyl chains.[5] Cations with relatively long alkyl chains are of more lipophilic nature than cations with shorter alkyl chains, and undergo stronger interactions with the cell membrane. The consequences are facilitated insertion of the cation into the phospholipid bilayer and (partial) disintegration of the membrane, as well as larger accumulation of the toxic molecules inside the cells.[3,6,7] Based on the data currently available, it seems that the cation head group has only moderate influence on the biocompatibility. However, more data are needed to confirm this observation.

As far as the effect of the anion is concerned, a general influence of the anion's potential to hydrolyze or undergo other degradation reactions on toxicity has been indicated. One example is $[BF_4]$, which frequently decreases the biocompatibility of the ionic liquid in comparison to other anions.[8–10] This increased toxicity is due to the decomposition of $[BF_4]$ in the presence of water, forming hydrofluoric acid.[11,12] In addition, as for the cation, a slight influence of lipophilic side chains of the anion on the solvent's toxicity has been postulated. This might also be the reason for the toxicity that was observed in some cases for the [NTF] moiety.[13–15] However, not all sources confirmed the detrimental effect of [NTF].[16,17]

The type of organism used is also, to some degree, of importance when considering the biocompatibility of an ionic liquid. Differences in the tolerance of the solvent have indeed been observed when using different cell types.[17–20] Still, to date, no reference was found describing that specific physiological properties provide particular tolerance or susceptibility to a given class of ionic liquids. The type of organism used must thus be a parameter of smaller impact than the composition of the ionic liquid itself. More pronounced effects have been observed when comparing immobilized biocatalyst with the corresponding free cells: Immobilization provides increased tolerance toward the solvent.[4]

The duration of the exposure and the concentration at which the solvent is present are also factors of major importance concerning the extent of the effect ionic liquids have on an organism. Other parameters of influence are the effect of the solvent on the pH of the cell suspension,[10] its water content,[15,21–23] and, in some cases, the ionic liquid's miscibility to water.[24]

7.1.2 Availability and Purity

To be considered for application in a given process, the solvent must of course be available not only at sufficient volume, but also at sufficient purity and affordable costs. Purity is a factor of major importance because impurities resulting from the synthesis, such as halides or unreacted organic salts, easily accumulate in the ionic liquid and might not only influence the physical properties of the solvent, but also affect the biocatalyst.[25] However, the most common contaminant is water, as even water-immiscible ionic liquids are hygroscopic and can absorb a significant quantity of up to a few weight percent.[12] Generally, this does not harm the biocatalyst, but it can significantly affect the viscosity as well as other physical properties,[26–28] and might interfere with the biotransformation under investigation.[29] Different degrees of purity and water content are thought to frequently be at the origin of variations observed in published data concerning the same topics.[30] Consistent purity is therefore also of high importance because only then reliable comparisons and analyses of existing data, as well as persistent process productivities, are possible.

7.1.3 Stability

The ionic liquid of choice needs to show minimal stability. Chemical stability of the solvent is a crucial feature and necessary to allow consistent quality of the solvent during the process, as well as storage before and after it. Depending on the process conditions, and even more on the downstream processing, thermal stability might also be decisive. Thermal stability of ionic liquids depends on the composing anions and cations, as well as on the presence of impurities, and might thus vary significantly from one case to another.[31] The trend is leading away from ionic liquids containing $[PF_6]$ or $[BF_4]$ anions, as these tend to hydrolyze over time when in contact with humidity, and even more under heating conditions, forming hydrofluoric acid.[11,25,27,32] Generally, however, ionic liquids are reported to present good thermal stability, showing little or no degradation up to temperature ranges of 150°C–200°C. Sometimes even values larger than 400°C are indicated. These indications should, however, be considered with precaution, as most of them are decomposition temperatures derived from fast scan thermogravimetric analysis (TGA), which often lead to temperatures much larger than long-term thermal stability studies.[25,31]

7.1.4 Process Design Criteria

Besides the previously cited factors, a range of process design criteria such as viscosity, density, corrosiveness, and water miscibility has to be taken into account when choosing a suitable ionic liquid. These properties will be part of determining which reaction setup is applied.

7.1.4.1 Viscosity, Density, and Corrosiveness. Ionic liquids generally show relatively large viscosity in comparison to their commonly used organic counterparts, with values roughly within the range of 15–500 mPa·s.[33] Viscosity should be taken into consideration when choosing a suitable solvent because too large viscosity will not only make transfer from one vessel to another difficult, but it might also lead to mass transfer limitations if sufficient stirring is not assured. However, it should be kept in mind that the viscosity of ionic liquids is dramatically decreased by the presence of already low amounts of water.[26,34,35] Hence, many of the current applications involving ionic liquids along with an aqueous phase should not suffer too severely from this phenomenon. Most ionic liquids available to date are denser than water, with typical values ranging from 1 g/cm^3 to 1.6 g/cm^3.[28,36] Density of the solvent is generally only a matter of concern for biphasic ionic liquid–aqueous phase systems. In fact, only in these cases, a minimal relative difference of density between the two phases is often desired, as this would permit and facilitate the phase separation at the end of the process by gravitational forces. A relative density difference of 20% is supposedly favorable in theses cases. Corrosiveness was mainly observed for ionic liquids containing halides.[37] With the development of new second- and third-generation ionic liquids, this problem seems to be widely resolved.

7.1.4.2 Water Miscibility. The miscibility of ionic liquids with water varies from fully miscible to virtually immiscible.[38] Assumed to largely depend on the combination of a given anion with a specific cation,[35] the miscibility of ionic liquids with water is still perceived as varying widely and unpredictably.[12] In contrast to organic solvents, where the concept of solvent polarity is relatively useful to predict the miscibility behavior with water, this concept might be too elusive to permit predictions concerning the water miscibility of ionic liquids.[39] In fact, solvent–solute interactions in ionic liquids have been described as obeying a dual interaction model, that is, that ionic liquids act like polar solvents toward polar molecules but display nonpolar characteristics toward nonpolar compounds.[40]

However, as water miscibility is such a decisive property in terms of process design, the development of structure–property relationships is a topic constantly attracting interest. Recently, Ranke et al. had a closer look at the subject, trying to distinguish the contribution of the composing entities to the overall water solubility by splitting the activity coefficient of ionic liquids in water into anion and cation contributions.[41] This was achieved by deriving the effect of the anion on overall water solubility by regression against experimentally determined cation hydrophobicity parameters. The derived anion hydrophobicity data, together with the measured cation hydrophobicity values, would then permit prediction of the water solubility for anion–cation combinations that have not yet been analyzed. The future will show if the proposed method will indeed facilitate the directed molecular design of ionic liquids and be accurate enough to match the expectations.

7.1.5 Monophasic versus Biphasic Reaction Mode

Depending on the solvent's miscibility with water, ionic liquids can be used either as solvents, as co-solvents, or as second phase in biphasic reaction modes. While the first reaction setup is not usual in whole-cell biocatalysis, the other two types of applications are common. The choice for one of these reaction modes is mainly a question of reaction design, and less a matter of productivity. At least to date, the published data do not permit to generally relate water miscibility to better or worse biocompatibility or to larger productivity at the end of the whole-cell biotransformation.[35,42] In the literature, examples of both types of reaction modes reaching good process productivities are found.

When used as a co-solvent, the reason for adding ionic liquid to whole-cell applications is generally its solubilizing effect, that is, that larger quantities of a given compound can be dissolved in the reaction system when water miscible ionic liquid is added to the aqueous buffer than without it. This permits larger substrate loading within one batch and thus increased productivities than in a purely aqueous setup. However, one of the disadvantages of using water-miscible ionic liquids is that the biocatalyst is exposed to increased molecular toxicity compared with biphasic reaction modes, where it is mainly affected by phase toxicity. Similarly, the recovery of the ionic liquid from the monophasic reaction system after the biotransformation is relatively complicated and more elaborate separation techniques are necessary, such as extraction by an organic solvent or nanofiltration. This not only involves a supplemental process step, but also generates larger costs and potentially increased environmental pollution.

In contrast, when relying on a biphasic reaction mode, the disadvantages of monophasic systems can be avoided, while larger substrate loading is still possible. This is why most of today's applications involve biphasic reaction systems with water-immiscible ionic liquids rather than monophasic setups. In fact, when choosing a water-immiscible solvent with sufficiently large partition coefficients (D) for the substrate and the product in the ionic liquid–buffer system, these compounds will preferably dissolve in the ionic liquid and only show up in the aqueous phase in very reduced concentrations. Used as second phase, the ionic liquid can thus act as a substrate reservoir and *in situ* product extractant if the solvent is chosen with care (Scheme 7.1).

This limits the exposure of the cell to the toxic substrate and/or product and thus still permits higher loading without risking inhibition and damage of the biocatalyst.

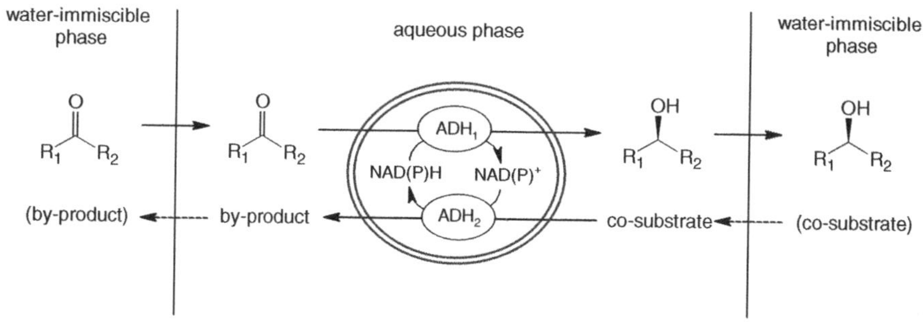

Scheme 7.1. Use of a water-immiscible ionic liquid for whole-cell biotransformations in a biphasic reaction setup, for example, the reduction of a ketone to the corresponding chiral alcohol with cell-intern cofactor regeneration.

Additionally, extracting the product from the reactive phase will shift the thermodynamic equilibrium toward the desired direction. As water-miscible ionic liquids do not form a second phase, monophasic setups expose the biocatalyst to the complete amount of substrate and product present in the reaction system. The biotransformation might thus be limited both because of toxic effects and for thermodynamic reasons.

When choosing water-immiscible ionic liquids for biphasic applications, solvents with partition coefficients in the ionic liquid–water system of log $D \geq 2$ for the substrate and the product in the ionic liquid–aqueous phase system are generally preferred.[17] Such values would roughly permit addition of substrate concentrations of the order of magnitude of several 100 mM to the ionic liquid, while still not surpassing substrate and product concentrations of only a few millimolar in the aqueous phase. This guarantees relatively low substrate concentrations in the aqueous phase, limiting the inhibition of the biocatalyst, and relatively large product concentration in the ionic liquid phase, making downstream processing more cost-efficient than in the corresponding monophasic reaction modes with relatively dilute product solutions at the end.

Immiscibility with water also permits an uncomplicated recovery of the ionic liquid from the rest of the biotransformation media, as simple phase separation through gravitational forces is possible. This facilitates the possible recycling of the ionic liquid and decreases material loss, not only reducing the process costs, but also limiting the impact on the environment. Product recovery is also facilitated by a biphasic reaction mode. Once the ionic liquid phase is isolated from the rest of the reaction broth, the product can be recovered from the solvent using one of the usual separation techniques, such as distillation (if the product is thermally stable), pervaporation, or extraction. These recovery methods are facilitated by the low vapor pressure and nonflammability of ionic liquids compared with their organic analogues, and the relatively high product concentrations found in the ionic liquid phase make these product isolation techniques even more cost-effective.

7.1.6 Hazard Potential

As more and more effort is invested in making the production of fine chemicals more sustainable, relying on biocatalysis and replacing volatile organic solvents by nonvolatile ionic liquids is very attractive. Still, it cannot be completely ruled out that some of the solvent will be lost during the process and will reach the environment, either by small quantities dissolving in the waste waters of the process, or while being disposed of when the solvent cannot be further used. Low hazard potential is therefore also a major criterion when choosing or designing an ionic liquid for a given application. Evaluations of the hazard potential of ionic liquids should include two factors: ecotoxicity and biodegradability of the solvent. The latter describes the persistence in the environment and thus on which time scale a spilling into the environment will have consequences, while the former gives an indication of the immediate severity of the spillage.

7.1.6.1 Ecotoxicity. Since ionic liquids receive increased attention for future applications, several studies on the toxicity of this class of solvents have been performed, investigating their effect on a large range of different organisms (e.g., mammalian cells IPC-81, zebrafish *Danio rerio*, luminescent marine bacteria *Vibrio fischeri*, limnic unicellular green algae *Scenedesmus vacuolatus*, wheat *Triticum aestivum*, cress *Lepidium sativum*, duckweed *Lemna minor*, as well as a soil invertebrate, the spring tail *Folsomia candida*).[43–48] The aim of these studies was to evaluate the influence of the structure of the cation and the anion on the ecotoxicity of the solvent. Even though not exactly the same organisms are observed in both types of studies, most considerations concerning biocompatibility are also found in ecotoxicity evaluations, as both are concerned with the effect of ionic liquids on living cells. The conclusions are thus roughly the same.

The choice of the cation has been found to significantly influence the tolerance of organisms toward the solvent. More precisely, a clear interdependency of the cation lipophilicity with its toxicity has been repeatedly shown.[43,46–49] This lipophilicity depends in a large measure on the length of the alkyl chains on the cation. The "side-chain effect" observed in biocompatibility evaluations with whole-cell biocatalysts is thus also valid for the assessment of ecotoxicity: The longer the alkyl chain, the more lipophilic the cation and the more toxic the solvent because it undergoes stronger interaction with the phospholipid bilayer composing the cell membrane. Confirming this observation is the fact that when introducing polar functions into the side chains, such as ether, hydroxyl, or nitrile functions, decreasing the lipophilicity of the cation, the cytotoxicity was reduced.[48] In contrast to the side-chain effect, only a low influence on the solvent's toxicity was observed for the head group of the cation, except for 4-dimethylaminopyridinium and quinolinium heads, which showed large toxicity toward several different microorganisms.[47,48]

Concerning the anion, the trends observed are as little consistent as those observed during biocompatibility studies. This is again due to the fact that the anions currently found in ionic liquids are of very diverse structures. As the potential toxic effect of each anion is intrinsic to its nature, this diversity makes it more difficult to distinguish structure–activity relationships. To date, the effect of the anion on the solvent's toxicity does not seem to be as consistent as the influence of the alkyl chains on the cation, except for [NTF], which is described as presenting relatively large hazard potential.[50] A general influence of the anion's potential to hydrolyze or undergo other degradation reactions on toxicity has also been indicated. Hence, reactive anions that might easily undergo hydrolysis or other degradation reactions should be avoided. In addition, as for the cation, a slight influence of lipophilic side chains on the anion on the solvent's toxicity has been postulated,[50] which might also be the reason for the large toxicity that was observed for the [NTF] moiety.[45]

From an ecotoxicological point of view, lipophilic solvents should thus generally be avoided. However, lipophilicity of the solvent is sometimes a desired feature, for example, when a biphasic reaction mode is wanted because this feature often also means low water solubility. In these cases, a compromise has to be reached between reaction design criteria on the one hand and biocompatibility and environmentally safe solvents on the other hand.

7.1.6.2 Biodegradability. In addition to ecotoxicity, estimations of the hazard potential ionic liquids represent should include evaluating the biodegradability of the solvents. Ionic liquids have received attention as potential solvents for industrial applications since only a few years ago, and although a large number of different toxicity studies have been published, investigations about their biodegradability have only appeared recently. Assessing the biodegradability of the ionic liquid includes determining the degradation of both the cation and the anion. Ideally, both entities should be entirely mineralized to CO_2 and biomass, yielding completely nontoxic products. This imposes more conditions on the molecular design of these solvents than just concentrating on either of the composing ions.

Evaluating biodegradability is performed mostly by using respirometry methods such as the so-called Closed Bottle or CO_2 Headspace (ISO 14593) tests, or by determining the dissolved organic carbon (DOC) in the assay. In these experiments, the substance under investigation is incubated in aerobic conditions either with commercially available freeze-dried microorganism mixtures, or with waste water microorganisms from activated sludge treatment plants. Then, the evolution of the DOC, molecular oxygen, or carbon dioxide is followed over time, and the observations made are compared with the starting conditions, or reported as a fraction of the theoretically possible maximum. The results of this kind of experiments are generally rated according to the OECD 301 guidelines stating that a substance can be considered "readily degradable" if the biodegradation level determined according to one of the above-mentioned tests is larger than 60% (Closed Bottle or CO_2 Headspace test) or 70% (DOC) within a 10-day frame within 28 days.[51] Most studies compared several different cation head groups with alkyl chains of varying lengths, combined with different anion entities. Among all of the ionic liquids tested to date, most could not be considered readily degradable, and they would thus persist in the environment longer than ideally wished for from an ecological point of view.

Concerning the head groups, pyridinium entities have shown slightly better biodegradability than imidazolium compounds.[52] In contrast, phosphonium ionic liquids showed generally relatively low biodegradability.[53]

The effect of the anion on biodegradability is generally not very pronounced, except for octyl sulfate, which permitted particularly good biodegradation levels of ionic liquids that include it.[54,55]

Generally, there is interdependency between the compound's chemical and thermal stability—wished for in industrial applications—and its biodegradability: More stable molecules show longer persistence in the environment, an undesired fact from an ecological point of view. Another conflict is arising between ecotoxicity and biodegradability: It could be shown that longer alkyl chains on the cation moiety provide higher biodegradability. These are, however, not recommendable from an ecotoxicological point of view because longer alkyl side chains lead to increased toxicity of the ionic liquid toward most tested microorganisms.

One possible strategy to increase biodegradability is to try to create supplemental sites for enzymatic attacks during the biodegradation process by modifying the side-chain structure. Including an ester function into the side chains could, for example, increase the degradation rate and level of some ionic liquids.[54–57] However, the biodegradability did not improve significantly by introducing amide functions.[55] In addition, it should be noted that when modifying the structure of an ionic liquid for ecological purposes, the variation of physicochemical properties that this might induce has to be taken into account, and it should be investigated afterward if the ionic liquid is still suitable for industrial applications.

Concerning the hazard potential of ionic liquids, the synthesis of the solvent under investigation might also be included in the previous considerations. Here, materials produced from nontoxic and biodegradable starting components would be preferable, for example, ionic liquids synthesized on the basis of naturally derived moieties such as lactic acid, acetic acid, sugars, or amino acids (see also Chapter 1).[58–63]

7.1.7 Recyclability

Recyclability of the solvent would not only limit waste production and thus be favorable from an ecological point of view, but it would also significantly reduce the process costs, especially considering the still very large expenses from ionic liquids. The conditions for reusing the same ionic liquid batch are stability of the material and that no impurities are accumulated in this phase over the different process cycles. Only then can constant productivity be guaranteed. The requirement of stability possibly includes high thermal stability as well as stability under reduced pressure if evaporation or vacuum distillation techniques are used to remove the product from the ionic liquid. As already mentioned, ionic liquids including [PF_6] or [BF_4] anions are known to be relatively unstable in the presence of water, but most ionic liquids have been described as having very good thermal stability. Accumulation of impurities other than degradation products from the ionic liquid might occur due to the process itself. This could be avoided by redesigning part of the process or by applying specific purification methods at the end of each cycle. To date, little data have been published about the recyclability of ionic liquids in whole-cell biocatalysis, although this is definitely a matter of great importance to industry. In fact, the large cost associated with ionic liquids is holding back their large-scale application, and proving that they can be reused over a large number of cycles without loss of productivity might help further trigger the interest in this class of materials.

7.1.8 Availability of Information

Finally, well-studied solvents are always preferred in applied research and industry. In fact, the availability of toxicological data as well as information about physical and chemical properties is of invaluable advantage when developing a given application of ionic liquids. This currently favors the use of second-generation ionic liquids, although more recently developed ionic liquids surely present advantages in comparison with the former. In fact, the trend is going to improve any possible parameter related to the use of ionic liquids, be it the use of more environmentally friendly starting materials in the synthesis of the solvent, modifying physical properties such as water (non)miscibility, viscosity or density, higher stability, greater biocompatibility, or better biodegradability and reduced environmental toxicity. Due to the paucity of known data, the establishment of unequivocal relationships between the structure of the ionic liquid and these parameters is still difficult. However, as more and more applications of ionic liquids are presented, the knowledge gained is increasing, making such conclusions easier and the directed design of ionic liquids for a given purpose more efficient.

7.2 BIOCOMPATIBILITY, TOLERANCE, AND ACCUMULATION IN THE CELL

When developing whole-cell processes involving ionic liquids, it is of major importance to know how the solvent will affect the biocatalyst. If the organism used does not tolerate the solvent at least to a minimal extent, the biocatalyst will rapidly be damaged and lose its activity. Biocompatibility of the solvent toward the cell is thus crucial for the process to be successful. In the following, methods for determining the tolerance of an organism toward a solvent, as well as the general conclusions of different biocompatibility studies will be presented. Following this, the interaction of ionic liquids with cells will be considered in more detail. In evaluations concerning the ecotoxicity of ionic liquids, much data has been presented on the effect of these solvents toward many aquatic organisms. As these organisms are not directly relevant for whole-cell biocatalysis, the corresponding results will however not be discussed here. The focus will lie exclusively on organisms used in whole-cell biocatalysis, and the effect ionic liquids have toward them.

7.2.1 Methods

There are many possible ways of evaluating the tolerance of an organism toward a given solvent. One of the simplest methods is determining the viability of cells incubated over a given duration in media containing the solvent. This is the least time-consuming method, as only one sample is taken for an end-point determination after a fixed amount of time. After the incubation, the viability of the cells is evaluated either by using staining techniques, which provide information about the integrity of the cell membrane (e.g., methylene blue staining for yeast cells, or using commercially available staining kits), or by determining the number of colony forming units (CFU) through plating on agar.[10,13,15–17,35,64]

Another method to assess the toxicity of a solvent on a microorganism is monitoring its growth at a given concentration of the solvent.[1,8,9,14,65,66] Comparing the resulting growth rate with the corresponding observations in purely aqueous reference conditions gives a relative measure of the tolerance of the cells toward the solvent. The samples are analyzed in terms of their cell concentrations, either by optical density (OD) measurements, or by counting the CFUs by plating of the volume taken at each point of time on agar. This method is slightly more time-intensive, as more than one sample has to be analyzed in order to determine the growth rate.

A very similar technique consists of evaluating the viability of the cells during or after the exposure to the ionic liquid by indirect means of consumption or production of compounds signaling cellular activity. Usually, the consumption of a nutriment such as glucose is chosen as an indicator,[4,22] or the production of a typical metabolite of the particular organism investigated, for example, lactic acid for lactobacilli strains.[2]

However, sometimes the activity of the cells after incubation is also determined using any compound accepted as substrate by the biocatalyst.[67]

Finally, a standard method in (eco)toxicity studies that has also been applied to microorganisms used as biocatalysts in whole-cell applications is the determination of the minimal inhibitory concentration (MIC), the minimal bactericidal concentration (MBC), or the half maximal effective concentration (EC_{50}).[18,20,24,60,68,69] These values are determined by registering the concentrations of a given ionic liquid at which no growth of the organism considered is observed or at which the number of CFUs is reduced by 50%. The results give very exact information on an absolute scale and are therefore very valuable information. In comparison to the other methods, which always give only relative information, knowledge of these values makes comparisons between different studies seemingly straightforward. Still, the data should be considered with care, as the exact proceeding must be identical to allow reliable comparisons.

7.2.2 Tolerance

7.2.2.1 Composition of the Ionic Liquid and Organism Type. As described earlier in this chapter, establishing general guidelines concerning ionic liquids is generally difficult due to the paucity of existing data. Concerning biocompatibility, this is made even more difficult by the large number of different possible ionic liquids and microorganisms. Additionally, the very diverse proceedings applied to evaluate the cells' tolerance toward the solvent make reliable comparisons almost impossible. The proceeding during the experimental evaluation significantly influences the outcome of the biocompatibility tests and would have to be very similar from one case to another to even allow only qualitative comparisons. Thus, there is only one consistent observation found in almost each of the biocompatibility studies: the side-chain effect, that is, increasing toxicity of the ionic liquid with increasing length of the alkyl chain on the cation. This effect was observed very clearly in the work of Lee et al.[24] The authors determined EC_{50} values for *Escherichia coli* with several 1-akly-3-methylimidazolium ionic liquids. Different counter-ions were used ([NTF], [PF_6], [BF_4], [$MeSO_4$], and [SbF_6]) and the alkyl chains were varied in length from ethyl up to octyl substituents. In each of these sets of homologue ionic liquids, a significant decrease of EC_{50} was observed with increasing chain length, signaling lower tolerance of ionic liquids with longer chains compared with solvents with shorter chains, independent of the type of anion. The only exception was the NTF containing set of solvents, which generally showed the lowest biocompatibility. Here EC_{50} values were as low as 150 mg/L for both tested variants ([BMIM][NTF] and [HMIM][NTF]). The best tolerated ionic liquids were [EMIM][BF_4] and [EMIM][CF_3SO_3], with EC_{50} values of 35,000 (±5000) mg/L and 20,000 (±10,000) mg/L, respectively.

Docherty and Kulpa reported biocompatibility tests with six different bromide ionic liquids containing either an 1-alkyl-3-methylimidazolium or an 1-alkyl-3-methylpyridinium cation.[1] The alkyl chains included here were butyl, hexyl, and octyl chains. The biocompatibility evaluations were performed with five different organisms, chosen specifically to cover a large variety of physiological and respiratory capabilities ranging from aerobe to facultative anaerobe strains, and including both gram-positive and gram-negative cell types: *Escherichia coli*, *Staphylococcus aureus*, *Bacillus subtilis*, *Pseudomonas fluorescens*, and *Saccharomyces cerevisiae*. In these experiments, the side-chain effect was observed for each of the ionic liquids and each of the microorganisms. The underlying phenomenon must thus be due to a common cellular structure or process.

The side-chain effect was also observed during an extensive set of studies concerning the toxicity of ammonium-based and phosphonium-based ionic liquids as well as pyridinium and benzimidazolium chlorides.[18,60,68,69] The experiments were performed on a large number of organisms, including several rods, cocci, and fungi, as well as one *Bacillus* strain. Throughout all the different combinations of ionic liquids with a microorganism, the toxicity increased with increasing alkyl chain length. The side-chain effect thus seems to be a truly general phenomenon, largely independent of the cation head group, the counter-ion used, and the organism involved. Other examples for this observation are found in the works by Matsumoto et al., Lou et al., Wang et al., and Yang et al.,[2,4,10,15,64] among others. This effect is strongly supposed to be due to the lipophilic interaction of the cation with the cell membrane. Due to its amphiphilic nature, the cation can insert partially into the membrane and provoke its local disruption. Hence, the longer the alkyl chains, the more lipophilic the cation, and the more toxic the ionic liquid, because the described interaction is more intense. This hypothesis is confirmed by the fact that the introduction of hydrophilic groups, such as hydroxyl or ether function, into the side chain on the cation decreases the overall toxicity of the ionic liquid.[10,64]

The influence of the cation head group on the biocompatibility of the ionic liquid has not been analyzed in detail. Only Docherty and Kulpa used a series of homologue ionic liquids solely differing by their respective head groups (pyridinium or imidazolium).[1] Differences were observed in some cases, but the observed trends were consistent neither within a given type of organisms nor within a set of homologue ionic liquids. Based on these results, it might be speculated that the cation head group is a factor of lesser importance for the biocompatibility of the solvent. Still, to support that assumption more data are clearly needed.

Concerning the anion, more data are available, as several authors investigated ionic liquids with the same cation but varying anions. A first trend frequently observed is that [PF_6] seems to be relatively biocompatible, at least in combination with [BMIM]. Ganske and Bornscheuer compared the effect of [BMIM][PF_6] and [BMIM][BF_4] on *E. coli*, *Pichia pastoris*, and *Bacillus cereus*.[8] [BMIM][BF_4] severely affected the growth of all three microorganisms tested starting at concentrations of 1% (v/v). The effect was strongest on *E. coli*, with an MIC of 0.7–1% (v/v) in the presence of the [BF_4] containing solvent. At concentrations of 4% (v/v) the growth of the other two microorganisms was also stopped. The tolerance of *E. coli* toward [BMIM][PF_6] was lower than for the [BF_4] containing ionic liquid, with an MIC determined at 0.3–0.7% (v/v). *B. cereus*, in contrast, tolerated [BMIM][PF_6] significantly better than the previous solvent, and *P. pastoris* was not affected by this ionic liquid at all, even at concentrations as large as 10% (v/v). Another yeast strain, *S. cerevisiae*, confirmed the good stability of this cell type toward [BMIM][PF_6].[9] At concentrations of 3% (v/v), its growth rate was only marginally lowered as compared with the reference assay without ionic liquid. However, as observed above, [BMIM][BF_4] was more toxic and significantly reduced the growth rate. For *Aureobasidium pullullans*, [BMIM][PF_6] was also less toxic than the corresponding ionic liquid with [BF_4].[22] It seems thus that [PF_6] is generally less toxic, at least in combination with this cation, than [BF_4]. However, the different water solubilities of both solvents might also interfere in the toxicity observed: While [BMIM][PF_6] is not miscible with water, [BMIM][BF_4] is completely miscible.

Similarly to the previous observation, a detrimental effect of [BF_4] was also observed by Lou et al.[10] Here, water-miscible ionic liquids were analyzed concerning their toxicity toward *Rhodotorula* sp. AS2.2241. After incubation of 1 hour only, the cell viability in the presence of 1-alkyl-3-methylimidazolium ionic liquids with [BF_4] had decreased to 23% of its initial value, and after 24 hours, 100% of the cells were dead. In contrast, the assay with the corresponding ionic liquids with [Br], [NO_3], or [Cl] anions still showed >80% cell viability after 24 hours.

Other ionic liquids showing relatively bad biocompatibility toward *Rhodotorula* sp. AS2.2241 and *Trigonopsis variabilis* were those with [CF_3SO_3] anions.[10,64] After 24 hours incubation with this ionic liquid at 10% (v/v), the cell viability had decreased to 40% of its initial value. Trihexyltetradecylphosphonium and methyltrioctylammonium chlorides were more toxic toward *E. coli* than the corresponding solvents with [NTF] anions.[14] The latter seemed to present relatively good biocompatibility, as the cell growth was almost not affected in comparison to the purely aqueous reference. The same observation was made by Pfruender et al.: The [NTF] containing ionic liquids [BMIM][NTF] and [Oc_3MeN][NTF] did not have any adverse effect on either *E. coli*, *Lactobacillus kefir*, or *S. cerevisiae*.[16,17] However, in comparison to [PF_6], Wang et al. observed slightly lower tolerance of [NTF] containing ionic liquids by *Rhodotorula* sp. AS2.2241.[15] This was also observed for *E. coli* by Bräutigam et al.[13] Here, the observed viabilities were somewhat lower than those observed in the work of Pfruender et al. When incubated over the same duration (5 hours) at ionic liquid concentrations of 20% (v/v), [BMIM][PF_6] affected *E. coli* more than shown above (relative membrane integrity <70%). Similarly, when comparing ionic liquids with [NTF] with those with [PF_6], the former induced a significant decrease in the relative membrane integrities of approximately 30%. However, the worst ionic liquids in terms of biocompatibility were those containing [E_3FAP] anions. The membrane integrities determined after 5 hours' incubation at 20% (v/v) were <10% for all the cation–[E_3FAP] combinations tested.

The microorganism investigated is also a factor of influence when evaluating the biocompatibility toward a given ionic liquid. Many different cell types have been analyzed in different studies comparing the tolerance of these organisms toward a given ionic liquid, including a series of common rods, fungi, cocci, and bacilli, as well as xenobiotic degrading bacteria.[2,8,15–20,35,64,66,68,69] In these works, it was repeatedly observed that different organisms show different tolerance of given solvents. Still, even when the range of tested organisms was specifically chosen to represent a large variety of physiological and respiratory capabilities as was done by Docherty and Kulpa, no consistent trend could be derived indicating that specific physical properties confer especially good tolerance of a given type of ionic liquids.[1] Based on the currently available data, it seems that biocompatibility would have to be evaluated for each cell type used. Careful extrapolations might only be possible for extremely similar microorganisms.

The observations described above are the trends that seem to be generally valid, but only on a relative scale. It should be kept in mind that these have been assembled on the basis of only a relatively small amount of data and that they include only very few of the different possible organisms, cations, and anions. It is not ruled out that these might no longer be applicable in very specific cases of, for example, an unfortunate combination of two nontoxic or only moderately toxic entities producing a relatively harmful solvent or of a moderately toxic ionic liquid with an exceptionally sensitive organism. *B. subtilis*, for example, was much more affected by [HMPyr][Br], [OMPyr][Br], and [OMIM][Br] than *E. coli* or *S. cerevisiae*,[1] and the relative difference in toxicity observed for [Cl] and [NTF] anions was by far larger when combined with [Oc_3MeN] than with [$P_{6,6,6,14}$].[14]

7.2.2.2 Other Factors of Influence. In addition to the factors described above, the concentration of the ionic liquid in the medium and the duration of exposure are factors of major influence when evaluating how well an organism tolerates a given solvent. Yang et al. analyzed the latter effect for immobilized baker's yeast cells with several 1-alkyl-3-methylimidazolium ionic liquids.[4] The toxicity of the solvents increased, unsurprisingly, with the exposure duration and when repeatedly acting on the encapsulated cells. Interestingly, when in the presence of ionic liquid without interruption, the decrease of activity with time was more important than when the cells were repeatedly exposed to it over the same total duration. This could be due to partial recovery of the cells during the intermission periods.

Then, not only the organism type is a determining factor, but its appearance is also of importance. Encapsulation in alginate beads was shown not only to be favorable from the point of view of reaction design—facilitating the recycling of the biocatalyst and the process workup—but also concerning biocompatibility: The immobilized *S. cerevisiae* cells were much less sensitive to a range of 1-alkylimidazolium ionic liquids than the corresponding free cells.[4]

The water content of the ionic liquid was also shown to have an impact on the viability of the cells in the solvent.[4] When using anhydrous [BMIM][PF_6] the relative activity retention of immobilized baker's yeast after 12 hours incubation was only ~50%. However, after 12 hours' incubation in water-saturated ionic liquid or in a biphasic ionic liquid–water system, the relative activity retention was still ~70%.

A property that might also influence the biocompatibility of the solvent is not only the (non)toxic effect intrinsically due to the composition of the ionic liquid, but also its water miscibility. Lee et al. compared two hydrophobic ionic liquids ([BMIM][PF_6] and [OMIM][$MeSO_4$]) and two hydrophilic ionic liquids ([EMIM][BF_4] and [EMIM][$MeSO_4$]) and their effect on solid and suspension cultures of a recombinant *E. coli* (DE3) strain.[24] It was shown that hydrophobicity had opposite effects depending on the type of cultures performed: Whereas the hydrophilic ionic liquids tested here showed lower toxicity in a suspension culture, the hydrophobic ionic liquids added were comparatively biocompatible with cells grown on solid media. This observation was explained by the different diffusivity of both types of ionic liquids in the hydrophilic solid media. Thus, the water miscibility of the solvent does determine how intense the contact of the microorganism with the solvent for a given culture mode will be, and if the toxic effect is predominantly due to molecular or phase toxicity. Nevertheless, this property is probably only of moderate influence, and the observed toxicity will still be largely dependent on the factors mentioned above, that is, anion and cation structure, as well as the organism type. In fact, when comparing the effect of [BMIM][PF_6] (hydrophobic) and [BMIM][BF_4] (hydrophilic) on a suspension culture, the trend observed was the same as that described above for *E. coli*, but opposite for *P. pastoris* and *B. cereus*, which tolerated the hydrophobic ionic liquid better.[8]

Finally, it should be noted that ionic liquids can be indirectly harmful to the biocatalyst by provoking the precipitation of the culture medium. One such case was reported with the ionic liquid AMMOENG™ 100 added to the growth medium.[66] The authors related the lack of growth in this system to the consequential shortage of essential nutriments for the cells. They also indicated that this phenomenon had also been observed with other ionic liquids and that it depended on the type of culture medium used.

7.2.2.3 Comparison with Organic Solvents. Often presented as a favorable alternative to organic solvents, the use of ionic liquids is frequently compared with this class of solvents, especially in terms of biocompatibility. In the pioneering work presented by Cull et al., the biocatalyst (*Rhodococcus* R312) displayed higher activity in [BMIM][PF_6]–water systems than in toluene–water systems.[67] Toluene thus seems to harm *Rhodococcus* R312 more than [BMIM][PF_6]. The biocompatibility studies performed by Pfruender et al. showed significantly higher toxicity for the organic solvents MTBE, diisopropylether, *n*-octanol, and *n*-decanol than for the ionic liquids ([BMIM][PF_6], [BMIM][NTF], and [Oc_3MeN][NTF]) tested.[16,17] Only *n*-decane performed relatively well in comparison to the ionic liquids. The viability retention was only slightly lower with this organic solvent than in the assays with ionic liquid. The relatively good biocompatibility of [BMIM][PF_6] was confirmed by Zhang et al., where the activity of *Aureobasidium pullulans* was significantly more affected by *n*-hexane and dibutylphthalate than by the former.[22] Other ionic liquids, however, reduced the relative activity retention to the same degree as the organic solvents. While the above works presented [BMIM][PF_6] as a relatively benign solvent in terms of biocompatibility, it performed less well than tetradecane with *Sporomusa termitida*.[70] The good biocompatibility of tetradecane in comparison to some ionic liquids was confirmed by the observations of Cornmell et al.[65] Like Cull et al., Hussain et al. found that all the ionic liquids tested ([EMIM][TOS], [BMIM][PF_6], [Oc_3MeN][NTF] [BMIM][BF_4], [BMIM][$OcSO_4$], [BMIM][$MDEGSO_4$], and [CABHEM][$MeSO_4$]) showed better biocompatibility than toluene, although the ionic liquids were added at larger concentrations than the organic solvent (50% [v/v] vs. only 10% [v/v]).[35,67]

In view of these observations, it seems thus that the conclusions reached depend largely on the ionic liquid and on the organic solvent chosen for the comparison, and to some degree on the organism. This was nicely illustrated by the work of Matsumoto et al. and Lee et al.[2,24]

When evaluating the biocompatibility of an ionic liquid, it should however be kept in mind that sometimes cell growth was reported to be more sensitive to solvents with low biocompatibility than the desired biotransformation that was to take place in the reaction media.[65,66] In fact, growth inhibitory ionic liquids can still benefit the reaction, as it was the case in several of the presented works. Generally, only moderately biocompatible ionic liquids should therefore not be directly excluded from applications in whole-cell biocatalysis, but be included with the nontoxic solvents into the screening procedure for their use in biotransformations. This also holds true for cases in which the tested organic solvent was seemingly more biocompatible than the ionic liquid: A possibly favorable effect of the ionic liquid on the biotransformation in spite of its reduced biocompatibility in comparison to some organic counterparts—in addition to the numerous advantages that these materials present in comparison to organic solvents—should still be sufficiently convincing to consider the application of the ionic liquid in whole-cell biocatalysis.

7.2.3 Interaction Mechanism

The mechanisms of action of ionic liquids on organisms are not yet fully elucidated. When organic solvents are concerned, there is a common differentiation between molecular toxicity and phase toxicity. Molecular toxicity is described as the toxicity caused by solvent molecules dissolved in the aqueous phase. It might result from enzyme inhibition, protein denaturation, and/or modifications in the membrane composition due to accumulation of solvent in the membrane.[71] Phase toxicity, on the other hand, is caused by the presence of a second phase and due to the direct contact between the cells and the solvent. This differentiation should also be applicable to ionic liquids, although no published work separately investigating either effect was found.

Common conclusions of experiments evaluating the biocompatibility of organic solvents are that solvents having a log $P < 3$–4 show severe toxicity toward most cells because they are polar solvents and therefore partition easily into the aqueous phase and from there into the cell membrane.[71–73] These guidelines might not be valid for ionic liquids, as the concepts of hydrophobicity and polarity cannot be applied to ionic liquids in the same way as to organic solvents.[2] Hydrophobic organic solvents have relatively large log P and are nonpolar, whereas hydrophobic ionic liquids might not be miscible to an aqueous phase and still show a polarity similar to that of short chain primary and secondary alcohols or secondary amides.[2,30]

However, effects observed on microorganisms due to interaction or contact with organic solvents might also appear when bringing the cells into contact with ionic liquids. These include accumulation of hydrophobic components of the solvent in the cytoplasmic membrane causing an increase of the surface and the thickness of the membrane, changes in the fatty acid composition, modification of the microviscosity, and/or damage of membrane structures or disintegration of the membrane.[74,75] As a consequence, cell functions may be destroyed or inhibited, and adverse effects on passive and active membrane transport systems may occur, harming the cell physiology. However, adaptation mechanisms to such changes have been observed to counteract these modifications. They include increases in unsaturated fatty acids resulting in a reduction of the membrane permeability or *cis–trans* conversions changing the fatty acid composition of the membrane in order to decrease the partition coefficient between the outside environment and the cell membrane. To date, it is neither confirmed nor ruled out that ionic liquids might provoke similar phenomena when acting on cells.

7.2.3.1 Effect on the Cell Membrane. The cell membrane is supposed to be the primary target of solvent toxicity[16] and interaction of ionic liquids with the phospholipid bilayer has been confirmed, as well as increased permeability of the membrane when the biocatalyst is exposed to these solvents.[10,13,14] To gain more information about this interaction, membrane-similar structures have been used to simulate the effect of ionic liquids on the cell membrane.[3,5,6] Evans et al. employed supported phospholipid bilayers (SPB) formed by 1,2-dielaidoylphosphocholine (DEPC), and 1,2-dimyristoylphosphoglycerol (DMPG) on different supports and large unilamellar vesicles (LUV) of DEPC. When exposing LUVs to different cations found in ionic liquids ([BMIM], [HMIM], [OMIM]), a clear increase in leakage from the vesicles and in the rate of leakage was observed with increasing alkyl chain length and with increasing cation concentration. Up to 500 mM, [BMIM] only produced minimal leakage (<6%), similar to the leakage provoked by NaCl in the same concentrations. At larger concentrations (1000 mM and 1500 mM), leakage and leakage rate increased, however without indications of bilayer disruption. It is supposed that the ionic liquid created small hole-like defects by inserting into the bilayer. In contrast, most vesicles were disrupted in the presence of [HMIM] at concentrations of 500 mM or higher, and 1500 mM [OMIM] even resulted in nearly total bilayer disruption. The cations of some ionic liquids clearly act as short chain surfactants, potentially disintegrating a phospholipid bilayer such as a cell membrane. The effect of different anions commonly found in ionic liquids was also tested on the bilayer. [BF_4] and [PF_6] did not induce any leakage, suggesting that these anions do not interact with the lipid bilayer. [NTF] on the other hand showed small and unstable defects in the lipid bilayer at concentrations of 500 mM.

Analyses of the effects of [OMIM], [NTF], and [BMPL][NTF] on silica-supported phospholipid bilayers formed of DEPC confirmed the previous results. [NTF] was shown to leave large stable defects with a pore-like interior in the bilayer at some positions, but leaving the surrounding bilayer intact. After exposure to [OMIM], the SPB presented increased roughness, and it could even be shown that this cation removes lipids from the bilayer and possibly redeposits them with entrapped water molecules. When exposed to [BMPL][NTF], a thin layer of adsorbed ionic liquid formed on the SPB, confirming the strong interaction with the cell membrane. Lipid removal could not be ruled out either in this case.

Cromie et al. approached the subject by making use of molecular dynamics simulations.[7] Based on an empirical force field model, they investigated the interaction between [BMIM][Cl] and [BMIM][NTF], respectively, with a neutral lipid bilayer made of cholesterol molecules. The latter should simulate the reaction of the cell membrane toward the ionic liquid exposure, even though it is conceded that the mechanisms of interaction for the neutral cholesterol might be slightly different from those for the ionic or zwitterionic phospholipids composing the cell membrane. Still, the results obtained by the simulations are consistent with the previously cited experiments involving phospholipids. The simulations performed with [BMIM][Cl] showed a partial incorporation of the butyl chain into the cholesterol bilayer, even though the good solubility of [Cl] in water limited this tendency. In contrast, [BMIM][NTF] almost exclusively accumulated at the water–cholesterol bilayer interface, giving rise to a nanometer-thick deposit of ionic liquid. Moreover, the anion here was found to also penetrate into the bilayer—even more than the cation—which is thought to be due to its higher lipophilicity in comparison to [Cl]. The authors finally pointed out that, even if the incorporation of the different ion remains quantitatively low (on average, ~0.2 ions per square nanometer of bilayer), this might still affect cell membrane properties such as surface tension, bending rigidity, and permeability.

7.2.3.2 Accumulation inside the Cell. Only few groups reported detection of ionic liquid accumulating inside the biocatalyst or on its surface. Gorman–Lewis and Fein determined the adsorption of [BMIM][Cl] to *B. subtilis* by UV-Vis spectroscopy at 211 nm.[76] The amount of adsorbed ionic liquid was determined by comparing the known initial ionic liquid concentration with the content found in solution after incubation and filtration of the cells. These measurements did not show any difference between both values, and the authors thus concluded that [BMIM][Cl] has no affinity to the surface of this gram-positive bacteria. It should be pointed out, however, that these experiments were performed under "realistic environmental conditions," employing concentrations of ionic liquid of less than 1 mM, that is, amounts much lower than those present, for example, in biphasic whole-cell biocatalyses.

Cornmell et al. relied on Fourier transform infrared (FT-IR) spectroscopy to analyze the chemical composition of *E. coli* K12 MG1655 cells after exposure to ionic liquids. The cells were incubated with four different ionic liquids: the relatively biocompatible [$P_{6,6,6,14}$][NTF] and [Oc_3MeN][NTF], and their respective supposedly toxic chloride salts, [$P_{6,6,6,14}$][Cl] and [Oc_3MeN][Cl].[14] The cells were added into assays composed of fresh medium containing 23% (v/v) ionic liquid. Within 24 hours, samples were taken for FT-IR analyses. These analyses showed that for all ionic liquids under investigation, both the cation and the anion—except for [Cl] not giving an FT-IR signal and thus not being detected—were found inside the cells or adsorbed so strongly to the surface of the cells that the intense washing before the analyses could not remove it. Over 24 hours, the toxic ionic liquids accumulated more rapidly in the cells than the more biocompatible ones. To localize where the ionic liquids accumulated exactly, cells incubated with [$P_{6,6,6,14}$][NTF] were fractionated into cytoplasmic and membrane fractions. Analysis by FT-IR revealed the presence of this ionic liquid only in the membrane fraction and not in the cytoplasmic fraction, indicating that it is sufficiently lipophilic to accumulate there.

Applying principal component analysis (PCA) and discriminant function analysis (DFA), the FT-IR analyses were also used to establish and compare the metabolic "fingerprints" of the different cell samples. It was hoped that using statistical analysis methods, differences in the spectra recorded could be attributed to differences in cell physiology. These analyses showed that the cells incubated with [Oc_3MeN][NTF] showed most similarities to the reference sample not exposed to any ionic liquid, suggesting that this ionic liquid induced the least perturbation of the cellular chemistry.

Lou et al. employed fluorescence microscopy to evaluate if ionic liquid had been taken up by *Rhodotorula* sp. AS2.2241.[10] This was possible because the ionic liquids used were dialkylimidazolium-based, which are fluorescent at an excitation wavelength of 320–420 nm. The cells were incubated for 24 hours in Tris-HCl buffer containing 10% (v/v) hydrophilic ionic liquid, and then washed with phosphate buffered saline (PBS) to remove ionic liquid adsorbed to the cell surface before analysis. The observations under the microscope clearly showed uptake of ionic liquid by the cells. However, by this means only, it was not possible to determine where the ionic liquid included was exactly localized. The authors acknowledge that this point, as well as how the uptake of ionic liquid influences the viability and the biocatalytic activity of the cell, needs further investigation.

Once inside the cell, ionic liquids will interact with any component present inside it. To date, not much work has investigated the underlying mechanisms of how ionic liquids affect the cell and the structures inside it after solvent molecules have crossed the protecting membrane. Concerning more precisely the effect on the enzymes inside the cell, further information might be found in the domain of enzymatic catalysis. In this area of catalysis, ionic liquids have also been applied, and the influence of ionic liquids on a given enzyme is a subject of major interest, which has received much attention in recent years. For details about this subject, please refer to Chapters 2, 5, and 6.

7.3 STATE OF THE ART

In 2000, Cull et al. were the first to publish the application of a room-temperature ionic liquid in whole-cell biocatalysis. The reaction considered was the reduction of 1,3-dicyanobenzene to 3-cyanobenzamide using *Rhodococcus* R312 (Scheme 7.2).[67]

This transformation is of industrial relevance for the production of acrylamide and is limited by the low water solubility of the substrate. To overcome this limitation, a biphasic reaction mode was chosen (20% v/v solvent), and the ionic liquid [BMIM][PF_6] was used to avoid the standard implication of an organic solvent for such applications (here, toluene). The initial reaction rate was lower in the reaction setup involving the ionic liquid, but the product concentration after 20 minutes was larger than in the organic reference. When evaluating the activity of the cells after incubation in water–toluene and water–[BMIM][PF_6] assays, it was found that *Rhodococcus* R312 tolerated

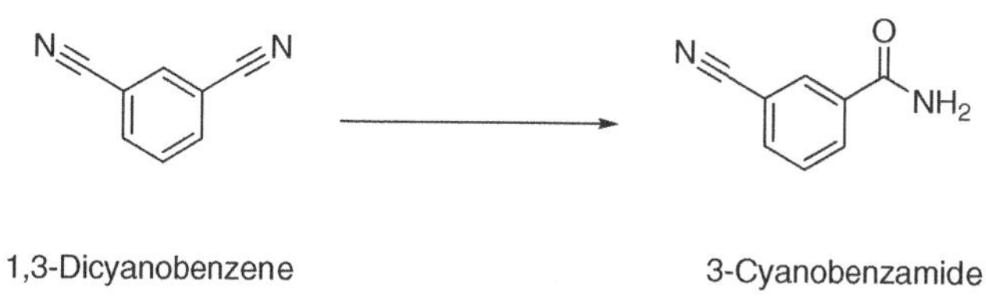

Scheme 7.2. Asymmetric reduction of 1,3 dicyanobenzene by *Rhodococcus* R312.

the ionic liquid better than the organic solvent. The decreased initial reaction rate could thus be attributed to mass transfer limitations due to the higher viscosity of the ionic liquid in comparison to toluene, and not to a relative loss of activity of the biocatalyst through interaction with [BMIM][PF_6]. This was the first demonstration of a successful application of ionic liquids in whole-cell biocatalysis.

Since then, several research groups have entered this promising field, and the number of applications of ionic liquids in whole-cell biocatalysis showed an impressive increase over the last decade. Most of the work involves biphasic reaction systems formed by a hydrophobic ionic liquid with an aqueous phase, as used by Cull et al. In fact, biphasic reaction systems present certain advantages over monophasic reaction modes, such as facilitated product isolation and recovery of the product and the ionic liquid, which make their use more attractive. Nevertheless, some applications also involve hydrophilic ionic liquids as a co-solvent with the aqueous phase.

Most of the reactions studied in whole-cell biocatalysis involving ionic liquids are asymmetric reductions of ketones by oxidoreductases for the production of chiral alcohols. Indeed, this class of enzymes takes particular advantage of the whole-cell approach due to its need of cofactors for performing the reduction. Oxidoreductases are also one of the enzyme classes most used in industry, and thus the subject of ongoing research (see also Chapter 6). Their large substrate spectrum involves toxic and low water soluble compounds, making biphasic reaction modes particularly attractive. In contrast, hydrolases, for example, the largest enzyme class involved in industrial applications, do not appear as frequently in whole-cell biocatalysis, since they do not need cofactors (see Chapter 5). Hence, they do not profit from the cell-intern cofactor regeneration as other enzymes would.

Still, applications involving ionic liquids are getting more and more diverse, already also including for example, deracemization of racemic alcohol mixtures,[77] hydrogenation of C–C double bonds,[70] degradation of xenobiotics,[19] or even biodiesel fuel production.[78] The cells used as biocatalysts cover a great diversity of microorganisms, ranging from the very commonly used (recombinant) *E. coli* and many different yeast species, to cocci, bacilli, or even strictly anaerobe organisms. In the following sections, an excerpt of the most important and interesting applications will be described.

7.3.1 Asymmetric Reductions by Whole Cells in Ionic Liquids

Following the first results published by Cull et al.,[67] Howarth et al. presented the reduction of a range of aliphatic and cyclic ketones to the corresponding alcohols in water–[BMIM][PF_6] mixtures (1:10) using immobilized baker's yeast (Scheme 7.3).[79]

The *S. cerevisiae* cells were immobilized in alginate beads in order to facilitate the workup procedure at the end of the transformation. The authors compared the yields and enantiomeric excesses observed here with literature data from biotransformations using organic solvents as alternative media. For most of the reactions, the resulting yields were larger in the reaction systems with [BMIM][PF_6] than in organic solvents by up to 20%, except for the reduction of ethyl acetoacetate (Scheme 7.3e) (70% vs. 78% with petroleum ether) and pentane-2,4-dione (Scheme 7.3d), which led to a significantly lower yield (22% vs. 90% in aqueous medium). Enantiomeric excesses were

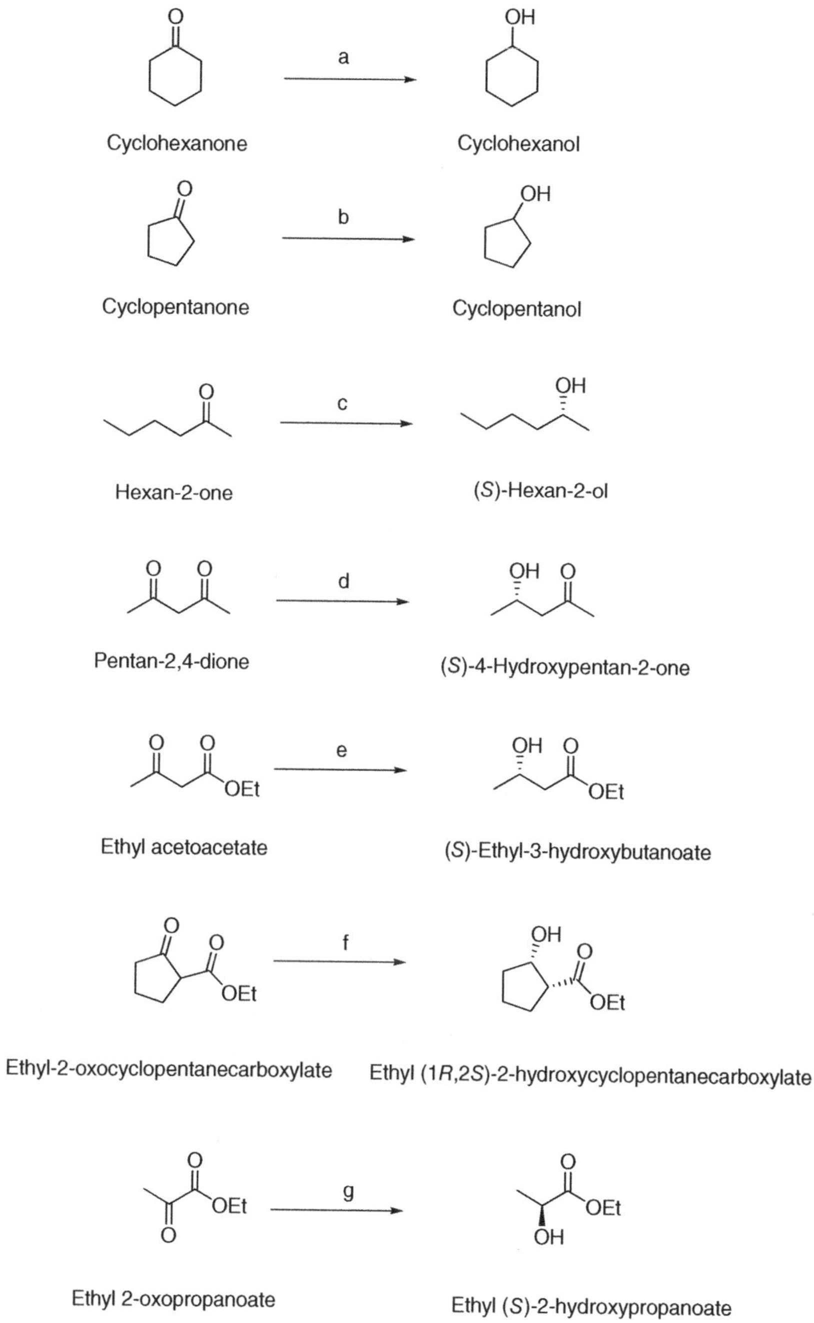

Scheme 7.3. Asymmetric reductions of aliphatic and cyclic ketones by immobilized *Saccharomyces cerevisiae*.

Scheme 7.4. Asymmetric reduction of 4-chloro-acetophenone by *Lactobacillus kefir* DSM20587.

generally lower by up to 15%, except for the two previously mentioned reactions, which showed similar and larger enantioselectivity, respectively. The reduction of 4-methoxyacetophenone and 4-bromoacetophenone in this setup was not successful. No product could be recovered for either of these transformations. Wolfson et al. later showed that the reduction of ethyl acetoacetate in very similar reaction setups was more efficient in water or glycerol media, leading to 100% (after 20 hours) and 99% (after 48 hours) conversions, respectively, both with enantiomeric excesses of 99%.[80]

Pfruender et al. analyzed the biotransformation of 4-chloroacetophenone to (*R*)-1-(4-chlorophenyl)ethanol with the gram-positive *Lactobacillus kefir* DSM20587 in biphasic ionic liquid–water systems (Scheme 7.4).[16]

The aim of this study was to demonstrate the possibility of increasing the yield of such asymmetric reductions by using hydrophobic ionic liquids instead of monophasic aqueous systems or biphasic reaction systems with organic solvents. The ionic liquids were hydrophobic solvents ([BMIM][NTF], [BMIM][PF_6], and [Oc_3MeN][NTF]), chosen on the basis of the results previously presented,[67,79] showing that whole cells could be active in such systems. First experiments were conducted in milliliter-scale assays composed of 20% (v/v) ionic liquid in phosphate buffer. The yields observed in the reaction systems with ionic liquid were twice as large as those reached by the purely aqueous reference system (88–93% vs. 46%), whereas the biphasic assay with MTBE led to a yield of only 4%, confirming the high toxicity of this organic solvent toward the biocatalyst.[16] All the ionic liquids tested led to very similar yields and enantiomeric excesses (ee > 99%), with [BMIM][NTF] performing slightly better than the other two ionic liquids (93% vs. 88% conversion). This same reaction setup was then used for a scale-up to a volume of 200 mL, carried out in a stirred tank reactor.[81] Here, the volumetric productivity was 20.4 g/L and a final product concentration of 81.6 g/L (*R*)-1-(4-chlorophenyl)ethanol was reached. Concerning the process workup, no emulsification of both phases with the cells was observed. The phase separation could thus be performed easily by centrifugation or sedimentation. Additionally, the absence of by-product formation and accumulation in the ionic liquid phase would make the recycling of this phase possible, an important characteristic for future applications in industry.

In 2006, this study was expanded by the same author, including *E. coli* K12 (DSM 498) and *S. cerevisiae* FasB His6 in addition to *L. kefir* as biocatalysts.[17] The suitability of ionic liquids for applications in whole-cell biocatalysis was tested using three model reactions: the reductions of 4-chloroacetophenone, *tert*-butyl-6-chloro-3,5-

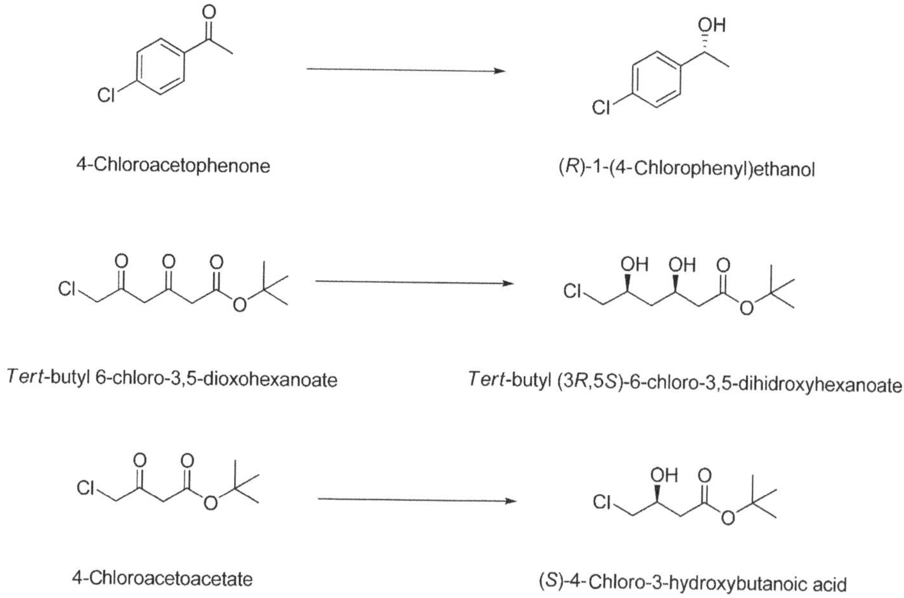

Scheme 7.5. Asymmetric reduction of ketones by *Escherichia coli* K12 (DSM 498), *Saccharomyces cerevisiae* FasB His6, and *Lactobacillus kefir* DSM20587.

dioxohexanoate, and 4-chloroacetoacetate to their corresponding alcohols (Scheme 7.5). The biotransformations were performed in the same biphasic reaction setups, using the hydrophobic ionic liquids mentioned above (20% v/v) in phosphate buffer and compared with assays involving *n*-butyl acetate (7% v/v) as second phase.

All three model reactions confirmed the beneficial effect of involving ionic liquids in the biphasic whole-cell biotransformations. The reduction of 4-chloroacetoacetate by *S. cerevisiae* showed only poor yields of 12%, 29%, and 17% for reaction systems with [BMIM][NTF], [BMIM][PF_6], and *n*-butyl acetate, respectively. However, these results could be significantly improved by increasing the biocatalyst concentration from 20 to 50 g/L. This permitted a yield of ~80% and enantiomeric excess of ~84% with 20% (v/v) [BMIM][PF_6]. In contrast, a similar reaction setup using the organic solvent *n*-butyl acetate gave only ~44% yield and ~62% enantiomeric excess for the (*S*)-alcohol. The reduction of *tert*-butyl-6-chloro-3,5-dioxohexanoate with *L. kefir* confirmed the favorable effect of the presence of a second ionic liquid phase. The biphasic reaction setup with [BMIM][NTF] led to a product yield of ~81% compared with ~48% in the monophasic aqueous reference. For each of the biphasic systems applied to the reduction of 4-chloroacetophenone with *L. kefir*, very good enantiomeric excesses (>99.5%) and yields >90% were observed, again proving the successful application of ionic liquids in whole-cell biocatalysis.

In addition to these studies, the authors pointed out that the larger viscosity of the ionic liquids in comparison to organic solvents or water could be the source of mass transfer limitations in biphasic reaction systems. To clarify the importance of this phenomenon, the initial reaction rates measured during the reduction of 4-chloroacetophenone with *L. kefir* were compared with the viscosities determined for each of the solvents added to the assays. These reaction rates, determined after 15 minutes' reaction duration, were 75 μM/s L ([BMIM][PF_6]), 62 μM/s L ([BMIM][NTF]), 55 μM/s L ([Oc_3MeN][NTF]), and 43 μM/s L (aqueous). When considering the viscosities evaluated for the water equilibrated ionic liquids—397 mPa s ([BMIM][PF_6]), 27 mPa s ([BMIM][NTF]), and 897 mPa s ([Oc_3MeN][NTF])—it becomes clear that larger viscosities do not necessarily induce lower reaction rates. It was concluded that mass transfer limitations could be excluded as possible reasons for the different reaction rates observed and that the large viscosity of the ionic liquids did not constitute an issue here.

Finally, the importance of the partition coefficients of the substrate in the biphasic reaction system was also underlined. A clear increase in the yields was registered when 100 μM cofactor (NADP) was added to the reduction of 4-chloroacetoacetate by *S. cerevisiae*. This led to the conclusion that the cell membranes were damaged during the biotransformation, a hypothesis confirmed by measuring the membrane integrity of samples taken after the reaction. The differences in the membrane integrities determined for each reaction system were consistent with the yields reached. The toxicity of the reaction systems here was, however, not inherent to the ionic liquid used, which had previously been proven not to deteriorate the cell membrane,[16] but to the different concentrations of toxic substrate found in the various aqueous phases due to different partition coefficients of the latter in the various buffer–ionic liquid system. This observation emphasizes the importance of this factor when choosing an appropriate ionic liquid.

Matsuda et al. demonstrated the importance of the presence of minimal quantities of water in whole-cell reaction systems with ionic liquids.[42] *Geotrichum candidum* IFO 5767 was used to reduce *o*-fluoroacetophenone to (*S*)-1-(*o*-fluorophenyl)ethanol in assays containing either the hydrophobic [BMIM][PF_6] or the hydrophilic [EMIM][BF_4] (Scheme 7.6).

When the dried cells were added into the pure ionic liquid, neither of the reaction systems showed conversion of the substrate after 16 hours. To improve the activity, it was decided to add water to the assays. Interestingly, the results were different for both

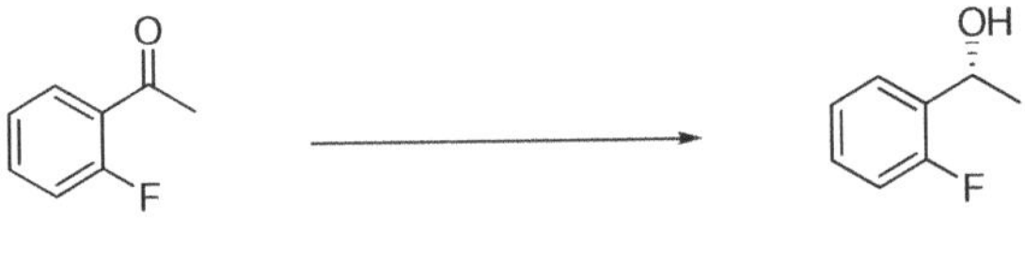

o-Fluoroacetophenone (*S*)-1-(*o*-Fluorophenyl)ethanol

Scheme 7.6. Asymmetric reduction of o-fluoroacetophenone by *Geotrichum candidum* IFO 5767.

ionic liquids tested: While the assay containing the water-immiscible ionic liquid [BMIM][PF_6] showed a dramatic increase of yield (84%), the assay with the water-miscible solvent [EMIM][BF_4] still led to only very small quantities of product (<1% yield). The authors concluded that a water layer around the cells is essential for the biocatalyst to work properly. In order to permit the formation of a water layer in the water-miscible ionic liquid, a water-absorbing polymer was used to immobilize the biocatalyst. This guaranteed the presence of a minimal quantity of water near the biocatalyst. Indeed, using this setup, the reaction system with [EMIM][BF_4] reached a very satisfying 96% yield. Additionally, even for the reaction system with water-immiscible ionic liquid, the yield increased from 84% to 92% when using the cells immobilized on the water-absorbing polymer. This observation was related to the increased surface area between the ionic liquid and the water layers. Finally, the polymer might also stabilize the pH and thereby create more favorable reaction conditions. Selectivity was constantly very satisfying, with enantiomeric excesses >99% whenever the reduction was observed.

To confirm the success of this reaction setup, other ionic liquids and substrates were tested. In addition to the two ionic liquids cited above, the following solvents were introduced: [BMIM][BF_4], [BMIM][NTF], [BMIM][CF_3SO_3], [EMIM][PF_6], [EMIM][NTF], [EMIM][CF_3SO_3], [EMIM][NO_3], [EMIM][$MeSO_4$], [EMIM][$EtSO_4$], and [OMIM][PF_6]. The assays with ionic liquids containing sulfur did not show any or hardly any reduction taking place even after 16 hours. Similarly, the ionic liquid with a nitrate anion did not permit any conversion. Solvents with [BF_4] and [PF_6] seemed most suitable (88–96% yield), whereas [NTF] led only to moderate yields (49–61%). However, the observed conversions could not be linked to the hydrophobic properties of the different solvents. The reduction of several acetophenone derivatives, as well as benzyl acetone, 2-hexanone, a β-keto ester, and a fluorinated epoxy ketone, showed that *G. candidum* immobilized on a water-absorbing polymer accepts a large variety of substrates with moderate to good activity and at excellent enantioselectivities (>99%).

Lou et al. performed an extensive study on the effect of several reaction engineering parameters on the biocatalytic production of chiral alcohols in reaction systems with ionic liquids.[9] The factors investigated were the ionic liquid content, the buffer pH, the reaction temperature, and the initial substrate concentration. As a model reaction, the transformation of acetyltrimethylsilane to (*S*)-1-trimethylsilylethanol by immobilized *S. cerevisiae* was chosen (Scheme 7.7).

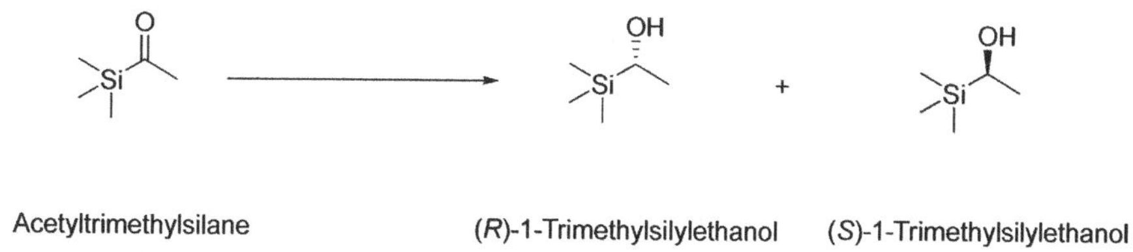

Scheme 7.7. Asymmetric reduction of acetyltrimethylsilane by immobilized *Saccharomyces cerevisiae*.

Two different ionic liquid–buffer systems were used, involving either the hydrophilic [BMIM][BF_4], added as co-solvent, or the hydrophobic [BMIM][PF_6], forming a second phase. Yield, selectivity, and reaction rate reached in the assays with ionic liquid while varying the aforementioned factors were compared with those obtained in reaction systems solely composed of buffer, or with *n*-hexane as second phase. This solvent was chosen based on previous experiments showing that this is an organic solvent well tolerated by the cells and suitable for the reaction considered.[82]

For each of the factors analyzed, variations strongly influenced the yield and the reaction rate, but mostly left the enantiomeric excess unaffected. The authors underlined the importance of the effect of pH and temperature on the biotransformation: Both parameters influence the activity and stability of all the enzymes present in the whole cell, and thereby also the cofactor regeneration, influencing the overall reaction rate. It is therefore necessary to determine those reaction conditions at which the desired enzymes show highest activity, and at which unwanted isoenzymes are least active or not active at all. The optimal temperature was 30°C for each of the systems, while the optimal pH varied from 7.3 for the biphasic ionic liquid systems, to 7.5 when the ionic liquid was used as co-solvent, to 8 for the biphasic organic system. The optimal initial substrate concentration varied largely. The use of a co-solvent or a second phase permitted larger substrate loading than in the purely aqueous reference. In the reaction systems with ionic liquid up to six times larger quantities of substrate could be added without negatively influencing the yield and the enantioselectivity, while the use of *n*-hexane allowed three times higher loading than the reference system. The reduced substrate loading when using an organic solvent in comparison to the biphasic ionic liquid–water system is explained by the different partition coefficients of the substrate in each system. Both reaction systems involving ionic liquids showed in optimized reaction conditions very good yield (>99%) after 12 hours, and excellent enantioselectivity (>99.9%). The biphasic organic reaction system performed slightly less well, with a final yield of 97.4%, however at a dramatically lower reaction rate. The reaction rate determined was only 1.45 Mm/h, slower than the reaction systems with [BMIM][PF_6] and [BMIM][BF_4] by more than a factor 40 and 50, respectively. The purely aqueous system showed a lower yield (84.1%), but was faster, with a reaction rate of 4.88 mM/h. The scale-up of the biotransformation to 154 mL was performed using the optimized [BMIM][PF_6]–buffer reaction system. High yield (95.1%) and excellent enantioselectivity (>99.9%) were reached in this setting.

The previous analysis was completed by investigating the operational stability of the immobilized biocatalyst in the different reaction setups. Best stability was observed in the [BMIM][PF_6] system, in which the biocatalyst still retained 81% of its initial activity after eight cycles of 12 hours each. The monophasic reaction system with [BMIM][BF_4] was second with 63% activity retention after these 96 hours of operation, versus only 51% for the purely aqueous setup. Again, the biphasic organic system with *n*-hexane was worst for the biocatalyst. Here, 96 hours of operation left only a residual activity of 48%. The biocatalyst might thus be recycled without severe loss of productivity over a small number of cycles in the reaction system containing ionic liquid as second phase. In addition, no emulsification of the biphasic system was observed, indicating a facilitated process workup and a possible recycling of the ionic liquid.

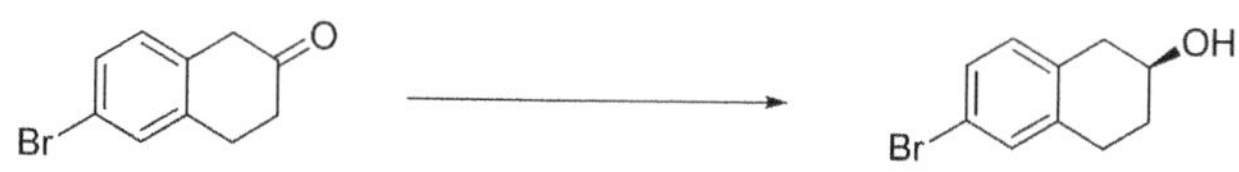

Scheme 7.8. Asymmetric reduction of 6-bromo-β-tetralone by *Trichosporon capitatum* MY1890 and *Rhodococcus erythropolis* MA7213.

Hussain et al. studied the reduction of a β-tetralone by yeast *Trichosporon capitatum* MY1890 and bacterium *Rhodococcus erythropolis* MA7213 in reaction systems involving co-solvents or a biphasic setup (Scheme 7.8).[35]

The authors evaluated some of the physical properties of the solvents thought to influence the outcome of the biotransformation, such as water miscibility, density, and viscosity, as well as biocompatibility and substrate solubility. The aim was to possibly link the observed conversions and reaction rates to these properties. Several reaction setups were compared, including water-miscible and immiscible ionic liquids (20% v/v), as well as the reference system composed of 10% (v/v) ethanol in the aqueous phase. The ionic liquids added as co-solvent to the assays were: [EMIM][TOS], [BMIM][BF_4], [BMIM][$OcSO_4$], [BMIM][$MDEGSO_4$], and [CABHEM][$MeSO_4$]. [BMIM][PF_6] and [Oc_3MeN][NTF] were introduced as second phase to the fermentation broth. The different ionic liquids were chosen to represent a large diversity of possible anions and cations. In the reference system with ethanol, *T. capitatum* showed complete conversion after 8 hours. This was only reached by one of the reaction systems including ionic liquids, namely [EMIM][TOS]. All other systems analyzed performed less well, [BMIM][$MDEGSO_4$] and [CABHEM][$MeSO_4$] reaching only 10% (w/w) and 11% (w/w) conversion, respectively. The hydrophobic ionic liquids led to intermediate yields of 40% (w/w) and 60% (w/w) for [BMIM][PF_6] and [Oc_3MeN][NTF], respectively. Their better performance in comparison to most of the hydrophilic ionic liquids was attributed to their water-immiscible nature. Indeed, this reduces the exposure of the biocatalyst to the solvents, preventing intense contact as occurring with water-miscible ionic liquids. The initial reaction rates were also determined, and these measurements followed the trends of the final conversions: Reaction systems with lower conversions also presented lower reaction rates than those with larger final conversions. The reasons for the low conversions observed are supposed to be association of the ionic liquid with the cell membrane, preventing transfer of substrate into the cell, damage of the biocatalyst through interaction with the ionic liquid, or possibly mass transfer limitations due to the larger viscosity of the solvents. The latter point, however, might only partially account for lower reaction rates, as the viscosities of the ionic liquids are dramatically lowered when even small quantities (~1% v/v) of water are added to the ionic liquid.[35] As *T. capitatum*'s growth is of filamentous nature, quantitative viability measurements in form of viable counts were not possible, and the former hypotheses could only be verified qualitatively. Selectivity is only indicated for [EMIM][TOS], which showed enantiomeric excesses of 88% (*S*), identical to the selectivity measured for the reference system with 10% v/v ethanol.

R. erythropolis generally showed much lower reduction of 6-bromo-β-tetralone. The reference assay with ethanol (10% v/v) reached only 14% conversion after 8 hours. The best performing solvent was again the hydrophilic [EMIM][TOS], with 28% conversion at significantly larger initial reaction rate than the reference system (18.7 vs. 4.7 mg (*S*)-6-bromo-β-tetralol/g_{cell} h). The use of this ionic liquid, permitting the twofold increase of productivity, is thus clearly improving the outcome of this application. One of the hydrophobic ionic liquids, [Oc_3MeN][NTF], showed fair performance in comparison to the reference, with a final conversion of 17% (w/w) and significantly larger initial reaction rate (15.0 mg (*S*)-6-bromo-β-tetralol/g_{cell} h), but all other ionic liquids led to less satisfying results in terms of final conversion.

The observations made correlated only partially with the viability measurements, indicating that the toxicity of the different solvents tested is probably one of the reasons for variations in biocatalytic efficacy, but not the only one. No predictable relationship between the physical properties evaluated and the activity data could be established. Cell viability data and substrate solubility are thus apparently not sufficient to predict the efficiency of the biotransformation. The lack of knowledge concerning the interaction of ionic liquids with the cells and the resulting effect on the reaction rate does not permit complete understanding of the mechanisms taking place. The authors concluded that screening of ionic liquids was currently still a more viable method for applications in industry than rational selection.

In 2007, Bräutigam et al. presented an extensive study including a large pool of the commonly available ionic liquids forming biphasic systems with aqueous phosphate buffer.[13] The goal of this study was to determine guidelines permitting the reduction of the number of ionic liquids that could be employed in biphasic whole-cell biocatalysis to the few most appropriate ones for a given reaction system. This should be achieved by applying a systematic procedure based on physical properties, as well as on the conversion reached in small-scale screening experiments. The authors demonstrated the applicability of this procedure using three model reactions, namely the asymmetric reduction of 4-chloroacetophenone, phenacyl chloride, and ethyl 4-chloroacetoacetate to their corresponding chiral alcohols (Scheme 7.9).

The biocatalyst used was a recombinant *E. coli* BL21 (DE3) with overexpressions for *L. brevis* alcohol dehydrogenase (LB ADH), executing the reduction of the ketone to the alcohol, and for *Mycobacterium vaccae* N10 formate dehydrogenase (MV FDH), ensuring the cell-intern cofactor regeneration by consumption of formate. First, nine solvents were chosen from the large pool of available ionic liquids based on criteria concerning their melting point (<30°C), their density (>1.2 g/cm^3), and their viscosity (<400 mm^2/s). These included [BMIM][PF_6], [HMIM][PF_6], [BMIM][NTF], [HMIM][NTF], [BMPL][NTF], [HMPL][NTF], [BMPL][E_3FAP], [HMIM][E_3FAP], and [EWTMG][E_3FAP]. To evaluate which solvent would be best suited for a given process, the ionic liquids were then rated according to their biocompatibility, the corresponding distribution coefficients of the substrate and the product in the biphasic buffer–ionic liquid systems, and the yields reached during biotransformations at the milliliter scale.

The biocompatibility of the different solvents was assessed by evaluating the membrane integrity of the cells after incubation over 5 hours in phosphate buffer with 20% (v/v) ionic liquid. The results of these experiments were discussed above (cf.

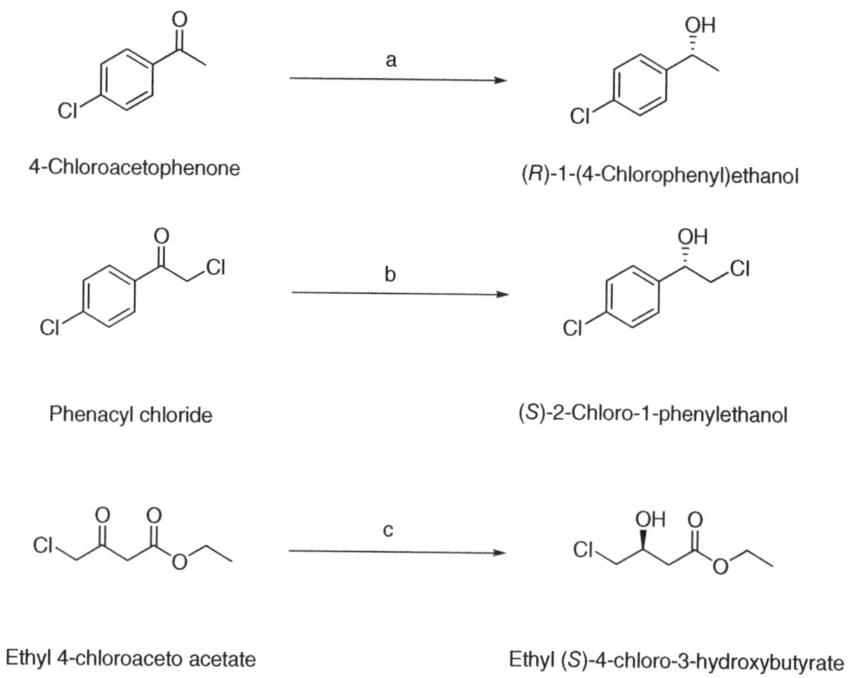

Scheme 7.9. Asymmetric reduction of ketones by *Escherichia coli*.

Section 7.2.2.1). The respective partition coefficients of the substrate and the product in the ionic liquid–buffer system were determined for each substrate/product pair. These coefficients are important because this factor indicates how much of the substances will be found in the aqueous phase near the biocatalyst at given concentrations in the ionic liquid phase. The partition coefficients in the ionic liquid–buffer system (indicated as log *D*) determined for 4-chloroacetophenone/1-(4-chlorophenyl)ethanol were between 2 and 2.4 for the product and between 2.6 and 3 for the substrate, except for the ionic liquids including [E_3FAP] anions, which showed partition coefficients lower than 2 for the product. Generally, for similar ionic liquids, the values showed a slight increase with the alkyl chain length on the cation. This observation was explained by the increasing surfactant character of ionic liquids with longer side chains than those with shorter substituents on the cation.

The biotransformations at the milliliter scale were performed with each of the different ionic liquids (20% v/v) and for each of the model reactions. The yields reached in the transformation of 4-chloroacetophenone to 1-(4-chlorophenyl)ethanol reflected the trends observed for the partition coefficients: The assays containing [E_3FAP] ionic liquids showed significantly lower yields after 1 hour than the rest of the reaction systems. Still, the yields reached in all other assays were very similar around 60%, except for [BMIM][PF_6], which gave slightly lower product concentrations. Enantiomeric excesses were very satisfying for each of the ionic liquids tested (≥99.5%). In comparison, the aqueous reference led to a yield of only 8% and an enantiomeric excess of ~96%. While the biphasic reaction mode with ionic liquids constituted a significant improvement, yields were still only moderate. This was not the case for both other reaction systems tested, including ethyl 4-chloroacetoacetate and phenacyl chloride as substrates. With the latter, a yield of 89% could be reached with [HMIM][PF_6] at an enantiomeric excess of 99.6%. The reduction of ethyl 4-chloroacetoacetate reached even larger yields (~96–99%) with each of the ionic liquids tested and also showed excellent enantioselectivity (ee > 99%). With [BMIM][PF_6], the productivity of (*S*)-α-chloro-3-hydroxybutyrate was thus increased 13-fold in comparison to the aqueous reference, to 20 g/L h.

These considerations should then all be included to determine which solvent would be most suitable for the different reactions. The results obtained also showed that the cell-intern cofactor regeneration was not negatively affected by the presence of this class of solvents and that the application of a biphasic reaction mode with ionic liquids was a success.

The systematic approach for choosing a suitable ionic liquid described above was again applied by Bräutigam et al. in 2009.[83] Here, the pool of ionic liquids screened was even larger, including the aforementioned nine ionic liquids and also the following twelve solvents: [(E2OH)MIM][NTF], [(EOE)MMO][NTF], [(MOP)MPI][NTF], [(P3OH)PYR][NTF], [(MOE)MPL][NTF], [(MOE)MPL][E_3FAP], [(NEMM)MOE][NTF], [(NEMM)MOE][E_3FAP], [(EO2E)MPL][NTF], [(EO2E)MPL][E_3FAP], [(NEMM)EO2E][NTF], and [(NEMM)EO2E][E_3FAP]. The ionic liquids tested can be divided into seven different cation classes (imidazolium, morpholinium, pyridinium, guanidinium, pyrrolidinium, piperidinium, and ammonium) (Scheme 7.10) and three different anion classes: hexafluorophosphate, bis(trifluoromethanesulfonyl)imide, and tris(pentafluorethyl)trifuorophosphate (Scheme 7.11).

It was expected that this would permit trends in the suitability of the solvents analyzed according to their composing entities to be distinguished. The reaction systems considered for the evaluation were here the asymmetric reduction of 4-chloroacetophenone

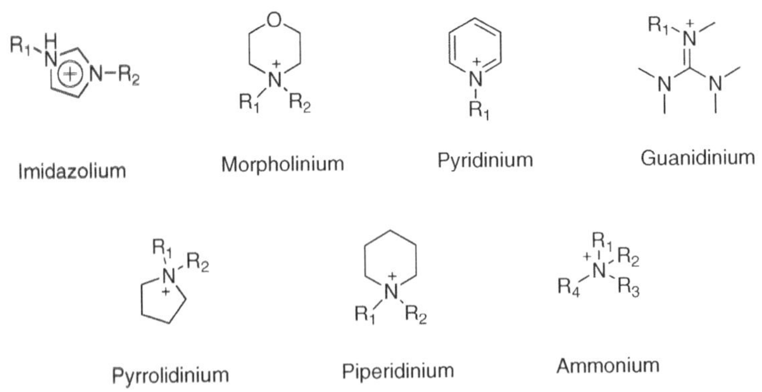

Scheme 7.10. Structures of the ionic liquid cations.

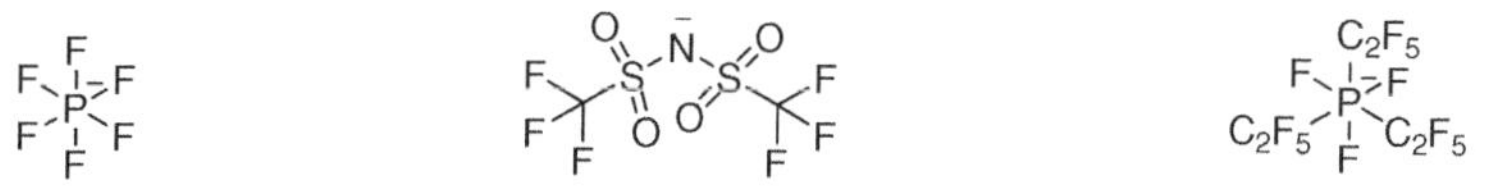

Hexafluorophosphate Bis(trifluoromethansulfonyl)imide Tris(pentafluoroethyl)trifluorophosphate

Scheme 7.11. Structures of the ionic liquid anions.

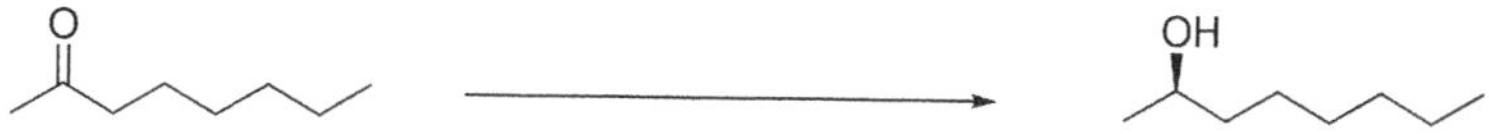

2-Octanone (*R*)-2-Octanol

Scheme 7.12. Asymmetric reduction of 2-octanone by *Escherichia coli*.

(Scheme 7.9a) or 2-octanone (Scheme 7.12), respectively, to the corresponding chiral alcohols.

The transformation was performed using a recombinant *E. coli* with overexpressions for *L. brevis* ADH, performing the asymmetric reduction, and *Candida boidinii* FDH, responsible for the regeneration of the cofactor (NADH). As previously, the partition coefficients were determined for both substrate/product pairs in the biphasic systems ionic liquid–buffer. For both reaction systems, the substrate (ketones) showed larger partition coefficients in each of the biphasic systems analyzed than the corresponding products (alcohols). As observed for 4-chloroacetophenone/1-(4-chlorophenyl)ethanol, the partition coefficients of 2-octanone/2-octanol were on average between log *D* of 2.5 and 3 for the substrate, and 2 and 2.4 for the product, although somewhat more distributed than for the former reaction system. For both substrates, [E_3FAP] anions led to larger partition coefficients than the corresponding ionic liquids with [NTF] anions. On the contrary, for 1-(4-chlorophenyl)ethanol, [PF_6] and [NTF] anions showed larger partition coefficients than [E_3FAP] anions. However, this did not hold true for 2-octanol, which distributed equally well independent of the anion. Considering these observations, it was pointed out that the choice in favor of some ionic liquids and against others should consider the partitioning of both the substrate and the product involved. If this cannot be conciliated, the final decision should be in favor of the ionic liquid leading to lower concentrations in the aqueous phase of the more toxic of the two substances.

The biotransformations at the milliliter scale again proved the advantage of using ionic liquids as second phase in comparison to the monophasic approach. For both reaction systems, only two ionic liquids ([(E2OH)MIM][NTF] and [(P3OH)PYR][NTF]) showed lower yields than the assay only containing buffer (<5% yield). These were completely unsuitable for the application considered and rejected from further consideration. Qualitatively, the same trends were observed for the production of (*R*)-2-octanol and (*R*)-1-(4-chlorophenyl)ethanol: Ionic liquids performing well in one reaction system also did in the other reaction system. All the assays showed clearly that ionic liquids containing [E_3FAP] anions lead to less favorable results in terms of yield. This confirms previous observations made with a different biocatalyst.[13] Qualitatively, however, the most favorable ionic liquids led to larger yields in the 2-octanone/2-octanol system than for the other reaction system analyzed, with yields of up to almost 80% after 1 hour of reaction. The enantiomeric excesses were >99.6% for almost all of the 21 assays.

Comparing the results observed for the milliliter biotransformations with the partition coefficients determined, it becomes clear that relatively favorable partition coefficients not necessarily lead to good yields and vice-versa. Thus, the outcome of the biotransformation can not only be related to this variable, but other parameters also have to be included, for example, biocompatibility.

Combining the results of all of the previous analyses permitted the identification of ionic liquids that seemed particularly well suited for this application. Following these evaluations, a scale-up of both reductions was performed (200 mL). For the reduction of 2-octanone, the biphasic reaction system with [HMPL][NTF] (20% v/v) showed a lower initial reaction rate than the monophasic setup, supposedly due to the larger viscosity of the ionic liquid. However, the final yield after 4 hours was significantly larger, with a yield >90% with ionic liquid versus only 55% in the assay containing solely phosphate buffer. Additionally, it should be noted that the yield in the monophasic assay only increased until a reaction time of 30 minutes, after which the reaction stopped because the biocatalyst was seriously damaged by the high concentrations of toxic substrate (and product). This demonstrates the advantage of a biphasic reaction mode, where the ionic liquid effectively protected the biocatalyst against the harmful interaction with the toxic substrate and product. The reduction of 4-chloroacetophenone in biphasic systems was compared for second phases (20% v/v) composed of [HMPL][NTF], [BMIM][NTF], and *n*-decane. The reaction systems containing ionic liquid led to significantly larger yields and initial reaction rates than the reaction system with *n*-decane. Concerning the enantiomeric excesses, the ionic liquids were more suitable and showed very satisfying results (~99.6%), whereas in *n*-decane, a decrease in selectivity was observed over time. This was attributed to the damaging effect of *n*-decane toward the membrane of the biocatalyst. Such a deterioration of the membrane could provoke a cofactor outflow, explaining the severe decrease in the reaction rate with time in this reaction system. The success of the biphasic ionic liquid systems used in comparison to the monophasic systems, or those involving organic solvents, proved the suitability of the selection process proposed.

Schroer et al. applied a water-immiscible ionic liquid to overcome thermodynamic limitations in the reduction of 2,5-hexadione to (2*R*,5*R*)-hexanediol and compared this approach with two other *in situ* removal techniques.[84] The reduction of the ketone should be executed using a recombinant *E. coli* containing overexpressions for *L. brevis* ADH. This enzyme necessitates the presence of a cofactor to be able to perform the reduction. In the present work, it was proposed to regenerate the cofactor by applying a substrate-coupled approach, in which the same enzyme oxidizes a co-substrate consuming the oxidized form of the cofactor and yielding its reduced form needed again for a subsequent reduction of the main substrate (Scheme 7.13).

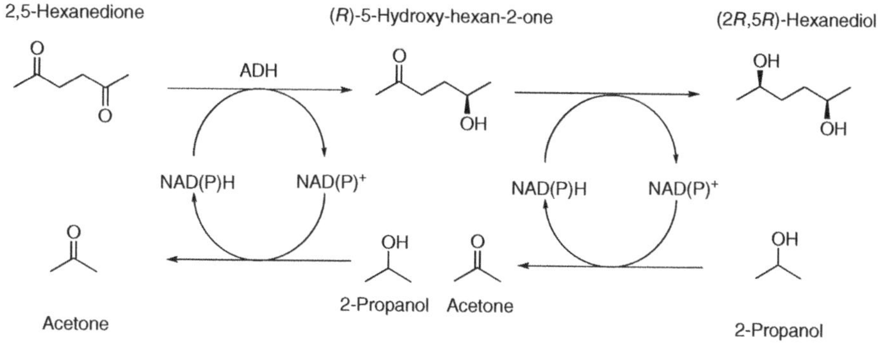

Scheme 7.13. Asymmetric reduction of 2,5-hexanedione to (2*R*,5*R*)-hexanediol by *Escherichia coli* with substrate-coupled cofactor regeneration using 2-propanol.

The co-substrate used in this setup was 2-propanol, which is transformed to acetone by *L. brevis* ADH. The thermodynamic limitation comes from the fact that these two reactions are in equilibrium, meaning that the accumulation of acetone will slow down the biotransformation gradually until it comes to a halt. Additionally, acetone might inhibit the enzyme and damage the biocatalyst. A removal of this substance during the reduction might thus not only favor the transformation from a thermodynamic point of view, but also productivity due to increased activity of the biocatalyst. In fact, the reduction of 2,5-hexanedione without *in situ* acetone removal led to a yield of only 55% for the wanted product after 7 hours, while the remaining 45% of the transformed substrate accumulated in form of the intermediate product (*R*)-5-hydroxyhexanone. The reaction initially evolved rapidly toward the desired product, but the reaction rate significantly decreased after only 1 hour. To overcome this thermodynamic limitation, three different *in situ* removal techniques were proposed to evacuate at least part of the acetone produced: stripping—as acetone was the most volatile substance present in the process—pervaporation, and a biphasic process setup in which the water-immiscible phase acts as an extraction medium for acetone. The stripping of acetone was very successful and permitted the acetone concentration to remain at a maximum of 75 mM in the initial phase of the biotransformation, with a further constant decrease to minimal values. In comparison to the 180 mM acetone accumulating in the reaction system without removal, this constituted a dramatic improvement, which was also confirmed in the yield now reaching values >95%. The results in terms of acetone concentration in the system obtained using pervaporation through a polydimethoxysiloxane membrane were very similar to the stripping procedure. However, the yields were slightly lower with values >90% and the membrane used was not 100% selective for acetone only, but eliminated 2-propanol at mass transfer rates 50% that of acetone. This means that 2-propanol had to be added, too, in order to keep the cofactor regeneration up. In addition, this removal technique is much more cost-intensive than the stripping approach. Thus, pervaporation probably does not constitute the method of choice if no additional advantages can be presented in comparison to the stripping approach.

Scheme 7.14. Asymmetric reduction of ethyl acetoacetate by *Pichia membranaefaciens* Hansen ZJPH07.

The *in situ* removal using a biphasic reaction system is based on the different partitioning of the substances concerned in the two phases involved. Here, the solvents chosen for the biphasic reaction mode were MTBE and the ionic liquid [BMIM][NTF]. 2-propanol, which should preferably not be extracted from the aqueous phase, showed a partition coefficient of 1 and 0.4 in the MTBE-buffer system and the ionic liquid–buffer system, respectively. The values indicated for acetone were 1.1 for the MTBE–buffer system and 2.0 for [BMIM][NTF]–buffer. The different partition coefficients are surely the explanation to the yields reached in both reaction systems: The assay with [BMIM][NTF] reached satisfying yields >95%, whereas the organic solvent led only to 24% yield.

To further compare the different methods tested, the stability of the biocatalyst in each system was assessed. Here, pervaporation performed best, with 82% activity retention of the biocatalyst. Stripping, however, seemed to harm the recombinant *E. coli* cells more, leading to only 25% relative activity. In the biphasic system with [BMIM][NTF], the biocatalyst showed intermediate stability, with 63% relative activity left.

He et al. performed an extensive study investigating the effect of buffer pH, temperature, initial substrate concentration, and biocatalyst concentration on the reduction of ethyl acetoacetate to ethyl (*R*)-3-hydroxybutyrate (Scheme 7.14), and added a further parameter seldom analyzed: the co-substrate type.[23] The goal was to optimize the reaction concerning these variables and observe the effect of different co-substrates on the reaction outcome.

The biotransformation was executed using *Pichia membranaefaciens* Hansen ZJPH07 cells in ionic liquid-containing systems. A first reaction system was formed by phosphate buffer (0.2 M, pH 8) to which the hydrophilic ionic liquid [BMIM][BF_4] was added, as well as 30 g/L glucose for the cofactor regeneration. While the selectivity increased with increasing ionic liquid content, the yields showed a maximum (32%) at 2.5% (v/v) ionic liquid in the buffer. This ionic liquid concentration was thus chosen for further experiments. In these reaction conditions, the use of glucose as standard co-substrate led to an enantiomeric excess of ~60%. When adding other co-substrates to the reaction, not only the yield of the biotransformation, but also the enantiomeric excess was affected. Ethanol, glycerol, and sucrose addition led to similar results to the reaction system without addition of co-substrate in terms of selectivity, and showed increased yields. However, all four performed less well than when adding glucose, in terms of both yield and enantioselectivity. Butanol addition caused the least satisfying reaction outcome. When using isopropanol as co-substrate, the highest yield was

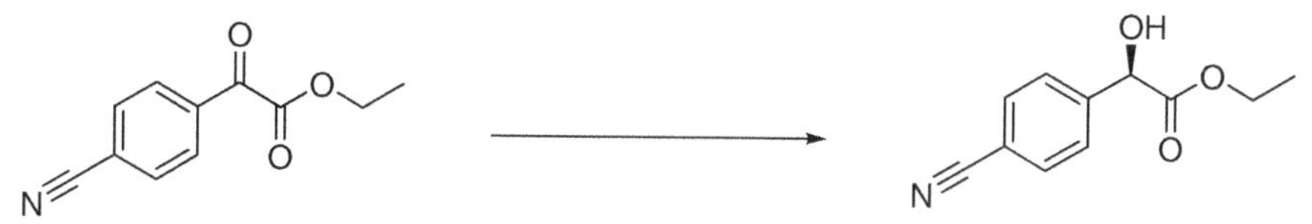

Ethyl 4-cyanobenzoylformate Ethyl (*R*)-4-cyanomandelate

Scheme 7.15. Asymmetric reduction of ethyl 4-cyanobenzoylformate by *Escherichia coli*.

observed (>70%), but the (*S*)-enantiomer was formed, instead of the desired (*R*)-molecule. In view of producing the (*R*)-enantiomer, glucose was thus the co-substrate of choice. The other experiments led to the following optimized reaction conditions: 0.55 M ethyl acetoacetate, 50 g/L glucose, and 240 g_{cww}/L biocatalyst in phosphate buffer of pH 8 containing 2.5% [BMIM][BF_4]. In these conditions, the yield reached was 77.8% and the enantiomeric excess was 73% after 12 hours of reaction. In comparison to the reaction setup without ionic liquid, these were significant improvements from 68.5% and 65.1% for the yield and enantioselectivity, respectively.

Kratzer et al. analyzed the synthesis of ethyl (*R*)-4-cyanomandelate through reduction of ethyl 4-cyanobenzoylformate, an aromatic α-keto ester (Scheme 7.15).[85]

This reaction is limited by the low solubility of the substrate, its low stability in the phosphate buffer used, and its toxicity toward the biocatalyst. Using a recombinant *E. coli* with overexpressions for *Candida xylose* reductase and *C. boidinii* FDH, the yield of the biotransformation in purely aqueous medium could be increased in comparison to what was achieved with a recombinant baker's yeast strain (*S. cerevisiae* XR2μ) from 45% to 82%, but due to the aforementioned reasons, no further increase in productivity could be reached. A biphasic reaction mode was supposed to help overcome these limitations. Several water-immiscible solvents were tested for this purpose, including ethyl acetate, butyl acetate, and hexane, as well as [BMIM][PF_6]. Even though performing by far better than the organic solvents analyzed in the same conditions (<10% yield), and leading to excellent enantioselectivity (>99.9%), the application of the ionic liquid (50% v/v) reached lower yields than the monophasic reaction mode, with only ~15% yield versus ~27% for the purely aqueous assay. The addition of cofactor (NAD^+) permitted only a slight improvement in the biphasic system to ~22%, at constant selectivity. Besides the only moderate improvement, the addition of such expensive compounds would be difficult to implement at large scale in industrial processes. Regrettably, only one volume ratio was tested for the buffer–solvent systems, and this was chosen relatively high (1:1). Testing a larger range of phase ratios would have provided more information about the reaction system, and whether the conclusions found for a 50% (v/v) are more generally valid may be evaluated.

Shi et al. analyzed the effect of different nonaqueous media on the reduction of 2-oxo-4-phenylbutanoate to 2-hydroxy-4-phenylbutanoate by *S. cerevisiae* (Scheme 7.16).[86]

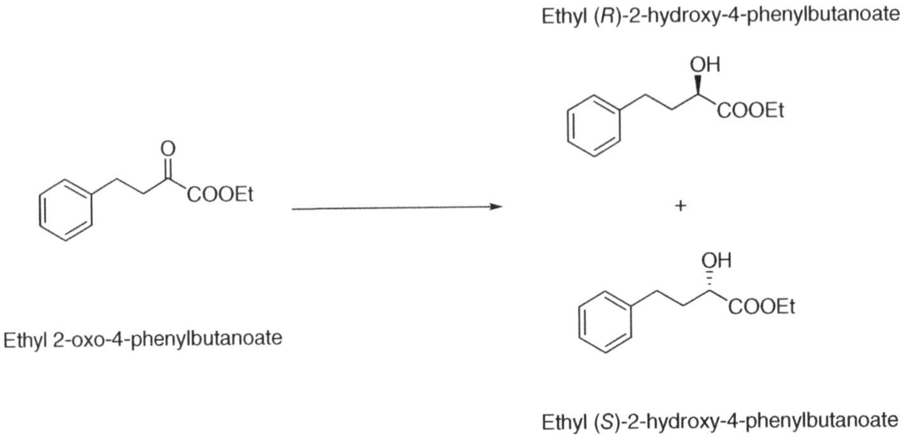

Scheme 7.16. Asymmetric reduction of ethyl 2-oxo-4-phenylbutanoate by *Saccharomyces cervisiae*.

The aim was to avoid water for this biocatalysis, as the (*R*)-enantiomer was decomposed to 3-phenylpropanol by hydrolytic enzymes also present in the whole yeast cell when in an aqueous environment. The media tested included a hydrophobic ionic liquid ([BMIM][PF_6]) and organic solvents, (ethyl ether, benzene, and toluene). The authors investigated the effect of addition of small volumes of water (1–3% v/v) to [BMIM][PF_6], toluene, and benzene on yield, conversion, and selectivity. For all three solvents, an increase in the water content up to 3% (v/v) significantly increased the yield. The increase in conversion with the water content was more significant for the organic solvents than for the ionic liquid, which showed already 60% conversion at only 1% (v/v) water. The enantioselectivity, however, was differently affected in the various systems: While the organic solvents showed a significant increase in enantioselectivity up to ~70% and ~80% (*R*) for ethyl ether and benzene, respectively, larger amounts of water led to increased formation of the (*S*)-enantiomer in the ionic liquid-containing system, with enantiomeric excesses of up to ~30% (*S*) at 3% (v/v) water. However, when water was added at a larger volume ratio (10:1 ionic liquid–water), a second phase formed, and the enantioselectivity again shifted to the (*R*)-side, giving a poor enantiomeric excess of only 6.6% (*R*). Conversion (87.4%) and yield (42.3%), however, increased by 10% and 2%, respectively, in comparison to the assay containing only 3% water. The enantiomeric excess was dramatically improved by addition of small volumes of short chain alcohol (methanol to butanol) to the previous setup ionic liquid–water (10:1 v/v). The addition of 1% (v/v) ethanol led to the best overall results at an enantiomeric excess of 82.5% (*R*), for slightly larger yields (45.8%) and conversion of the substrate (89.6%) slightly larger than without additives. These improvements were explained by the utilization of ethanol by the cell as co-substrate for the cofactor regeneration, thereby favoring the desired reduction performed by the oxidoreductase. This work nicely demonstrated that additives and the volume ratio used can have a signifi-

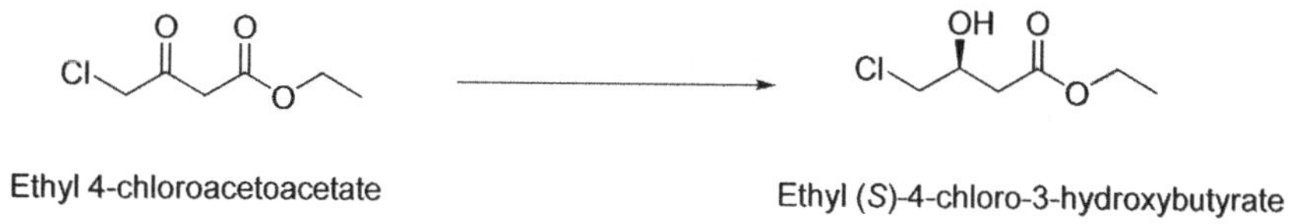

Scheme 7.17. Asymmetric reduction of ethyl 4-chloroacetoacetate by *Aureobasidium pullulans* CGMCC 1244.

cant influence on the enantioselectivity of a given biphasic biotransformation, and showed the importance of choosing the correct reaction conditions.

Zhang et al. tested a range of biphasic and monophasic reaction systems involving ionic liquids and organic solvents for the asymmetric reduction of ethyl 4-chloroacetoacetate to ethyl (*S*)-4-chloro-3-hydroxybutyrate by *Aureobasidium pullulans* CGMCC 1244 (Scheme 7.17).[22]

A. pullulans had already been shown to be able to reach a conversion of 94.5% for the reduction considered, and produce up to 56.8 g/L ethyl (*S*)-4-chloro-3-hydroxybutanoate with an enantiomeric purity of 97.7% in reaction media composed of phosphate buffer (pH 6.5) and dibutylphthalate.[23] Here, the goal was to try to further increase the regioselectivity and the conversion by combining ionic liquids instead of organic solvents with the aqueous reaction phase. The ionic liquids tested were [BMIM][PF_6], [BMMIM][PF_6], [BMIM][BF_4], [OPy][BF_4], and [BMIM][CF_3SO_3]. These reaction systems were compared with assays containing the organic solvents dibutylphthalate or *n*-hexane. To determine the most suitable reaction medium, the authors proceeded similarly to Bräutigam et al., evaluating the metabolic activity retention of the cells in the different media proposed, and determining the partition coefficients of the substrate and the product in both phases present. In assays composed of 50% (v/v) solvent in phosphate buffer, metabolic activity retention was largest for [BMIM][PF_6], surpassing even the value determined for the aqueous reference not containing any solvent. [BMMIM][PF_6] was also well tolerated by the cells, leading to residual activity as high as in the reference. Dibutylphthalate showed also relatively good biocompatibility (>80% relative activity), while *n*-hexane and the other ionic liquids led to relative activities of around 60% after 24 hours' incubation. Biotransformations were performed in assays of the same composition (50% v/v solvent) with each of the proposed solvents. Enantioselectivity was very similar for all the reaction systems analyzed, between 96.1% and 97.3%. The final conversions, however, varied significantly between 32% for the system containing [BMIM][BF_4] and 94.2% when using [BMIM][PF_6]. The assay with *n*-hexane reached only 38.7% and was thus only slightly better than the aqueous reference (35.9%). [BMMIM][PF_6] showed intermediate conversion of 50.1%, while dibutylphthalate was among the best performing with 90.0% conversion. The conversions observed were not completely predictable on the basis of only the partition coefficients and the biocompatibility of each reaction system, indicating that these are not the only factors influencing the outcome of a given biotransformation. As [BMIM][PF_6] led to the highest conversion among all the solvents tested, this ionic liquid was chosen for further analyses concerning the optimization of the phase ratio, mixing rate, reaction temperature, pH, and initial substrate concentration. The variation of these factors influenced both yield and conversion but left the enantioselectivity unaffected. In optimized reaction conditions (50% (v/v) ionic liquid, 292 mM substrate, pH 6.6, and 30°C), the conversion reached after 8 hours was 95.6% at an enantiomeric excess of 98.5%. As the substrate inhibited the biocatalyst, at concentrations of more than 292 mM, it was decided to add the substrate during the reaction. This permitted a further increase in the conversion to 97.5%, leading to a product concentration of 75.1 g/L. Compared with the same reaction conditions in the aqueous reference system (~40% conversion), the ionic liquid did not only permit an increase in the conversion by more than a factor of two, but also showed a larger reaction rate due to reduced substrate inhibition of the biocatalyst in the biphasic system. The results obtained here were very similar to those obtained for a range of hydrophobic ionic liquids during the reduction of the same substrate by a recombinant *E. coli* in terms of yield and enantioselectivity, but showed lower productivity.[13] The investigation was completed by evaluating the recyclability of the ionic liquid phase. When the same batch of ionic liquid was reused six times, a decrease in the conversion from initially 96.7% to 92.1% was observed, as well as a reduced enantioselectivity of 96.4% in comparison to 98.4% during the first cycle.

Dipeolu et al. investigated the use of water-miscible ionic liquids to deliver water-insoluble substrates in a whole-cell biotransformation.[66] As a model reaction, the reduction of nitrobenzene to aniline using *Clostridium sporogenes* was chosen (Scheme 7.18), a transformation limited by the relatively low water solubility of the substrate.

The effect of five water-miscible ionic liquids on the growth of the biocatalyst and on the biotransformation was thus assessed and compared with the results observed when using common organic solvents such as ethanol and heptane. The ionic liquids considered were [BMIM][BF_4], AMMOENG 100, DMEAA, [$EtOHNMe_3$][Me_2PO_4], and [EMIM][$EtSO_4$].

To evaluate which of the solvents would be most suitable, their respective biocompatibility was first assessed by determining the growth rate of *C. sporogenes* cells in anaerobic conditions in the presence of different amounts of solvent. These analyses showed that two of the ionic liquids—DMEAA and [$EtOHNME_3$][Me_2PO_4]—actually favored growth, as the rates observed were larger than those determined in the purely aqueous reference without solvent. The authors suppose that the cells might have been able to metabolize these ionic liquids or that the solvents increased the availability of other nutrients in the medium. The addition of [BMIM][BF_4] resulted in severely reduced growth at 0.25% (v/v), and completely inhibited it at 0.5% (v/v). This is thought

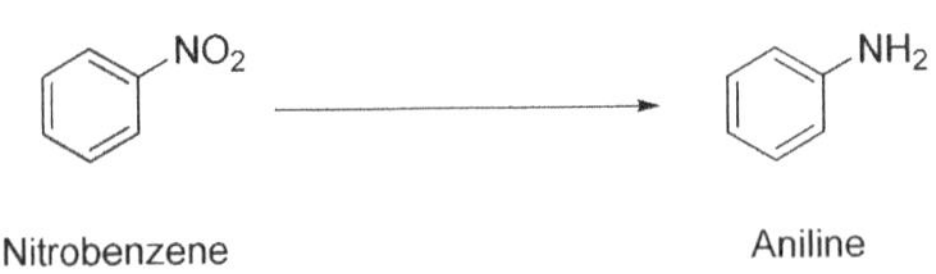

Scheme 7.18. Asymmetric reduction of nitrobenzene by *Clostridium sporogenes*.

to be not necessarily inherent to the ionic liquid itself, but due to the hydrolysis of the anion, producing hydrofluoric acid, which is highly toxic to cells. AMMOENG 100 almost completely inhibited growth, even at concentrations as low as 0.1%, because its addition to the growth medium provoked immediate precipitation of the culture medium. The apparent growth inhibition was thus possibly due to the lack of available nutrients. [EMIM][$EtSO_4$] was intermediate between the other ionic liquids. Up to 1% (v/v), growth was reduced by 23% in comparison to the reference system. However, at a concentration of 2% (v/v), it became severely inhibiting, and relative growth was only 42%. The only organic solvent tested, ethanol, was only slightly inhibiting, even at a concentration of 2% (v/v). Based on the biocompatibility tests, the choice was made for [EMIM][$EtSO_4$] to be used in further experiments, even though it was not the most biocompatible solvent. DMEAA and [$EtOHNME_3$][Me_2PO_4] were not further considered because it was not clear what exactly led to the increase of growth when these solvents were added to the cell suspension. The biotransformations showed that the application of ionic liquids can be very favorable for nitro reductions: When the solvents were added to a concentration of 4% (v/v), the yield reached in [EMIM][$EtSO_4$] was 45%, significantly larger in comparison to ethanol (8%) and still larger than that observed in heptane (39%). It is worth noting that [EMIM][$EtSO_4$] could still benefit the reaction, even though it was only moderately biocompatible. The cell growth was in fact more sensitive to the solvent than the desired biotransformation which was to take place in the reaction media. Thus, only moderately biocompatible ionic liquids should not be directly excluded from the pool of possibly suitable solvents for use in biotransformations but included in screening experiments.

Recently, several authors investigated the asymmetric reduction of 4-methoxyacetophenone to (*S*)-1-(4-methoxyphenyl)ethanol in ionic liquid-containing reaction systems (Scheme 7.19).

Earlier, Howarth et al. had not been successful in converting this substrate using immobilized baker's yeast in water–[BMIM][PF_6] mixtures (volume ratio of 1:10). Wang et al., in contrast, used immobilized *Rhodotorula* sp. AS2.2241 cells to catalyze the reduction under consideration.[15] As inhibition by substrate and product represented a serious problem in monophasic aqueous systems,[87] hydrophobic ionic liquids were considered to be used as second phase. Indeed, a biphasic reaction mode chosen appropriately should reduce the substrate and product concentrations in the cells' environment by acting as a reservoir for both. The solvents tested were [RMIM][PF_6], where

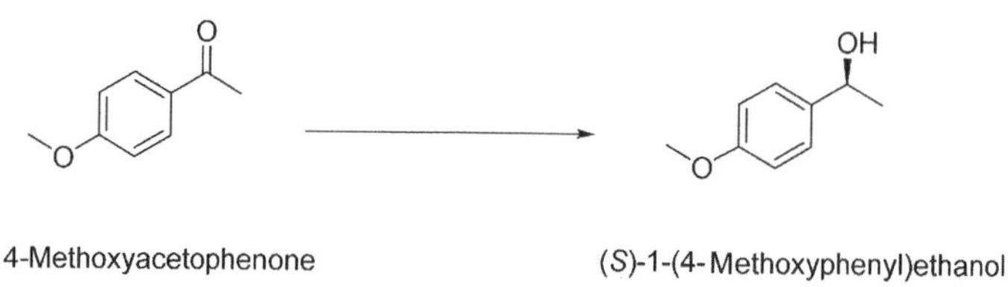

Scheme 7.19. Asymmetric reduction of 4-methoxyacetophenone by immobilized *Rhodotorula* sp. AS2.2241and *Trigonopsis variabilis* AS2.1611.

"R" represents the alkyl chain varying from butyl to heptyl, [EMIM][NTF], and [BMIM][NTF]. All of the biotransformations performed in assays containing 20% (v/v) ionic liquid showed excellent enantiomeric excesses (>99%). Substrate conversion was best for [BMIM][PF_6] (69.5%), and decreased with increasing chain length within the set of [RMIM][PF_6] ionic liquids to only 38.2% conversion for the heptyl chain on the cation. [EMIM][NTF] and [BMIM][NTF] led to 66.4% and 58.3% conversions, respectively. Biocompatibility of the different reaction systems was tested with and without the presence of the substrate. The same "side-chain effect" was observed, as increasing chain length went with decreasing biocompatibility for the [RMIM][PF_6] ionic liquids. However, [BMIM][NTF] seemed to be better tolerated than [EMIM][NTF]. When the substrate was added, biocompatibility generally decreased. The effect was, as expected, most dramatic for the monophasic aqueous system, with a drop of cell viability from >95% to <60%. The drop in cell viability was significantly lower for the ionic liquid-containing systems, with a loss of generally only 5–10%. The ionic liquids thus truly acted as substrate reservoirs, which was confirmed by the partition coefficients in the solvent–buffer systems determined for the substrate and the product: For the substrate, the partition coefficients were between 31 and 46, whereas for the product the partition coefficients were between 6 and 10 in the ionic liquid–buffer systems.

Using the best performing ionic liquid, [BMIM][PF_6], the reaction system was optimized with regard to buffer pH, reaction temperature, phase ratios, and substrate concentration. In optimized conditions, the biphasic reaction system containing 20% (v/v) showed a clearly lower initial reaction rate (1.6 μmol/h vs. 5.7 μmol/h), which was attributed to the lower substrate concentration found in the aqueous phase of the biphasic system in comparison to the amounts found in the monophasic system, as well as to possible mass transfer limitations. However, more importantly, the biphasic system again led to more efficient substrate conversion (95.5%) within 50 hours reaction time than the monophasic aqueous system (90.5%), at additionally larger initial substrate concentrations. Product yield in the biphasic reaction system was thus significantly increased in comparison to the monophasic assay composed of buffer phase only.

Finally, the authors also investigated the operational stability of the immobilized biocatalyst in the different reaction systems. In comparison to the aqueous system, where stability decreased rapidly to only 25% after eight batches, the biocatalyst showed good operational stability in the [BMIM][PF_6]-containing system, with only a slight decrease after the same number of cycles (relative activity retention >90%).

Lou et al. extended the previous work by testing a second range of different solvents for the same application as those studied by Wang et al., this time using hydrophilic ionic liquids as co-solvents.[10] The ionic liquids evaluated here included many [C_2OHMIM] ionic liquids with different anions, as well as several [RMIM][BF_4], [RMIM][NO_3], and [RMIM][Br] ionic liquids where "R" stands for different alkyl chains varying in length from ethyl to heptyl. The biotransformations were performed in assays containing 10% (v/v) solvent in Tris-HCl buffer, over different reaction durations. The yields resulting from these reductions varied greatly, strongly depending on the ionic liquid involved. The most striking result was observed for ionic liquids including the [BF_4] anion. These solvents all led to extremely poor yields, <10%, even though the reaction was left going on up to 80 hours. For the set of [Br]-containing ionic liquids, yields decreased with increasing alkyl chain length. These ionic liquids led to moderate yields of between ~30% and ~60%. This is also the range of values covered by the yields reached in systems involving [NO_3] anions. The best performing ionic liquids were the set of [C_2OHMIM] ionic liquids, except when containing [BF_4], with yields of around 70–80%. The biocompatibility tests were performed in two different setups, without substrate addition and in the presence of substrate, and qualitatively reflected the yields observed previously during the biotransformations. Generally, the viability was lower when substrate was added into the reaction system, showing the inherent toxicity of 4-methoxyacetophenone toward the biocatalyst. Interestingly, one ionic liquid, [C_2OHMIM][NO_3], led to a particularly well-tolerated reaction system within the set of experiments in the presence of the substrate.

Using the most suitable ionic liquid among those tested, the ionic liquid content, buffer pH, reaction temperature, and substrate concentration were optimized. Hereby, the effect of the ionic liquid on the buffer pH was underlined, showing a clear influence that should be taken into consideration in such evaluations of the reaction conditions. Each of the factors analyzed affected both yield and conversion considerably, with only a minor effect on the enantioselectivity. In optimized reaction conditions, the asymmetric reduction in buffer containing 5% (v/v) [C_2OHMIM][NO_3] reached a very satisfying yield of 98.3% at excellent enantiomeric excesses >99%. In comparison to the aqueous reference (90.5% yield), and to the results obtained by Wang et al. using [BMIM][PF_6] (20% v/v) (95.5% yield),[15] this reaction system has brought further improvement. This was also reflected by the initial reaction rates measured in optimized conditions, which were largest for the [C_2OHMIM][NO_3] system (9.8 $\mu mol/g_{cell}$ h), followed by the aqueous reference (7.1 $\mu mol/g_{cell}$ h), while the system with [BMIM][PF_6] was slowest (2.0 $\mu mol/g_{cell}$ h). In addition, the use of the [C_2OHMIM][NO_3]–buffer system permitted significantly larger substrate loading (12 mM), with an optimal initial substrate concentration more than twice the optimal amount determined for the aqueous reference (5 mM).

Finally, the stability of the immobilized biocatalyst was investigated. In the current system, a decrease in relative activity to 93% was observed after 10 batches (i.e., 250 hours of operation). This was comparable to the stability when using [BMIM][PF_6] (95% relative activity),[15] and a considerable improvement in comparison to the aqueous reference (49% relative activity).

The same authors evaluated the same range of water-miscible ionic liquids for application with another biocatalyst, immobilized *T. variabilis* AS2.1611, for the reduction of 4-methoxyacetophenone (Scheme 7.19).[64] The biotransformations in the presence of 5% (v/v) ionic liquid showed qualitatively the same trends concerning the different ionic liquids tested, but on an absolute scale less satisfying results than with immobilized *Rhodotorula* sp. AS2.2241 cells. This microorganism thus seems to be more sensitive than *Rhodotorula* sp to the solvents added. Enantiomeric excesses were by far not as good as in the previous study, and the yield reached was also lower for each of the reaction systems tested. The only ionic liquid that gave satisfying results was [C_2OHMIM][NO_3], with 95% yield and enantiomeric excess >99%. Interestingly, *T. variabilis* performed better in the purely aqueous system than did *Rhodotorula* sp., leading to an 80% yield for the former versus 68.6% for the latter. It thus seems that *T. variabilis* tolerates the substrate a little better than *Rhodotorula* sp.[10] However, this might be due in part to the slightly larger biocatalyst concentration (0.4 g/mL *T. variabilis* and 0.32 g/mL *Rhodotorula* sp.). The biocompatibilities determined for *T. variabilis* in the different reaction systems gave a very similar picture to that reported for *Rhodotorula* sp. The optimization of ionic liquid content, buffer pH, reaction temperature, and substrate concentration significantly improved the efficiency of the process. In optimized conditions (2.5% (v/v) [C_2OHMIM][NO_3], pH 8.5, 30°C, and substrate concentrations of 15 mM), the yield reached was 97.2% at excellent enantioselectivity (99.6%). In comparison to the yield and the enantioselectivity reached by the purely aqueous reaction system (86.4% and 87%, respectively), the involvement of this ionic liquid was thus clearly beneficial, even leading to a larger reaction rate (7.1 μmol/h vs. 4.7 μmol/h). However, the previous study using *Rhodotorula* sp. showed still slightly better performance with values of 9.8 μmol/h, 98.3%, and >99% for the reaction rate, product yield, and enantiomeric excess, respectively. Interestingly, *T. variabilis* tolerated not only larger substrate concentrations in the reaction system, but also large temperatures (>30°C) better when ionic liquid was added as co-solvent than in a purely aqueous medium. A scale-up to a volume of 250 mL was also performed, leading to good yields (95.7%) and enantiomeric excesses (>99%).

Finally, the stability of the biocatalyst in the ionic liquid-containing buffer was relatively good, still showing 85% relative activity after 12 subsequent batches. As no emulsification of the medium with the cells was observed and no by-products accumulated in the reaction medium, the ionic liquid-based co-solvent system might thus be recycled.

7.3.2 Other Whole-Cell Biotransformations in Ionic Liquids

Cornmell et al. were the first to investigate the oxidation of aromatic compounds by whole-cell biocatalysis in biphasic reaction media with ionic liquids.[65] The biotransformation considered was the dihydroxylation of toluene to toluene *cis*-glycol using a recombinant *E. coli* K12 strain (Scheme 7.20).

The biphasic reaction mode was chosen here due to the strong substrate toxicity.[88] With the second phase acting as a substrate reservoir, the biphasic approach should protect the biocatalyst from its toxic effect toward the cell. The authors turned to ionic liquids for this application because organic solvents had previously shown only moderate success.[88,89] The ionic liquids to be tested for their possible application as second

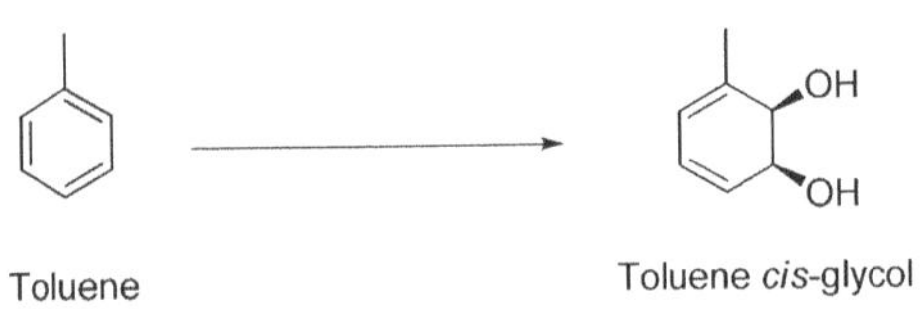

Scheme 7.20. Dihydroxylation of toluene by *Escherichia coli* K12.

phase were [BMIM][Cl], [EMIM][Cl], [OMIM][Cl], [Oc_3MeN][Cl], [$P_{6,6,6,14}$][Cl], and [NDecPy][Cl], as well as their corresponding analogues containing the [NTF] anion.

The authors first assessed the biocompatibility of the solvents by measuring the growth of the biocatalyst in the presence of the ionic liquids. Only two of them permitted growth at concentrations of 0.23% (v/v), with growth rates of 0.219/h and 0.276/h for [Oc_3MeN][NTF] and [$P_{6,6,6,14}$][NTF], respectively, although it was slightly reduced in comparison to the aqueous reference (0.357/h) and the reference with tetradecane (0.336/h). These two ionic liquids were thus used for further investigations. Optimal substrate concentrations had been determined at 3.76 mM for monophasic biotransformations at the 50 milliliter scale. In these conditions, the presence of [Oc_3MeN][NTF] and [$P_{6,6,6,14}$][NTF] did not have any adverse effects on the biotransformation, as both the reaction rate and the final yield (87–90%) reached were unaffected by the presence of the solvents after 20 hours and the same as in the reference assay without solvent addition. The viable cell count, however, was again slightly lower in the presence of the ionic liquids than in the aqueous reference. Cell growth thus seemed more sensitive to the inhibition due to the ionic liquids than the biotransformation. Following these observations, the maximum toluene concentration tolerated by the biocatalyst in each reaction system was determined. In the aqueous system, only 9.4 mM toluene could be added, while the ionic liquids permitted a loading of ~75 mM. This eightfold increase in toluene concentration in the ionic liquid-containing systems showed that these solvents effectively protected the biocatalyst from the toxic effect of the substrate and truly acted as toluene reservoirs. This resulted in specific yields of toluene *cis*-glycol of 2.89 mmol/g_{DCW} and 2.68 mmol/g_{DCW} for [Oc_3MeN][NTF] and [$P_{6,6,6,14}$][NTF], respectively, a dramatic improvement from the 0.83 mmol/g_{DCW} observed in the aqueous reference. The reduced yields reached in these biphasic systems (~27% yield vs. 88.4% yield in the monophasic system) were attributable to limited oxygen availability in the sealed reaction vessels. This was verified by performing experiments in a bioreactor at a larger scale (1.25 L) that permitted air sparging, during which toluene was fed to the reaction medium. Here, larger mean product concentrations were reached in the biphasic system with [$P_{6,6,6,14}$][NTF] (55.5 mM vs. 20.7 mM at the 50-mL scale), as well as in the monophasic system (18.2 mM vs. 8.3 mM at the 50-mL scale). The specific product yield of these larger scale experiments clearly showed the success of the biphasic ionic liquid system, where 11.7 mmol/g_{DCW} was produced, in comparison to only 3.46 mmol/g_{DCW} in the aqueous reaction system. Yields could only be estimated because of the stripping off of toluene due to the air sparging through the reactor. Based on the total quantity of toluene added to the reactor, minimal yields of 36% and 15% for the ionic liquid-containing system and the aqueous reference, respectively, were estimated, but actual yields were supposed to be significantly larger.

Baumann et al. used biphasic ionic liquid–buffer systems with three xenobiotic degrading bacteria.[19] More precisely, the degradation of phenol was investigated using *Pseudomonas putida*, *Achromobacter xylosoxidans*, and *Sphingomonas aromaticivorans*. Phenol being already toxic to the cells at concentrations <1 g/L,[90] a biphasic reaction mode, protecting the cells from large concentrations of phenol would still permit relatively high loading of the xenobiotic and thus increase the process efficiency.

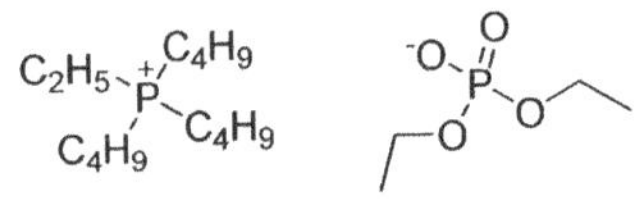

Scheme 7.21. Tributyl(ethyl)phosphonium diethylphosphate.

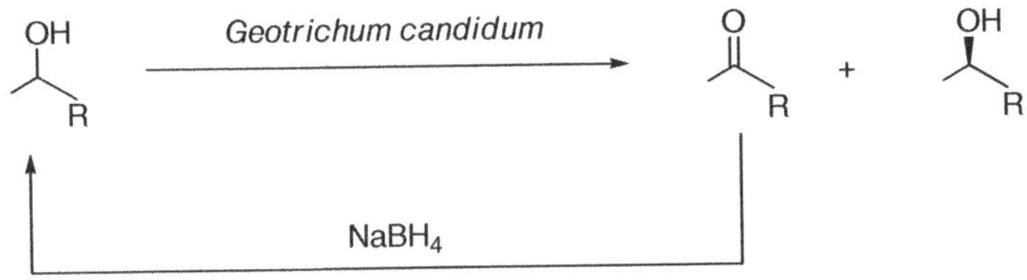

Scheme 7.22. Optical resolution of racemic secondary alcohols with *Geotrichum candidum* NBRC 5767 and $NaBH_4$.

Six ionic liquids were tested in terms of their suitability as second phase. Among them, only one was completely biocompatible with all three microorganisms: tributyl(ethyl)phosphonium diethylphosphate (Scheme 7.21).

This ionic liquid was thus used for the biodegradation under consideration, based on its good biocompatibility, but also because of its larger density relative to water compared with other ionic liquids evaluated. The choice of one biocatalyst was made for *P. putida*, as this was the microorganism the authors had the most experience with for the degradation of phenol.[90,91] The use of a second ionic liquid phase permitted a phenol loading of a total of 1580 mg/L in the reaction system. Due to the favorable partitioning of phenol in the two phases present, this resulted in an aqueous concentration of 400 mg/L, which is below the toxicity limit of *P. putida*. If the same quantity of phenol had been added to a monophasic reaction system, this value would have been largely exceeded, causing severe damage to the biocatalyst. In the biphasic setting involving ionic liquids, however, the biocatalyst was active and complete degradation of the xenobiotic was observed after 12 hours. The use of ionic liquids for this type of applications can thus be considered a success. The authors indicate that the performance found here is very similar to what was observed during previous biodegradations in the presence of second organic phases.[90,91] Concerning the operating conditions, the larger density of the second ionic liquid phase did not create any difficulties in terms of mixing and dispersion, and the use of this type of solvents caused fewer security issues than common organic solvents.

Tanaka et al. applied water-miscible ionic liquids in the optical resolution of racemic secondary alcohols.[77] *G. candidum* NBRC 5767 was used to oxidize the (*S*)-form of the alcohol to the corresponding ketone, leaving the (*R*)-enantiomere unchanged (Scheme 7.22).

To guarantee the necessary presence of given amounts of water,[42] the biocatalyst was immobilized on a water-absorbing polymer. First, experiments were conducted in buffer containing [EMIM][BF_4] at a concentration of 50% (v/v), using a large range of different aromatic and aliphatic alcohols as substrates. The asymmetric oxidation produced very pure alcohol, with enantiomeric excesses of mostly >94% (*R*), but with only moderate yields for the alcohol, between 39% and 67%. To increase the yield, reductive agents such as $NaBH_4$ or $NaBH_3CN$ were added. These should complete the reaction by reducing the ketone again to its corresponding alcohol, and thereby reintroduce it into the deracemization procedure. The use of $NaBH_3CN$ did not prove very successful, but good results could be achieved with $NaBH_4$, for example, for the optical resolution of 1-phenylethanol. At concentrations of 6.4 or 9 equivalent $NaBH_4$, alcohol yields of 95% and 96%, respectively, could be reached after 24 hours, at excellent enantiomeric excesses of 99%. The authors proved that the deracemization could be applied to a large number of substrates, although not all had comparable success to 1-phenylethanol.

Arai et al. presented a novel application of ionic liquids as second phases in the production of biodiesel fuels by whole-cell biocatalysis.[78] In the reaction considered, fatty acid methyl esters were produced by transesterification of triglycerides with methanol in biphasic reaction systems. The triglycerides were provided by soy bean oil, and the transesterification was performed by different lipase-producing biocatalysts immobilized on biomass support particles (BSP): wild-type *Rhizopus oryzae* (w-ROL), and three different recombinant *Aspergillus oryzea* expressing *Fusarium heterosporum* lipase (r-FHL), *Candida antarctica* lipase B (r-CALB), and mono- and diacylglycerol lipase from *A. oryzea* (r-mdlB). A biphasic reaction mode was applied to avoid deactivation of the biocatalyst by one of the substrates, methanol. Indeed, ionic liquids solve methanol, but they dissolve neither fatty acid methyl esters nor soy bean oil, forming a second phase with the substrate/product mix containing the immobilized cells, and protecting the biocatalyst from toxic contact with the short-chain alcohol. In addition to their protective effect, ionic liquids can dissolve glycerol, a by-product of the process having a negative effect on biodiesel production and on lipase activity.[92,93]

The ionic liquids tested were [BMIM][BF_4], [EMIM][BF_4], and [EMIM][CF_3SO_3]. When comparing the production of methyl esters in the different solvent–oil systems (at volume ratios of 1:1) containing 4 M equivalent methanol, the results were fairly unequivocal: The ionic liquid-free reaction system showed very poor yields for almost all of the biocatalysts tested. As the methanol did not dissolve in soy bean oil, it formed a second organic phase that was toxic to the biocatalyst. Only r-mdlB showed low activity. Methanol-dissolving solvents should thus effectively reduce this toxicity and lead to more satisfying yields. When adding [BMIM][BF_4], [EMIM][BF_4], and [EMIM][CF_3SO_3], the activity of r-mdlB was not affected. This biocatalyst showed neither an increase nor a decrease in activity in comparison to the monophasic setup. r-CALB was still the least productive, with only minimal yield in the presence of ionic liquids, but this nevertheless constituted an improvement in comparison to the complete lack of activity in the monophasic aqueous system. The activity of r-FHL also increased by on average ~10% and thus showed fair yields similar to what was reached with r-mdlB. The most impressive increase of activity was, however, registered for w-ROL. While producing <5% methyl ester content in the monophasic reaction mode, it led to ~45% methyl ester content in the [BMIM][BF_4]- and [EMIM][BF_4]-containing systems, and performed only slightly less well in [EMIM][CF_3SO_3]. The use of a second ionic liquid phase thus permitted a clear improvement of the reaction outcome for three of the four biocatalysts.

The protective effect of the ionic liquid was confirmed when varying concentrations of methanol were added to the system with w-ROL. While the ionic liquid-free reaction system still showed the largest methyl ester content (32%) at a molar ratio of methanol to oil of 1, the activity of w-ROL dramatically decreased with increasing ratios and showed almost no activity anymore at molar ratio of 8. In contrast, the ionic liquid-containing systems showed a significant increase up to a molar ratio of 4 (~45% methyl ester content). The authors concluded that the ionic liquids truly act as a reservoir for methanol, thereby reducing the toxic effect on the biocatalyst. It was only at molar ratios of 8 that the activity decreased again to a methyl ester content of <30% after 24 hours' reaction time.

As this would permit interesting cost reductions, the recyclability of the immobilized biocatalyst was also evaluated. Using a combination of w-ROL and r-mdlB, the final methyl ester content decreased dramatically to less than 20% after the second cycle, and further again during the third cycle. The stability of this mixed biocatalyst was improved by gluteraldehyde cross-linking. Here, the decrease in the methyl ester content was still significant but reduced in comparison to the untreated biocatalyst mixture.

7.4 PROSPECTS

Although the first applications of ionic liquids in biocatalysis appeared only a decade ago, past studies and applications point to several trends that will receive increased attention in the next few years.

One of the most urgent matters and a subject that will continue to attract attention will be the establishment of structure–property relationships. This is due to the large benefit that such data would provide, be it in biocatalysis, chemical catalysis, electrochemistry, or any other domain where ionic liquids will find application. The estimation of given properties based on the structure of the composing entities would significantly reduce the time and cost investments necessary to distinguish solvents that might be suitable for a given application, and avoid having to rely on trial-and-error methods when establishing new processes. Similarly, based on such knowledge, existing ionic liquids could be modified to fit precise purposes, for example, reduced water solubility or viscosity, or improved biocompatibility and increased biodegradability. These modifications could help improve existing processes on technical, economical, and ecological levels. Similarly, models to predict physical properties such as viscosity, density, and hydrophobicity have started appearing.[41,94,95] The future will permit validation or further improvement of these models, as well as development of new ones. While waiting for progress in these areas, the development of commercially available screening kits for ionic liquids covering a large range of different physicochemical properties could also help distinguish suitable solvents for given applications among the multitude of possible combinations.[96]

Another domain of ongoing research is the quest for safer, cheaper, and more ecologically friendly ionic liquids. Hence, while new ionic liquids are constantly developed, especially large efforts are made with regard to their ecotoxicity, biodegradability, and biocompatibility. There have already been significant improvements from the so-called first-generation ionic liquids composed of haloaluminate salts to the second generation, air- and moisture-stable ionic liquids allowing a comparably good ease of handling (see also Chapter 1).[97] Today, the development of new ionic liquids is further veering away from fluoride-containing anions, which still show a potential to hydrolyze, to anions such as [$EtSO_4$] and [$OcSO_4$], which are more desirable from an (eco) toxicological point of view.[34,46] However, new "bio-based" ionic liquids are also attracting increased attention. The idea is to build ionic liquids on the basis of components of known biodegradable and toxicological properties. Using naturally derived moieties would make the currently still largely synthetic ionic liquids bio-renewable and reduce their hazard potential.[98] Syntheses using lactic acid, acetic acid, sugars and sugar substitutes, or amino acids have been presented.[58–63] In particular, the latter are of great interest in the production of chiral ionic liquids due to their inherent chirality and the fact that they constitute the largest natural source of quaternary nitrogen precursors.[99] While these bio-based ionic liquids show good biodegradability and biocompatibility, their physical properties and their suitability for given processes still have to be evaluated, in particular their water miscibility.

Another type of solvents that should also be mentioned are the so-called deep eutectic solvents (DES). They are mixtures of ammonium or metal salts with hydrogen-bond donors showing low melting points. Reported to have low vapor pressure and low flammability, DESs are claimed to be a potentially safer, cheaper, and more biocompatible alternative to ionic liquids.[93,100] Typical solvents are composed of choline chloride combined with amines, carboxylic acids, or alcohols.[93,101,102] While they have been applied in chemical[103,104] and enzymatic catalysis,[100] to date no applications in whole-cell biocatalysis have been presented. The future will show if these solvents might also be suitable for applications in whole-cell biotransformations. First steps in this direction have been made by developing a procedure to suspend microorganisms in pure DESs.[105] After suspension and incubation of up to 24 hours in a mixture of glycerol and choline chloride, the *E. coli* cells still showed good viability.

From the point of view of reaction design, a subject that will probably gain much attention is the recycling of the ionic liquid phase. Reusing ionic liquid phase over several process cycles would not only be ecologically favorable, but also of large financial interest, considering the still very large cost of ionic liquids in comparison to commonly used organic solvents. This would help to significantly reduce environmental exposure to this class of solvents, considerably limiting potential toxic effects, and render industrial applications of ionic liquids economically feasible. While several investigations of the recyclability of ionic liquids have been carried out, the number of cycles has been limited to mainly less than 10, and the processes considered were either chemical[106–108] or enzymatic catalysis.[109–111] Another interesting application is the use of supercritical fluids. To date, this trend has only been observed in chemical and enzymatic catalyses, but it might, in future, also appear in whole-cell biocatalyses.

As was described above, biphasic reaction systems facilitate the product recovery. Due to the low vapor pressure of ionic liquids, it is common to recover the product from the ionic liquid by distillation. Although this is a very attractive approach for thermally stable molecules with low boiling points, thermally unstable products will be lost during this procedure due to the elevated temperatures applied. In these cases, extraction of the product from the ionic liquid is a common alternative. A standard method here is the use of organic solvents immiscible with the ionic liquid. However, this technique does not only make a further purification step necessary after the extraction, but it also leads back to what scientists have been seeking to avoid since the beginning by applying ionic liquids: the use of volatile, flammable organic solvents presenting safety issues and having a major impact on the environment. An interesting and highly efficient alternative has been proposed which makes use of supercritical fluids—mostly supercritical CO_2—for the extraction of organic products from the ionic liquid.[112,113] The success of this procedure is based on two sets of properties: first, the different miscibilities of ionic liquids and supercritical CO_2, which permits the formation of a biphasic extraction system; and second, the good solubility of supercritical CO_2 in ionic liquids, which is controlled by pressure, and the insolubility of ionic liquids in supercritical CO_2, which allows an efficient extraction without loss of ionic liquid in the extractant.[30,114] In contrast to organic solvents used for extraction, supercritical CO_2 is nontoxic, nonflammable, and in most cases nonreactive. Additionally, the separation of the product-loaded supercritical fluid from the ionic liquid after the biotransformation is very simple and can be made conveniently by simple variation of pressure. The same procedure is applied to recover the extracted product from the supercritical fluid. Finally, the supercritical fluid can be easily recovered after the last step and reused in a subsequent downstream processing cycle, allowing its frequent recycling. The only drawback of this procedure is the relatively high cost of the equipment needed to produce and control the supercritical conditions of the fluid (see Chapter 4).

As already mentioned, to date, supercritical fluids have mostly been applied in chemical[115–118] and enzymatic catalyses,[119–122] where the method benefits from the stability and negligible solubility of the catalysts in the supercritical fluids. Still, it is not ruled out that these solvents will also find application in whole-cell biocatalysis, first and foremost in syntheses involving poorly volatile or thermally unstable products.

REFERENCES

1 K. M. Docherty, C. F. Kulpa, Jr., *Green Chem.* **2005**, *7*, 185–189.

2 M. Matsumoto, K. Mochiduki, K. Kondo, *J. Biosci. Bioeng.* **2004**, *98*, 344–347.

3 K. O. Evans, *Colloids Surf. A* **2006**, *274*, 11–17.

4 Z. Yang, R. Zeng, Y. Wang, X. Li, Z. Lv, B. Lai, S. Yang, J. Liao, *Food Technol. Biotechnol.* **2009**, *47*, 62–66.

5 K. O. Evans, *Int. J. Mol. Sci.* **2008**, *9*, 498–511.

6 K. O. Evans, *J. Phys. Chem. B* **2008**, *112*, 8558–8562.

7 S. R. T. Cromie, M. G. Del Pópolo, P. Ballone, *J. Phys. Chem. B* **2009**, *113*, 11642–11648.

8 F. Ganske, U. T. Bornscheuer, *Biotechnol. Lett.* **2006**, *28*, 465–469.
9 W. Lou, M. Zong, T. J. Smith, *Green Chem.* **2006**, *8*, 147–155.
10 W. Lou, W. Wang, R. Li, M. Zong, *J. Biotechnol.* **2009**, *143*, 190–197.
11 R. P. Swatloski, J. D. Holbrey, R. D. Rogers, *Green Chem.* **2003**, *5*, 361–363.
12 F. van Rantwijk, R. A. Sheldon, *Chem. Rev.* **2007**, *107*, 2757–2785.
13 S. Bräutigam, S. Bringer-Meyer, D. Weuster-Botz, *Tetrahedron Asymmetry* **2007**, *18*, 1883–1887.
14 R. J. Cornmell, C. L. Winder, G. J. T. Tiddy, R. Goodacre, G. Stephens, *Green Chem.* **2008**, *10*, 836–841.
15 W. Wang, M. Zong, W. Lou, *J. Mol. Catal., B Enzym.* **2009**, *56*, 70–76.
16 H. Pfruender, M. Amidjojo, U. Kragl, D. Weuster-Botz, *Angew. Chem. Int. Ed. Engl.* **2004**, *43*, 4529–4531.
17 H. Pfruender, R. Jones, D. Weuster-Botz, *J. Biotechnol.* **2006**, *124*, 182–190.
18 J. Pernak, *Eur. J. Med. Chem.* **2001**, *36*, 313–320.
19 M. D. Baumann, A. J. Daugulis, P. G. Jessop, *Appl. Microbiol. Biotechnol.* **2004**, *67*, 131–137.
20 A. Cieniecka-Rosłonkiewicz, J. Pernak, J. Kubis-Feder, A. Ramani, A. J. Robertson, K. R. Seddon, *Green Chem.* **2005**, *7*, 855–862.
21 Y. Li, X. Shi, M. Zong, C. Meng, Y. Dong, Y. Guo, *Enzyme Microb. Technol.* **2007**, *40*, 1305–1311.
22 F. Zhang, Y. Ni, Z. Sun, P. Zheng, W. Lin, P. Zhu, N. Ju, *Chin. J. Catal.* **2008**, *29*, 577–582.
23 J. He, L. Zhou, P. Wang, L. Zu, *Process Biochem.* **2009**, *44*, 316–321.
24 S. Lee, W. Chang, A. Choi, Y. Koo, *Korean J. Chem. Eng.* **2005**, *22*, 687–690.
25 P. J. Scammells, J. L. Scott, R. D. Singer, *Aust. J. Chem.* **2005**, *58*, 155–169.
26 K. R. Seddon, A. Stark, M. Torres, *Pure Appl. Chem.* **2000**, *72*, 2275–2287.
27 J. G. Huddleston, A. E. Visser, W. M. Reichert, H. D. Willauer, G. A. Broker, R. D. Rogers, *Green Chem.* **2001**, *3*, 156–164.
28 K. Marsh, *Fluid Phase Equilib.* **2004**, *219*, 93–98.
29 M. H. Ghatee, A. R. Zolghadr, *Fluid Phase Equilib.* **2008**, *263*, 168–175.
30 S. Cantone, U. Hanefeld, A. Basso, *Green Chem.* **2007**, *9*, 954–971.
31 M. Kosmulski, *Thermochim. Acta* **2004**, *412*, 47–53.
32 H. Ohtani, S. Ishimura, M. Kumai, *Anal. Sci.* **2008**, *24*, 1335–1340.
33 R. A. Mantz, P. C. Trulove, in *Ionic Liquids in Synthesis* (Ed. P. Wasserscheid and T. Welton), Wiley-VCH, Weinheim, Germany, **2003**, pp. 56–68.
34 J. Jacquemin, P. Husson, A. A. H. Padua, V. Majer, *Green Chem.* **2006**, *8*, 172–180.
35 W. Hussain, D. J. Pollard, G. J. Lye, *Biocatal. Biotransformation* **2007**, *25*, 443–452.
36 S. Zhang, N. Sun, X. He, X. Lu, X. Zhang, *J. Phys. Chem. Ref. Data* **2006**, *35*, 1475–1518.
37 J. F. Brennecke, E. J. Maginn, *AIChE J.* **2001**, *47*, 2384–2389.
38 C. Roosen, P. Müller, L. Greiner, *Appl. Microbiol. Biotechnol.* **2008**, *81*, 607–614.
39 J. Dupont, R. F. de Souza, P. A. Z. Suarez, *Chem. Rev.* **2002**, *102*, 3667–3692.
40 D. W. Armstrong, L. He, Y. Liu, *Anal. Chem.* **1999**, *71*, 3873–3876.

41 J. Ranke, A. Othman, P. Fan, A. Müller, *Int. J. Mol. Sci.* **2009**, *10*, 1271–1289.

42 T. Matsuda, Y. Yamagishi, S. Koguchi, N. Iwai, T. Kitazume, *Tetrahedron Lett.* **2006**, *47*, 4619–4622.

43 J. Ranke, K. Molter, F. Stock, U. Bottin-Weber, J. Poczobutt, J. Hoffmann, B. Ondruschka, J. Filser, B. Jastorff, *Ecotoxicol. Environ. Saf.* **2004**, *58*, 396–404.

44 C. Pretti, C. Chiappe, D. Pieraccini, M. Gregori, F. Abramo, G. Monni, L. Intorre, *Green Chem.* **2006**, *8*, 238–240.

45 M. Matzke, S. Stolte, K. Thiele, T. Juffernholz, J. Arning, J. Ranke, U. Welz-Biermann, B. Jastorff, *Green Chem.* **2007**, *9*, 1198–1207.

46 J. Ranke, S. Stolte, R. Störmann, J. Arning, B. Jastorff, *Chem. Rev.* **2007**, *107*, 2183–2206.

47 S. Stolte, M. Matzke, J. Arning, A. Böschen, W. Pitner, U. Welz-Biermann, B. Jastorff, J. Ranke, *Green Chem.* **2007**, *9*, 1170–1179.

48 S. Stolte, J. Arning, U. Bottin-Weber, A. Müller, W. Pitner, U. Welz-Biermann, B. Jastorff, J. Ranke, *Green Chem.* **2007**, *9*, 760–767.

49 J. Ranke, A. Müller, U. Bottin-Weber, F. Stock, S. Stolte, J. Arning, R. Störmann, B. Jastorff, *Ecotoxicol. Environ. Saf.* **2007**, *67*, 430–438.

50 S. Stolte, J. Arning, U. Bottin-Weber, M. Matzke, F. Stock, K. Thiele, M. Uerdingen, U. Welz-Biermann, B. Jastorff, J. Ranke, *Green Chem.* **2006**, *8*, 621–629.

51 OECD Chemical group, *OECD Guidelines for the Testing of Chemicals*, OECD, Paris, France, **1993**.

52 K. M. Docherty, J. K. Dixon, C. F. Kulpa Jr., *Biodegradation* **2006**, *18*, 481–493.

53 F. Atefi, M. T. Garcia, R. D. Singer, P. J. Scammells, *Green Chem.* **2009**, *11*, 1595–1604.

54 M. T. Garcia, N. Gathergood, P. J. Scammells, *Green Chem.* **2005**, *7*, 9–14.

55 N. Gathergood, M. T. Garcia, P. J. Scammells, *Green Chem.* **2004**, *6*, 166–175.

56 J. R. Harjani, R. D. Singer, M. T. Garcia, P. J. Scammells, *Green Chem.* **2008**, *10*, 436–438.

57 J. R. Harjani, R. D. Singer, M. T. Garcia, P. J. Scammells, *Green Chem.* **2009**, *11*, 83–90.

58 S. T. Handy, M. Okello, G. Dickenson, *Org. Lett.* **2003**, *5*, 2513–2515.

59 E. B. Carter, S. L. Culver, P. A. Fox, R. D. Goode, I. Ntai, M. D. Tickell, R. K. Traylor, N. W. Hoffman, J. H. Davis, Jr., *Chem. Commun.* **2004**, *6*, 630–631.

60 J. Pernak, I. Goc, I. Mirska, *Green Chem.* **2004**, *6*, 323–329.

61 K. Fukumoto, M. Yoshizawa, H. Ohno, *J. Am. Chem. Soc.* **2005**, *127*, 2398–2399.

62 J. Kagimoto, K. Fukumoto, H. Ohno, *Chem. Commun.* **2006**, *21*, 2254–2256.

63 G. Ou, M. Zhu, J. She, Y. Yuan, *Chem. Commun.* **2006**, *44*, 4626–4628.

64 W. Lou, W. Wang, T. J. Smith, M. Zong, *Green Chem.* **2009**, *11*, 1377–1384.

65 R. J. Cornmell, C. L. Winder, S. Schuler, R. Goodacre, G. Stephens, *Green Chem.* **2008**, *10*, 685–691.

66 O. Dipeolu, E. Green, G. Stephens, *Green Chem.* **2009**, *11*, 397.

67 S. G. Cull, J. D. Holbrey, V. Vargas-Mora, K. R. Seddon, G. J. Lye, *Biotechnol. Bioeng.* **2000**, *69*, 227–233.

68 J. Pernak, K. Sobaszkiewicz, I. Mirska, *Green Chem.* **2003**, *5*, 52–56.

69 J. Pernak, M. Smiglak, S. T. Griffin, W. L. Hough, T. B. Wilson, A. Pernak, J. Zabielska-Matejuk, A. Fojutowski, K. Kita, R. D. Rogers, *Green Chem.* **2006**, *8*, 798–806.
70 A. Lenourry, J. M. Gardiner, G. Stephens, *Biotechnol. Lett.* **2005**, *27*, 161–165.
71 R. Bar, *J. Chem. Technol. Biotechnol.* **1988**, *43*, 49–62.
72 A. Rajagopal, *Enzyme Microb. Technol.* **1996**, *19*, 606–613.
73 Y. Sardessai, *Res. Microbiol.* **2002**, *153*, 263–268.
74 T. Hahn, K. Botzenhart, in *Handbook of Solvents* (Ed. G. Wypych), Chemtec, Toronto, Canada, **2001**, pp. 865–872.
75 S. Isken, J. A. de Bont, *Extremophiles* **1998**, *2*, 229–238.
76 D. J. Gorman-Lewis, J. B. Fein, *Environ. Sci. Technol.* **2004**, *38*, 2491–2495.
77 T. Tanaka, N. Iwai, T. Matsuda, T. Kitazume, *J. Mol. Catal., B Enzym.* **2009**, *57*, 317–320.
78 S. Arai, K. Nakashima, T. Tanino, C. Ogino, A. Kondo, H. Fukuda, *Enzyme Microb. Technol.* **2010**, *46*, 51–55.
79 J. Howarth, P. James, J. Dai, *Tetrahedron Lett.* **2001**, *42*, 7517–7519.
80 A. Wolfson, C. Dlugy, D. Tavor, J. Blumenfeld, Y. Shotland, *Tetrahedron Asymmetry* **2006**, *17*, 2043–2045.
81 D. Weuster-Botz, *Biochem. Eng. J.* **2002**, *11*, 69–72.
82 W. Lou, *Enzyme Microb. Technol.* **2004**, *35*, 190–196.
83 S. Bräutigam, D. Dennewald, M. Schürmann, J. Lutje-Spelberg, W. Pitner, D. Weuster-Botz, *Enzyme Microb. Technol.* **2009**, *45*, 310–316.
84 K. Schroer, E. Tacha, S. Lütz, *Org. Process Res. Dev.* **2007**, *11*, 836–841.
85 R. Kratzer, M. Pukl, S. Egger, B. Nidetzky, *Microb. Cell Fact.* **2008**, *7*, 37.
86 Y. Shi, Y. Fang, Y. Ren, H. Wu, H. Guan, *J. Ind. Microbiol. Biotechnol.* **2008**, *35*, 1419–1424.
87 W. Yang, J. Xu, Y. Xie, Y. Xu, G. Zhao, G. Lin, *Tetrahedron Asymmetry* **2006**, *17*, 1769–1774.
88 P. Phumathon, *Enzyme Microb. Technol.* **1999**, *25*, 810–819.
89 J. Tsai, L. Wahbi, G. Dervakos, G. Stephens, *Biotechnol. Lett.* **1996**, *18*, 241–244.
90 H. A. Vrionis, A. M. Kropinski, A. J. Daugulis, *Biotechnol. Bioeng.* **2002**, *79*, 587–594.
91 L. Collins, A. Daugulis, *Biotechnol. Tech.* **1996**, *10*, 643–648.
92 V. Dossat, *Enzyme Microb. Technol.* **1999**, *25*, 194–200.
93 A. P. Abbott, D. Boothby, G. Capper, D. L. Davies, R. K. Rasheed, *J. Am. Chem. Soc.* **2004**, *126*, 9142–9147.
94 R. L. Gardas, J. A. Coutinho, *Fluid Phase Equilib.* **2008**, *266*, 195–201.
95 J. O. Valderrama, A. Rea′tegui, R. E. Rojas, *Ind. Eng. Chem. Res.* **2009**, *48*, 3254–3259.
96 P. Kim, D. Pollard, J. Woodley, *Biotechnol. Prog.* **2007**, *23*, 74–82.
97 G. A. Baker, S. N. Baker, S. Pandey, F. V. Bright, *Analyst* **2005**, *130*, 800–808.
98 Y. Fukaya, Y. Iizuka, K. Sekikawa, H. Ohno, *Green Chem.* **2007**, *9*, 1155–1157.
99 G. Tao, L. He, N. Sun, Y. Kou, *Chem. Commun.* **2005**, *28*, 3562–3564.
100 J. T. Gorke, F. Srienc, R. J. Kazlauskas, *Chem. Commun.* **2008**, *10*, 1235–1237.
101 A. P. Abbott, G. Capper, D. L. Davies, R. K. Rasheed, V. Tambyrajah, *Chem. Commun.* **2003**, *1*, 70–71.

102 H. G. Morrison, C. C. Sun, S. Neervannan, *Int. J. Pharm.* **2009**, *378*, 136–139.

103 A. P. Abbott, T. J. Bell, S. Handa, B. Stoddart, *Green Chem.* **2006**, *8*, 784–786.

104 C. A. Nkuku, R. J. LeSuer, *J. Phys. Chem. B* **2007**, *111*, 13271–13277.

105 M. C. Gutiérrez, M. L. Ferrer, L. Yuste, F. Rojo, F. del Monte, *Angew. Chem. Int. Ed. Engl.* **2010**, *122*, 2204–2208.

106 W. Chen, L. Xu, C. Chatterton, J. Xiao, *Chem. Commun.* **1999**, *13*, 1247–1248.

107 V. Le Boulaire, R. Grée, *Chem. Commun.* **2000**, *22*, 2195–2196.

108 S. T. Handy, X. Zhang, *Org. Lett.* **2001**, *3*, 233–236.

109 S. Nara, J. Harjani, M. Salunkhe, *Tetrahedron Lett.* **2002**, *43*, 2979–2982.

110 N. M. T. Lourenço, S. Barreiros, C. A. M. Afonso, *Green Chem.* **2007**, *9*, 734–736.

111 E. Fehér, V. Illeova, I. Kelemenhorvath, K. Belafibako, M. Polakovic, L. Gubicza, *J. Mol. Catal., B Enzym.* **2008**, *50*, 28–32.

112 L. A. Blanchard, J. F. Brennecke, *Ind. Eng. Chem. Res.* **2001**, *40*, 287–292.

113 M. C. Kroon, J. van Spronsen, C. J. Peters, R. A. Sheldon, G. Witkamp, *Green Chem.* **2006**, *8*, 246–249.

114 L. A. Blanchard, D. Hancu, E. J. Beckman, J. F. Brennecke, *Nature* **1999**, *399*, 28–29.

115 A. Bösmann, G. Franciò, E. Janssen, M. Solinas, W. Leitner, P. Wasserscheid, *Angew. Chem. Int. Ed. Engl.* **2001**, *40*, 2697–2699.

116 R. A. Brown, P. Pollet, E. McKoon, C. A. Eckert, C. L. Liotta, P. G. Jessop, *J. Am. Chem. Soc.* **2001**, *123*, 1254–1255.

117 F. Liu, M. B. Abrams, R. T. Baker, W. Tumas, *Chem. Commun.* **2001**, *5*, 433–434.

118 M. F. Sellin, P. B. Webb, D. J. Cole-Hamilton, *Chem. Commun.* **2001**, *8*, 781–782.

119 P. Lozano, T. D. Diego, D. Carrié, M. Vaultier, J. L. Iborra, *Chem. Commun.* **2002**, *7*, 692–693.

120 M. T. Reetz, W. Wiesenhöfer, G. Franciò, W. Leitner, *Chem. Commun.* **2002**, *9*, 992–993.

121 M. Reetz, W. Wiesenhöfer, G. Franciò, W. Leitner, *Adv. Synth. Catal.* **2003**, *345*, 1221–1228.

122 P. Lozano, T. Diego, M. Larnicol, M. Vaultier, J. L. Iborra, *Biotechnol. Lett.* **2006**, *28*, 1559–1565.

8

NONSOLVENT APPLICATIONS OF IONIC LIQUIDS IN BIOTRANSFORMATIONS

Pablo Domínguez de María and Christina Kohlmann

8.1 INTRODUCTION

As extensively discussed in this book, one of the most attractive properties of ionic liquids (ILs) is their versatility to be finely tuned in cation and anion composition. Obviously, physical–chemical properties of the IL widely change from one derivative to another, and therefore significant room for designing is feasible. As a consequence of this variability ILs offer a broad portfolio of applications, since one can smartly design hydrophilic or hydrophobic ILs, miscible or immiscible with water, and so on, to adapt the IL properties for certain, specific application. Therefore, the range of possibilities derived from ILs can be significantly enhanced, in relation to the options that conventional organic solvents may provide. In the field of biotransformations, examples of ILs as pure solvents (either hydrophobic or hydrophilic), or as co-solvents (e.g., with buffer systems) have been extensively reported, despite the topic took off just a decade ago (see previous chapters).

Remarkably, ILs are not only solvents. ILs are composed of salts, and it is well known that anions and cations may play an influence in proteins, by interacting with amino acids that are present in the peptide scaffold. Conclusively, when ILs are used

Ionic Liquids in Biotransformations and Organocatalysis: Solvents and Beyond, First Edition. Edited by Pablo Domínguez de María.

in biotransformations, they will not act as *simple* solvents, but will also influence (positive or negative) at the molecular level of the protein, modifying microenvironments, removing water molecules in some cases, and so on. These properties are especially relevant when ILs are used as co-solvents in aqueous solutions. In these cases, the IL often loses its structure, and the result is an aqueous solution of ions, where apparently the Hofmeister series might rationalize their influence in some cases, albeit not in all of them (see Chapters 2 and 5 for extended discussions on this aspect). As stated in Chapter 3, it is questionable whether these aqueous solutions of ions could be formally regarded as "ionic liquids," although in the literature this is commonly assumed. Importantly for this chapter, since these ions actually influence the catalytic behavior of enzymes, other (nonsolvent) uses of ILs can be envisaged.[1]

Likewise, again by modulating the IL properties, other applications (e.g., as membrane or in process development) can be set, thus facilitating the workup conditions or the biocatalyst recyclability, and so on. These applications (especially the combinations with other solvents or supercritical fluids) are extensively treated in Chapter 4 of this book, and thus in this chapter only brief insights into them will be given.

Finally, being composed of ions, ILs may provide a beneficial work frame in the field of electro(bio)chemistry. Some interesting nonsolvent applications have also been reported in that specific field.

In the following sections of this chapter, different nonsolvent applications of ILs in biotransformations will be discussed.

8.2 IONIC LIQUIDS AS ADDITIVES IN BIOTRANSFORMATIONS

Probably the *simplest* way to apply ILs in biotransformations is their use as mere additives in aqueous solutions, for example, adding catalytic amounts of ILs to explore the effects of the ions of the IL in enzymatic performances. One example of this was the remarkable enhanced enantioselectivity and acceleration of the enzymatic rate observed in the hydrolytic desymmetrization of 2,2-disubstituted prochiral malonic acid diesters catalyzed by pig liver esterase (PLE).[2–4] A proper combination of a co-solvent (*iso*propanol, 10% v/v) together with catalytic amounts of some ILs (<1% v/v) led to this outstanding observation (Table 8.1).

As observed (Table 8.1), both cation and anion of the IL played a role in the enzymatic outcome, in terms of enantioselectivity and reaction rate. Remarkably, PLE immobilized onto *Eupergit*® could be reused for several cycles without showing any loss in the catalytic activity. Therefore, not only an enhanced stereoselection and rate were observed, but also a proper stability to conciliate enzyme costs with productivities was reached. Other commonly used ILs such as [BMIM][BF_6] did not lead to improvements in the enzymatic performance, emphasizing again the importance that each ion composing ILs may have in the final outcome of a certain biocatalytic reaction.

Analogous positive effects have been observed in other hydrolases (namely lipases). For instance, lipase from *Mucor javanicus* shows enhanced stability and activity in hydrolytic reactions when pretreated with [EMIM][BF_4].[5] Also, *Rhizomucor*

TABLE 8.1. Overview of the PLE-Catalyzed Desymmetrization of 2-methyl-2-phenyl-malonic Acid Diester in the Presence of Catalytic Amounts of Some Ionic Liquids[1–4]

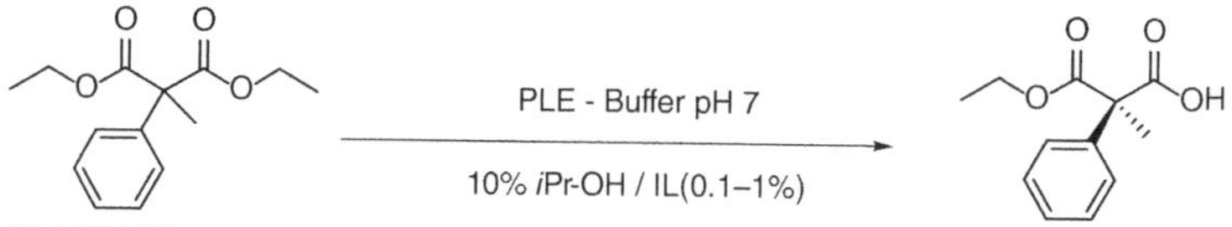

Ionic liquid (0.1–1%)	Time (min)	Conversion (%)	ee (%) (*R*)
Buffer	95	>90	78
Buffer / *i*Pr-OH (10% v/v)	200	>90	95
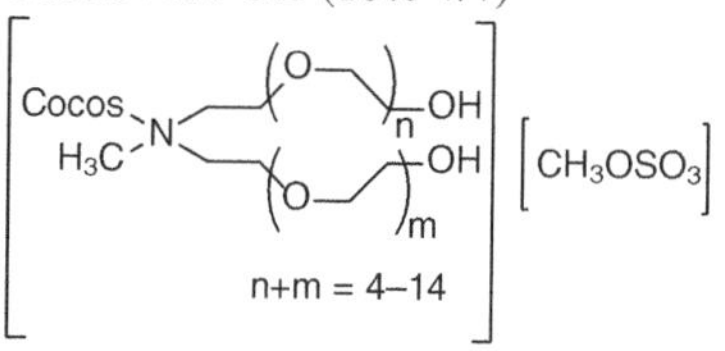 **Ammoeng™ 100**	55	>90	97
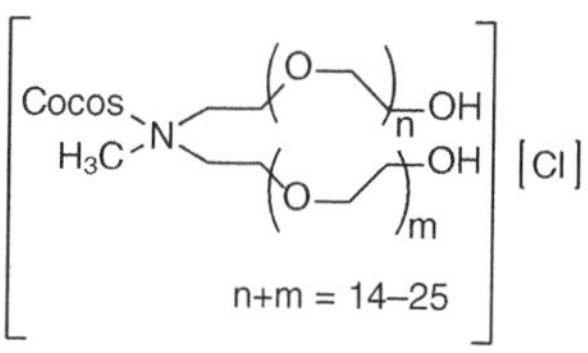 **Ammoeng™ 101**	65	>90	97
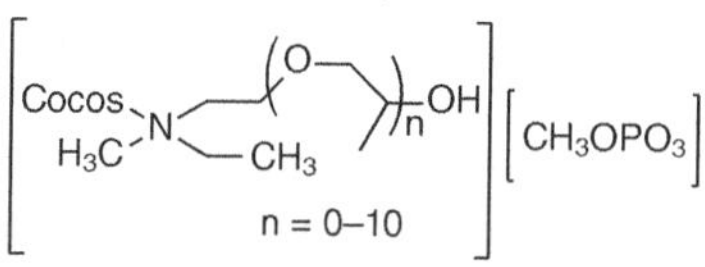 **Ammoeng™ 112**	95	>90	95
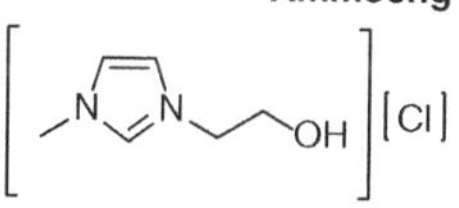	105	>90	92

miehei lipase displays higher rates in the regioselective hydrolysis of peracetylated disaccharides in the presence of some amounts of different ILs.[6] Another example that clearly addresses how relevant the subtle changes in the IL structure can be is the work reported by Itoh et al.[7,8] studying the effect on the enantioselectivity (*E*) that various different ILs had in kinetic resolutions of racemic alcohols, by means of lipase-catalyzed transesterifications. Therein ILs were added as additives (~10 mol% compared with substrate) to the applied nonconventional media (*i*Pr_2O, organic solvent). Excellent improvements in enantioselectivities were observed when 1-butyl-2,3-dimethylimidazolium [BDMIM] was used to form the IL, regardless of the anion.

Conversely, low enantioselectivities were observed when ILs based on 1-butyl-3-methylimidazolium [BMIM] were added (Scheme 8.1). As a plausible explanation for these differences, the presence of a 2-proton in the imidazolium ring [BMIM] with high acidity was pointed out, since then [BMIM] might act as a Brønsted acid catalyst in transesterifications, therefore causing the observed reduction in enantioselectivity.[7,8]

8.3 IONIC LIQUIDS FOR COATING ENZYMES: THE ILCE CONCEPT

Another important nonsolvent option of ILs in biocatalysis is to coat the enzyme with different ILs, leading to improved properties of enantioselectivities, activities, or stabilities. One example, recently reported, was to graft the surface of formate dehydrogenase (FDH) from *Candida boidinii* with different chaotropic cations in the surface, leading to an enhanced stability of the enzyme (see Scheme 6.1, Chapter 6).[9] In an analogous approach, immobilization of *Pseudomonas cepacia* lipase was conducted by coating it with 1-(3′-phenylpropyl)-3-methylimidazolium hexafluorophosphate, [PPMIM][PF_6], IL with a melting point of 53°C.[10] Thus, IL was firstly melted, lipase was aggregated, and the stirred mixture was cooled. Solid material was then cut into pieces, serving as more stable immobilized catalyst (reused for several times), and displaying often a remarkable higher enantioselectivity. This is the so-called ionic liquid-coated-enzyme (ILCE) concept. Analogously, other authors have reported similar results with the same enzyme but with different IL coating agents.[7,8] A summary of most relevant results is summarized in Table 8.2. In some cases enantioselectivity was not affected, but the enzymatic reaction resulted much faster than experiments performed with a native enzyme. Experiments reported with MALDI-TOF mass spectrometric analysis of the ILCE derivatives showed that the ionic liquid-coated agent actually binds the enzyme, thus modifying the microenvironment of the biocatalyst.[8] Albeit much more research is needed, results resemble the experiment reported on grafting FDH with chaotropic cations (Scheme 6.1, Chapter 6),[9] showing that ILs may represent a fantastic tool to study/assess protein interactions at the molecular level, by means of these nonsolvent applications.

Analogous approaches of the ILCE concept were reported for *Candida antarctica* lipase B (CALB) at high temperatures (95°C) and in either hexane or solvent-free conditions, providing a higher stability of the biocatalyst under these harsh reaction conditions.[11,12] Similar strategies for enhanced enzyme stabilities have been reported when lipases are co-immobilized with ILs in sol–gel derivatives.[13] Other enzymes such laccases have also proven to be successfully co-immobilized with ILs.[14] Some other remarkable examples include the combination of ILs and polyethyleneglycol (PEG) for the co-lyophilization of enzymes, providing more robust biocatalysts.[15–18]

Combining the aforementioned properties of ILs for coating enzymes, together with the promising electrochemical properties that some ILs may bring, the ILCE concept has also been successfully reported in electro(bio)chemical reactions catalyzed by horseradish peroxidase (HRP) (see Section 8.6 of this chapter).[19,20]

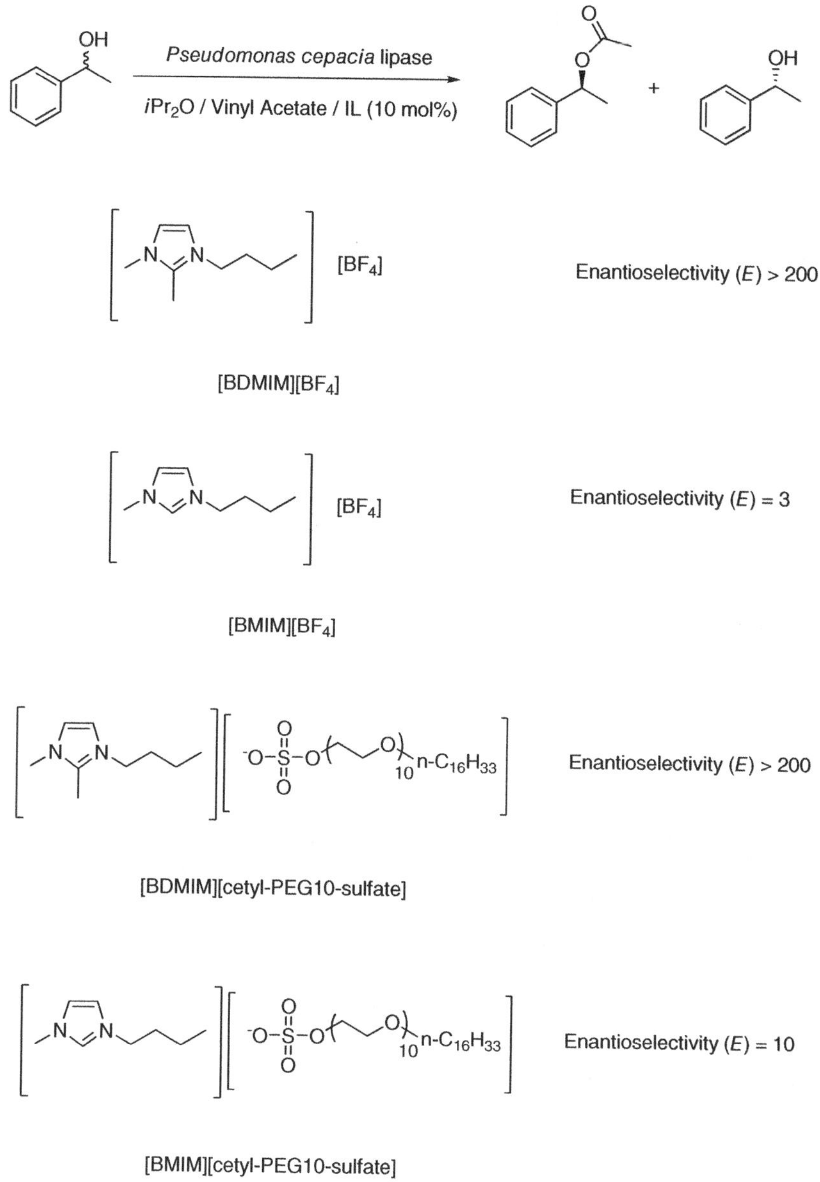

Scheme 8.1. Influence of the IL in the enantioselectivity (*E*) of *Pseudomonas cepacia* lipase-catalyzed kinetic resolution of racemic alcohols.[7,8]

TABLE 8.2. Some Selected Examples on the ILCE Concept, Reported in the Literature for the Lipase-Catalyzed Kinetic Resolution of Alcohols[1,7,8,10]

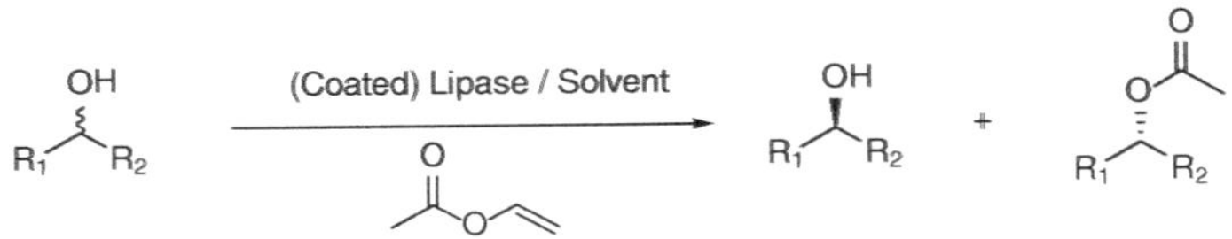

Substrate	Ionic liquid coating agent	Enantioselectivity (E)	Reference
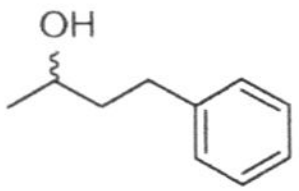	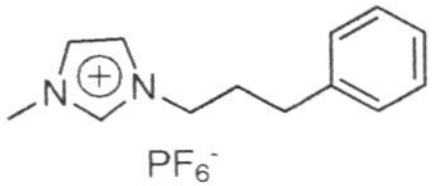	Native: 265 Coated: 532	10
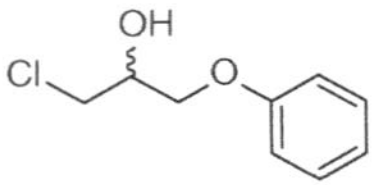	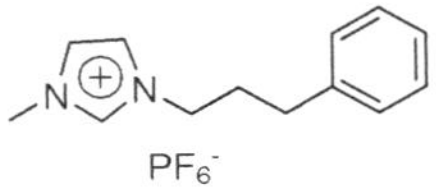	Native: 293 Coated: 574	10
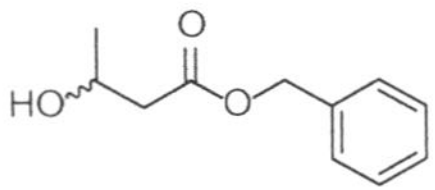	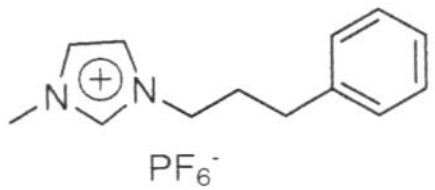	Native: 107 Coated: 156	10
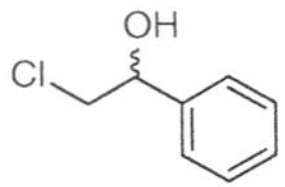	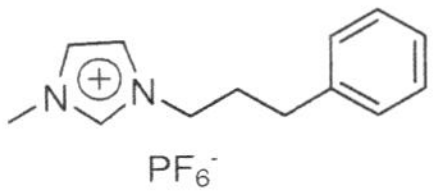	Native: 198 Coated: 176	10
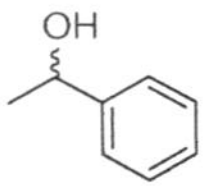	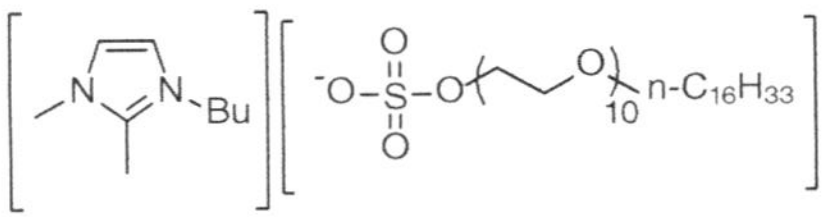	Native: 17 Coated: 96	7

8.4 IONIC LIQUIDS COMBINED WITH MEMBRANES AND BIOTRANSFORMATIONS

Another interesting nonsolvent application of ILs is their use as membranes, with different physical–chemical properties depending on the ion composition of the IL. Applied to biocatalysis, one example was to disperse and/or coat lipases with ILs, and subsequently adsorb them onto ceramic membranes. Some of the derivatives remained stable and active during months.[21] This is a further promising proof-of-principle on how ILs may provide many applications in different areas of chemistry.

Another important lead that ILs may bring is that of supported ionic liquid membranes (SLMs).[22–24] Herein ILs act as a separative phase of two additional phases. Importantly, these two phases can be two organic phases, two aqueous phases, or any other combination of them. This approach may lead to novel concepts in process compartmentalization. In biocatalysis, a remarkable example has been reported (also discussed in other chapters of this book, from different viewpoints).[25–27] In that concept, two lipases and three different phases were used for the kinetic resolution of (*R,S*)-Ibuprofen (Scheme 8.2).

In the first compartment, lipase from *Candida rugosa* catalyzes the enantioselective esterification of (*R,S*)-Ibuprofen with ethanol. High ethanol concentrations (65% v/v) were needed to decrease the water amount and therefore shift the reaction to ester products. Later on, only (*S*)-ibuprofen–ester diffuses selectively over the IL-based membrane to the second compartment. There lipase from porcine pancreas catalyzes the hydrolysis of ester to afford (*S*)-ibuprofen. As it becomes obvious, many other applications or derivations of that concept may be envisaged (e.g., applied to dynamic kinetic resolution strategies).

8.5 IONIC LIQUIDS ANCHORING SUBSTRATES

Another interesting application of ILs derives from the high functionality of the ions commonly employed. This offers a huge number of possibilities for the substrate anchoring to that ILs, facilitating the workup and purification of products, together with the recovery of nonprocessed substrates. This is the so-called ionic liquid-phase organic synthesis (IoLiPOS) concept.[28–30] Several examples of this concept have also been reported in biocatalysis. The kinetic resolution of (*R,S*)-Ibuprofen was reported by anchoring the substrate to an IL and thus conducting the lipase-catalyzed enantioselective hydrolysis, leading to a 41% yield and 86% enantiomeric excess under nonoptimized conditions (proof-of-concept) (Scheme 8.3).[31]

A similar approach combines ILs as solvents, either [BMIM][PF_6] or [BMIM][BF_4], in which a task-specific IL served as a substrate for a kinetic resolution and further separation of enantiomers (Scheme 8.4).[32]

CALB was used as the biocatalyst, and over a sequential two-step transesterification, both enantiomers were separated. The approach simplifies significantly the downstream processing, and thus laborious chromatography separations could be bypassed. Furthermore, since ILs used as solvents were enzyme friendly, the biocatalyst could be

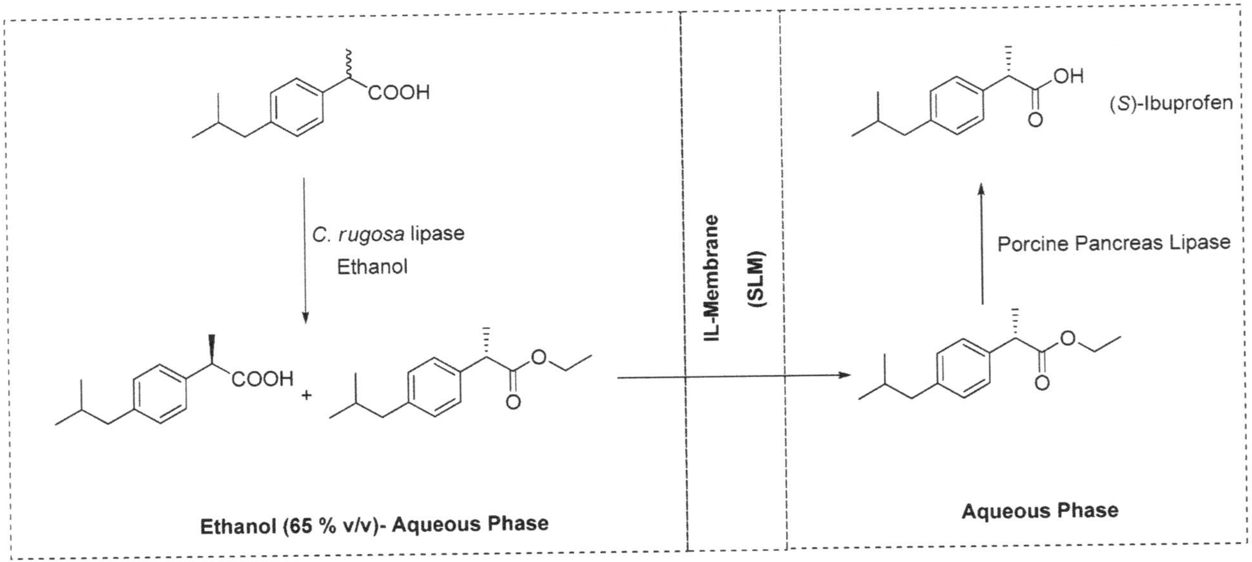

Scheme 8.2. SLM-based concept for the lipase-catalyzed kinetic resolution of (*R*,*S*)-Ibuprofen.[25–27]

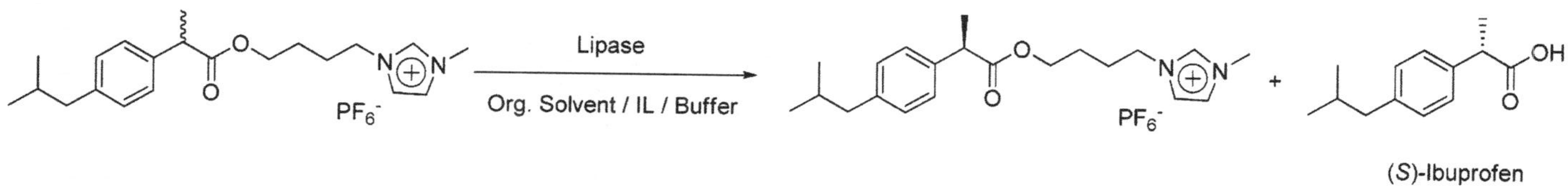

Scheme 8.3. IL-anchoring (*R,S*)-Ibuprofen for the lipase-catalyzed kinetic resolution.[31]

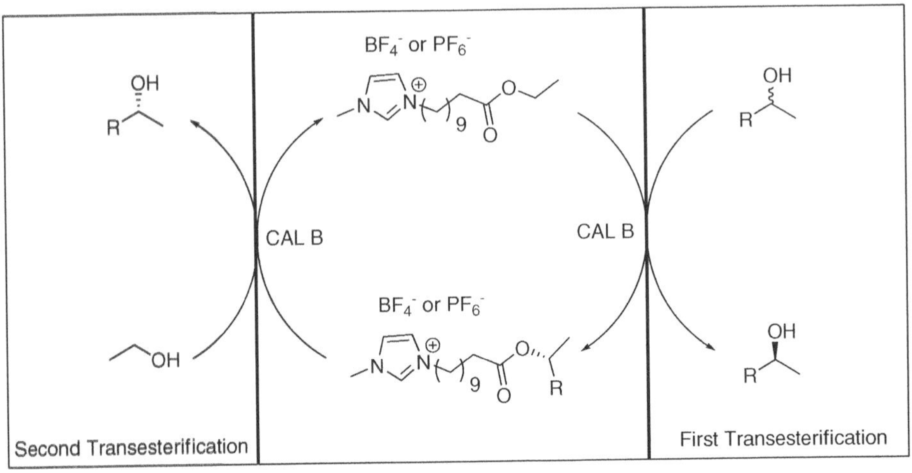

Scheme 8.4. Lipase-catalyzed kinetic resolution using ionic liquids as anchoring substrates.[32]

reused for several times without apparent loss of activity.[32] Analogous anchoring strategies have been reported by others as well.[33]

8.6 IONIC LIQUIDS AND BIOELECTROCHEMISTRY

As stated, ILs consist entirely of ions and therefore show outstanding electrochemical properties, such as high conductivities up to 10 S/m and broad electrochemical windows, that is, electrochemical potential range in which a compound is stable against oxidation or reduction at the electrodes, for example, 1.2 V for water and more than 6 V, depending on the used IL.[34] Consequently, the first successful applications of ILs were in electrochemistry, and thus they also found their way into bioelectrochemistry nowadays.[35] Besides the electrochemical properties, the good compatibility of some ILs with proteins, enzymes, and even whole cells is also beneficial for bioelectrochemistry, as even improved activities or stabilities of some biocatalysts can be obtained.[36,37] Additionally, excellent properties as solubilizers and their potential to stabilize nicotinamide cofactors[38] turn ILs into promising performance additives for bioelectrochemical applications such as biosensors or electroenzymatic syntheses.

Electroenzymatic synthesis is the combination of an oxidoreductase-catalyzed reaction with electrochemical reactant supply on a preparative scale. Unfortunately, to render those reactions in a competitive manner, some challenges need to be tackled. First, the productivity of those reactions is restricted by the conductivity of the aqueous reaction medium, resulting in the need for biocompatible conducting salts. Second, the applied biocatalysts suffer from deactivation at the electrode surface, leading to insufficient biocatalyst utilization. Thus, it would be helpful to stabilize the catalyst under process conditions. Finally, the conversion of hardly water-soluble substrates is difficult, as conventional methods to overcome this limitation are not effective. The addition of an organic co-solvent would cause a decrease in the conductivity of the reaction medium, resulting in even lower productivities and biocatalyst utilizations, whereas the use of a second organic phase would complicate tremendously the reaction setup. In this area, ILs proved to be effective performance additives for different electroenzymatic syntheses (Figure 8.1), being able to overcome those limitations and improve electroenzymatic systems synergistically.[39] For instance, the addition of 2% to 10 vol% of [EMPY][$EtSO_4$] caused a linear increase in the obtainable space–time yield (STY = amount of product per reaction volume and time) for the electrochemical generation of NADPH (Figure 8.1A) and the turnover frequency (TOF = number of regeneration cycles within a defined time frame) of the utilized rhodium mediator, giving rise to 2.7-fold enhanced results for 10 vol% IL. These findings prove the limitation of obtainable productivities by the conductivity of the reaction medium. Furthermore, the IL showed a high potential to stabilize the reduced nicotinamide cofactor in solution, resulting in a three times longer half-life in contrast to pure buffer. For a D-amino acid oxidase-catalyzed resolution of a methionine racemate with electrochemical regeneration of flavin adenine dinucleotide (FAD) by ferrocene carboxylic acid (Figure 8.1B) the obtained STY, $TOF_{\text{ferrocene carboxylic acid}}$, and turnover number of the biocatalyst (TON = produced amount of product in mol per mol biocatalyst) showed also a linear dependence on the amount of IL. At best, STY and TOF were improved by factor 1.5, whereas the TON was enhanced 2.4 times for 10 vol% IL; thus, the IL did not only act as a conducting salt, but also as a stabilizer for the biocatalyst. In a further example, ILs were used as additives for the chloroperoxidase-catalyzed synthesis of (*R*)-phenylmethylsulfoxide with electrochemical generation of H_2O_2 (Figure 8.1C), since the performance of this reaction was not only limited by conductivity and biocatalyst stability issues, but also due to poor solubility of the substrate thioanisol in aqueous media. As even small amounts of IL were able to act as solubilizers, reactions were conducted in the presence of 2 vol% of different IL, always resulting in increased STY and TON. Best STY was obtained with [BMIM][$MeSO_4$], while maximum TTN was achieved with [EMIM][$EtSO_4$]; for both an improvement by a factor of 4.2 was obvious.

Besides being an attractive performance additive for electroenzymatic syntheses, IL can also be utilized for the development of biosensors. Electrochemical biosensors unify the specificity and selectivity of biocatalytic reactions with high-performance electroanalytical techniques. The performance of the sensors is highly dependent on adequate immobilization of the biological components at the electrode surface, as the biochemical reaction is responsible for the generation of an electrical signal that is correlated with the concentration of the detectable analyte. Therefore, special focus was on the development of effective immobilization techniques to gain durable and reliable sensors. To outline the potential of ILs as additives for the preparation of biosensors, some successful examples will be summarized in the following paragraphs.

A straightforward simple method is the direct immobilization of a biocatalyst by an IL film on an electrode surface. For instance, as reported by Yu et al., HRP could be fixed to a glassy carbon electrode by immersion of the electrode in a [BMIM][Cl]-containing aqueous solution of the enzyme, or by pipetting a small amount of the same solution directly onto the surface and subsequent drying.[40] Preparing a biosensor in

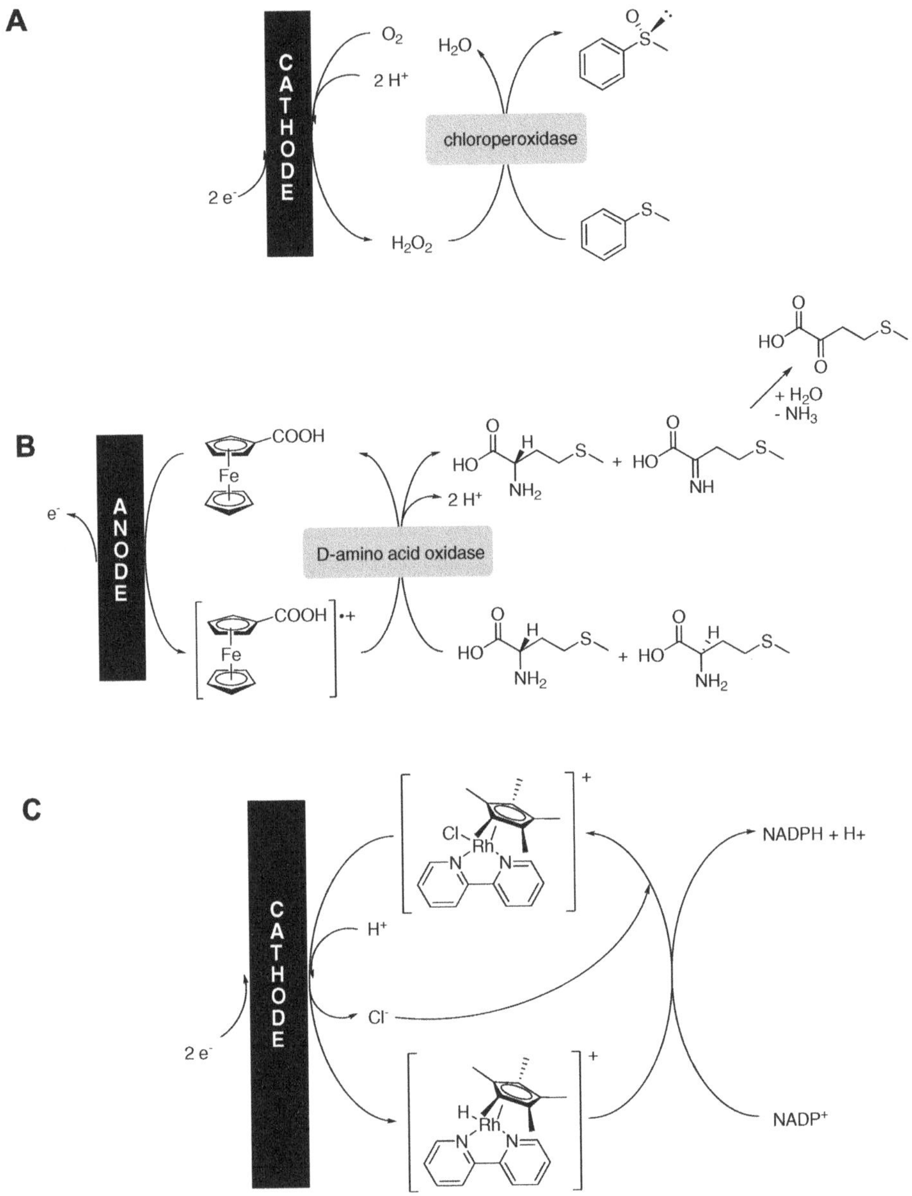

Figure 8.1. Electroenzymatic syntheses tested with ILs as performance additives. (A) Electrochemical generation of NADPH via a rhodium mediator. (B) D-amino acid oxidase-catalyzed resolution of a methionine racemate with electrochemical regeneration of enzyme-bound FAD by ferrocene carboxylic acid. (C) Chloroperoxidase-catalyzed synthesis of (*R*)-phenylmethylsulfoxide with electrochemical generation of H_2O_2.

those manners showed a clear redox peak, while for an electrode without addition of IL no electrochemical reaction could be detected. Thus, the IL enables direct electron transfer between the biocatalyst and the electrode. The electrode showed both high operational and storage stability, as only slight differences in the cyclic voltammogramm were apparent after 50 consecutive cycles or storage in distilled water for one week. Furthermore, while applied in the reduction of O_2, the peak maximum was shifted to less negative potentials, indicating accelerated electron transfer.

López et al. constructed an amperometric biosensor entrapping glucose oxidase in polymerized IL.[40] Therefore, they dispersed the enzyme in an aqueous mixture of 1-vinyl-3-ethylimmidazolium bromide and bisacrylamide as cross-linker. Subsequent emulsion polymerization resulted in microparticles. The biosensor was prepared by covering a platinum electrode with a thin Nafion film to prevent influences of anions present in biological media; the IL microparticles were placed on top and hold back via a dialysis membrane. The obtainable detection limit for glucose of the sensor was dependent on the amount of microparticles on the surface and reached 5.9 μM at best. The biosensor showed high precision and recoveries for monitoring glucose in human serum. Additionally, it exhibited a life time longer than 150 days and could be applied for measuring nonaqueous samples.

In biosensor technology, also the immobilization of the biological component in silica glasses derived from sol–gel is a very common technique.[41] Positive features of those materials are their simple and easy synthesis, their porosity, and their chemical as well as thermal stability.[42] Nevertheless, alcohols formed as by-products during the preparation, as well as diffusion limitations, can hamper the performance of biosensors.[43] Liu et al. determined the influence of IL as additives for the preparation of such a matrix and found improved biocompatibility besides excellent biosensor properties.[44,45] The polymer was synthesized by hydrolysis of tetraethyl orthosilicate in the presence of [BMIM][BF_4]; subsequent immobilization of HRP was achieved by mixing with the IL–sol–gel. In contrast to a gel prepared without addition of IL, the biocatalyst possessed a 30-fold increased activity, which was attributed to the higher porosity of the material. Furthermore, due to the viscosity of the IL, cracking of the silica glass could be prevented. To fabricate a biosensor, in a first step, ferrocene solution working as mediator was pipetted onto the surface of a glassy carbon electrode and subsequently dried. In the next step, the electrode was covered with the HRP–IL–sol–gel. The resulting sensor showed a high sensitivity for H_2O_2 with a detection limit of 1.1 μM and a high stability over 20 days.

Besides silica glasses, chitosan, a linear polysaccharide composed of randomly distributed β-(1-4)-linked D-glucosamine and *N*-acetyl-D-glucosamine is frequently used for the construction of biosensors.[46–49] Chitosan exhibits ideal characteristics as support enzyme immobilization due to its hydrophilicity, biocompatibility, biodegradability, and antibacterial properties. Furthermore, it can be produced from natural recourses since it is derived from the natural polymer chitin. Consequently, different examples of chitosan–IL biosensors are available. For example, Lu et al. developed an electrode comparable to the one from Yu et al. (as described above) with addition of chitosan, applying [BMIM][BF_4] instead of [BMIM][Cl].[49] Yu et al. reported an enzyme surface concentration five times higher than the theoretical monolayer coverage and fivefold increased electroactive HRP. When used in the detection of H_2O_2, the biosensor exhibited a sensitivity of 184 mA/cm^2 M and a detection limit of 0.25 μM. Zheng and coworkers achieved the immobilization of glucose oxidase on a nanogold electrode via electrodeposition of a chitosan–[BMIM][BF_4]–enzyme mixture on the surface of the electrode.[50] The emerging biosensor possessed a detection limit of 1.5 μM for glucose as well as high precision and could therefore be applied successfully in the monitoring of the latter from human serum yielding satisfying recovery.

Moreover, IL can be applied as a binder for the preparation of biosensors, in which the biological component is immobilized in the presence of a conductive support, for example, carbon composite materials. Vieira and coworkers published the preparation of different biosensors using IL as a binder for graphite powder. For instance, a biosensor for the detection of chlorogenic acid was prepared by mixing polyphenol oxidase immobilized in cross-linked chitosan with iridium nanoparticles dispersed in [BMIM][PF_6] and graphite powder.[51] In contrast to sensors where pure nujol (a conventional binder) or nujol–IL mixtures are used, the application of pure ILs showed highest sensitivities. The biosensor exhibited a detection limit of 0.915 μM for chlorogenic acid. When utilized in the determination of the latter in coffee, the biosensor showed results comparable to capillary electrophoresis. A further biosensor for the detection of rosmarinic acid was prepared in a similar way by the same working group.[52] In this case a peroxidase was immobilized on chitosan cross-linked with citrate; the immobilisate and graphite powder were merged, followed by dispersion in a mixture of [BMIM][$(CF_3SO_2)_2N$] and nujol. The obtained biosensor showed a detection limit of 0.0725 μM and excellent recovery when measuring samples from plant extracts. Besides carbon powders (or nanoparticles), carbon nanotubes also gained considerable attention, as they possess large surface areas, excellent electronic properties, as well as chemical and thermal stability. Kachoosangi et al. prepared an electrode of multiwall carbon nanotubes (MWCNT) and *n*-octylpyridinium hexafluorophosphate and found best electrochemical properties for a composition of 10 mass% MWCNT and 90 mass% IL.[53] In contrast to electrodes prepared with conventional binders, the [OPY][PF_6]-based electrode showed superior performance: The electrochemical signal exhibited by the Teflon electrode was negligible, which can be attributed to the low conductivity of Teflon. However, when using mineral oil as a binder, an effective electrode was composed; nevertheless, the [OPY][PF_6] electrode showed a 10 times higher signal. It was possible to incorporate glucose oxidase in the MWCNT–IL–composite by simple mixing. The resulting biosensor was successfully applied in the detection of glucose. While the biosensor prepared with mineral oil as a binder showed a response of only 0.003 μA/mM for glucose, the [OPY][PF_6]-based sensor exhibited a response of 5 μA/mM.

Altogether, ILs exhibit excellent properties to be used in bioelectrochemical applications. For biosensors, the utilization of IL broadens the field of possible manufacturing procedures and effectively contributes to the formation of robust and efficient biosensors. For electroenzymatic syntheses, ILs are promising performance additives, as they do not only act as conducting salts, but also further proved to stabilize reduced nicotinamide cofactors and enzymes in solution, and are able to dissolve hardly water soluble substrates.

REFERENCES

1 P. Domínguez de María, *Angew. Chem. Int. Ed. Engl.* **2008**, *47*, 6960–6968.

2 S. Wallert, K. H. Drauz, I. Grayson, H. Gröger, P. Domínguez de María, C. Bolm, *Green Chem.* **2005**, *7*, 602–605.

3 S. Wallert, I. Grayson, K. H. Drauz, I. Grayson, H. Gröger, C. Bolm, P. Domínguez de María, F. Chamouleau, WO 2006/005409 A1. **2006**.

4 P. Domínguez de María, C. A. García Burgos, G. Bargeman, R. W. van Gemert, *Synthesis* **2007**, *10*, 1439–1452.

5 D. T. Dang, S. H. Ha, S. M. Lee, W. J. Chang, Y. M. Koo, *J. Mol. Catal., B Enzym.* **2007**, *45*, 118–121.

6 M. Filice, J. M. Guisán, J. M. Palomo, *Green Chem.* **2010**, *12*, 1365–1369.

7 T. Itoh, S. Han, Y. Matsushita, S. Hayase, *Green Chem.* **2004**, *6*, 437–439.

8 (a) Y. Abe, K. Yoshiyama, Y. Yagi, S. Hayase, M. Kawatsura, T. Itoh, *Green Chem.* **2010**, *12*, 1976–1980; (b) T. Itoh, Y. Matsushita, Y. Abe, S. H. Han, S. Wada, S. Hayase, M. Kawatsura, S. Takai, M. Morimoto, Y. Hirose, *Chem. Eur. J.* **2006**, *12*, 9228–9237.

9 M. Bekhouche, B. Doumeche, L. J. Blum, *J. Mol. Catal., B Enzym.* **2010**, *65*, 73–78.

10 J. K. Lee, M. J. Kim, *J. Org. Chem.* **2002**, *67*, 6845–6847.

11 P. Lozano, R. Piamtongkam, K. Kohns, T. De Diego, M. Vaultier, J. L. Iborra, *Green Chem.* **2007**, *9*, 780–784.

12 P. Lozano, T. De Diego, T. Sauer, M. Vaultier, S. Gmouh, J. L. Iborra, *J. Supercrit. Fluids* **2007**, *40*, 93–100.

13 S. H. Lee, T. T. Ngoc Doan, S. H. Ha, Y. M. Koo, *J. Mol. Catal., B Enzym.* **2007**, *45*, 57–61.

14 M. B. Turner, S. K. Spear, J. D. Holbrey, R. D. Rogers, *Biomacromolecules* **2004**, *5*, 1379–1384.

15 K. Nakashima, J. Okada, T. Maruyama, N. Kamiya, M. Goto, *Sci. Technol. Adv. Mater.* **2006**, *7*, 692–698.

16 K. Nakashima, T. Maruyama, N. Kamiya, M. Goto, *Chem. Commun.* **2005**, 4297–4299.

17 T. Maruyama, H. Yamamura, T. Kotani, N. Kamiya, M. Goto, *Org. Biomol. Chem.* **2004**, *2*, 1239–1244.

18 T. Maruyama, S. Nagasawa, M. Goto, *Biotechnol. Lett.* **2002**, *24*, 1341–1345.

19 F. Zhao, X. Wu, M. Wang, Y. Liu, L. Gao, S. Dong, *Anal. Chem.* **2004**, *76*, 4960–4967.

20 Y. Liu, M. Wang, J. Li, Z. Li, P. He, H. Liu, J. Li, *Chem. Commun.* **2005**, 1778–1780.

21 M. Mori, R. Gómez García, M. P. Belleville, D. Paolucci-Jeanjean, J. Sánchez, P. Lozano, M. Vaultier, G. Ríos, *Catal. Today* **2005**, *104*, 313–317.

22 C. P. Mehnert, *Chem. Eur. J.* **2005**, *11*, 50–56.

23 M. H. Valkenberg, C. de Castro, W. F. Hölderich, *Green Chem.* **2002**, *4*, 88–93.

24 L. C. Branco, J. G. Crespo, C. A. M. Afonso, *Angew. Chem. Int. Ed. Engl.* **2002**, *41*, 2771–2773.

25 E. Miyako, T. Maruyama, N. Kamiya, M. Goto, *Chem. Commun.* **2003**, 2926–2927.

26 E. Miyako, T. Maruyama, N. Kamiya, M. Goto, *Biotechnol. Lett.* **2003**, *25*, 805–808.

27 E. Miyako, T. Maruyama, N. Kamiya, M. Goto, *J. Biosci. Bioeng.* **2003**, *96*, 370–374.

28 S. T. Handy, M. Okello, *J. Org. Chem.* **2005**, *70*, 2874–2877.

29 H. Hakkou, J. J. Eynde, J. Hamelin, J. P. Bazureau, *Synthesis* **2004**, 1793–1798.

30 W. Miao, T. H. Chan, *Org. Lett.* **2003**, *5*, 5003–5005.

31 P. U. Naik, S. J. Nara, J. R. Harjani, M. M. Salunkhe, *J. Mol. Catal., B Enzym.* **2007**, *44*, 93–98.

32 N. M. T. Lourenço, C. A. M. Afonso, *Angew. Chem. Int. Ed. Engl.* **2007**, *46*, 8178–8181.

33 M. Brossat, T. S. Moody, S. J. C. Taylor, J. W. Wiffen, *Tetrahedron Asymmetry* **2009**, *20*, 2112–2116.

34 F. Endres, *Z. Phys. Chem.* **2004**, *218*, 255–283.

35 P. Wasserscheid, T. Welton, *Ionic Liquids in Synthesis*, Vol. 1, 2nd ed., Wiley-VCH, Weinheim, **2008**.

36 F. van Rantwijk, R. A. Sheldon, *Chem. Rev.* **2007**, *107*, 2757–2785.

37 C. Roosen, P. Müller, L. Greiner, *Appl. Microbiol. Biotechnol.* **2008**, *81*, 607–614.

38 C. Kohlmann, L. Greiner, S. Lütz, DE 10 2008 061 866.7. **2008**.

39 C. Kohlmann, L. Greiner, W. Leitner, C. Wandrey, S. Lütz, *Chem. Eur. J.* **2009**, *15*, 11692–11700.

40 P. Yu, Y. Q. Lin, L. Xiang, L. Su, J. Zhang, L. Q. Mao, *Langmuir* **2005**, *21*, 9000–9006.

41 M. S.-P. Lopez, D. Mecerreyes, E. Lopez-Cabarcos, B. Lopez-Ruiz, *Biosens. Bioelectron.* **2006**, *21*, 2320–2328.

42 M. E. Tess, J. E. Cox, *J. Pharm. Biomed. Anal.* **1999**, *19*, 55–68.

43 Y. Liu, L. Shi, M. Wang, Z. Li, H. Liu, J. Li, *Green Chem.* **2005**, *7*, 655–658.

44 Y. Liu, M. Wang, Z. Li, H. Liu, P. He, J. Li, *Langmuir* **2005**, *21*, 1618–1622.

45 S. C. Fernandes, I. R. W. Z. Oliveira, I. C. Vieira, *Enzyme Microb. Technol.* **2007**, *40*, 661–668.

46 S. C. Fernandes, I. R. W. Z. Oliveira, O. Fatibello-Filho, A. Spinelli, I. C. Vieira, *Sens. Actuators B Chem.* **2008**, *133*, 202–207.

47 I. R. W. Z. Oliveira, S. C. Fernandes, I. C. Vieira, *J. Pharm. Biomed. Anal.* **2006**, *41*, 366–372.

48 T. J. Castilho, M. P. T. Sotomayor, L. T. Kubota, *J. Pharm. Biomed. Anal.* **2005**, *37*, 785–791.

49 X. Lu, Q. Zhang, L. Zhang, J. Li, *Electrochem. Commun.* **2006**, *8*, 874–878.

50 X. D. Zeng, X. F. Li, L. Xing, X. Y. Liu, S. L. Luo, W. Z. Wei, B. Kong, Y. H. Li, *Biosens. Bioelectron.* **2009**, *24*, 2898–2903.

51 S. C. Fernandes, S. K. Moccelini, C. W. Scheeren, P. Migowski, J. Dupont, M. Heller, G. A. Micke, I. C. Vieira, *Talanta* **2009**, *79*, 222–228.

52 K. dos Santos Maguerroski, S. C. Fernandes, A. C. Franzoi, I. C. Vieira, *Enzyme Microb. Technol.* **2009**, *44*, 400–405.

53 R. T. Kachoosangi, M. M. Musameh, I. Abu-Yousef, J. M. Yousef, S. M. Kanan, L. Xiao, S. G. Davies, A. Russell, R. G. Compton, *Anal. Chem.* **2009**, *81*, 435–442.

PART III

IONIC LIQUIDS IN ORGANOCATALYSIS

9

IONIC LIQUIDS AS (CO-) SOLVENTS AND CO-CATALYSTS FOR ORGANOCATALYTIC REACTIONS

Štefan Toma and Radovan Šebesta

9.1 NONTRADITIONAL MEDIA IN ORGANOCATALYSIS

Research in alternative reaction media for organic synthesis is an important part of green chemistry.[1] Ionic liquids, for their adjustable properties, represent the most promising candidates. Convenience of ionic liquids as replacement media in organic synthesis has been demonstrated for almost all kinds of reactions.[2] Furthermore, ionic liquids can be tailor-made for special applications, and such *task-specific ionic liquids* offer exciting possibilities for achieving desired properties or functions. A range of catalysts and reagents has been supported in the ionic liquid phase, often resulting in enhanced reactivities and selectivities of numerous important transformations.[3]

Development of organocatalysis led to the identification of many efficient transformations. Achieved progress was summed up in two books[4] and several reviews.[5] Proline and other amino acid-derived organocatalysts or their analogues are acting via formation of enamines[6] or iminium salts[7] with aldehydes or ketones. A special type of organocatalysts is chiral urea or thiourea derivatives, which are acting as hydrogen-bond catalysts[8] or chiral Brønsted acids.[9] The most frequently used solvents for organocatalytic reactions are dimethyl sulfoxide (DMSO), dimethylformamide (DMF),

Ionic Liquids in Biotransformations and Organocatalysis: Solvents and Beyond, First Edition. Edited by Pablo Domínguez de María.

acetonitrile chloroform, or alcohols. Water, sea-water, or brine is also an attractive, environmentally benign, reaction medium for organocatalytic reactions.[10] Some organocatalytic reactions benefit from presence of water; however, this issue is somewhat ambiguous (see also Chapter 10).[11]

One of the main drawbacks of organocatalytic reactions in common solvents is that reactions are rather slow in spite of high catalyst loading. The amount of organocatalyst often reaches as much as 20 mol% or even 30 mol%, and this creates difficulties with product isolation and purification. This factor indicates that atom economy of such reactions is very low. It is not surprising therefore that there were attempts to solve these problems by finding more suitable solvents for organocatalytic reactions and ionic liquids are promising candidates.

9.2 EARLY ORGANOCATALYTIC REACTIONS IN IONIC LIQUIDS

Three research groups started independent investigation of the application of ionic liquids as reaction media for organocatalytic reactions. The first organocatalytic reaction performed in ionic liquids was an aldol reaction of aromatic aldehydes with acetone and alicyclic ketones.[12] In the reaction depicted in Scheme 9.1, we found that using ionic liquids enabled lowering the amount of (*S*)-proline from 30 mol% to 5 mol% and still the product was isolated in 89% yield and with 74% ee. Reaction worked well also with only 1 mol% of the catalyst. Yield then decreased to 74%, but the enantioselectivity was not affected. Even more important was the fact that the reaction medium could be reused in the next reaction (after 2 hours drying *in vacuo* at 40°C and without addition of the catalyst). In the third cycle, the product was isolated in 74% yield with 63% ee. Similar results were achieved with a range of benzaldehydes having both electron-donating and electron-withdrawing substituents. Enatioselectivity of the reaction was higher in ionic liquid than in DMSO. Interesting results were achieved also at aldol reactions with alicyclic ketones. The results are summarized in Table 9.1.

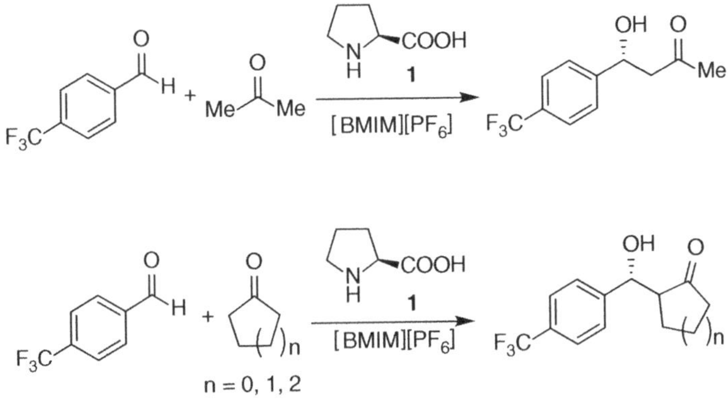

Scheme 9.1. (*S*)-Proline catalyzed an aldol reaction of 4-trifluomethylbenzaldehyde with acetone and cyclic ketones in ionic liquid.

TABLE 9.1. (S)-Proline-Catalyzed Aldolization of 4-Trifluorbenzaldehyde with Alicyclic Ketones

n	Yield (%)	*anti* : *syn*	ee *anti* (%)	ee *syn* (%)
1	94	1:1	32	86
2	91	20:1	93	n.d.
0	53	6:1	10	8

n.d., not determined.

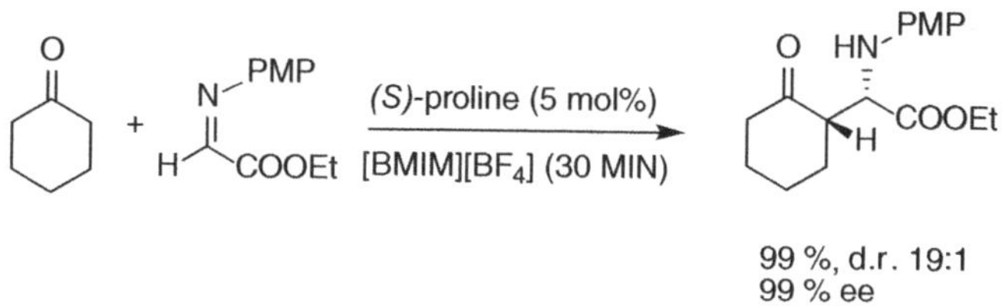

Scheme 9.2. (*S*)-Proline-catalyzed Mannich reaction in ionic liquid.

Practically at the same time, Loh and coworkers published a similar investigation of aldol reaction in ionic liquids.[13] The authors tested four different ionic liquids, among them [BMIM][PF_6]. They used 30 mol% of (*S*)-proline as the catalyst, and prolonged reaction time (up to 48 hours). However, isolated yields of aldol products were slightly lower (60–75%) and reaction mixture contained up to 34% of the elimination product. Enantioselectivity of the reactions ranged between 58% and 71%.

In the third work, Barbas described a Mannich reaction of *N*-PMP-protected α-imino ethyl glyoxylate with cyclohexanone.[14] The reaction, using 5 mol% of (*S*)-proline, was 4–50 times faster in [BMIM][BF_4] than in DMSO. Interestingly, diastereoselectivity as well as enantioselectivity of the reaction was excellent (d.r. 19:1, 99% ee) (Scheme 9.2). Ionic liquid containing the catalyst was used in four consecutive reaction cycles with only a slight decrease in yield and constant enantioselectivity. Authors also noted poor performance of hydroxyacetone in such a Mannich reaction in ionic liquid.

9.3 IONIC LIQUIDS AS SOLVENTS FOR ORGANOCATALYTIC REACTIONS

9.3.1 Aldol Reactions

Development of (*S*)-proline-catalyzed aldol reactions in ionic liquids continued with auto-aldol and cross-aldol reactions between aliphatic aldehydes.[15] Scheme 9.3 shows that the best results in these reactions were achieved using a 1.5:1 mixture of ionic liquid and DMF.

Product yields were only 65% in the neat ionic liquid, and oligomers of propionaldehyde were formed. After extraction of the reaction product, the remaining reaction

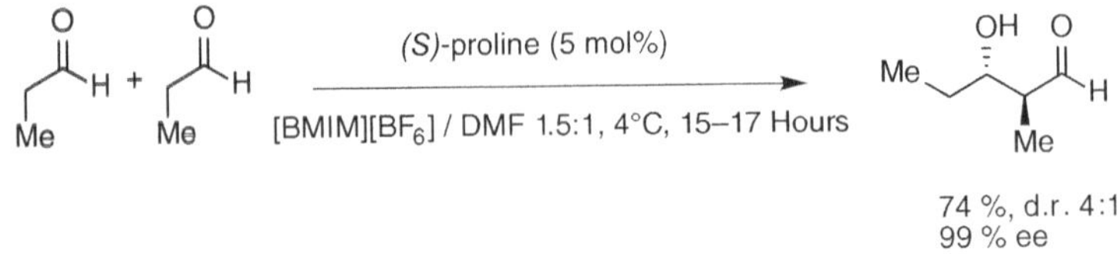

Scheme 9.3. (S)-Proline-catalyzed auto-aldol reaction in ionic liquid.

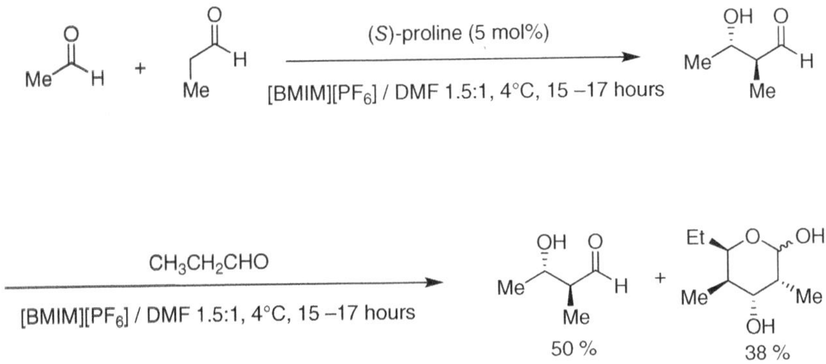

Scheme 9.4. (S)-Proline-catalyzed synthesis of pyranose sugars in ionic liquid.

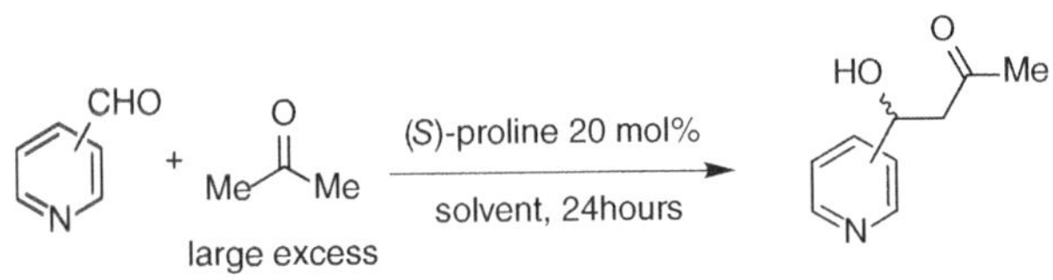

Scheme 9.5. (S)-Proline-catalyzed aldol reactions of pyridinecarbaldehydes with acetone.

medium was dried *in vacuo* and used in subsequent reaction. Five cycles were possible without lowering yields or enantioselectivity. Cross-aldol reactions between acetaldehyde (the acceptor group) and other aliphatic aldehydes ran well, and in the case of isovaleraldehyde and cyclohexanecarbaldehyde, a high diastereoselectivity (d.r. 19:1) was observed. Authors attempted also on synthesis of polyketide sugars. Reaction was started by mixing acetaldehyde and propionaldehyde, and later on the reaction mixture was transferred via a syringe pump to a solution of propionaldehyde in [BMIM][PF_6]/DMF mixture. Dimer as well as pyranose as single diastereomer with an anomeric ratio of 1:2 (α:β) was isolated (Scheme 9.4).

The next paper described (*S*)-proline-catalyzed reaction with pyridinecarbaldehydes[16] (Scheme 9.5). The authors tested several polar solvents (Table 9.2) and found that reactions gave usually low product yields, with the exception of water, in which, however, rather low enantioselectivity was observed. Surprisingly, reaction ran consid-

TABLE 9.2. (S)-Proline-Catalyzed Aldol Reactions of 4-Pyridinecarbaldehyde with Acetone

Entry	Solvent	Time (hour)	Yield (%)	ee (%)
1	DMSO	24	<10	70
2	NMP	24	<10	70
3	H_2O	12	84	18
4	DMSO + 50 mol % H_2O	12	41	5
5	[BMIM][BF_4]	6	98	70
6 (2nd cycle)	[BMIM][BF_4]	6	98	70
7 (3rd cycle)	[BMIM][BF_4]	6	98	70
8 (4th cycle)	[BMIM][BF_4]	6	96	68

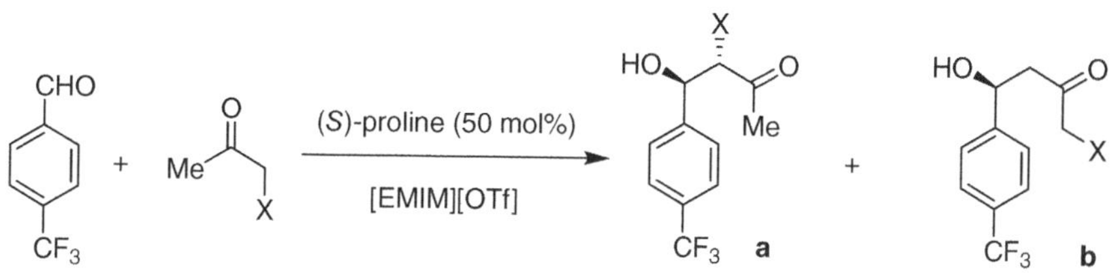

Scheme 9.6. Aldol reactions of 4-trifluoromethylbenzaldehyde with substituted acetones.

TABLE 9.3. (S)-Proline-Catalyzed Aldol Reactions of 4-Trifluormethylbenzaldehyde with Substituted Acetones in Ionic Liquid

Entry	X	Yield **a** (%)	Yield **b** (%)	d.r. **a**
1	Me	68	29	99:1
2 (2nd cycle)	Me	64	39	99:1
3	Cl	68		83:17
4 (2nd cycle)	Cl	86		85:15
5	F	41	41	78:22
6	OMe	71	20	75:25
7	OH	91		50:50

erably faster in [BMIM[BF_4] (6 hours) and high yields of products with high enantioselectivity were obtained. Furthermore, a reaction medium could be reused four times.

Similar results were achieved with pyridine-2- and 3-carbaldehydes. The beneficial effect of ionic liquid was evident, in spite of the fact that authors used a large excess of acetone (1 mL of acetone for 1 mmol of aldehyde), which means that the reaction was carried out in fact in a [BMIM][BF_4]/acetone 1:1 mixture. Kitazume have studied (*S*)-proline-catalyzed aldol reactions of benzaldehydes with substituted acetones in ionic liquids (Scheme 9.6).[17] The best results were achieved with 4-trifluoromethylbenzaldehyde (Table 9.3).

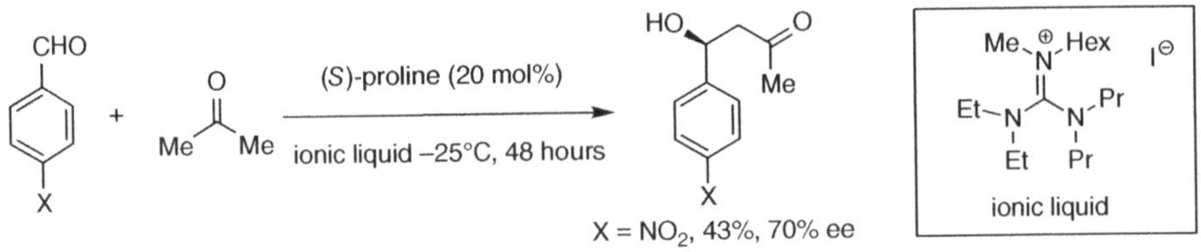

Scheme 9.7. Aldol reactions of aromatic aldehydes with acetone in guanidine ionic liquids.

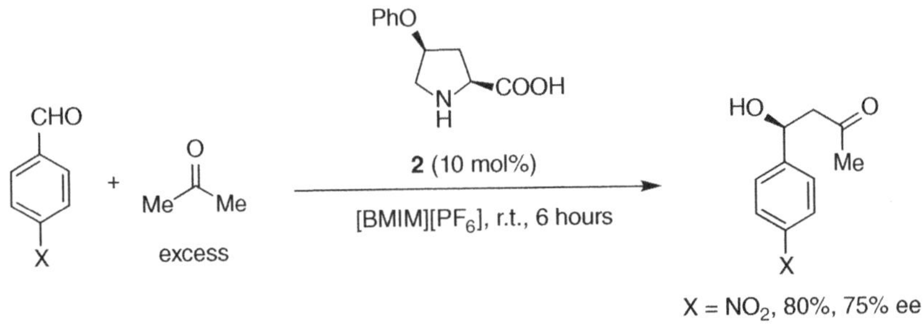

Scheme 9.8. Aldol reactions of aromatic aldehydes with acetone catalyzed by **2**.

The authors found that secondary chloroderivatives under Et_3N catalysis in ionic liquid underwent intramolecular nucleophilic substitution and epoxides were formed. Aldol reaction catalyzed with aldolase antibody 38C2 was also described in Kitazume's paper. The reaction proceeded only with hydroxyacetone and reaction time was rather long—14 days. It is of interest that product yields increased from 21% to 89% in the second cycle and 66% in the third cycle, but the d.r. was low (40:60). Guanidine ionic liquids were also tested in (*S*)-proline-catalyzed aldol reactions of aromatic as well aliphatic aldehydes with acetone (Scheme 9.7).[18]

Noticeable fact is that reactions ran well also with 10 mol% and even with 5 mol% of (*S*)-proline. Addition of water lowered the enantioselectivity of the reaction. Aromatic as well as aliphatic aldehydes reacted well. The highest enantioselectivity (99%) was observed at reaction with 4-methoxybenzaldehyde and 4-bromobenzaldehyde. Again the problem was a large excess of acetone (1 mL of acetone for 1 mmol of aldehyde) used in the study. Also (*S*)-*cis*-4-phenoxyproline (**2**) proved to be an excellent catalyst for aldol reactions of aromatic aldehydes with acetone (Scheme 9.8).[19]

It is of interest that practically the same results were obtained with 1 mol% of the catalyst, just the reaction time had to be extended to 16 hours. Experiments with catalyst loadings 5 mol% revealed that the reaction medium could be reused four times without any effect on the yields or enantioselectivity. On the other hand, the effect of ionic liquid is questionable because the authors were working with a large excess of acetone and results of the reactions in neat acetone were essentially the same as in the acetone–ionic liquid mixture.

Guo et al.[20] have used the prolinamide **3** as the catalyst at reaction of aromatic as well as aliphatic aldehydes with acetone in [BMIM][BF_4]. They found that using 20 mol% of the catalyst at 0°C leads to yields of up to 84% with 90–99% ee. The highest enantioselectivity was observed at reactions with aliphatic aldehydes and reaction medium could be recycled three times without any adverse effect on yields as well as on enantioselectivity. However, again more that 10 mol excess of acetone was used.

Prolinamide catalyst **4** was tested in aldol reaction of 4-nitrobenzaldehyde with acetone and other ketones in ionic liquids.[21] This reaction was much faster than that catalyzed by (*S*)-proline. The authors tested four ionic liquids ([BMIM]PF_6], [BMIM][NTf_2], [BMIM][BF_4], and [BDMIM][BF_4]) and found that yields of the product (70–84%) as well as enantioselectivities (47–50% ee) were practically the same and the reaction medium could be recycled three times. In a subsequent paper[22] authors examined the effect of addition of water to [BMIM][BF_4] and found the best results (reaction time 8 hours, 99% yield, *anti*/*syn* 85:15) with a 50:50 mixture of ionic liquid and water. On the other hand addition of water slightly lowered enantioselectivity from 93% ee (*anti*) and 26% ee (*syn*) to 82% ee (*anti*) and 40% ee (*syn*).

Gryko and Lipinski prepared a range of (*S*)-prolinethioamides and tested them in an aldol reaction with acetone in different solvents.[23] Results with the best catalyst **5** (Figure 9.1) are summarized in Table 9.4.

The best enantioselectivity was observed in [BMIM][BF_4]; however the product yield was rather low. When 0.5 equivalent of acetone was used, bis(aldol) product was isolated in 33% yield. Other amino acids or its derivatives were used as the catalysts

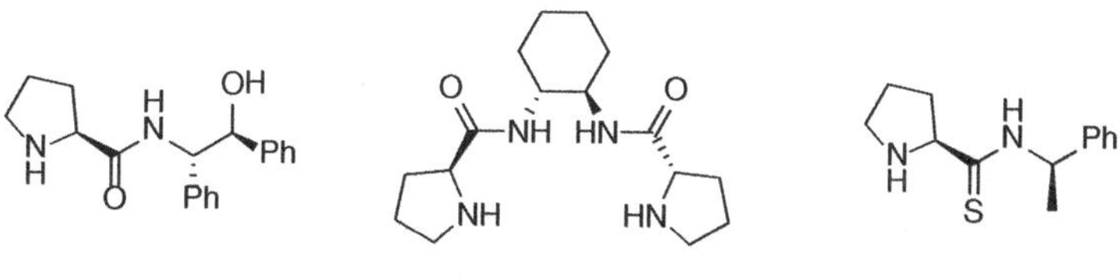

3 **4** **5**

Figure 9.1. Prolinamide organocatalysts.

TABLE 9.4. Prolinthioamide **5**-Catalyzed Aldol Reactions of 4-Formylbenzonitrile with Acetone

Entry	Solvent	T (°C)	Time (hour)	Yield (%)	ee (%)
1	Acetone	r.t.	24	67	72
2	Acetone	4	72	74	86
3	Acetone	−18	96	57	93
4	Acetone	−78	10	21	100
5	*i*-PrOH/H_2O	4	68	76	23
6	Acetone/H_2O	4	68	70	68
7	[BMIM][BF_4]	4	68	32	100

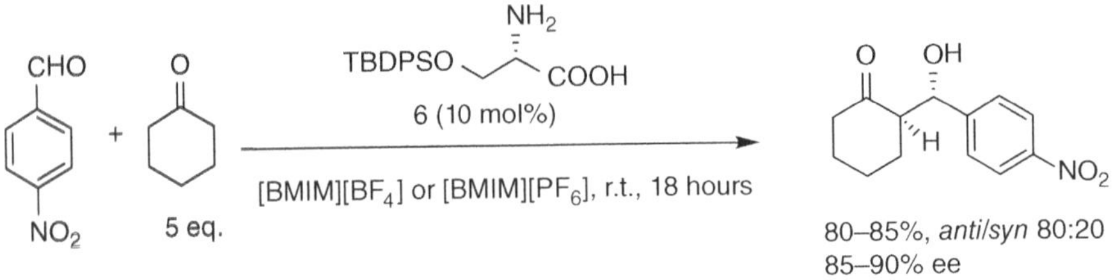

Scheme 9.9. Silyloxyserine-catalyzed reaction of 4-nitrobenyaldehyde with cyclohexanone in ionic liquids.

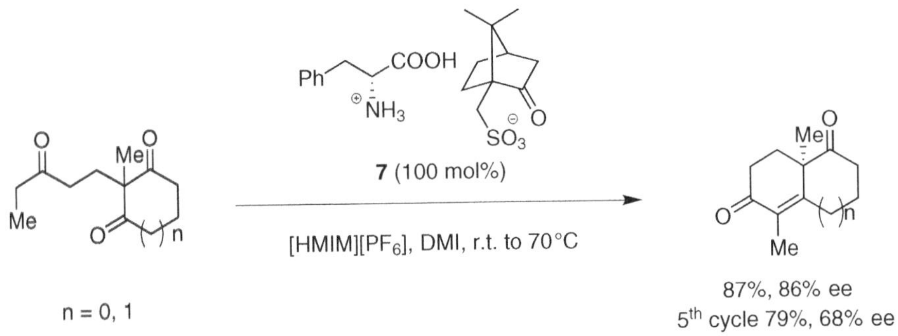

Scheme 9.10. Aldol reaction catalyzed by **7** in ionic liquid leading to Wieland–Mischer ketone analogue.

for aldol reactions in ionic liquids. Teo et al. used silyloxyserine organocatalyst **6**[24] and the results are given in Scheme 9.9. Reaction medium could be recycled four times with identical results; just reaction time should be prolonged to up to 60 hours.

Lombardo and coworkers used protonated arginine (arginine.PTSA) or protonated lysine (lysine.PTSA) as the catalyst for aldol reaction of benzaldehyde with cyclohexanone.[25] They tested two ionic liquids ([BMIM][N(CN)$_2$] and [BMPY][OTf]). Results achieved in [BMPY][OTf] (95% yield, *anti*/*syn* 70:30, and 94% *ee*), were comparable or better than in the DMSO/water 20:1 mixture. Reaction in [BMIM][N(CN)$_2$] gave a lower yield as well as enantioselectivity, but better diasteroselectivity (90:10).

Catalyst **7**, (*R*)-phenylalanine protonated by camphorsulfonic acid, is effective for the aldol reaction leading to Wieland-Mischer ketone analogue.[26] Reactions were carried out in [BMIM][PF$_6$] and [HMIM][PF$_6$] with or without an additive. The best additive proved to be *N*,*N*-dimethylpyrrolidinone (DMI). With this additive it was possible to recycle the reaction medium five times without dramatic effect on enantioselectivity (Scheme 9.10).

Kidwai et al. have tested different amino acids as catalysts at aldol reactions leading to xanthene derivatives.[27] From the amino acid tested, (*S*)-histidine, (*S*)-lysine, (*S*)-alanine, and glycine, the last one was the best catalyst. The same yields were achieved in the fourth recycle of the reaction medium.

9.3.2 Mannich Reactions

Barbas has returned to Mannich reactions after publishing his seminal paper.[14] He checked 18 different organocatalysts at the reaction of iso-pentanal with α-imino ethyl glyoxylate and found that (*S*)-proline was the best one and that reaction in [BMIM][BF_4] gave the better results (90% yield, 93% ee) than in other six organic solvents.[28] In the next paper he thoroughly studied solvent effect on the reaction of acetone with α-imino ethyl glyoxylate.[29] Eight solvents were tested and reaction in [BMIM][BF_4] worked very well (80% yield and 97% ee). Better results were achieved only in neat acetone. The possibility of reaction medium recycling was also studied on the reaction of cyclohexanone with α-imino ethyl glyoxylate. The first experiment resulted in a 99% yield, *syn*/*anti* 19:1, 99% ee, and the fourth one in the same reaction medium gave a 84% yield of the product, with the same diastereo- and enantioselectivity. Barbas also studied a one-pot three-component Mannich reaction (Scheme 9.11). He found that the reaction worked well also in ionic liquid and that aliphatic aldehydes are more reactive than aromatic ones.

During development of an *anti*-Mannich reaction, Tanaka and Barbas found that β-amino acid catalysts, such as *trans*-5-methylpyrrolidine-3-carboxylic acid (**8**), were less active in ionic liquid than in other media.[30] In [BMIM][BF_4] the product was isolated in 35% yield and 77% ee compared with 85% and 99% ee in DMSO (Scheme 9.12).

A similar reaction with 2-phenylpropionaldehyde in ionic liquid [BMIM][BF_4] was also described by Barbas,[31] but both yield and enantioselectivity was significantly lower

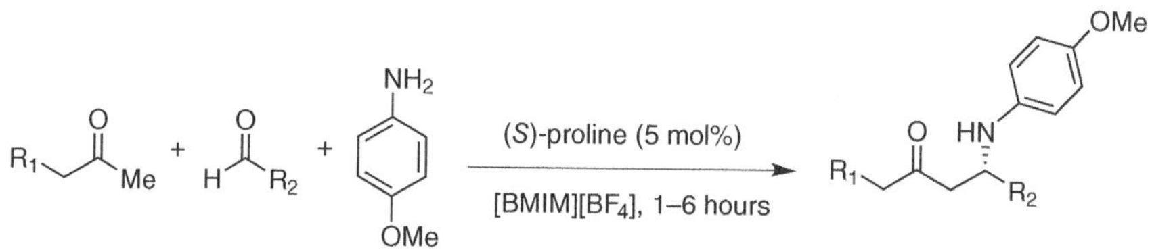

Scheme 9.11. The one-pot three-component, (*S*)-proline-catalyzed Mannich reaction in ionic liquids.

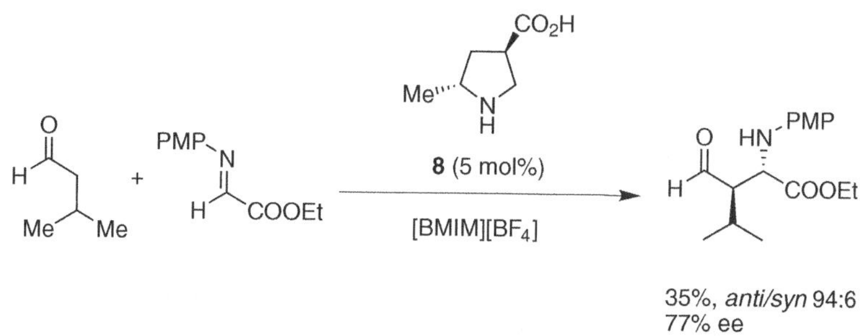

Scheme 9.12. *anti*-Selective Mannich reaction in ionic liquid.

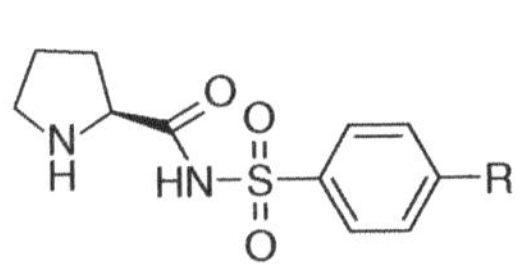

Figure 9.2. *N*-Arylsulfonyl proline amides.

TABLE 9.5. The Effect of Catalyst and Ionic Liquid on the Mannich Reaction between Cyclohexanone and α-Imino Ethyl Glyoxylate

Entry	Catalyst	Solvent	Yield (%)	*syn* : *anti*	ee (%) *syn/anti*
1	**9**	[EMIM][$EtSO_4$]	32	77:23	80/47
2	**10**	[EMIM][$EtSO_4$]	69	70:30	68/8
3	**11**	[EMIM][$EtSO_4$]	75	70:30	81/82
4	Pro	[EMIM][$EtSO_4$]	39	75:25	91/65
5	**9**	[BMIM][$C_8H_{17}SO_4$]	47	37:63	31/34
6	**10**	[BMIM][$C_8H_{17}SO_4$]	54	47:53	20/18
7	**11**	[BMIM][$C_8H_{17}SO_4$]	98	47:53	27/29
8	Pro	[BMIM][$C_8H_{17}SO_4$]	73	44:56	28/24
9	**9**	[HMIM][NTf_2]	34	60:40	72/35
10	**10**	[HMIM][NTf_2]	97	58:42	48/9
11	**11**	[HMIM][NTf_2]	62	71:29	72/77
12	Pro	[HMIM][NTf_2]	60	57:43	90/78

than in DMSO (39% yield and 27% ee vs. 66% yield and 86% ee). Berkessel-type sulfonamide catalysts **9–11** were tested in a Mannich reaction of α-imino ethyl glyoxylate with cyclic ketones in different ionic liquids (Figure 9.2).[32] From the results given in Table 9.5, it follows that [EMIM][$EtSO_4$] and [HMIM][NTf_2] were good reaction media for this transformation.

9.3.3 α-Amination and Aminoxylation of Carbonyl Compounds

Addition of proline-formed enamine to diazadicarboxylate leads to α-amination of aldehyde. Ionic liquids proved to be useful for this reaction too.[33] Several catalysts have been tested and again simple (*S*)-proline was found to be the best (65 minutes, 85%, 84% ee). (*S*)-Tiazolidin-4-carboxylic acid was even more enantioselective, but the reaction was considerably slower (8 hours, 17%, 92% ee). Several ionic liquids were tested, but the best results were achieved in the simplest and cheapest: [BMIM][BF_4] and [BMIM][PF_6] (Scheme 9.13). The worst ionic liquids were AMMOENG™ 100 (44% yield 11% ee) and phosphonium ionic liquid CYPOSIL 101 from which it was very difficult to isolate the product (45% yield and practically 0% ee).

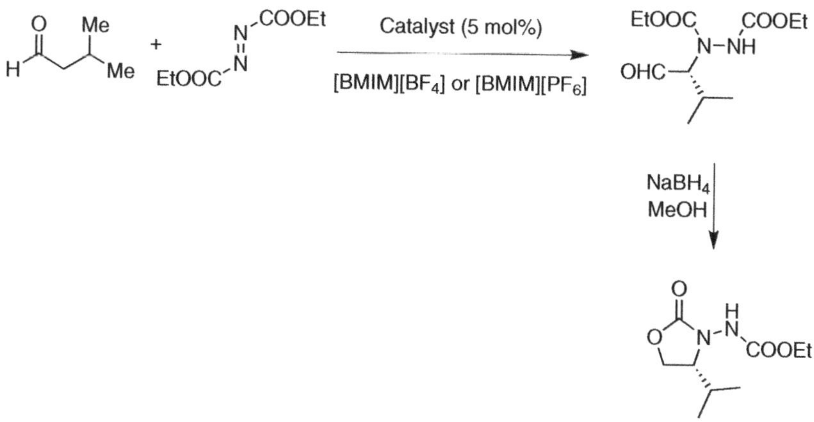

Scheme 9.13. α-Amination of aldehydes in ionic liquids.

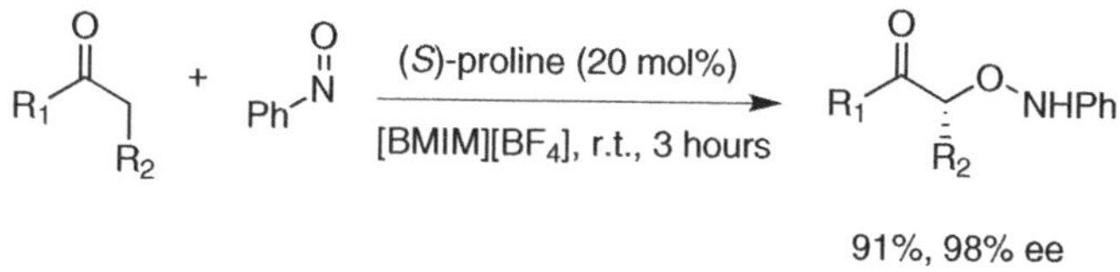

Scheme 9.14. α-Aminoxylation of aldehydes in ionic liquid.

Results achieved with aldehydes in ionic liquids are comparable or even better than those in classical solvents. The reaction medium could be recycled three times without decrease in enantiomeric excesses. The reaction worked well also with cyclic C_4–C_7 ketones.

Related, proline-catalyzed, addition to N=O bond leads to valuable optically active α-hydroxyaldehydes and ketones. Huang and coworkers described the enantioselective aminoxylation of carbonyl compounds in ionic liquids (Scheme 9.14).[34] They found that both aldehydes and ketones underwent this reaction and yields of products were higher in ionic liquids ([BMIM][BF_4], [BMIM][PF_6]) than in conventional solvents. Enantioselectivities were also excellent (95–99% ee). The same reaction with similar results was reported independently.[35] Ionic liquid containing (*S*)-proline was recycled four times without a decrease in yield and enantioselectivity.

Practically the same results were achieved also using just 5 mol% of (*S*)-proline. Lowering catalyst loading to 1 mol% resulted in a 64% yield of the product, but its optical purity remained high (98% ee).

9.3.4 Michael Additions

The study of Michael addition in ionic liquids started with our work on reactions of aliphatic aldehydes and ketones to β-nitrostyrene (Scheme 9.15).[36] The addition of isovaleraldehyde to β-nitrostyrene in [BMIM][BF_4] afforded product in 95% yield, but

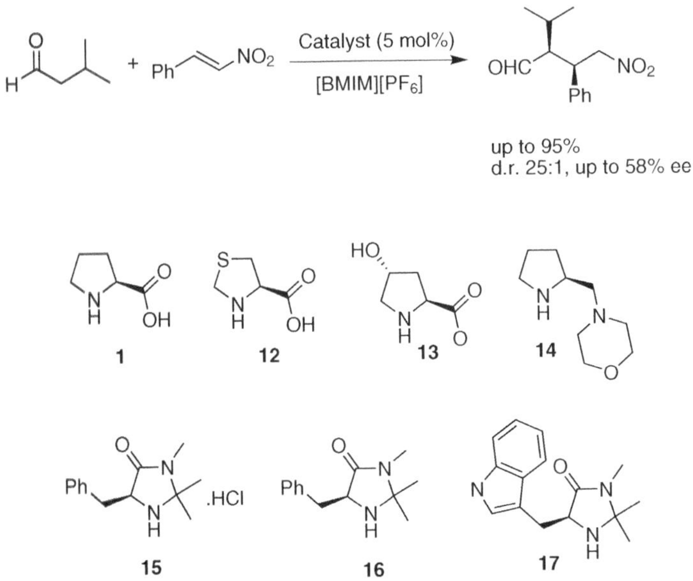

Scheme 9.15. Michael addition of aldehydes to β-nitrostyrene in ionic liquids.

only with 43% ee. The highest enantioselectivity (58% ee) was observed in [BMIM][laurate]. We evaluated several organocatalysts and the most efficient were amino acids **1**, **12**, and **13** and amine **14**. MacMillan's catalysts **15–17** were not active in this reaction.

Michael addition of ketones ran slightly worse, but better than that of β-diketones. Rasalkar et al. studied the same reaction and used 20 mol% of simple (*S*)-proline as a catalyst.[37] They tested several ionic liquids, the best one being [MOEMIM][OMs] (1-methyl-3-(2-methoxyethyl)-imidazolium mesylate). The authors observed high *syn* selectivity (*syn*/*anti* 95:5) and enantioselectivities of up to 75% ee. The catalytic system was used three times with the yield remaining constant but the enantiomeric excess of the product decreasing. Luo et al. examined the structure of ionic liquid in the same reaction.[38] They found that the reaction is faster in [HMIM][Cl] than in [OMIM][Br], with reaction times of 3 and 7 hours, respectively. Reaction in [BMIM][BF_4] was much slower (72 hours). Yields and diastereoselectivities were in all cases high. In the same reaction Xu used chiral pyrrolidine–pyridine-based catalyst **18**.[39] The best results were achieved under solvent-free conditions (12 hours, 99%, d.r. 94:6, 89% ee), then in tetrahydrofuran (THF) and *i*PrOH. The reaction in [BMIM][PF_6] was rather slow and gave a low yield (72 hours, 40%).

Tsogoeva et al. catalyzed Michael addition of 2-nitropropane to cyclohex-2-enone with several tripeptide organocatalysts, such as **19** (Figure 9.3).[40] Interestingly, the addition was slower in [BMIM][PF_6], compared with the reaction in organic solvents, such as $CHCl_3$, DMF, or DMSO.

Hagiwara et al. studied Michael additions of simple aldehydes to metyl vinyl ketone in [BMIM][PF_6].[41] They tested several proline-derived diamine catalysts **20–22**

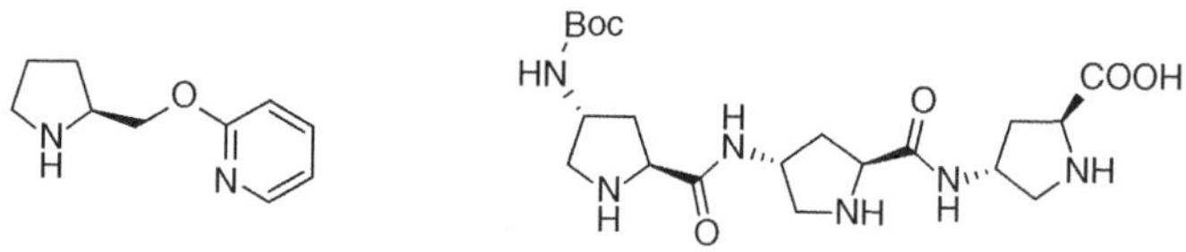

Figure 9.3. Pyrrolidine–pyridine catalyst **18** and tripeptide catalyst **19**.

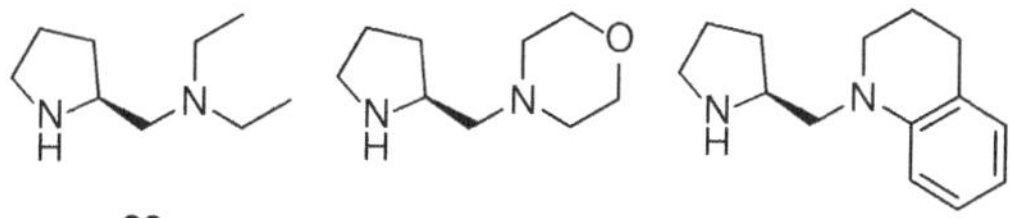

Figure 9.4. Diamine organocatalysts **20–22** and proline amide **23**.

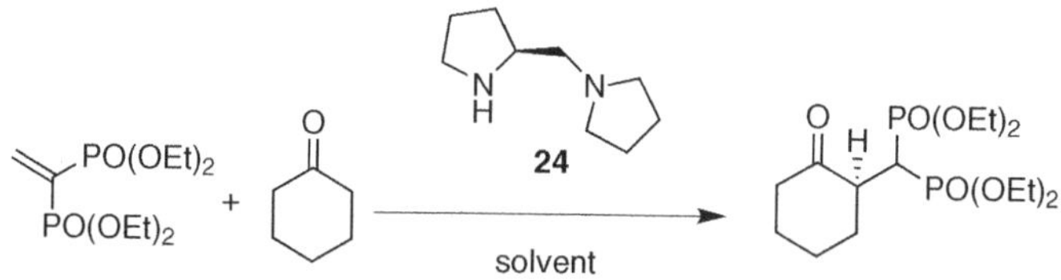

Scheme 9.16. Diamine **24**-catalyzed Michael addition of cyclohexanone to vinyldiphosphonate.

(Figure 9.4). However, the results were not very encouraging because the reactions were slow (r.t. 48 hours) and afforded moderate yields (44–72%) of the product with low enantiomeric purity (18–48% ee).

Almasi et al. studied the addition of ketones to β-nitrostyrene catalyzed by several proline amides.[42] With the best one, catalyst **23** (Figure 9.4), solvent effect was tested on the reaction course. Results achieved in [BMIM][PF_6] (3d, 90%, d.r. 87:13, 62% ee) were comparable with results achieved in MeOH, $CHCl_3$, or NMP. The study of the solvent effect was problematic, as 10 equivalents of ketone (pentan-3-one) were always used.

Barros et al. studied the Michael addition of cyclohexanone to vinyldiphosphonate (Scheme 9.16).[43] With diamine catalyst **24** the reaction worked in several solvents, with

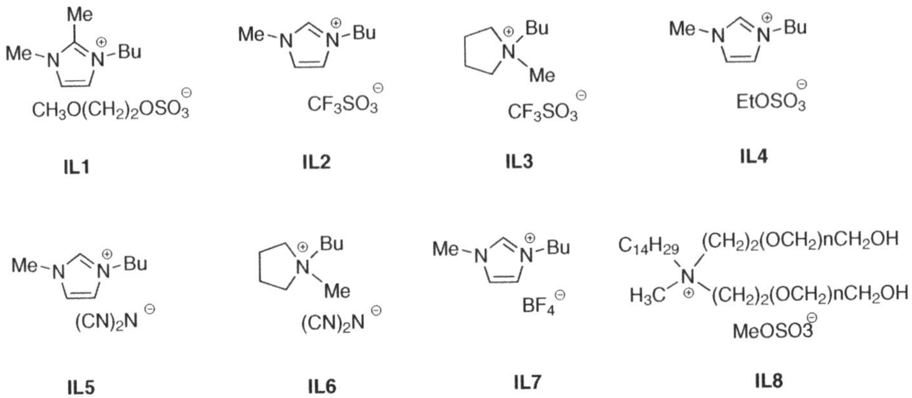

Figure 9.5. Structures of ionic liquids used for Michael addition of aldehydes to nitrostyrene.

full conversion but with mediocre enantioselectivity. The highest enantioselectivity (46%) was achieved in [BMIM][PF_6].

We were interested also if organocatalyzed Michael addition of thiols to chalcone is possible in ionic liquids, and found[44] that the reaction proceeded very well but nearly without any enantioselectivity. Control experiments without organocatalysts proved that the nucleophilic activity of thiophenols in ionic liquid was so high that no catalyst was necessary.[45] Likewise, we assessed whether the Michael addition of 3-methylbutanal to (*E*)-β-nitrostyrene can also be catalyzed by *N*-toluenesulfonylproline amide **9**. Therefore, we decided to test this reaction in several ionic liquids as reaction media (Figure 9.5).[46]

We made three interesting observations: (1) no reaction was observed in ionic liquid with substituted 2-position, for example, **IL1**; (2) addition of water (10 equiv.) was necessary to reach good yields in reasonable time; and (3) the structure of major diasteroisomer depended on the structure of the ionic liquid. All used ionic liquids should, at least in principle, be neutral. However, we found that the pH of their 10% aqueous solution varies from 1.58 to 8.32. Acido-basic properties of the resulting solution had a dramatic effect on the studied reaction. The results are summarized in Table 9.6.

The pH of aqueous ionic liquid solutions can perhaps explain why no reaction was observed in **IL1**. This ionic liquid is so acidic that (*S*)-proline moiety is protonated and no enamine of the aldehyde can be formed. The effect of pH on the stereochemical result of the reaction is intriguing, but at present we do not have any plausible rationalization of it. The acidity of the solutions is likely a result of partial anion hydrolysis or the remaining acid. On the other hand, the fact that several aqueous solutions of ionic liquids were basic, was puzzling. The presence of residual amounts of *N*-methylimidazole from the synthesis of ionic liquids was already suggested as a possible explanation. This prompted us to study kinetics of *N*-methylimidazole-catalyzed addition of malonodinitrile to chalcone in different solvents (Scheme 9.17).[46]

TABLE 9.6. Michael Addition of 3-Methylbutanal to (E)-β-Nitrostyrene Catalyzed by **2**

Ionic liquid	pH of ionic liquid	Yield (%)	ee (%)
[BMIM][$C_8H_{17}OSO_3$]	7.89	98	42 (+)
IL1	1.58	–	n.d.
IL2	2.58	68	18 (–)
IL3	3.09	38	30 (–)
IL4	4.33	83	42 (–)
IL5	7.09	98	28 (+)
IL6	7.09	89	20 (+)
IL7	8.32	98	70 (+)
IL8	7.01	95	28 (+)

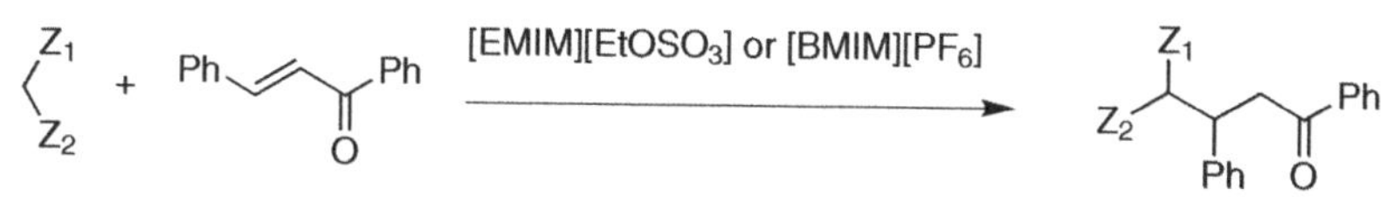

Scheme 9.17. Michael addition of active methylene compounds in ionic liquids.

TABLE 9.7. Michael Addition of C-nucleophiles to Chalcone

Donor	Solvent	Catalyst	Yield (%)
Dimethyl malonate	[EMIM][OSO_3Et]	(*S*)-Pro	0
Dimethyl malonate	[EMIM][OSO_3Et]	Piperidine	59
Dimethyl malonate	CH_2Cl_2	(*S*)-Pro	0
Dimethyl malonate	CH_2Cl_2	Piperidine	0
2-Nitropropane	[BMIM][PF_6]	(*S*)-Pro	31
2-Nitropropane	[BMIM][PF_6]	Piperidine	90
2-Nitropropane	[EMIM][OSO_3Et]	(*S*)-Pro	25
2-Nitropropane	[EMIM][OSO_3Et]	Piperidine	95
2-Nitropropane	CH_2Cl_2	(*S*)-Pro	0
2-Nitropropane	CH_2Cl_2	Piperidine	0

During investigation of this reaction, we found that addition of malononitrile proceeded well even without addition of any basic catalyst. Less reactive dimethyl malonate needed a catalyst, either (*S*)-proline or piperidine. However (*S*)-proline probably acts only as a base, because no stereoselectivity was observed in this case. Piperidine was an even more efficient catalyst for this transformation in [EMIM][SO_4Et] than in CH_2Cl_2. A similar situation was observed with less acidic 2-nitropropane (Table 9.7).[47]

The reaction medium has a profound effect on this transformation. Our study revealed that several ionic liquids contain, from their preparation, small amounts of unreacted *N*-methylimidazole. The amount can be sufficient for the reaction to work (Table 9.8). We also proved that *N*-methylimidazole as well as piperidine is a stronger

TABLE 9.8. Solvent Effect on the Reaction Rate of the Michael Addition of Malonodinitrile to Chalcone Catalyzed by N-Methylimidazole (5 mol%)

Solvent	K ($\times10^{-4}$) [s^{-1}]	$t_{1/2}$ [s]
CH_2Cl_2	no reaction	–
CH_3CN	0,30	23,105
THF	1,26	5501
[EMIM][SO_4Et]	1,80	3851
[BMIM][PF_6]	1,70	4077
[BMIM]PF_6][a]	18,9	366
[BMMIM][PF_6]	5,50	1260
[BMIM][$N(CN)_2$]	3,69	1879
[EMPYR][SO_4Et]	0,61	11,438
[EMPYR][SO_4Et][b]	0,26	27,182

[a] 1 mol% of piperidine was used as a catalyst.
[b] 5 mol% of pyridine was used as a catalyst.

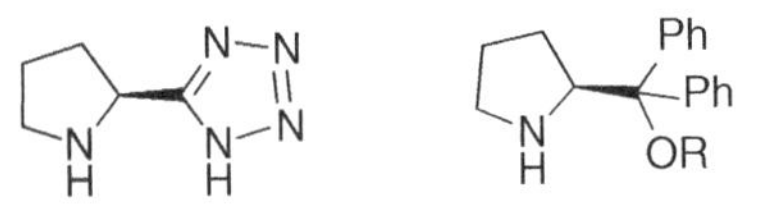

25

26, R = H
27, R = SiMe3

Figure 9.6. Tetrazole analogues of proline **25** and prolinol organocatalysts **26** and **27**.

base in ionic liquids than in classical solvents. This is in agreement with results presented recently by D'Anna and coworkers.[48]

Later on we decided to make a thorough testing of catalysts, solvents, and also temperature effects on organocatalyzed Michael addition of simple aldehydes to β-nitrostyrene.[49] Besides proline (**1**) and prolinesulfonamides **9** and **10**, organocatalysts **25**–**27** were tested (Figure 9.6). The following ionic liquids were examined: [BMIM] BF_4], [EMIM][EtOSO_3], [BMIM][CF_3SO_3], [BMIM][$C_8H_{17}OSO_3$], and [BMIM] [$N(CN)_2$]. The results are summarized in Table 9.9.

From the results given in Table 9.9, it follows that additions catalyzed by (*S*)-proline (**1**), (*S*)-*N*-toluensulfonylproline amide (**9**), (*S*)-*N*-(4-nitrophenyl)sulfonylproline amide (**10**), and (*S*)-5-(2-pyrrolidinyl)-1*H*-tetrazole (**25**) proceeded well and the addition product was isolated in moderate to excellent yields after 20–48 hours at room temperature (r.t.). High *syn*/*anti* selectivities (up to 98:2) were observed, but enantioselectivities were only moderate (up to 54% ee). Interestingly, catalyst **9** afforded product with opposite sense of asymmetric induction than other catalysts, when used in [EMIM] [SO_4Et] and [BMIM][CF_3SO_3] (Table 9.9, entries 7 and 8). This result is probably due to the interaction of the reaction intermediate with the cation and anion of the ionic liquid. The product was isolated with 45% ee instead of 32% (15 mol% of **25**). A further

TABLE 9.9. Michael Addition of Isovaleraldehyde to β-Nitrostyrene

Entry	Cat.	Ionic liquid	Yield (%)	d.r. (*syn*/*anti*)[a]	ee (%)[b]
1	**1**	[BMIM][BF_4]	85	98:2	38
2	**1**	[BMIM][BF_4][c]	51	94:6	33
3	**1**	[EMIM][SO_4Et]	68	95:5	32
4	**1**	[BMIM][CF_3SO_3]	60	96:4	35
5	**1**	[BMIM][$SO_4C_8H_{17}$]	72	97:3	47
6	**1**	[BMPYR][$N(CN)_2$][f]	81	95:5	35
7	**9**	[EMIM][SO_4Et]	80	92:8	38 (−)
8	**9**	[BMIM][CF_3SO_3]	68	95:5	36 (−)
9	**9**	[BMIM][$SO_4C_8H_{17}$]	98	95:5	45
10	**9**	[BMPYR]$N(CN)_2$][f]	96	94:6	42
11	**10**	[BMIM][BF_4]	38	96:4	14
12	**10**	[EMIM][SO_4Et]	45	98:2	28
13	**10**	[BMIM][CF_3SO_3]	72	95:5	46
14	**10**	[BMIM][$C_8H_{17}SO_4$]	51	88:12	46
15	**10**	[BMPYR][$N(CN)_2$][f]	64	95:5	54
16	**25**	[BMIM][BF_4]	68	95:5	32
17	**25**	[BMIM][BF_4][c]	36	97:3	45
18	**25**	[BMIM][BF_4][d]	47[e]	97:3	36
19	**25**	[EMIM][SO_4Et]	74	95:5	12
20	**25**	[EMIM][SO_4Et][c]	45	96:4	8
21	**25**	[BMIM][CF_3SO_3]	64	86:14	40
22	**25**	[BMIM][$SO_4C_8H_{17}$]	51	93:7	34
23	**25**	[BMPYR][$N(CN)_2$][f]	70	93:7	34
24	**26**	[BMIM][BF_4]	<5	60:40	nd
25	**26**	[BMIM][CF_3SO_3]	10[f]	94:6	nd
26	**27**	[BMIM][BF_4]	47	71:29	8
27	**27**	[EMIM][SO_4Et]	21	75:25	53

[a] Determined by ^1H NMR spectra.
[b] Reported values refer to the *syn* isomer and were determined by chiral HPLC.
[c] Reactions were performed with 5 mol% of the catalyst.
[d] Reactions were performed with 2.5 mol% of the catalyst.
[e] Reactions proceeded 48 hours.
[f] 1-Butyl-1-methylpyrrolidinium dicyanamide.

decrease of catalyst loading to 2.5 mol% led to lower enantioselectivities (36% ee, Table 9.1, entries 16, 17, and 18). The concentration of the reaction mixture had no influence on the reaction course.

(*S*)-α,α-Diphenyl-2-pyrrolidinemethanol (**26**) as well as (*S*)-α,α-diphenyl-2-pyrrolidinemethanol trimethylsilyl ether (**27**) are efficient catalysts for the Michael addition of aldehydes to nitroalkanes in organic solvents, such as hexane, dioxane, and CH_3CN.[50] Our results show that catalysts **26** and **27** are considerably less effective in ionic liquids. Catalyst **26** was inactive in [BMIM][BF_4] as well as [BMIM][CF_3SO_3] when only small amounts of the adduct were isolated or detected by ^1H NMR (Table 9.9, entries 24 and 25). Somewhat better results were achieved with catalyst **27** (Table

TABLE 9.10. Temperature Effect on the Michael Addition

Entry	Cat.	Ionic liquid	Temp. (°C)	Yield (%)	d.r. (*syn/anti*)[a]	ee (%)[b]
1	**1**	[BMIM][BF_4]	5	64	96:4	40
2	**1**	[BMIM][BF_4]	20	85	96:4	38
3	**1**	[BMIM][BF_4]	30	90	95:5	40
4	**1**	[BMIM][BF_4]	40	94	86:14	42
5	**1**	[BMIM][BF_4]	50	96	83:17	40
6	**1**	[BMIM][BF_4]	60	94	67:33	54
7	**1**	[BMIM][BF_4]	70[c]	78	75:25	50
8	**1**	[BMIM][BF_4]	80[c]	64	67:33	46
9	**25**	[BMIM][BF_4]	5	55	94:6	42
10	**25**	[BMIM][BF_4]	20	68	95:5	32
11	**25**	[BMIM][BF_4]	40	85	91:9	34
12	**25**	[BMIM][BF_4]	60	89	80:20	40
13	**25**	[BMIM][BF_4]	80[c]	68	67:33	40
14	**1**	[EMIM][$EtSO_4$]	5	47	95:5	33
15	**1**	[EMIM][$EtSO_4$]	20	61	95:5	32
16	**1**	[EMIM][$EtSO_4$]	40	68	83:17	34
17	**1**	[EMIM][$EtSO_4$]	60	80	75:25	38
18	**1**	[EMIM][$EtSO_4$]	80[c]	62	75:25	38
19	**1**	DMSO	5	34	95:5	36
20	**1**	DMSO	20	55	93:7	36
21	**1**	DMSO	40	47	86:14	38
22	**1**	DMSO	60	36	80:20	40
23	**1**	DMSO	80[c]	51	80:20	40

[a] Determined by ^1H NMR spectra.
[b] Reported values refer to the *syn* isomer and were determined by chiral HPLC.
[c] Reactions proceeded 2 hours.

9.9, entries 26 and 27), which afforded product in 47% yield in [BMIM][BF_4] and 21% yield in [EMIM][SO_4Et], respectively. Diastereoselectivities were very similar, 71:29 ([bmim][BF_4]), and 75:25 ([EMIM][SO_4Et]), although considerably lower than with other catalysts. The reaction medium had a strong influence on enantioselectivities with this catalyst. While in [BMIM][BF_4] the product was isolated with 8% ee, in [EMIM][SO_4Et] it was isolated with 53% ee.

In the second part of our work we studied the temperature effect on the organocatalyzed Michael addition of isovaleraldehyde to β-nitrostyrene in ionic liquids. For comparison, experiments were also performed in DMSO. Results are shown in Table 9.10.

From the data in Table 9.10, it follows that increasing the reaction temperature from 5 to 80°C resulted, generally, in a decrease in the diastereoselectivity of the reaction in ionic liquids as well as in DMSO. While *syn/anti* ratio was from 93:7 to 95:5 at 5°C and r.t., it decreased to 67:33–80:20 at 80°C. Similar observations were made when organocatalyzed Michael addition of aldehydes to nitrostyrene was performed in $CHCl_3$, CH_2Cl_2, THF, hexane, and CCl_4, respectively. On the other hand, the enantioselectivity of the addition was not reduced when the temperature increased from 5 to

80°C. Interestingly, the enantioselectivity of the reaction went through a maximum at 60°C. When tetrazole derivative **25** was used as the catalyst, the enantioselectivity decreased from 42% (5°C) to 32% (20°C) and rose again to 40% ee at 60 and 80°C. A slight increase in the enantioselectivity was observed in all studied reactions when the reaction temperature increased from 20 to 80°C.

9.3.5 Miscellaneous Reactions

Apart from aldol and Mannich reactions, α-functionalization of aldehydes, and Michael addition, cycloaddition reactions are intensively studied. Park et al. investigated the ionic effect in Diels–Alder reactions. They studied the reaction of cyclohexan-1,3-diene with acrolein in different solvents (Scheme 9.18).[51] Results are summarized in Table 9.11.

From the results given in Table 9.11, it follows that the best ionic liquid was [BMIM]PF_6], which can be reused two times without having a deteriorating effect on both the yields and enantioselectivity. It was also proved that practically the same results were achieved using only 5 mol% of the catalyst (Table 9.11, entry 5), and

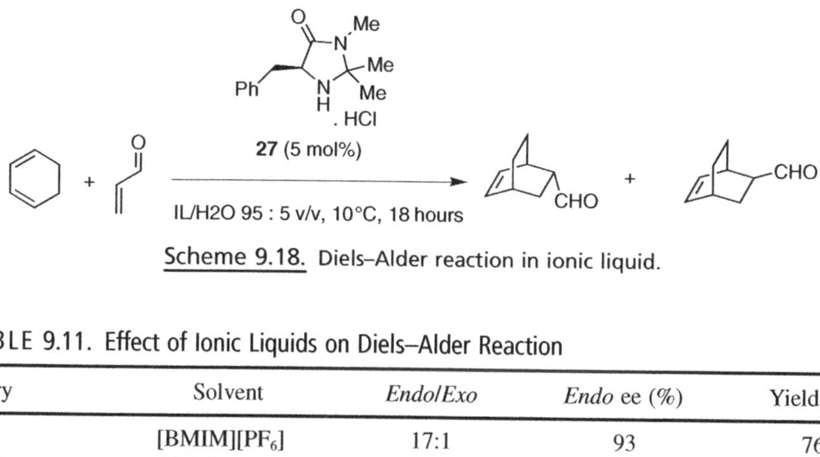

Scheme 9.18. Diels–Alder reaction in ionic liquid.

TABLE 9.11. Effect of Ionic Liquids on Diels–Alder Reaction

Entry	Solvent	*Endo*/*Exo*	*Endo* ee (%)	Yield (%)
1	[BMIM][PF_6]	17:1	93	76
2 1st reuse	[BMIM][PF_6]	17:1	91	72
3 2nd reuse	[BMIM][PF_6]	17:1	87	70
4	[BMIM][SbF_6]	17:1	74	92
5	[BMIM][PF_6][a]	17:1	93	85
6	[BMIM][PF_6][b]	13:1	88	70
7	[BMIM][PF_6][c]	13:1	86	43
8	[BMIM][OTf]	17:1	0	7
9	[BMIM][OTf][d]	18:1	54	22
10	[BMIM][BF_4]	17:1	0	5
11	CH_3CN	14:1	94	82

[a] 5 mol % of the catalyst was used.
[b] 2.5 mol % of the catalyst was used.
[c] 1 mol % of the catalyst was used.
[d] Additional 15% of water was added to the ionic liquid.

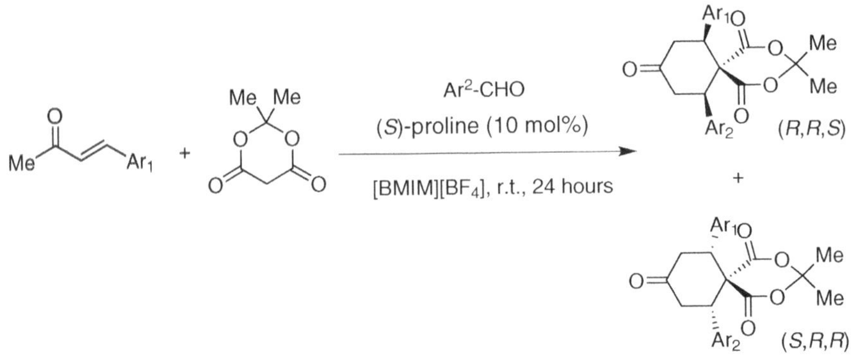

Scheme 9.19. Domino reaction in ionic liquid.

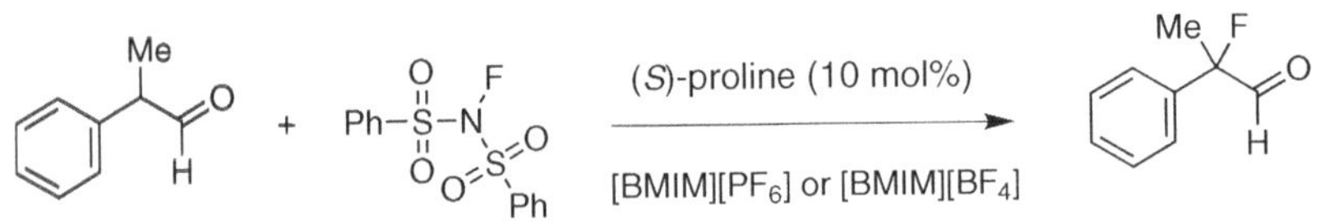

Scheme 9.20. Proline-catalyzed fluorination of aldehydes.

further lowering the catalyst amount to 2.5 mol% as well as 1 mol% (Table 9.11, entries 6 and 7) lowers mostly the reaction yields. It was proved also that addition of water could have a beneficial effect on the reaction course (Table 9.11, entry 9). The best enantioselectivity was achieved in acetonitrile, but kinetic experiments proved that reaction in acetonitrile is much slower than in [BMIM]PF_6]. Similar results were achieved at the Diels–Alder reaction of cyclopentadiene with cinnamic aldehyde, except that the *endo*/*exo* ratio was close to 1:1.

Barbas and coworkers published a series of three papers devoted to domino Knoevenagel/Diels–Alder reactions.[52] (Scheme 9.19). In the first paper they tested nine different solvents. In dichloromethane a 35% yield was achieved, (*R,R,S*):(*S,R,R*) was 100:0, and enantioselectivity was 59% ee. Results in methanol were similar: 92% yield, (*R,R,S*):(*S,R,R*) 12:1, and 60% ee. The reaction in [BMIM]PF_6] was much faster (24 hours instead of 72 or 48 hours, respectively).The chemical yield was very high, but enantioselectivity was low (95%, (*R,R,S*):(*S,R,R*) was 100:0, 6% ee). A similar reaction with indan-1,3-dione, instead of Meldrum's acid, was described. In [BMIM][PF_6] or [BMIM][BF_4]. it gave lower yields of the product and a much lower isomer ratio (33:66).

Barbas et al. also described the organocatalytic fluorination of aldehydes in ionic liquids (Scheme 9.20).[53] Reaction in common organic solvents such as acetonitrile, DMF, THF, and N-methylpyrrolidone (NMP) gave higher product yields with similar enantioselectivities. Recycling of the solvent was not tested.

9.4 IONIC LIQUIDS AS CO-CATALYSTS FOR ORGANOCATALYTIC REACTIONS: TOWARD NEW REACTIVITIES AND SELECTIVITIES

Ionic moieties, often structurally similar to ionic liquids, have been incorporated into many organocatalysts in order to improve the recyclability of catalysts and/or reusability of whole reaction media. An interesting feature of such ionically tagged organocatalysts is that they can be efficiently used in other reaction media than ionic liquids.[3a,54] However, the following section is aimed only at catalytic systems used in ionic liquids (see also next chapter for further information).

A special case is Hu et al.'s work.[55] The authors prepared a proline-derived chiral ionic liquid, namely (2-hydroxyethyl)-trimethylammonium (*S*)-2-pyrrolidine carboxylic acid salt ([Choline][Pro]) and used it as the catalyst (co-solvent) at the aldol reaction of aromatic aldehydes with excess of acetone and cyclic ketones in water. Both mono- and bis-aldol product was isolated. The best results are depicted in Scheme 9.21. The enantioselectivity of the reaction was, unfortunately, very low, as less than 10% ee was obtained in all experiments.

Lombardo and Trombini prepared ionically tagged organocatalysts **28** and **29** based on 4-hydroxyproline (Figure 9.7).[56] These catalysts performed well in direct aldol reaction in ionic liquids: Yields were up to 85% and enantioselectivities were in the 80–85% range. Catalysts performed even better in aqueous conditions.[57]

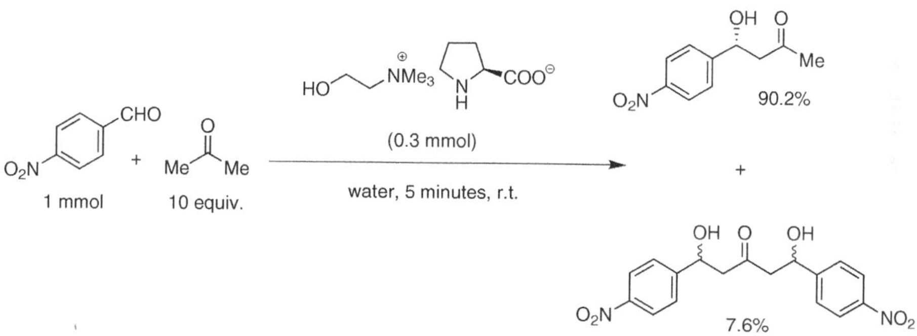

Scheme 9.21. Aldol reaction of 4-nitrobenzaldehyde with acetone catalyzed by chiral ionic liquid.

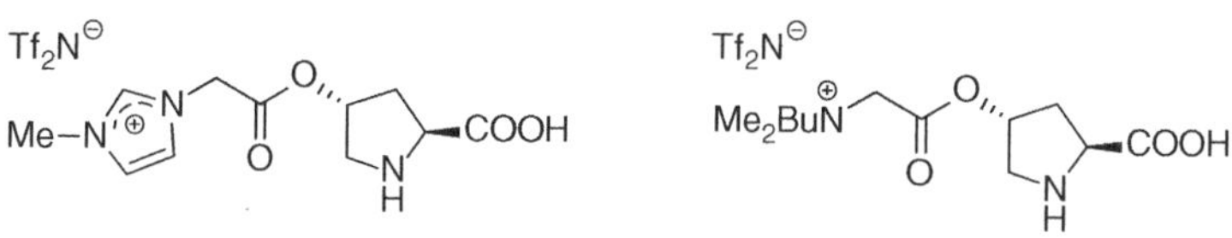

28 **29**

Figure 9.7. Immobilized 4-hydroxyproline-derived organocatalysts.

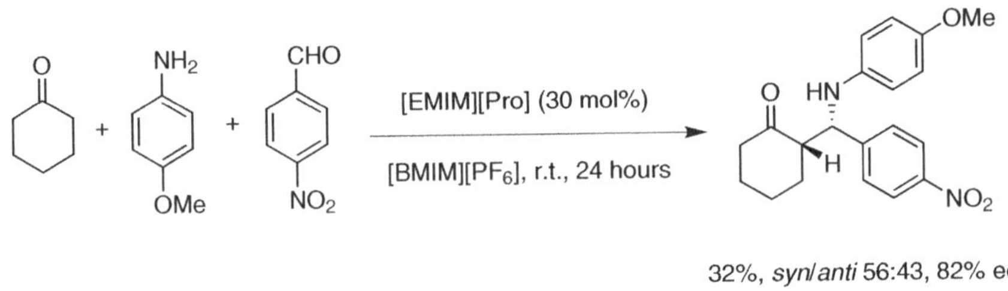

Scheme 9.22. The one-pot three component Mannich reaction in ionic liquid catalyzed by chiral ionic liquid.

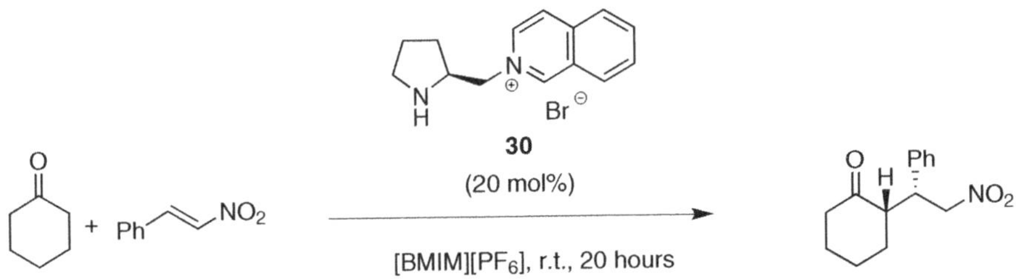

Scheme 9.23. Michael addition catalyzed by ionically tagged pyrrolidine organocatalyst **30**.

TABLE 9.12. Results of the Michael Addition of Cyklohexanone to β-Nitrostyrene in Various Solvents

Solvent	Yield (%)	d.r.	*ee* (%)
MeOH	94	92:8	0
CH_2Cl_2	94	92:8	59
THF	27	93:7	27
DMSO	37	91:9	24
–	91	90:10	63
[BMIM]BF_4	94	94:6	99

Gruttadauria and Noto developed a silica gel modified with ionic liquid. Proline used on this material efficiently catalyzed the aldol reaction between acetone and benzaldehyde.[58]

A one-pot three-component Mannich reaction was also performed in ionic liquid catalyzed by chiral ionic liquid, [EMIM][Pro], the structure of which was mentioned in the aldol section.[59] It was found that reaction in [BMIM][PF_6] went reasonably well (Scheme 9.22), but much better yields were achieved in DMF as well as in DMSO.

A Michael addition of cyclohexanone to β-nitrostyrene was described also by Xu et al.[60] The authors used a functionalized ionic liquid as catalyst and performed the reaction in several solvents (Scheme 9.23). The results are given in Table 9.12. The reaction in ionic liquid proceeded with a high yield (94%) and selectivity (d.r. 94:6,

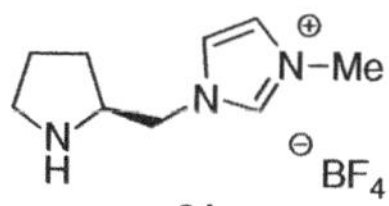

31

Figure 9.8. Ionically tagged pyrrolidine.

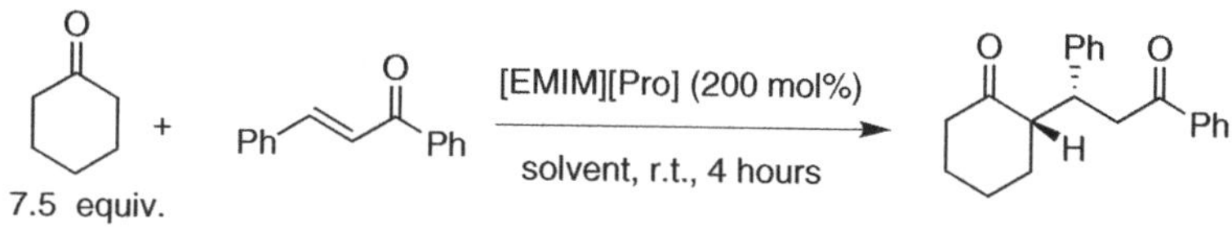

CH_3OH	98%, d.r. 20:80, ee 86%
[BMIM][PF_6]	75%, d.r. 20:80, ee 46%
[EMIM] [Pro]	91%, d.r. 21:79, ee 60%
cyclohexanone	89%, d.r. 14:86, ee 62%

Scheme 9.24. Michael addition in chiral ionic liquid [EMIM][Pro].

99% ee) using only 5–10 mol% of the catalyst, while in classical solvents catalyst loading had to be increased up to 20 mol%. The reaction medium was recycled four times without any deleterious effect on yields and selectivity.

Xu et al. also used a similarly immobilized (*S*)-proline derivative **31** for the Michael addition of cyclohexanone to (*E*)-β-nitrostyrene (Figure 9.8).[61] The reaction was performed in various solvents ([BMIM][PF_6], DMSO, DMF, or *i*PrOH), but no significant differences were observed between them. The catalytic system in ionic liquid was recycled three times with constant yields and selectivity, but the reaction time had to be prolonged from 24 to 100 hours.

Qian et al. studied the Michael addition of cyclohexanone to chalcone and used proline-derived ionic liquid [EMIM][Pro] as catalyst. They used 200 mol% of the catalyst in different solvents, including ionic liquids.[62] Results are given in Scheme 9.24. It is very difficult to discuss the solvent effect as the authors used 7.5 equivalents of cyclohexanone. It is not surprising therefore that the best results were achieved when cyclohexanone was used as the solvent.

9.5 KEY FACTORS IN CHOOSING IONIC LIQUIDS FOR ORGANOCATALYSIS AND PROSPECTS

In this chapter, we showed that ionic liquids can be used as reaction media for organocatalyzed reactions, especially for reactions proceeding via enamine intermediates. An advantage of such experiments is that a smaller amount of the catalyst can often be used than in common organic solvents and reaction media can be recycled several times. On the other hand, several reactions performing worse in ionic liquids than in common solvents were also shown. The advantage of organocatalytic reactions, especially in industrial applications, could be that the reaction medium can be recycled several times without addition of a new catalyst, or by addition of a small amount of the catalyst, which is more probable. It is necessary to note that some serious recycling studies are necessary.[63]

Only few organocatalytic reactions working via iminium ion intermediates have been investigated and more studies are certainly necessary. Our research showed that in such reactions amine impurity in ionic liquids can act as a strong base. This can diminish the stereoselectivity of the reaction. It is therefore advisable to measure the pH of the aqueous solution of the ionic liquid (if it is soluble in water) and check the reaction in the ionic liquid without addition of organocatalyst.

No investigations of SOMO or hydrogen bonding catalysis have been published so far, and research in this area seems to be highly needed. Practically no serious investigations of domino reactions in ionic liquids have yet been carried out. In the Barbas papers,[52] only one of the reactions was stereoselective. Our own attempts[64] to perform an Enders-type domino reaction[65] have not yet been successful. The desired product, polysubstituted cyclohexene, had a low yield, and it was not possible to isolate it from a complicated reaction mixture. We assume that it was due to the low solubility of the Hayashi–Jőrgensen catalyst **27** in ionic liquids in which the reactions were performed.

To the best of our knowledge, no hypothesis has been published yet which would explain the influence of ionic liquids on organocatalytic reactions. Furthermore, the fact that no uniform effect of ionic liquids is observed makes such unified hypothesis unlikely. We can only speculate if effect similar to that of water on some organic reactions, which enhances hydrophobic interactions,[66] cannot operate also in ionic liquids. Perhaps the suggested domain structure of ionic liquids, with altering hydrophobic and polar regions, is also responsible for the effect of ionic liquids on some organocatalytic reactions.[67] Further research in this area will hopefully shed more light on this topic.

Combining organocatalysis with ionic liquids has become a flourishing area of green chemistry. Further research on the application of ionic liquids, both in areas of organocatalysis already discussed and in those that are still unexplored, will undoubtedly bring many new exciting discoveries.

REFERENCES

1 R. A. Sheldon, *Green Chem.* **2005**, *7*, 267.

2 (a) P. Wasserscheid, T. Welton, *Ionic Liquids in Synthesis*, 2nd ed., Wiley-VCH, Weinheim, **2008**.; (b) R. D. Rogers, K. R. Seddon, *Ionic Liquids as Green Solvents. Progress and Progress*, Oxford University Press USA, Washington, DC, **2003**.

3 (a) R. Šebesta, I. Kmentová, Š. Toma, *Green Chem.* **2008**, *10*, 484.; (b) W. Miao, T. H. Chan, *Acc. Chem. Res.* **2006**, *39*, 897.

4 (a) A. Berkessel, H. Gröger, *Asymmetric Organocatalysis*, Wiley-VCH, Weinheim, Germany, **2005**.; (b) P. I. Dalko, *Asymmetric Organocatalysis: Reactions and Experimental Procedures*, Wiley-VCH, Weinheim, **2007**.

5 (a) P. I. Dalko, L. Moisan, *Angew. Chem. Int. Ed.* **2004**, *43*, 5138.; (b) J. Seayad, B. List, *Org. Biomol. Chem.* **2005**, *3*, 719.; (c) B. List, *Chem. Commun.* **2006**, 819.; (d) H. Kotsuki, H. Ikishima, A. Okuyama, *Heterocycles* **2008**, *75*, 493.; (e) H. Kotsuki, H. Ikishima, A. Okuyama, *Heterocycles* **2008**, *75*, 757.; (f) A. Dondoni, A. Massi, *Angew. Chem. Int. Ed.* **2008**, *47*, 4638.; (g) P. Melchiorre, M. Marigo, A. Carlone, G. Bartoli, *Angew. Chem. Int. Ed.* **2008**, *47*, 6138.

6 (a) B. List, *Acc. Chem. Res.* **2004**, *37*, 548.; (b) W. Notz, F. Tanaka, C. F. Barbas, *Acc. Chem. Res.* **2004**, *37*, 580.; (c) S. Mukherjee, J. W. Yang, S. Hoffmann, B. List, *Chem. Rev.* **2007**, *107*, 5471.

7 A. Erkkila, I. Majander, P. M. Pihko, *Chem. Rev.* **2007**, *107*, 5416.

8 (a) P. R. Schreiner, *Chem. Soc. Rev.* **2003**, *32*, 289.; (b) P. M. Pihko, *Angew. Chem. Int. Ed.* **2004**, *43*, 2062.; (c) Y. Takemoto, *Org. Biomol. Chem.* **2005**, *3*, 4299.; (d) A. G. Doyle, E. N. Jacobsen, *Chem. Rev.* **2007**, *107*, 5713.; (e) X. Yu, W. Wang, *Chem. Asian J.* **2008**, *3*, 516.; (f) S. J. Connon, *Chem. Commun.* **2008**, 2499.

9 (a) S. J. Connon, *Angew. Chem. Int. Ed.* **2006**, *45*, 3909.; (b) T. Akiyama, *Chem. Rev.* **2007**, *107*, 5744.

10 (a) J. Mlynarski, J. Paradowska, *Chem. Soc. Rev.* **2008**, *37*, 1502.; (b) J. Paradowska, M. Stodulski, J. Mlynarski, *Angew. Chem. Int. Ed.* **2009**, *48*, 4288.

11 D. G. Blackmond, A. Armstrong, V. Coombe, A. Wells, *Angew. Chem. Int. Ed.* **2007**, *46*, 3798.

12 P. Kotrusz, I. Kmentová, B. Gotov, Š. Toma, E. Solčániová, *Chem. Commun.* **2002**, (8), 2510.

13 T.-P. Loh, L.-C. Feng, H.-Y. Yang, J.-Y. Yang, *Tetrahedron Lett.* **2002**, *43*, 8741.

14 N. S. Chowdari, D. B. Ramachary, C. F. III Barbas, *Synlett* **2003**, 1906.

15 A. Córdova, *Tetrahedron Lett.* **2004**, *45*, 3949.

16 K. R. Reddy, L. Chakrapani, T. Ramani, C. V. Rajasekhar, *Synth. Commun.* **2007**, *37*, 4301.

17 T. Kitazume, Z. Jiang, K. Kasai, Y. Mihara, M. Suzuki, *J. Fluor. Chem.* **2003**, *121*, 205.

18 J. Shah, H. Blumenthal, Z. Yacob, J. Liebscher, *Adv. Synth. Catal.* **2008**, *350*, 1267.

19 Y.-H. Liu, Y.-W. Zhang, Y.-P. Ding, Z.-X. Shen, X.-Q. Luo, *Chin. J. Chem.* **2005**, *23*, 634.

20 H.-M. Guo, L.-F. Cun, L.-Z. Gong, A.-Q. Mi, Y.-Z. Jiang, *Chem. Commun.* **2005**, 1450.

21 A. S. Kucherenko, D. E. Siyutkin, V. O. Muraviev, M. I. Struchkova, S. G. Zlotin, *Mendeleev Commun.* **2007**, *17*, 277.

22 A. Kucherenko, D. Syutkin, S. Zlotin, *Russ. Chem. Bull.* **2008**, *57*, 591.

23 D. Gryko, R. Lipinski, *Eur. J. Org. Chem.* **2006**, *2006*, 3864.

24 Y.-C. Teo, G.-L. Chua, *Tetrahedron Lett.* **2008**, *49*, 4235.

25 M. Lombardo, S. Easwara, C. T. Filippo Pasia, D. D. Dhavale, *Tetrahedron* **2008**, *64*, 9203.

26 M. Nozawa, T. Akita, T. Hoshi, T. Suzuki, H. Hagiwara, *Synlett* **2007**, 0661.

27 M. Kidwai, K. Singhal, S. Kukreja, *Can. J. Chem.* **2008**, *86*, 799.

28 W. Notz, F. Tanaka, S.-I. Watanabe, N. S. Chowdari, J. M. Turner, R. Thayumanavan, C. F. Barbas, *J. Org. Chem.* **2003**, *68*, 9624.

29 W. Notz, S.-I. Watanabe, N. S. Chowdari, G. Zhong, J. M. Betancort, F. Tanaka, C. F. Barbas III, *Adv. Synth. Catal.* **2004**, *346*, 1131.

30 H. Zhang, S. Mitsumori, N. Utsumi, M. Imai, N. Garcia-Delgado, M. Mifsud, K. Albertshofer, P. H. Y. Cheong, K. N. Houk, F. Tanaka, C. F. Barbas, *J. Am. Chem. Soc.* **2008**, *130*, 875.

31 N. S. Chowdari, J. T. Suri, C. F. Barbas, *Org. Lett.* **2004**, *6*, 2507.

32 E. Veverková, J. Štrasserová, R. Šebesta, Š. Toma, *Tetrahedron Asymmetry* **2010**, *21*, 58–61. doi:10.1016/j.tetasy.2009.12.013.

33 P. Kotrusz, S. Alemayehu, Š. Toma, H. G. Schmalz, A. Adler, *Eur. J. Org. Chem.* **2005**, 4904.

34 K. Huang, Z. Z. Huang, X. L. Li, *J. Org. Chem.* **2006**, *71*, 8320.

35 H.-M. Guo, H.-Y. Niu, M.-X. Xue, Q.-X. Guo, L.-F. Cun, A.-Q. Mi, Y.-Z. Jiang, J.-J. Wang, *Green Chem.* **2006**, *8*, 682.

36 P. Kotrusz, Š. Toma, H. G. Schmalz, A. Adler, *Eur. J. Org. Chem.* **2004**, 1577.

37 M. S. Rasalkar, M. K. Potdar, S. S. Mohile, M. M. Salunkhe, *J. Mol. Catal. A Chem.* **2005**, *235*, 267.

38 S.-P. Luo, L.-P. Wang, H.-D. Yue, Z.-G. Le, W.-L. Yang, D.-Q. Xu, Z.-Y. Xu, *Acta Chim. Sin.* **2006**, *14*, 1483.

39 D.-Z. Xu, S. Shi, Y. Wang, *Eur. J. Org. Chem.* **2009**, 4848.

40 S. B. Tsogoeva, S. B. Jagtap, Z. A. Ardemasova, V. N. Kalikhevich, *Eur. J. Org. Chem.* **2004**, 4014.

41 H. Hagiwara, T. Okabe, T. Hoshi, T. Suzuki, *J. Mol. Catal. A Chem.* **2004**, *214*, 167.

42 D. Almasi, D. A. Alonso, E. Gómez-Bengoa, Y. Nagel, C. Nájera, *Eur. J. Org. Chem.* **2007**, 2328.

43 M. T. Barros, A. M. F. Phillips, *Eur. J. Org. Chem.* **2008**, 2525.

44 M. Mečiarová, Š. Toma, P. Kotrusz, *Org. Biomol. Chem.* **2006**, *4*, 1420.

45 (a) M. Mečiarová, Š. Toma, *Lett. Org. Chem.* **2006**, *3*, 794.; (b) M. Mečiarová, K. Hubinská, Š. Toma, B. Koch, A. Berkessel, *Monatsh. Chem.* **2007**, *138*, 1181.

46 M. Mečiarová, M. Cigán, Š. Toma, A. Gáplovský, *Eur. J. Org. Chem.* **2008**, 4408.

47 M. Mečiarová, Š. Toma, *Chem. Eur. J.* **2007**, *13*, 1268.

48 F. D'Anna, V. Frenna, S. La Marca, R. Noto, V. Pace, D. Spinelli, *Tetrahedron* **2008**, *64*, 672.

49 M. Mečiarová, Š. Toma, R. Šebesta, *Tetrahedron Asymmetry* **2009**, *20*, 2403.

50 (a) Y. Hayashi, H. Gotoh, T. Hayashi, M. Shoji, *Angew. Chem. Int. Ed.* **2005**, *44*, 4212.; (b) Y. Hayashi, T. Itoh, M. Ohkubo, H. Ishikawa, *Angew. Chem. Int. Ed.* **2008**, *47*, 4722.; (c) P. García-García, A. Ladépęche, R. Halder, B. List, *Angew. Chem. Int. Ed.* **2008**, *47*, 4719.

51 J. K. Park, P. Sreekanth, B. M. Kim, *Adv. Synth. Catal.* **2004**, *346*, 49.

52 (a) D. B. Ramachary, N. S. Chowdari, C. F. Barbas III, *Angew. Chem. Int. Ed.* **2003**, *42*, 4233.; (b) D. B. Ramachary, N. S. Chowdari, C. F. III Barbas, *Synlett* **2003**, 1910.; (c) D. B. Ramachary, C. F. Barbas III, *Chem. Eur. J.* **2004**, *10*, 5323.

53 D. D. Steiner, N. Mase, C. F. Barbas III, *Angew. Chem. Int. Ed.* **2005**, *44*, 3706.

54 S. Luo, L. Zhang, J.-P. Cheng, *Chem. Asian J.* **2009**, *4*, 1184.

55 S. Hu, T. Jiang, Z. Zhang, A. Zhu, B. Han, J. Song, Y. Xie, W. Li, *Tetrahedron Lett.* **2007**, *48*, 5613.

56 M. Lombardo, F. Pasi, S. Easwar, C. Trombini, *Adv. Synth. Catal.* **2007**, *349*, 2061.

57 M. Lombardo, F. Pasi, S. Easwar, C. Trombini, *Synlett* **2008**, 2471.

58 M. Gruttadauria, S. Riela, C. Aprile, P. Lo Meo, F. D'Anna, R. Noto, *Adv. Synth. Catal.* **2006**, *348*, 82.

59 X. Zheng, Y.-B. Qian, Y. Wang, *Eur. J. Org. Chem.* **2010**, 515.

60 D.-Q. Xu, B.-T. Wang, S.-P. Luo, H.-D. Yue, L.-P. Wang, Z.-Y. Xu, *Tetrahedron Asymmetry* **2007**, *18*, 1788.

61 D. Xu, S. Luo, H. Yue, L. Wang, Y. Liu, Z. Xu, *Synlett* **2006**, 2569.

62 Y. Qian, S. Xiao, L. Liu, Y. Wang, *Tetrahedron Asymmetry* **2008**, *18*, 1515.

63 B. Wu, W. Liu, Y. Zhang, H. Wang, *Chem. Eur. J.* **2009**, *15*, 1804.

64 M. Mečiarová, Š. Toma, Unpublished results, 2010.

65 (a) D. Enders, M. R. M. Huttl, C. Grondal, G. Raabe, *Nature* **2006**, *441*, 861.; (b) D. Enders, M. R. M. Hüttl, G. Raabe, J. W. Bats, *Adv. Synth. Catal.* **2008**, *350*, 267.

66 S. Otto, J. B. F. N. Engberts, *Org. Biomol. Chem.* **2003**, *1*, 2809.

67 A. A. H. Pádua, F. Costa Gomes, J. N. A. Canongia Lopes, *Acc. Chem. Res.* **2007**, *40*, 1087.

10

"NONSOLVENT" APPLICATIONS OF IONIC LIQUIDS IN ORGANOCATALYSIS

Michelangelo Gruttadauria, Francesco Giacalone, Paola Agrigento, and Renato Noto

10.1 INTRODUCTION

The beginning of the current century witnessed the paramount development of two research fields: organocatalysis and ionic liquids. A few years after the rediscovery of the proline-catalyzed aldol reactions[1] and MacMillan's imidazolidinone-catalyzed reactions,[2] these two fields merged. As first and natural approach, ionic liquids were employed as alternative reaction media in which organocatalytic reactions could be performed.[3] Owing to their tunable structures, the exploration was focused on the screening of several ionic liquids, as new reaction media for the above reactions, with the aim of obtaining higher stereoselectivity and/or for the recovery and reuse of the organocatalyst. Many studies were carried out demonstrating that enhanced catalytic activity can be reached but, in several cases, associated with limited reusability (see Chapter 9).[3]

The relatively high catalytic loading in many of the organocatalytic processes has led many researchers to investigate the catalyst recycling in order to increase both the economical and environmental benefits of the organocatalytic processes.[4] The search for a higher reusability of the ionic liquid phase and for higher active and stereoselective organocatalysts prompted many researchers to develop new synthetic strategies.

Ionic Liquids in Biotransformations and Organocatalysis: Solvents and Beyond, First Edition. Edited by Pablo Domínguez de María.

These efforts led to the development of "nonsolvent" use of ionic liquids in organocatalysis. Two new important approaches were widely applied: (1) immobilization of ionic liquids and organocatalysts, and (2) anchorage of organocatalyst to ionic liquids.

These approaches mainly regarded the use of ionic liquids for the immobilization or anchorage of proline-based organocatalysts for aldol, Michael, and Baylis–Hillman reactions. In addition to these organocatalytic reactions, ionic liquids were immobilized on solid supports to act, directly, as organocatalysts in useful synthetic processes such as Knoevenagel reactions and cycloaddition of epoxides with CO_2. In parallel, ionic liquids were used as organocatalysts also under homogeneous conditions. It is important to stress that whereas in the second approach ionic liquids are liquids, in the first approach the "ionic liquid" moiety, when used as support or linker or catalyst, loses its classical ionic liquid nature (i.e., m.p. < 100°C) because it is anchored to a solid support. Also in the second approach, the ionic liquid-anchored organocatalyst may not be liquid. Then, in both cases, we will refer to an "ionic liquid" moiety or to "ionic liquid-anchored organocatalyst" even if these are solid materials.

Then, this chapter is organized in three sections (see Figure 10.1):

2. Immobilization of ionic liquids and organocatalysts
   - 2.1 Covalently attached "ionic liquid" moieties as supports
   - 2.2 Covalently attached "ionic liquid" moieties as linkers
   - 2.3 Covalently attached "ionic liquid" moieties as organocatalysts

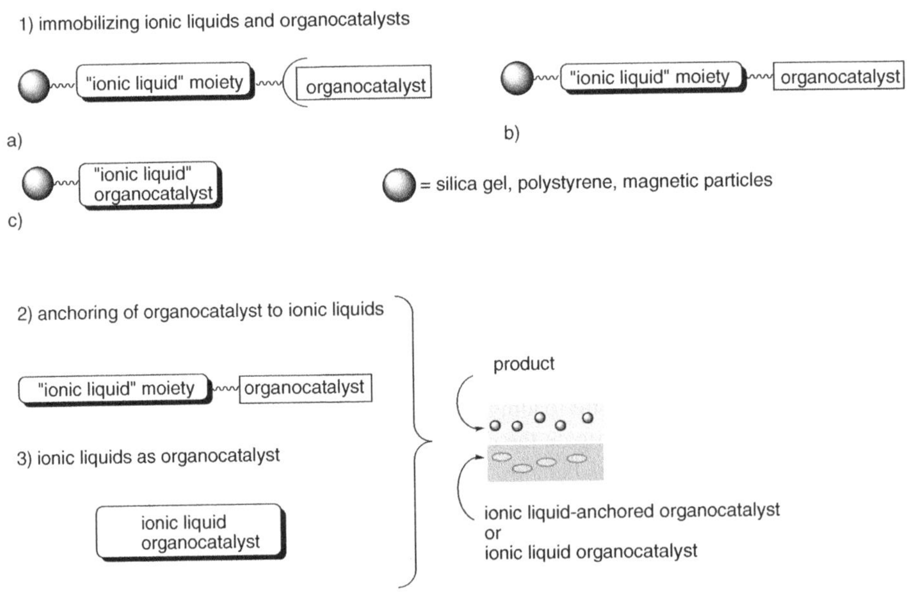

Figure 10.1. General approaches for "nonsolvent" applications of ionic liquids in organocatalysis.

3. Anchorage of organocatalyst to ionic liquids
   3.1 Aldol reactions
   3.2 Michael reactions
   3.3 Morita–Baylis–Hillman reaction and Claisen–Schmidt reaction
4. Ionic liquids as organocatalysts

10.2 IMMOBILIZING IONIC LIQUIDS AND ORGANOCATALYSTS

Three strategies can be followed in order to immobilize ionic liquids and organocatalysts: (1a) the "ionic liquid" moiety is covalently attached to an insoluble support, such as silica gel, polystyrene, or magnetic particles, then organocatalysts are adsorbed on their surfaces; (1b) organocatalysts are covalently attached to the support through an "ionic liquid" spacer. In the last approach (1c) the covalently attached "ionic liquid" moiety does not serve as support or linker but acts as organocatalyst.

10.2.1 Strategy 1a: Covalently Attached "Ionic Liquid" Moieties as Supports

In 2002 Mehnert described the supported ionic liquid catalysis concept, which combines the advantages of ionic liquids with those of heterogeneous supported materials.[5] This concept involves the treatment of a monolayer of covalently attached ionic liquid on the surface of silica gel with additional adsorbed ionic liquid. The preference for heterogeneous catalytic systems is motivated by the advantage of easy separation and the ability to use fixed-bed reactors. The supported ionic liquid phase (SILP) serves as the reaction phase in which the homogeneous catalyst is dissolved. Since ionic liquids are still expensive, it is desirable to minimize the amount of utilized ionic liquid in a process allowing at the same time an easy recovery of the catalyst. Moreover, it is of great interest to have a material behaving as bulk ionic liquid even when it is covalently attached as a monolayer or adsorbed multilayer on a surface. This new class of advanced materials could share the properties of true ionic liquids and the advantages of a solid support. With this idea in mind, two years after the first use of ionic liquids as solvents in organocatalysis,[6,7] the first example of supported ionic liquid asymmetric catalysis was reported.[8] The use of ionic liquids as solvents provided better results compared with dimethylsulfoxide (DMSO). Then, the question was whether proline immobilized on supported ionic liquid silica gels could act as proline in homogeneous ionic liquids. Three types of silica gels modified with 1,2-dimethyl-3-propyl-imidazolium tetrafluoroborate or hexafluorophosphate or chloride moieties (**1–3**) were prepared by reaction between silica gel and the corresponding trialcoxysilane derivatives (Figure 10.2).

These materials were tested in the aldol reaction between acetone and several aldehydes. To these modified silica gels were also adsorbed two ionic liquids, 1-butyl-3-methylimidazolium tetrafluoroborate [BMIM][BF_4] or 1-butyl-3-methylimidazolium hexafluorophosphate [BMIM][PF_6]. The organocatalyst, proline, was simply adsorbed on the surface of these ionic liquid-modified silica gels. In this manner, three types of

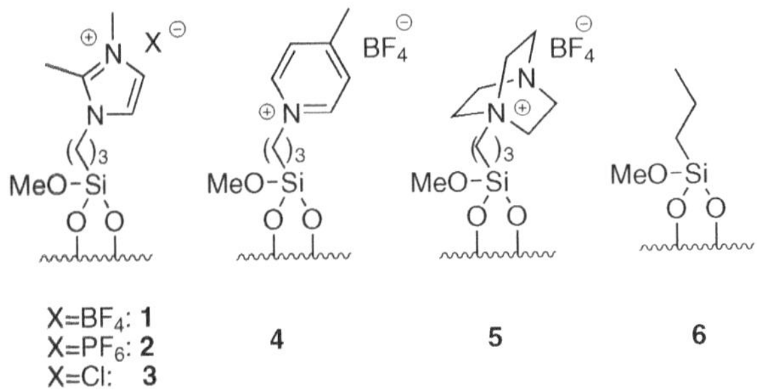

Figure 10.2. Structure of ionic liquid-modified silica gels.

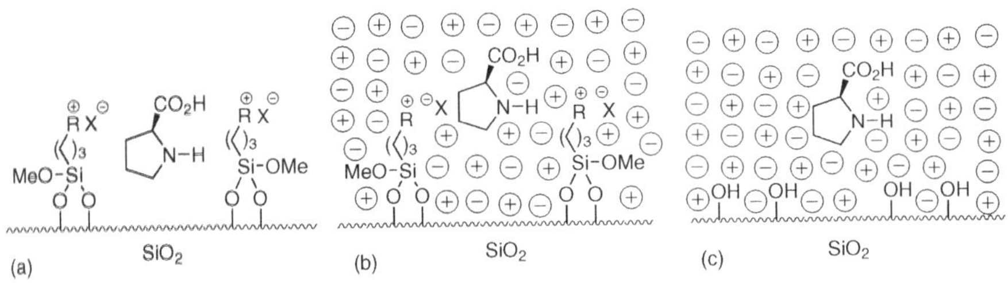

Figure 10.3. (a) Proline (30 mol%) adsorbed on ionic liquid-modified silica gels; (b) proline (30 mol%) adsorbed on ionic liquid-modified silica gels with additional adsorbed ionic liquid; (c) proline (30 mol%) adsorbed on unmodified silica gel containing adsorbed ionic liquid.

catalytic materials were prepared (Figure 10.3): type (a) constituted by proline adsorbed on silica gels modified with covalently attached ionic liquid moieties; type (b) constituted by proline adsorbed on silica gel modified with covalently attached ionic liquid moieties and adsorbed ionic liquids; type (c) constituted by proline adsorbed on silica gels modified with adsorbed ionic liquids.

Preliminary investigations carried out with these materials, using the reaction between acetone and benzaldehyde as model reaction, gave significant results. Proline immobilized on unmodified silica gel cointaining adsorbed ionic liquid [BMIM][BF_4] (type c) afforded the aldol product in low optical purity. On the other hand, proline immobilized on covalently modified silica gel with or without adsorbed ionic liquid (type b or a) gave a higher enantiomeric excess (ee) value. In the case of material **1**/[BMIM][BF_4]/pro (silica **1** containing adsorbed ionic liquid plus proline), a better yield was obtained. These results indicated that the surface of silica gel must be modified by covalently attached ionic liquid for a better outcome of the reaction. Probably, free acidic hydroxy groups on the surface of silica gel (see type c) participate in the coordination of the aldehydes, then playing the same role of the intramolecular hydrogen bond furnished by the proline carboxylic group. Further tests with a small set of aldehydes indicated that material **1**/[BMIM][BF_4]/pro gave good yields and *ee* values and could be used up to four consecutive runs.

In addition to the former modified silica gels **1-3**, 4-methylpyridinium tetrafluoroborate **4**, DABCO tetrafluoroborate **5**, and propyl **6** modified silica gels were prepared

(Figure 10.2).[9] The use of the last two modified silica gels (**5-6**) as support for proline gave catalytic materials that afforded aldol products in low optical purity while 4-methylpyridinium tetrafluoroborate-modified silica gel (**4**) gave a high ee value. On a whole, the above results indicated that, in order to have high ee values, the surface of silica gel must be modified with a covalently attached aromatic ionic liquid phase.

Recycling studies were also carried out using two catalytic systems, **4**/pro and **1**/[BMIM][BF_4]/pro. Proline was adsorbed both from acetonitrile/water and methanol solution. Better results were observed when methanol was employed. The catalytic system **4**/pro was not highly recyclable while the **1**/[BMIM][BF_4]/pro system was successfully used for six cycles. Moreover, the support **1** was easily recovered and recharged with fresh [BMIM][BF_4]/proline. Thus, the regenerated catalytic system was used up to 13 cycles with unchanged results (Scheme 10.1).

Further studies were carried out using new ionic liquid-modified silica gels **7–10** (Scheme 10.2).[10] These supports were prepared by radical reaction between the

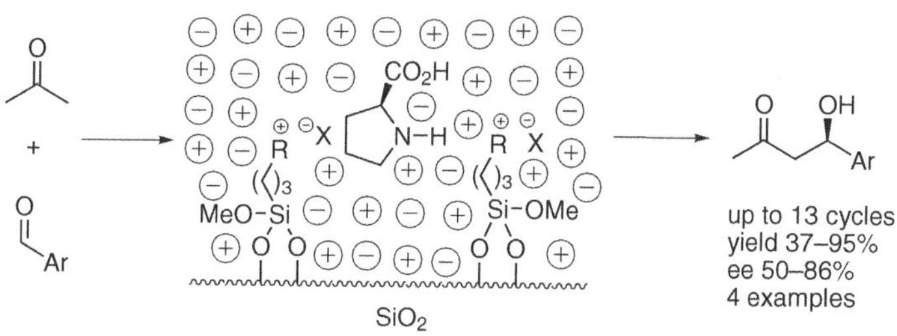

Scheme 10.1. Aldol reactions performed with **1**/[BMIM][BF_4]/pro system.

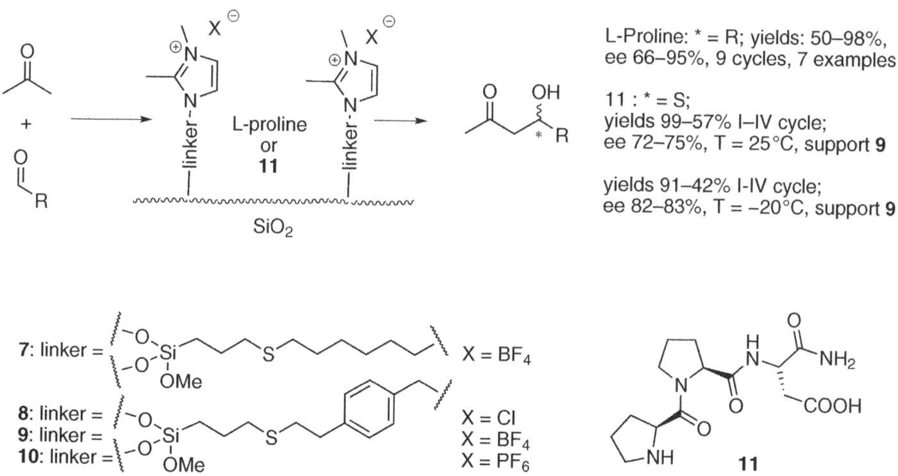

Scheme 10.2. Aldol reactions performed with **7–10**/pro and **9**/**11** systems.

mercaptopropyl-modified silica gel and the corresponding alkene (support **7**) or styrene (supports **8–10**) derivatives.

Proline (30 mol%) was adsorbed on supports **7–10** from a methanol solution. Differently from previous investigations, no additional adsorbed ionic liquid was used. Support **9** resulted in the best one, being recyclable up to nine times without the need of regeneration. Aldol products were obtained in 50–98% yield and 66–95% ee. It is worth noting that the enantioselectivity observed with some aldehydes were comparable to those obtained in pure ionic liquid and were higher than those obtained in pure acetone. These results suggested that the covalently attached monolayer of ionic liquid behaved as a bulk ionic liquid medium, but with the advantage of avoiding the need of expensive ionic liquids as solvents. Tripetide H-Pro-Pro-Asp-NH_2 **11**[11] (5 mol%) was supported on silica gels **8–10** and the catalytic materials tested in the reaction between acetone and 4-nitrobenzaldehyde both at room temperature and −20°C. The catalytic material **9/11** gave the best results, being used in four consecutive runs with slightly decreased enantioselectivity with respect to the unsupported **11** when used at room temperature. Reproduction of good enantioselectivity was observed in all reaction cycles, but usually decreased conversions were observed in the third and fourth cycles.

Two examples of covalently attached ionic liquid for enzyme immobilization were also reported. Although the catalysts were enzymes and not simple organocatalysts, these reports deserve a description because of the role played by the supports (see also Chapter 4 for a broader discussion).

The first example regarded monolith-supported ionic liquid materials.[12] The main idea of this approach was to have an SILP material in which the ionic liquid properties were transferred to the solid phase, which then behaved as a bulk ionic liquid. Moreover, in order to expand the greenness of the process, supercritical carbon dioxide ($scCO_2$) was used as solvent. Supercritical CO_2 has been described as a solvent with a strong deactivating effect on enzymes; however, the best results for enzymes catalysis in $scCO_2$ were obtained when enzymes were immobilized in ionic liquids.[13] The overall green aspect of the process is due to the following points: (1) advantages of SILP, (2) enzyme catalysis, (3) use of a supercritical fluid, and (4) advantages of a continuous flow process.

Two types of supports were prepared as depicted in Scheme 10.3. The monolithic SILP **13** was obtained by alkylation of the polymer **12** with butylimidazole, while material **15** was prepared by polymerization of **14** with ethylene dimethacrylate (EDMA) and 2-hydroxyethyl methacrylate (HEMA).

The microenvironment of the ionic liquid-modified monolith **13** was investigated by means of steady-state fluorescence spectroscopy, using pyrene as probe.[14] This methodology allowed a semi-quantitative assessment of the polarity of the SILPs. The pyrene value I_I/I_{III} measured for **13** indicated a completely different microenvironment with respect to resin **12**. The pyrene I_I/I_{III} value for **13** was 1.44, similar to that observed in methanol (I_I/I_{III} = 1.33) and higher than that measured for **12** (I_I/I_{III} = 1.01).

Candida antarctica lipase B (CALB) was immobilized by adsorption of an aqueous solution of the enzyme. The resulting catalytic material was used for the continuous flow synthesis of citronellyl propionate in sCO_2 (Scheme 10.4). The catalytic activity of these bioreactors remained unchanged for seven operational cycles of 5 hours each.

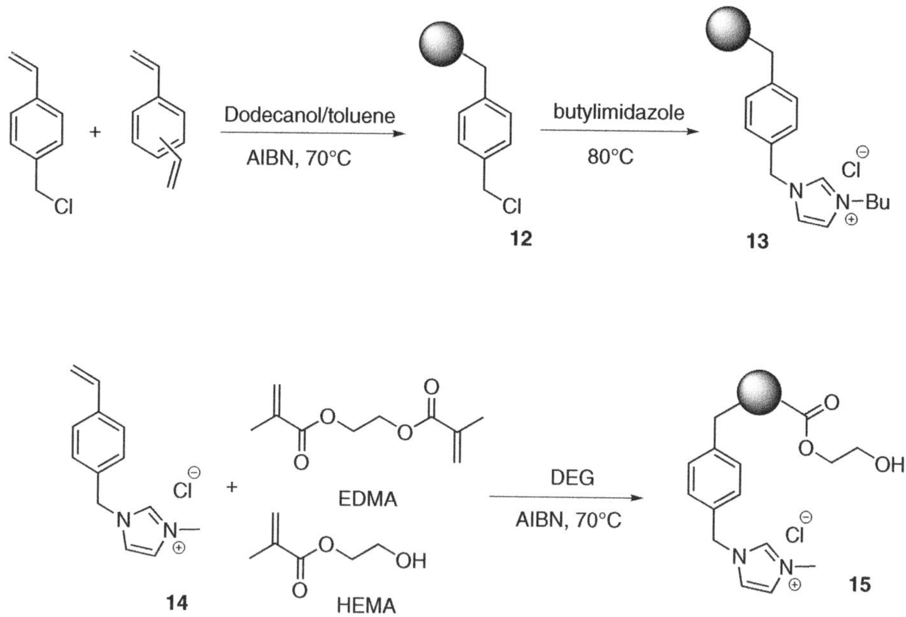

Scheme 10.3. Synthesis of monolith-supported ionic liquid materials **13** and **15**.

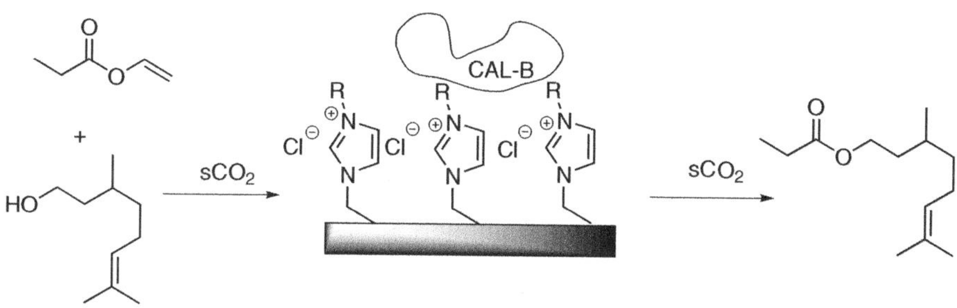

Scheme 10.4. Synthesis of citronellyl propionate in sCO_2 catalyzed by CALB immobilized on monolith-supported ionic liquid material **13**.

The best results were obtained with the most hydrophobic support **13** at 80°C and 10 MPa reaching a total turnover number of 35.8×10^4 mol product/mol enzyme.

The second example regarded the use of supported ionic liquid on magnetic silica nanoparticles for *Candida rugosa* lipase immobilization.[15] Imidazole, having different chain lengths (C_1, C_4, and C_8), was used as cation, while Cl^-, BF_4^-, PF_6^- were used as anions. Magnetic nanoparticles were treated with tetraethyl orthosilicate in order to allow the silica shell to grow on the surface of the nanoparticles. This silica support was then treated with the proper 1-alkyl-3-(triethoxysilylpropyl)-imidazolium salt in toluene at 90°C to give supports **16–20** (Scheme 10.5).

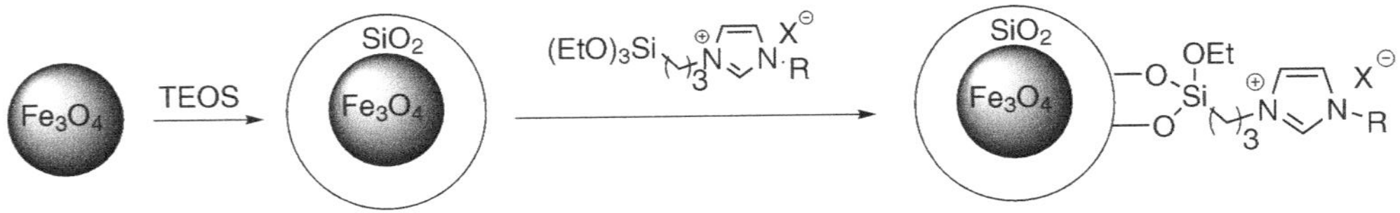

16: R=CH_3, X=Cl
17: R=CH_3, X=BF_4
18: R=CH_3, X=PF_6
19: R=C_4H_9, X=PF_6
20: R=C_8H_{17}, X=PF_6

Scheme 10.5. Synthesis of supported ionic liquid on magnetic silica nanoparticles.

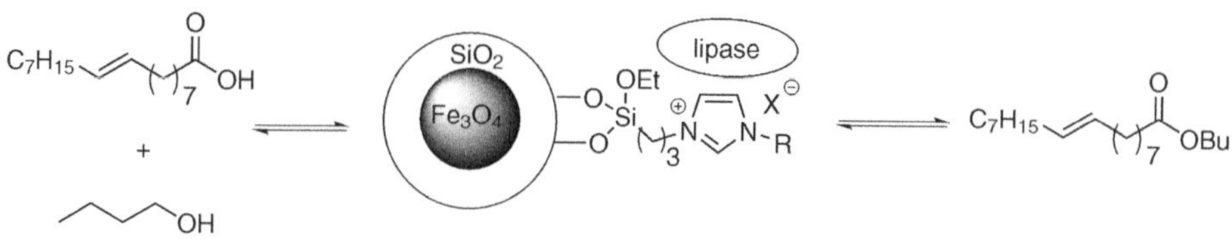

Scheme 10.6. Esterification reaction catalyzed by *Candida rugosa* lipase immobilized on supported ionic liquid on magnetic silica nanoparticles.

Candida rugosa lipase was supported from a phosphate buffer (pH 7.0), then the supported lipase was removed from the solution by a magnet. A large amount of lipase (63.89 mg/100 mg carrier) was loaded on the support through ionic adsorption, and the activity was tested in the ester synthesis between oleic acid and butanol (Scheme 10.6). For the catalytic materials having the same cation ($R{=}CH_3$), the lipase activity followed the trend $Cl^- < BF_4^- < PF_6^-$. For the catalytic materials having the same anion (PF_6), the lipase activity followed the trend $CH_3 < C_4H_9 < C_8H_{17}$. These results suggested that the immobilized enzyme activity was related both to the coordinating ability of anions and to the hydrophobicity of the ionic liquid moiety.

The role of the ionic liquid moiety was also evident if a comparison with the native lipase is made. The activity of bound lipase was 118.3% compared with that of native lipase. Moreover, immobilized lipase maintained 60% of its activity after eight repeated reaction cycles, while no activity was detected after six cycles for the native enzyme.

10.2.2 Strategy 1b: Covalently Attached "Ionic Liquid" Moieties as Linkers

Soon after the first report on SILP, organocatalysts were covalently modified with ionic liquid tags (see Section 10.3). As a consequence of this approach, the corresponding supported organocatalysts were also developed. Pyrrolidine-based chiral ionic liquids supported on polystyrene[16] (Merrifield resin) or silica gel[17] were prepared from pyrrolidine **21** and the proper support, followed by deprotection (Scheme 10.7). These catalytic materials were used as recyclable organocatalysts for asymmetric Michael additions to nitrostyrene.

Using the Michael addition between cyclohexanone and *trans*-β-nitrostyrene as model reaction, catalyst **25** was tested under different conditions. Several solvents were used for the above reaction, ranging from less polar (hexane, diethyl ether) to more polar (dimethylformamide [DMF], water). The stereochemical outcome of the reaction was insensitive to the polarity of the solvent, being always very high (98–99% ee; 98/2–99/1 diastereomeric ratio [d.r.]). However, different yields were observed, being higher in polar solvents such as CH_3CN, EtOH, H_2O, and DMF and also in nonpolar solvents such as hexane and toluene. Neat conditions, at room temperature, were, at last, chosen as optimal reaction condition. Michael products were obtained in high

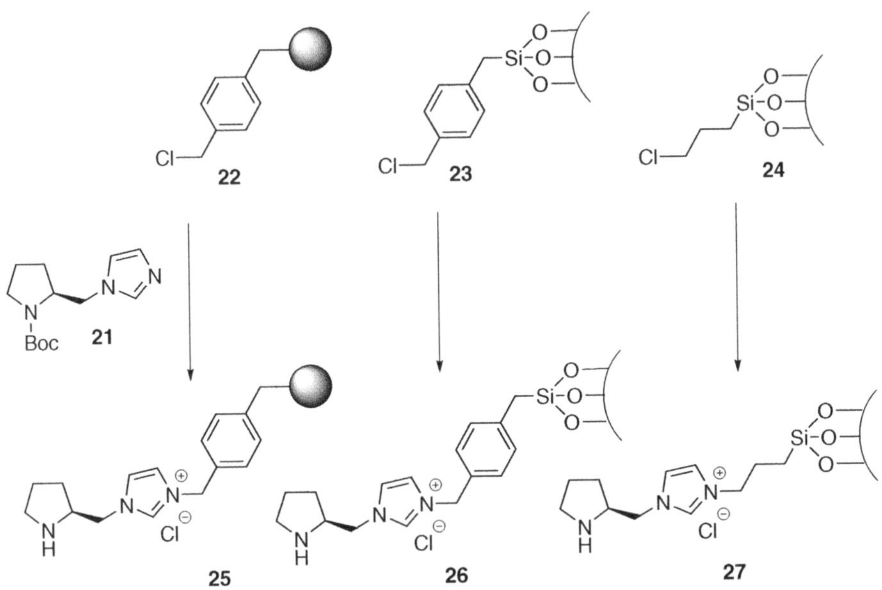

Scheme 10.7. Synthesis of pyrrolidine-based chiral ionic liquids supported on polystyrene (Merrifield resin) or silica gel.

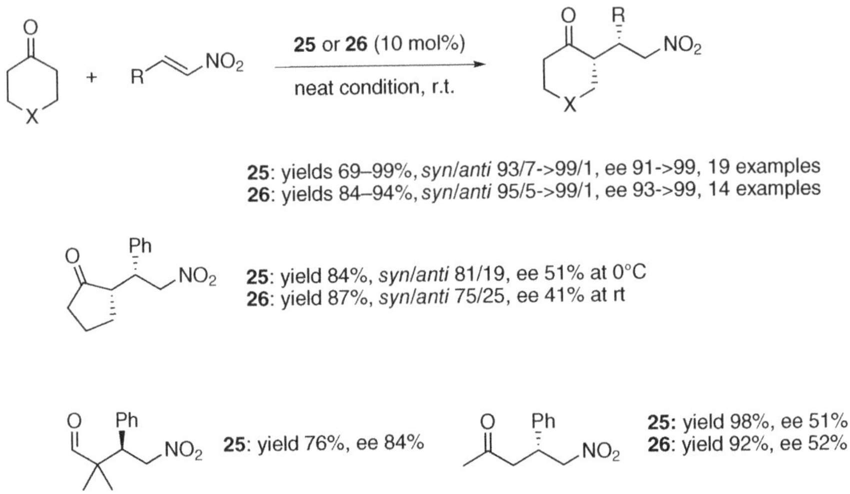

Scheme 10.8. Aldol reactions catalyzed by supported chiral ionic liquids **25** and **26**.

yields and stereoselectivity. Also silica-supported catalysts **26** and **27** were tested in the above reaction under different conditions and a similar behavior with catalyst **25** was observed (Scheme 10.8). Again neat condition was chosen. Silica-supported catalyst **26** gave a better result than catalyst **27**. These catalysts were also recyclable and used up to eight times without significant loss in activity and stereoselectivity.

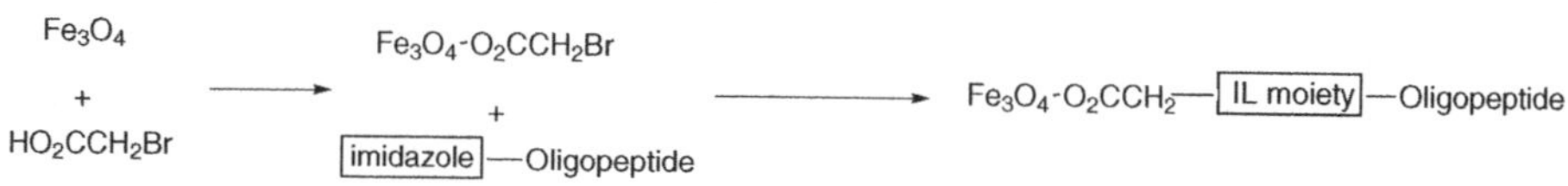

Scheme 10.9. Synthesis of supported oligopeptides on magnetic particles through an ionic liquid spacer.

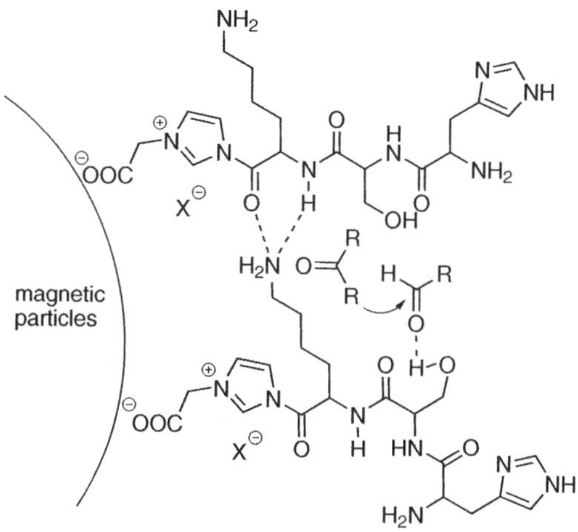

Figure 10.4. Aldol reactions catalyzed by oligopeptide supported on magnetic particles through an ionic liquid spacer.

Magnetic particles were used as support for the immobilization of oligopeptides through an ionic liquid spacer.[18] These materials were prepared as outlined in Scheme 10.9. The imidazole–oligopeptides were purified by HPLC, and bromoacetic acid was used as coupling reagent between the magnetic particles and the imidazole on the terminals of oligopeptide chain. These catalytic materials were tested in the asymmetric aldol reaction.

Several parameters were changed such as cation, anion, oligopeptides, and solvents. As cation, imidazolium containing C(2) hydrogen improved the enantioselectivity compared with the imidazolium containing C(2)-methyl. Both yield and enantioselectivity increased with the hydrophobicity of the anion ($Cl^- < BF_4^- < PF_6^-$). Probably, the hydrophobic environment of PF_6^- anion pushes the aldehyde acceptor into the catalytic helix, which facilitates its condensation with the ketone donor by exclusion of water. Moreover, the strong H-bonding ability of Cl^- ion with the ε-amino acid group of lysine may perturb the pK_a of the Lewis base lysine group, leading to a low enantioselectivity value. Variation of the amino acid sequence of the oligopeptide showed a limited effect on the yield and enantioselectivity, but introducing a series of lysine groups close to the imidazolium ring significantly increased the output (Figure 10.4). The oligopeptide of choice was L-Lys-L-Ser-L-His with a catalytic loading of 10 mol%.

Several solvents were screened, both for the unsupported and for the supported oligopeptide. It was observed that yield and enantioselectivity were less dependent on the solvent polarity in free oligopeptide, while the enantioselectivity increased with the supported oligopeptide with the polarity of solvent. This was ascribed to the incompatibility of the ionic liquids units on the magnetite-supported oligopeptide with nonpolar solvents, which induced aggregation leading to poor catalytic efficiency and enantioselectivity. Using magnetite–imidazole(PF_6)-L-Lys-L-Ser-L-His catalyst in DMSO at 30°C, several aldol reactions were carried out affording products in 35–93% ee. The recyclability of the catalyst was also considered. After five cycles a small decrease both in yield and in enantioselectivity was observed; however, the unsupported oligopeptide showed a more marked decrease in enantioselectivity.

10.2.3 Strategy 1c: Covalently Attached "Ionic Liquid" Moieties as Organocatalysts

While in the first two strategies the covalently attached "ionic liquid" moieties served as supports or linkers, covalently attached "ionic liquid" moieties may also play the role of catalysts in several reactions. Ionic liquid-modified silica gels were found to be useful recyclable catalysts for Knoevenagel condensation. Two imidazolium cation-based ionic liquid-modified silica gels were prepared on two different supports (Scheme 10.10). Catalyst **28** was prepared by co-condensation of tetraethoxysilane and 1-methyl-3-(3-triethoxysilylpropyl)imidazolium chloride to give an amorphous silica,[19] while catalyst **29** was prepared in the presence of P123 as template to give a SBA-15 mesoporous material.[20] Analogously, catalyst **30** was prepared following the procedure for **29**. The imidazolium-based catalysts gave better results than the pyridinium-based catalyst. High yields and good recyclability were observed (**29** was used up to 10 cycles).

Ionic liquid-anchored catalyst **29** was also used in the reaction of propylene oxide with CO_2, in the aza-Michael additions of amines to αβ-unsaturated compounds in water and for the Biginelli reaction (Scheme 10.11).[21] Recycling (five cycles) in the Biginelli reaction showed no loss in activity.

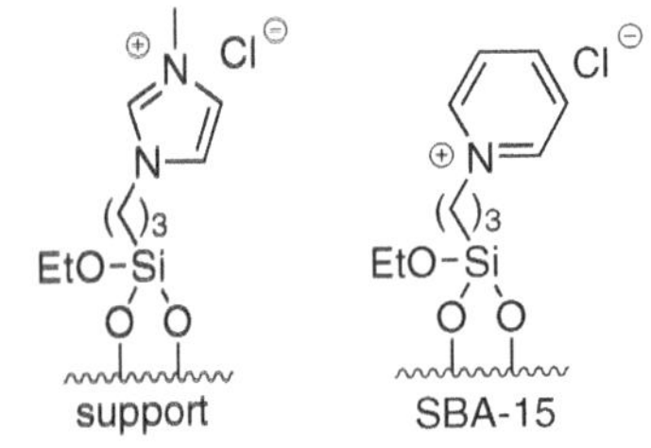

28: support = SiO_2
29: support = SBA-15
30

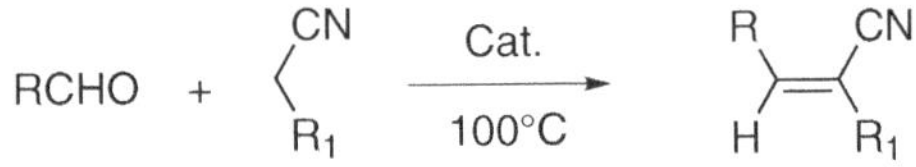

28: R_1 = COOEt: yields 60–98%, 7 examples
R_1 = CN: yields 93–97%, 7 examples

29: R_1 = COOEt: yields 75–90%, 6 examples
R_1 = CN: yields 87–93%, 6 examples

30: R_1 = COOEt: yields 82–84%, 2 examples
R_1 = CN: yields 89–92%, 2 examples

Scheme 10.10. Knoevenagel condensations catalyzed by supported ionic liquids **28–30**.

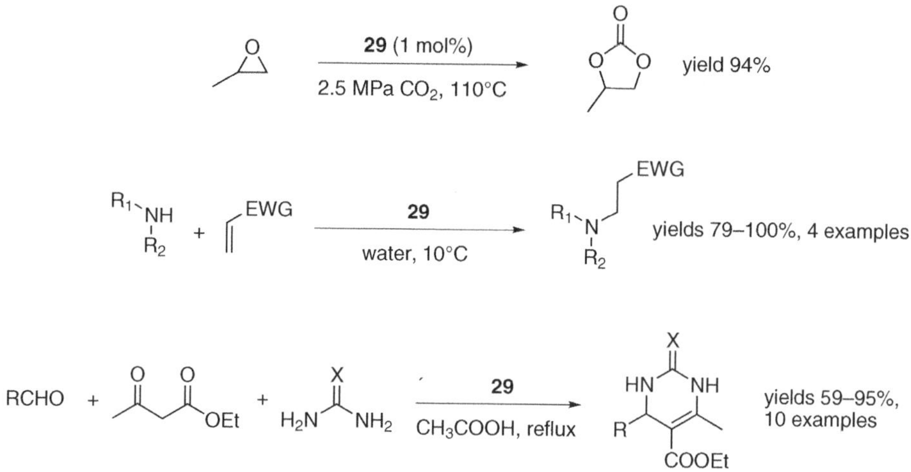

Scheme 10.11. Reactions catalyzed by supported ionic liquid **29**.

Co-condensation procedures, in the presence of 1-cetyl-3-methylimidazolium chloride as templating agent, were also employed for the preparation of imidazolium- and dihydroimidazolium-supported ionic liquids **33-34** and **36-37**, in different dilutions, as recyclable organocatalysts for Knoevenagel condensation (Scheme 10.12).[22]

The synthesis was carried out as reported in Scheme 10.12 by using the monosilylated and disilylated monomers **32** and **35**. For each monomer two materials with different amounts of organic loading were prepared.

Better results were obtained with material **36** having higher catalytic loading, showing that the predominant factor was the concentration of the ionic liquid moiety rather than the porosity in the materials. Moreover, the catalyst was easily recovered and used for five consecutive cycles without loss in activity (Scheme 10.13).

The monosilylated dihydroimidazolium chloride **38** bearing a C-8 alkyl chain was grafted on silica gel. Then, the corresponding *N*-octyldihydroimidazolium hydroxide **39** was obtained after treatment with aqueous NH_3 (Scheme 10.14).[23]

This organocatalyst was employed both in the cyanosilylation of several carbonyl compounds and in the epoxidation of cyclic ketones, with good results. Only one recycle was carried out with decreased activity (yield from 99 to 90%). Knoevenagel condensations were also carried out with basic ionic liquids supported on hydroxyapatite-encapsulated γ-Fe_2O_3 nanocrystallites.[24] Four organocatalysts (**40–43**) having different alkyl chains were tested. Catalyst **40** gave slightly higher yields and was used for several reactions with malononitrile (Scheme 10.15). Recovery and reuse showed a small decrease in isolated yields after five cycles.

Polystyrene-supported imidazolium hydroxide **44** and hydrogen carbonate **45** (Scheme 10.16) were tested as catalysts in the synthesis of 1,2-propylene glycol by hydrolysis of propylene carbonate. Catalyst **45** gave excellent yield and was used in six consecutive runs.[25] On the other hand, polystyrene-supported imidazolium hydroxide **44** was found to be a good recyclable (up to six cycles) organocatalyst for the

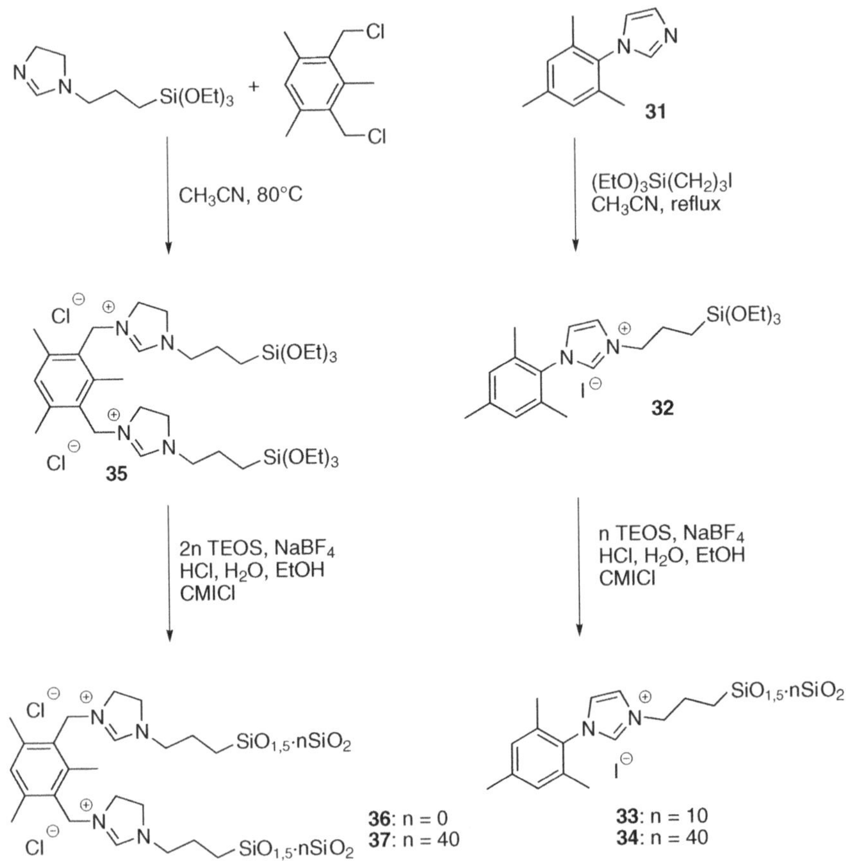

Scheme 10.12. Synthesis of imidazolium- and dihydroimidazolium supported ionic liquids **33-34** and **36-37**.

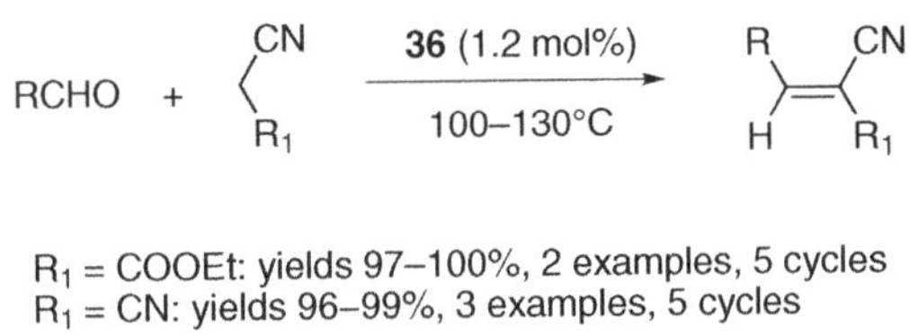

Scheme 10.13. Knoevenagel condensations catalyzed by supported ionic liquid **36**.

aldol-type coupling of aldehydes with ethyl diazoacetate (Scheme 10.16).[26] These supported organocatalysts worked nicely in water, affording products in good to high yields.

As the last example of supported Brønsted basic ionic liquid, simple choline hydroxide on MgO was used for aldol condensation reactions to afford α,β-unsaturated compounds in good yields.[27]

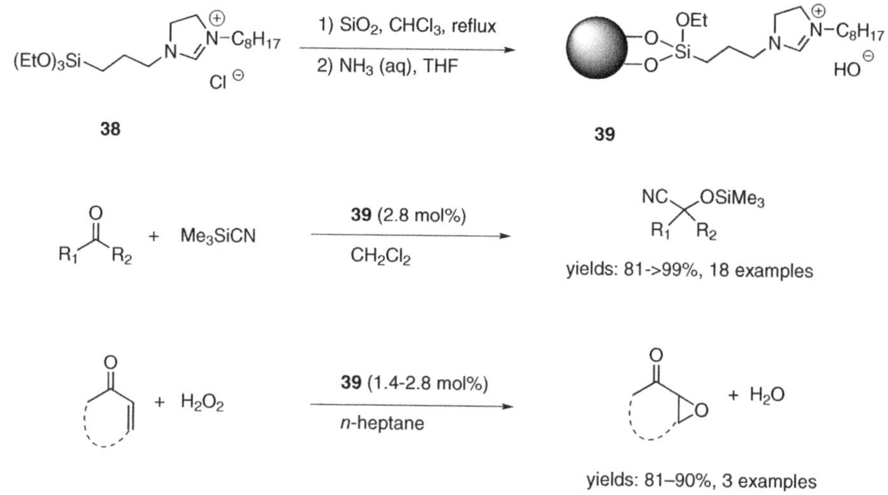

Scheme 10.14. Synthesis of supported *N*-octyldihydroimidazolium hydroxide **39** and its use in the cyanosilylation and epoxidation reactions.

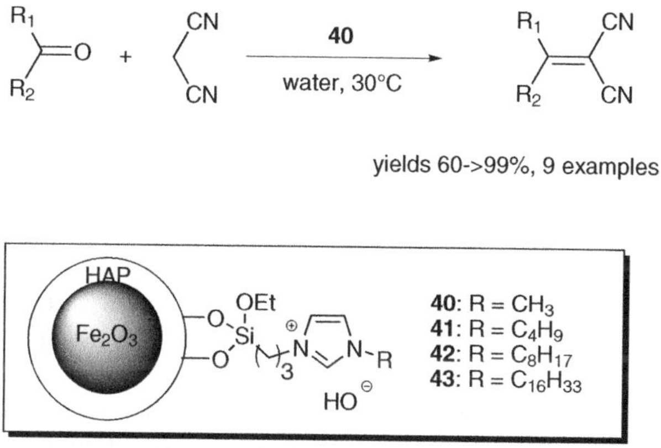

Scheme 10.15. Knoevenagel condensations catalyzed by supported ionic liquid **40**.

Several efforts were devoted to the cycloaddition of epoxides with CO_2 by using ionic liquid catalysts anchored to cross-linked polystyrenes or silica. Cross-linked polymer **46** obtained by copolymerization of divinylbenzene and 1-butyl-3-vinyl-imidazolium chloride was a good catalyst for the formation of cyclic carbonates from epoxides (Scheme 10.17).[28]

Catalyst **46** was much more active than the catalyst obtained by polymerization of 1-butyl-3-vinyl-imidazolium chloride without cross-linker. In the latter polymer, the density of the active sites was much larger than that in **46**, which resulted in insufficient use of the active sites. Organocatalyst **46** was used in five consecutive runs without loss in activity.

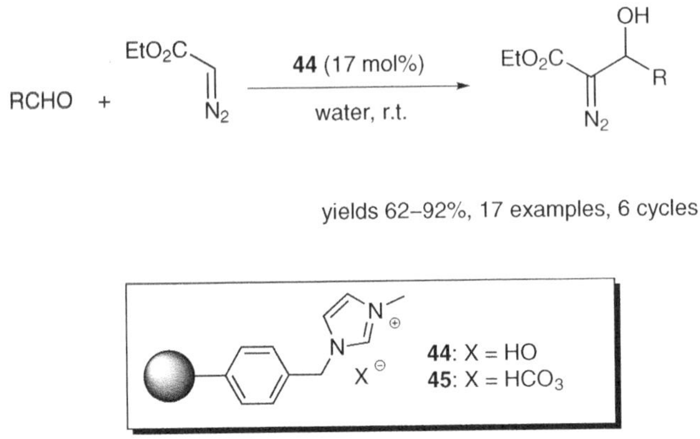

Scheme 10.16. Aldol-type coupling of aldehydes with ethyl diazoacetate catalyzed by **44**.

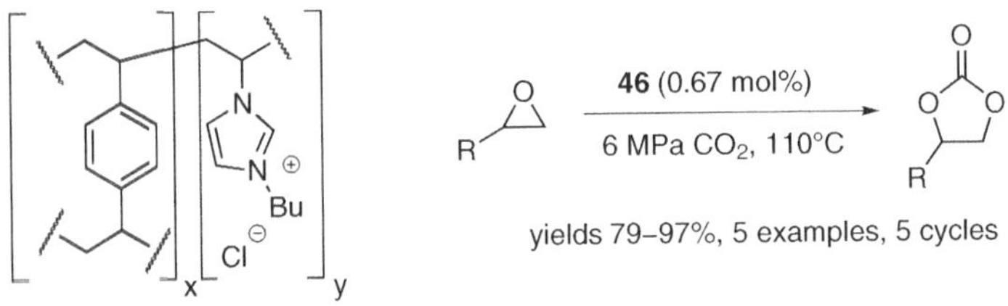

46

Scheme 10.17. Cycloaddition reactions of epoxides with CO_2 catalyzed by ionic liquid catalyst **46**.

Highly cross-linked chloromethylated polystyrene was used as starting material for the synthesis of ionic liquid-anchored organocatalysts **47–49**, as reported in Scheme 10.18.[29]

Organocatalysts **48-49** having the more nucleophilic anions were the more active in the formation of cyclic carbonates from epoxides (**48**: 1.6 mol%, 2.5 MPa CO_2, 115–125°C, yields 80–99%, seven examples). Moreover, the crucial role of the OH group was established since organocatalyst **50** was less active than **48**. In the proposed mechanism, such role was ascribed to the hydrogen bond between the OH group and the oxygen atom of the epoxide, which activated the epoxide to the nucleophilic attack of Br^- anion. Organocatalyst **48** was used in six consecutive runs without loss in activity.

Polystyrene-supported ionic liquid **51** (Figure 10.5) was employed as a recyclable phase transfer catalyst, although in high loading (50 mol%), in substitution reactions such as chlorination, bromination, iodination, and acetoxylation of a mesylated primary alcohol and azidation of a brominated primary alcohol, at 100°C.[30]

Six silica supported imidazolium ionic liquids (**52–57**, Figure 10.5) were prepared by grafting on commercial silica gel of the corresponding triethoxysilane derivatives.[31]

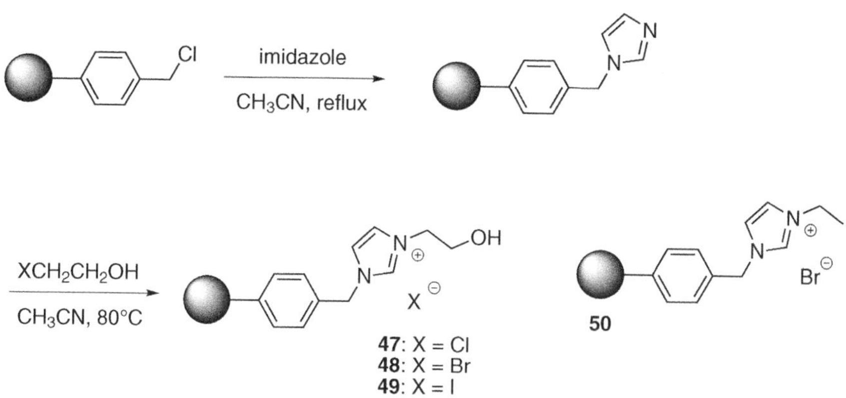

Scheme 10.18. Polystyrene-supported ionic liquids **47–50**.

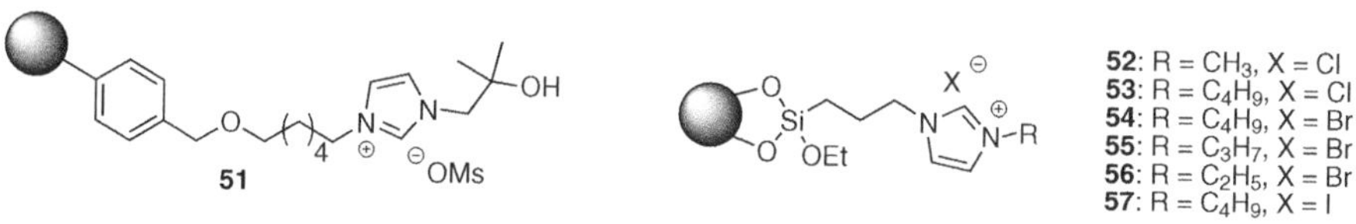

Figure 10.5. Polystyrene-supported ionic liquid **51** and silica gel supported ionic liquids **52–57**.

Materials with different alkyl chains and anions were investigated for the cycloaddition of carbon dioxide to allyl glycidyl ether. The immobilized ionic liquids with longer alkyl chain length and more nucleophilic anion showed higher activity (up to 77.8% conversion). The result was ascribed to the bulkiness of alkyl chain that forced the halide ions away from the cation more easily. The same reaction was also carried out with catalyst **54** prepared by template-free condensation under strong acidic conditions. The yield reached 91.7%.[32]

In addition to covalently attached ionic liquids, adsorbed ionic liquids ([BMIM][BF_4], [BMIM][PF_6], and [BMIM][Br]) on silica gel were also used as catalysts for the synthesis of cyclic carbonates from carbon dioxide.[33] [BMIM][BF_4] on SiO_2 was the catalyst of choice and used with good results (catalyst loading 1.8 mol%, 8 MPa CO_2, 160°C, yields 78–98%, five examples). After four cycles a minor decrease in activity was observed. Since the ionic liquid was not covalently attached, leaching of the adsorbed ionic liquid was determined (40 ppm).

Lower CO_2 pressure (1 MPa) was needed when magnetic nanoparticle-supported ionic liquid catalyst **58** was employed (Scheme 10.19).[34] This supported ionic liquid was successfully used for five different epoxides to afford carbonates in high yields. Imidazolium units having longer alkyl chain (C4 and C6) gave lower yields, whereas a comparable result was observed with [BMIM][Br], highlighting the advantage of the supported organocatalyst. Catalyst **58** was recovered by a magnet and used directly in

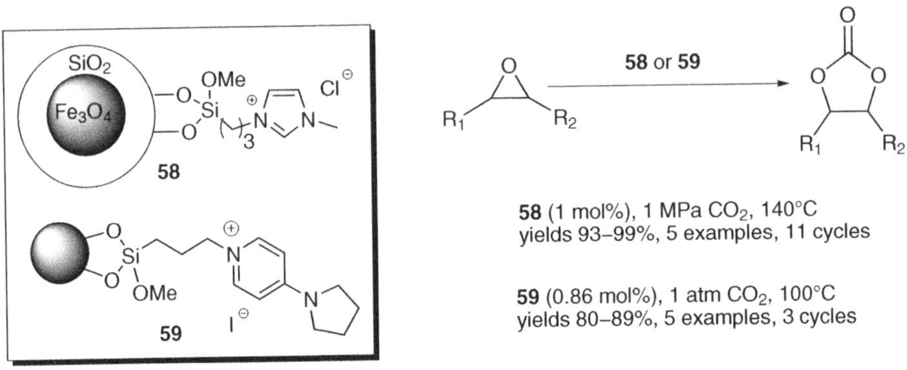

Scheme 10.19. Cycloaddition reactions of epoxides with CO_2 catalyzed by **58** or by **59**.

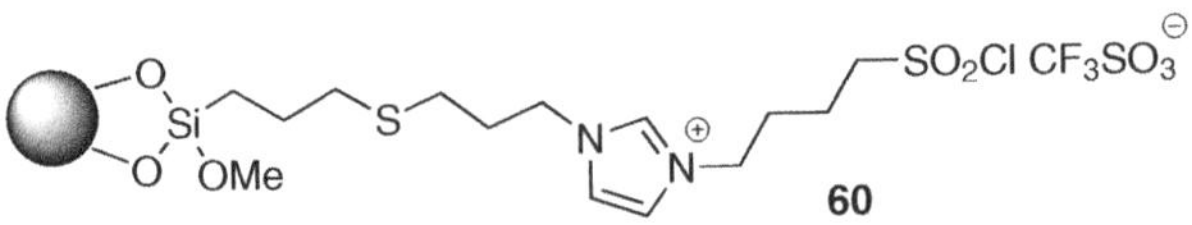

Figure 10.6. Silica gel supported ionic liquid **60**.

the next run after washing with CH_2Cl_2 and drying. After 11 cycles the activity was maintained. Even lower CO_2 pressure (1 atm) was used when the reaction was carried out in the presence of catalyst **59** (Scheme 10.19).[35]

Supported Brønsted acidic ionic liquids have also been reported. Supported ionic liquid catalyst **60** (Figure 10.6), obtained by radical chain transfer reaction of the corresponding 1-allylimidazolium-based ionic liquid on mercaptopropyl-modified silica gel, was used in the esterification reactions of simple acids with ethanol or C8 or C10 alcohols at 80–100°C.[36]

The supported ionic liquid **61** was synthesized by grafting the reaction between the ionic liquid 1-methyl-3-(triethoxysilylpropyl)imidazolium chloride on a silica support with an extensive system of meso- and macropores. The obtained material was then treated with sulfuric acid to afford the hydrogensulfate catalyst **61**.[37]

This acid catalyst was used for the Baeyer–Villiger reaction of several cyclic ketones to give the corresponding lactones in good yields (Scheme 10.20). The reusability of the catalyst was tested in four consecutive runs. No decrease in activity was observed.

10.3 ANCHORING OF ORGANOCATALYST TO IONIC LIQUIDS

In this approach a chiral (or nonchiral) unit is covalently tethered to an ionic liquid moiety, with the former serving as a catalytic site and the latter as a phase tag. The use of ionic liquids as phase tags for organocatalysts can play up to three roles (Figure

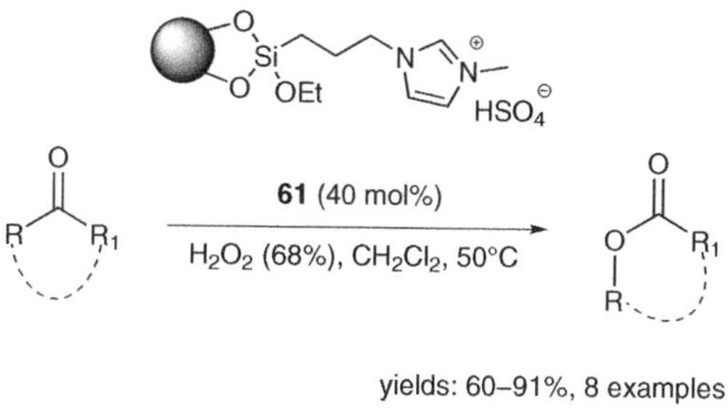

Scheme 10.20. Baeyer–Villiger reactions catalyzed by **61**.

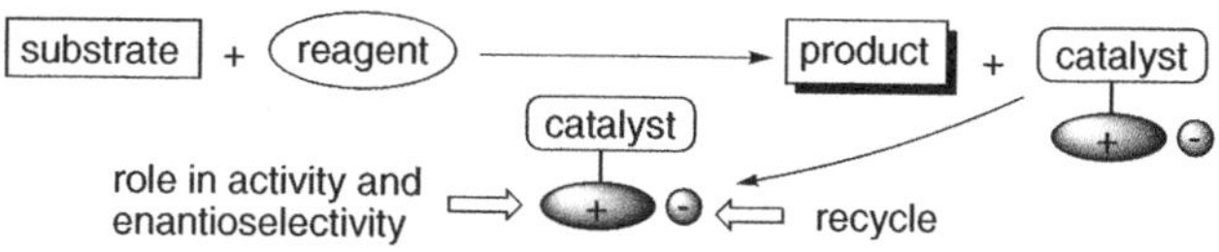

Figure 10.7. Roles of ionic liquid tags.

10.7): (1) they can facilitate catalyst recycling; (b) they can offer enhanced reactivity; and (3) they can offer enhanced enantioselectivity acting as chiral-induction groups.

By choosing different cations and anions the solubilities of ionic liquids can be readily tuned allowing phase separation from organic as well as aqueous media. This approach can also be applied to ionic liquid-anchored organocatalysts in order to separate the catalytic molecule from the product and to allow its reuse. Ionic liquid-anchored organocatalysts may be more active with respect to the supported ionic liquid organocatalysts because of their homogeneous nature, but recovery of the catalyst requires precipitation or extraction, which may not be quantitative.

Previously, it has been reported that reactions in ionic liquids are sometimes faster and cleaner than in conventional organic solvents,[38] and ionic liquid itself has been shown to possess weak Lewis acidity that can serve as catalyst for organic reactions.[39]

The role of ionic liquid moiety as a chiral-induction group may be rationalized by considering that (1) the bulky and planar organic cation may impart space shielding to the reaction intermediate and (2) the proximity of the ionic liquid unit to the active site may create a microenvironment that is favorable for the reaction. Indeed, the high polarity and ionic character of the ionic liquids exert synergistic effects on many organic reactions.[40]

The ionic liquid-anchored organocatalysts were employed for several types of reactions such as aldol reactions, Michael additions, and Baylis–Hillman reactions.

10.3.1 Aldol Reactions

In 2006, after the development of the supported ionic liquid asymmetric catalysis,[8,9] the first example of an organocatalyst anchored to an ionic liquid was reported.[41]

Catalyst **64** was easily prepared from the ionic liquid carboxylic acid **62** and (*2S, 4R*)-4-hydroxyproline Cbz-protected **63** (Scheme 10.21a).

In addition to compound **64**, ionic liquid-anchored proline **65** (Scheme 10.21b) was prepared in order to assess whether the imidazolium moiety can replace the role of the acidic proton since the C-2 hydrogen atom of the cation is fairly acidic. However, the latter compound gave a poor yield and enantioselectivity in the aldol reaction between 4-cyanobenzaldehyde and acetone.

Aldol reactions performed in pure ketone employing organocatalyst **64** (30 mol%) gave comparable results to those obtained in DMSO (Scheme 10.22). Moreover, catalyst **64** showed higher activity and gave better enantioselectivities in neat acetone when compared with native L-proline.

These results suggested that the ionic liquid-supported proline **64** was a more efficient and stereoselective organocatalyst than proline. In addition, the recyclability of catalyst **64** was also examined. After the reaction was completed, the mixture was concentrated, rinsed twice with dichloromethane, centrifuged, and then decanted. The dichloromethane solutions were concentrated to give the aldol product, whereas the residue contained the catalyst **64**. Four cycles were carried out with almost unchanged yield and enantioselectivity.

Whereas compound **65** was reported to be inefficient, other authors claimed that ionic liquid-anchored proline **66**, containing CF_3COO as anion instead of BF_4 in compound **65**, was an efficient catalyst for the aldol reaction between acetone and several aldehydes (Scheme 10.23).[42] Aldol products were obtained in good yields and good to high enantioselectivity values. Moreover, the catalyst was recovered after evaporation of the acetone and extraction with diethyl ether, which afforded the aldol product, leaving the catalyst as residue. The catalyst was used four times, with unchanged yield and enantioselectivity.

The ionic liquid moiety was also anchored to 4-hydroxyproline with an ether linkage.[43] Compound **68** was prepared from imidazolium bromide **67** and (*2S,4R*)-4-hydroxyproline Cbz-protected **63**, followed by deprotection (Scheme 10.24). Catalyst **68** was used in ionic liquid as solvent, thus combining the advantages of a homogeneous phase with the opportunity of exploiting solubility differences in the workup step.

The amount of catalyst **68** was decreased with respect to **64** being used in 10 mol%, and the reactions were carried out in [BMIM][BF_4]. Reactions between acetone and several substituted benzaldehydes afforded the aldol products in good to high yields and high enantioselectivity (Scheme 10.24). Moreover, catalyst **68** and [BMIM][BF_4] were recovered and reused for six cycles, with minor decreases in yields but always with reproducible enantioselectivity values.

Further studies were carried out by using ionic liquid-supported prolines **69** and **70** (Figure 10.8).[44] These compounds were prepared following a slightly modified procedure with respect to the previous one. Chloroacetate **71**, readily obtained from the corresponding protected 4-hydroxyproline, was used as starting material. Moreover, whereas catalyst **64** was employed in large amounts (30 mol%), catalysts **69** and **70** were screened in lower amounts (1–5 mol%).

Four ionic liquids were used as solvents, two imidazolium-based ionic liquids, [BMIM][Tf_2N] and [BMIM][TfO], and two butylmethylpyrrolidinium-based ionic

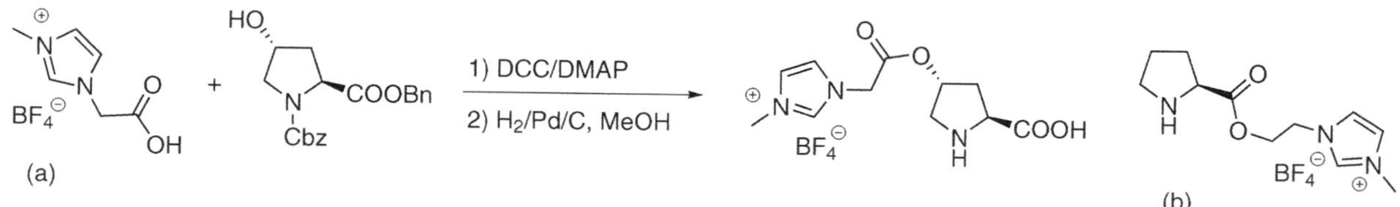

62 63 64 65

Scheme 10.21. Synthesis of ionic liquid-supported proline **64** and structure of ionic liquid-anchored proline **65**.

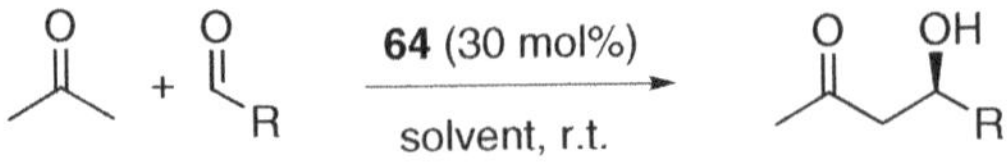

in acetone: yields 40–92%, ee 64–85%, 8 examples
in DMSO: yields 46–83%, ee 60–87%, 8 examples

Scheme 10.22. Aldol reactions catalyzed by **64**.

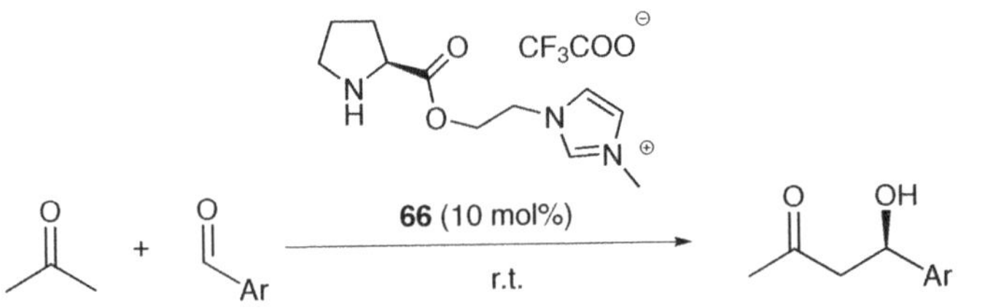

yields 54–94%, ee 67–92%
6 examples

Scheme 10.23. Aldol reactions catalyzed by **66**.

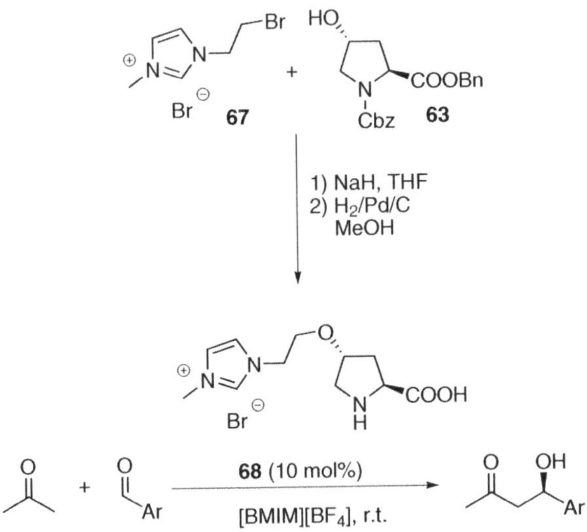

yields 53–94%, ee 64–93%, 14 examples

Scheme 10.24. Synthesis of catalyst **68** and aldol reactions catalyzed by it.

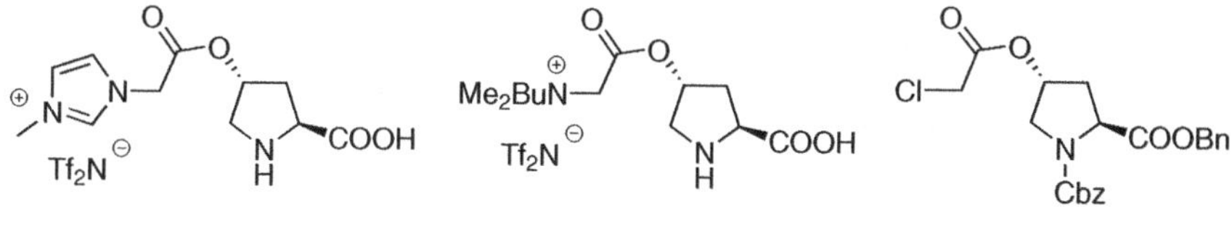

Figure 10.8. Structure of compounds **69–71**.

liquids, [BMPyrr][Tf_2N] and [BMPyrr][TfO]. Preliminary tests were carried out on the reaction between acetone and 4-nitrobenzaldehyde. The best result was obtained when catalyst **70** was used in 5 mol% with 10 equivalents of acetone in [BMIM][Tf_2N] (Scheme 10.25). However, the drawback of this procedure was represented by the difficult recovery and reuse of the catalyst at the concentration level used. After three cycles, a dramatic drop in yield was observed. Several reactions were carried out affording aldol products in better enantioselectivity values with respect to native proline in DMSO.

Although recycling experiments were unsatisfactory, this work demonstrated that the ionic liquid-anchored organocatalysts can be used in lower catalytic amounts than the "classical" 30 mol% usually used in proline-catalyzed aldol reactions.

In order to improve the stereoselectivities observed using catalyst **70** in ionic liquid, further studies were presented using catalyst **69** both in ionic liquid and in water as reaction medium. Enantio- and diastereoselectivities increased significantly when catalyst **69** was used in water with respect to its use under homogeneous conditions in [BMIM][Tf_2N].[45]

Indeed, water plays an important role in stereoselective organocatalytic reactions.[46] It has been suggested that the role of water is to prevent deactivation rather to promote activity.[47] Further studies were also presented about the role of water. Such studies, carried out on proline-catalyzed reaction between acetone and 2-chloro-benzaldehyde, led to the hypothesis about the conflicting role of water. Water increases the total catalyst concentration due to suppression of unproductive species (**72** and **73**, Figure 10.9a) and decreases the relative concentration of productive intermediates by shifting the iminium ion back to proline.[48]

When organocatalysts such as prolines with hydrophobic substituents[49] or linked to a hydrophobic support such as a polystyrene backbone[50] in the 4-position are employed in water, high levels of enantio- and diastereoselectivity and high yields are reached.

The aqueous biphasic environment can simulate the hydrophobic pocket of class I aldolases. As a consequence, hydrophobic reactants are forced into the hydrophobic pocket, resulting in increased activity and stereoselectivity of the catalyst (Figure 10.9b). A confirmation of this hypothesis is demonstrated by the fact that water-miscible ketones afforded moderate yield and stereoselectivity.[49a,51] Then, the use of a massive amount of water is supposed to favor the hydrolysis of the enamine, unless the reaction takes place in a hydrophobic pocket (such as in the enzyme or in polystyrene-supported

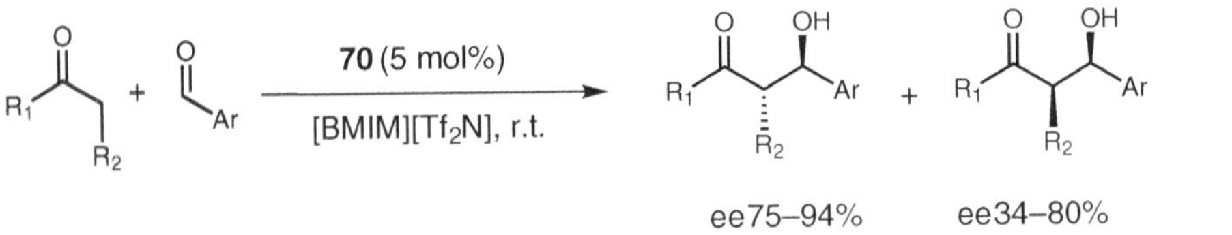

yields 35–78%, *anti/syn* 67/33–80/20, 7 examples

Scheme 10.25. Aldol reactions catalyzed by **70**.

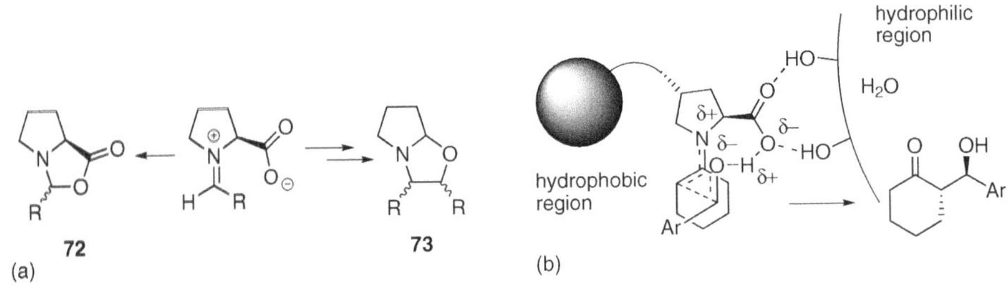

Figure 10.9. (a) Structure of unproductive species **72** and **73**; (b) proposed transition state model for the major stereoisomer in the aldol reactions catalyzed by 4-substituted-L-prolines.

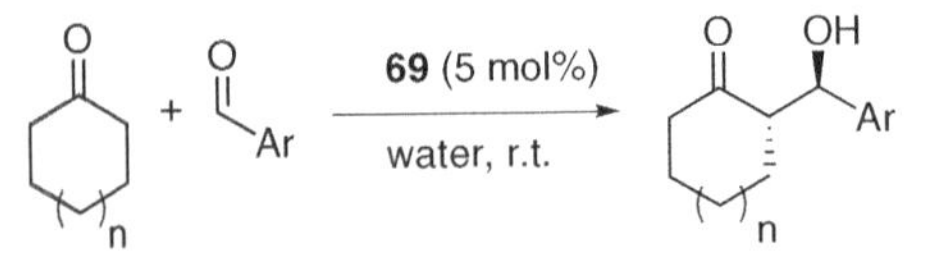

n = 1: yields 41–98%, *anti/syn* 91/9–98/2, ee 96–99%, 9 examples
n = 2: yield 25%, *anti/syn* 91/9, ee 98%, 1 example
n = 0: yield 91%, *anti/syn* 58/42, ee 80%, 1 example

Scheme 10.26. Aldol reactions catalyzed by **69**.

proline derivatives) separated from the aqueous phase. Compound **69**, having a hydrophobic ionic liquid tag, can simulate the proposed mechanism.

Several aldol reactions were carried out affording products in high stereoselectivities in the case of cyclohexanone and cycloheptanone, whereas cyclopentanone gave the aldol product in lower stereoselectivity (Scheme 10.26).

Aldol reactions between cyclohexanone and aliphatic aldehydes (isobutyraldehyde and cyclohexanecarboxaldehyde) were carried out using only a small amount of water (~4% v/v). The aldol products were obtained in good yields and excellent stereoselectivities. Catalyst **69** was recovered after removal of the water/ketone mixture and extraction. After five cycles a 10% decrease in yield was observed while stereoselectivities remained unchanged.

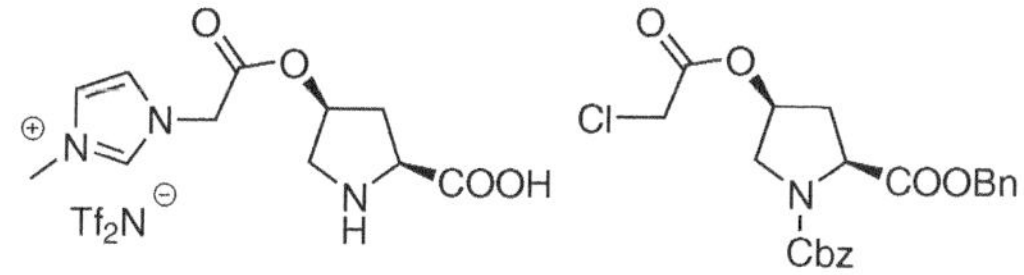

74 **75**

Figure 10.10. Structure of compounds **74–75**.

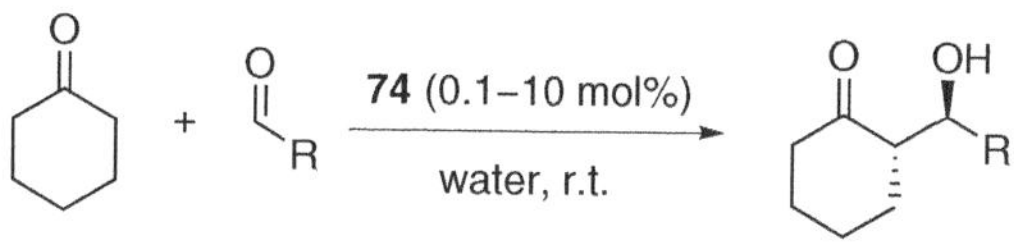

yields 44->99%, *anti/syn* 80/20->99/1
ee 70->99%, TON 7-930, 13 examples

Scheme 10.27. Aldol reactions catalyzed by **74**.

The possibility of having an organocatalyst that could work in a lower catalytic amount prompted the same authors to search for new structures and new reaction conditions than those previously reported. They noticed that ionic liquid-anchored proline **74** (see Figure 10.10 for structure) having a 1,4-*cis*-configuration different from compound **69**, which has a 1,4-*trans*-configuration, was a powerful catalyst for the aldol reaction.[52] Moreover, water was again found to be a much better reaction medium.

Compound **74** was prepared by nucleophylic substitution with methylimidazole on the chloro ester **75** followed by anion exchange with lithium(bis-trifluoromethylsulfonyl) imide and deprotection. Chloro ester **75** was prepared following a standard Mitsunobu protocol on the protected *trans*-4-hydroxy-L-proline **63**.

Catalyst **74** was synthesized following these considerations: Rotation around the C-4-O bond can produce two effects: (1) orienting the ionic tag toward the concave face of the octahydropentalene-like structure of side products **72** and **73**, thus destabilizing them and ensuring a higher amount of available catalytically active species; (2) bringing the ionic tag and its Tf_2N^- counter-ion in spatial proximity to the reactive centers during the rate-determining enamine addition step. If an internal electrostatic stabilization of the transition state was in action, a reaction rate enhancement should be observed. Authors referred to this hypothesis as a "*cis* effect."

Using the reaction between cyclohexanone and 4-nitrobenzaldehyde as model reaction, catalyst **74** was tested employing different catalytic loadings (from 5 to 0.1 mol%) and different amounts of water (0.25, 1.2, 3.2, and 32 equiv.) (Scheme 10.27). A comparison between catalysts **69** and **74** in 0.5 mol% catalytic loading and in the presence of 1.2 equivalents of water was made. After 5 hours, both catalysts afforded the aldol product in excellent optical purity (ee > 99%, *anti/syn* 92/8-94/6), but with markedly different activities (yield: **69**, 37%; **74**, 87%). These results indicated

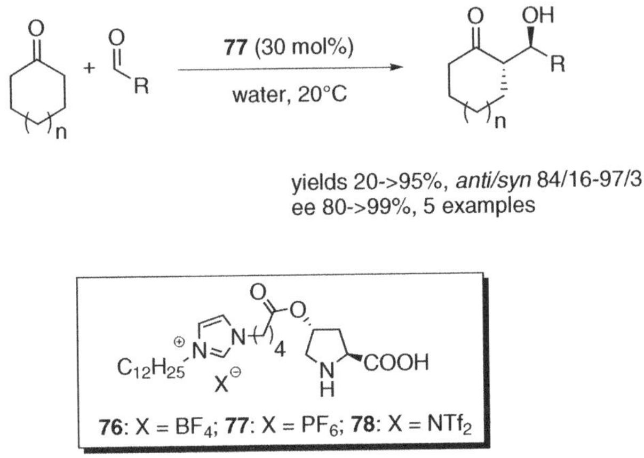

Scheme 10.28. Aldol reactions catalyzed by **77**.

that ionic liquid-anchored catalyst **74** was highly efficient. It was then used in the aldol reaction with several aromatic aldehydes in 0.1–2 mol% and with aliphatic aldehydes in 10 mol%, with excellent results.

Since in many cases the use of water as a reaction medium leads to products with higher optical purity, it might be expected that more hydrophobic structural analogues of catalysts **64** would possess a much higher activity and stereoselectivity when used in water. To test this hypothesis, two new ionic liquid-anchored proline derivatives bearing long-chain hydrocarbon groups at the imidazolium nitrogen atoms were prepared.[53] Compounds **76** and **77** had quite different solubilities in water. Tetrafluoroborate **76** gave a clear 5% aqueous solution, whereas hexafluorophosphate **77** gave a suspension under the same condition. In the presence of the water-soluble catalyst **76** the reaction did not take place. This result resembles those of aldol reactions catalyzed by water-soluble amino acids in water, low activity, and stereoselectivity. On the other hand, good results were obtained with water-insoluble catalyst **77** (Scheme 10.28). These reactions afforded aldol products in low to high yield and high stereoselecivity, but disappointingly the activity of catalyst **77** was not high. Indeed, it was used in 30 mol%.

Recycling experiments were carried out. The aldol products were recovered by extraction with diethyl ether and the residue was reused. After five cycles, catalyst **77** retained its activity and stereoselectivity.

Other highly hydrophobic ionic liquid-anchored compounds were reported by the same authors and used as catalysts for the aldol reaction in water. Such catalysts were based on pyridinium cations bearing proline, serine, or threonine moieties and PF_6^- anion (**79–82**, Figure 10.11). In addition, compound **78** carrying the (bis-trifluoromethylsulfonyl)imide anion was prepared and tested.[54]

Catalysts **76–82** were prepared with the same strategy, nucleophilic displacement of the bromine atom, followed by anion exchange and deprotection, as depicted in Scheme 10.29.

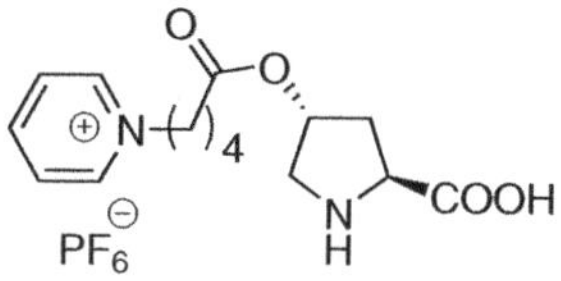

79

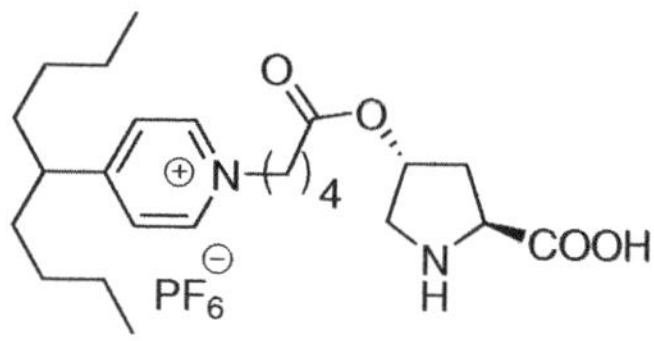

80

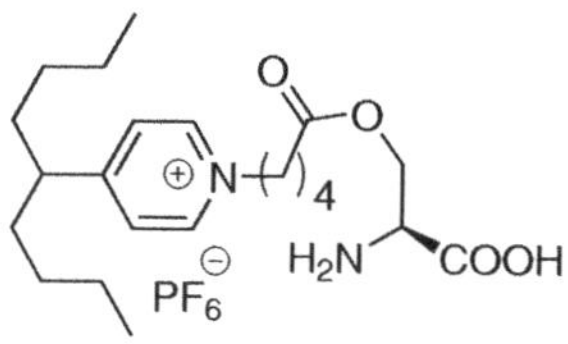

81

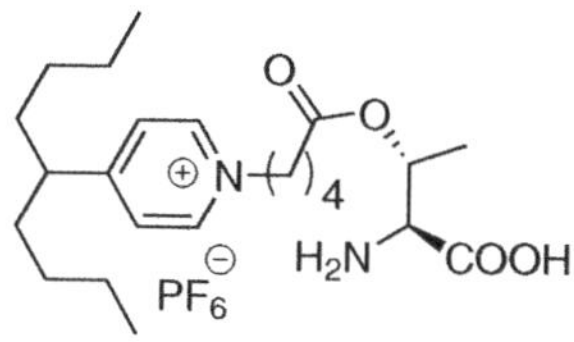

82

Figure 10.11. Structure of compounds **79–82**.

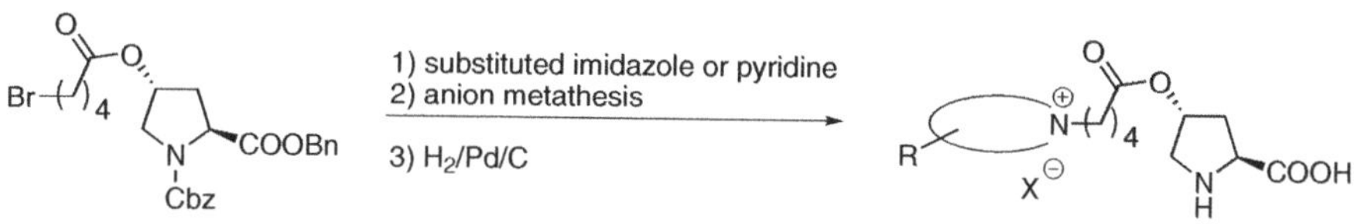

76–82

Scheme 10.29. Synthesis of ionic liquid-anchored prolines **76–82**.

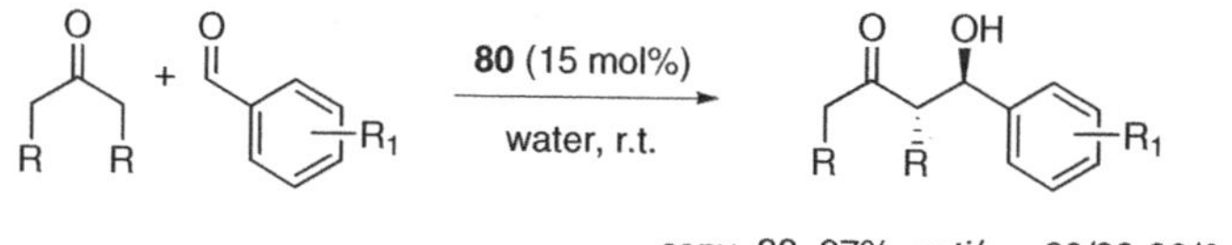

conv. 38–97%, *anti/syn* 80/20-98/2
ee 92->99%, 10 examples

Scheme 10.30. Aldol reactions catalyzed by **80**.

Catalysts **78–82** (15 mol%) were screened using the reaction between cyclohexanone (3 equiv.) and 4-nitrobenzaldehyde. Catalyst **78** gave excellent results in terms of yield and stereoselectivity, even though a dramatic drop in activity was observed in the third cycle. Compound **79**, despite the presence of the hydrophobic PF_6 anion, was soluble in water and did not catalyze the aldol reaction. Compound **80** gave excellent results and was used in eight consecutive runs without loss in activity and stereoselectivity (procedure not reported). Ionic liquid-anchored serine **81** and threonine **82** were used in 20 mol% and gave the corresponding aldol product in lower yield and stereoselectivity. Then, catalyst **80** was employed in a set of reactions with good results (Scheme 10.30).

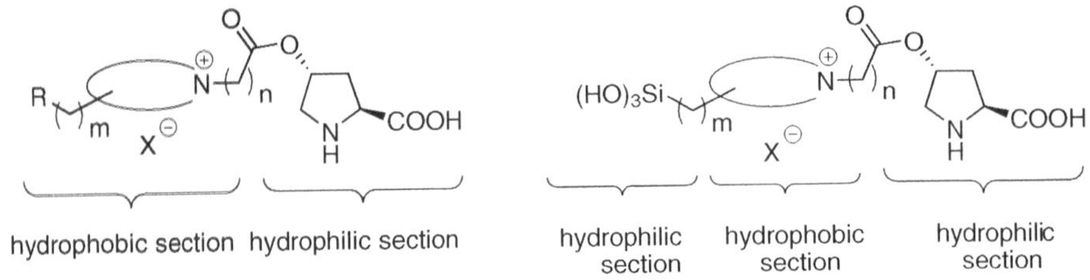

83 **84**

Figure 10.12. General structure of molecules of type **83** and **84**.

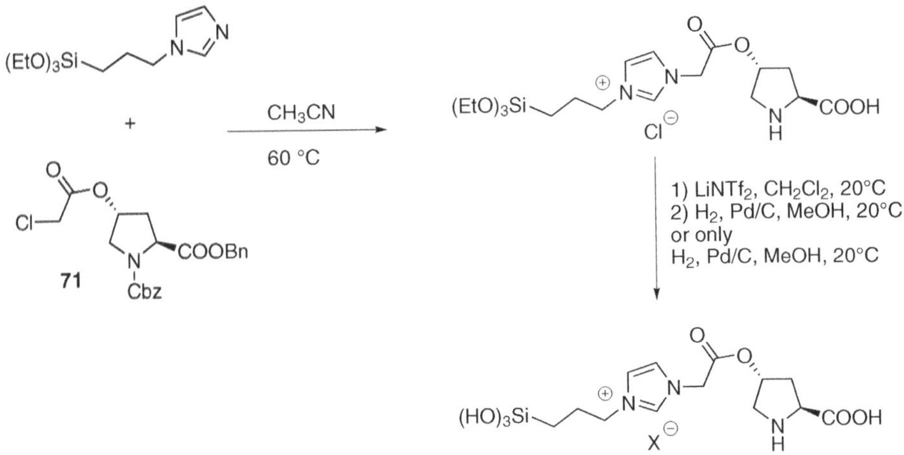

85: X = Cl; **86**: X = NTf_2

Scheme 10.31. Synthesis of ionic liquid-anchored prolines **85-86**.

Compared with catalyst **74**, which was used in low amount (2–0.1 mol%) with excellent results, catalyst **80** was much less active (15 mol%). It could be interesting to investigate if inversion of configuration at C-4 could enhance its activity.

Based on the data reported in the literature, the most useful ionic liquid-anchored proline derivatives for aldol reactions in water could be represented in the general formula **83** (Figure 10.12). The key factor for the successful application of these molecules is the hydrophobicity of the ionic liquid moiety. However, it has been questioned that maximizing the hydrophobicity of this part of the molecule could provide an inefficient recycling procedure at the extraction stage since the catalyst could be extracted in the same solvent. In order to overcome this possible drawback, a new type of ionic liquid-anchored organocatalyst with the correct amphiphilicity profile should be designed.

It has been proposed that molecules of type **84** (Figure 10.12) could help in the recycling procedure.[55] Two molecules of type **84** (**85** and **86**) were prepared following the synthetic pathway reported in Scheme 10.31.

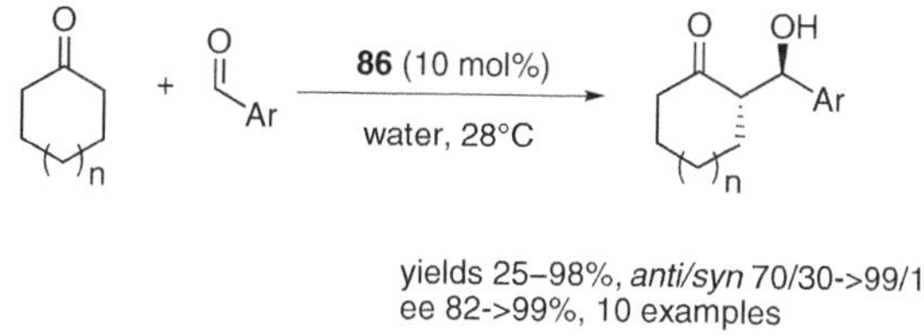

Scheme 10.32. Aldol reactions catalyzed by **86**.

Molecule **85** was completely soluble in water and no aldol reaction between cyclohexanone and 4-nitrobenzaldehyde occurred when it was used as catalyst. As expected, change of chloride anion with the more hydrophobic NTf_2 anion furnished a more active and stereoselective catalyst. Catalyst **86** was used in 10 mol% in water to give aldol products in good yields and high to excellent stereoselectivity (Scheme 10.32). While the hydrophobic core of catalyst **86** made it active and stereoselective for its use in water, the hydrophilic sections made its extraction in water possible.

Indeed, at the end of the reaction, excess of ketone and water were removed under reduced pressure and the solid residue was partitioned between water and diethyl ether. Catalyst **86** was polar enough to be miscible with water, while aldol products were extracted in the organic phase. The aqueous phase containing catalyst **86** was directly used in the next run. However, this approach did not give excellent results; in the fourth and fifth cycles the yield decreased, and in the fifth cycle enantioselectivity decreased as well.

Since supported prolinamides showed high performances as catalysts in aldol reactions in water,[51a,b] ionic liquid-anchored prolinamides **87** and **88** were synthesized and tested in the above reactions (Scheme 10.33).[54] Preliminary tests indicated that organocatalyst **88** was more active and stereoselective than **87**. Indeed, bromide **87** produced a clear 7% aqueous solution at room temperature, whereas the hexafluorophosphate **88**/H_2O mixture was a suspension under the same conditions. Using catalyst **88** in water at 3°C several aldol reactions with cyclic ketones were carried out affording products in high yields and stereoselectivities. Several methyl alkyl ketones were employed in the reaction with 4-nitrobenzaldehyde at room temperature to give aldol products in high enantioselectivity (Scheme 10.33). Catalyst **88** was used up to four cycles showing unchanged stereoselectivity but decreased activity.

Ionic liquid-anchored pyrrolidine derivatives **89–98** were prepared and tested as reusable organocatalysts for aldol reactions.[56] These molecules were synthesized starting from L-prolinol and following the straightforward procedure reported in Scheme 10.34. The ionic liquid-anchored derivatives **89–98** were soluble in chloroform, dichloromethane, and methanol but insoluble in solvents such as diethyl ether, ethyl acetate, and hexane. Compounds **89–98** were tested in the aldol reaction between acetone and 4-nitrobenzaldehyde.

Yields were not satisfactory; only catalysts **95** and **98** gave yields higher than 60%, but in all cases enantioselectivities were poor (~10% ee). Since the hydroxyl containing

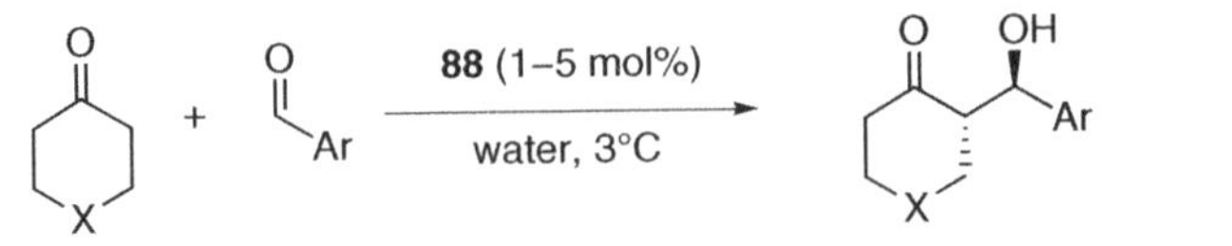

yields 64–99%, *anti*/*syn* 62/38-99/1
ee 92–99%, 8 examples

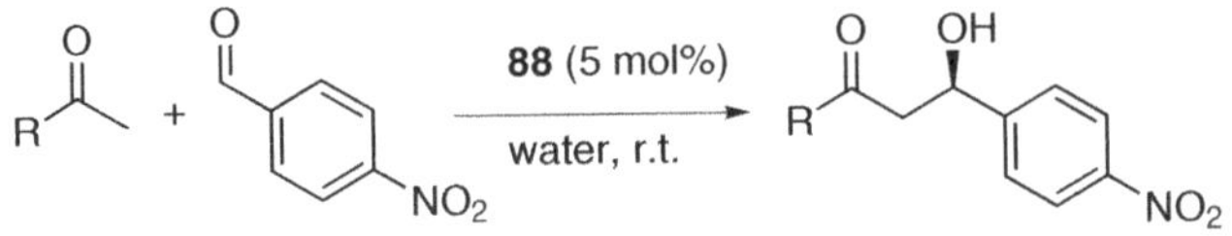

yields 48–95%, ee 82–97%, 5 examples

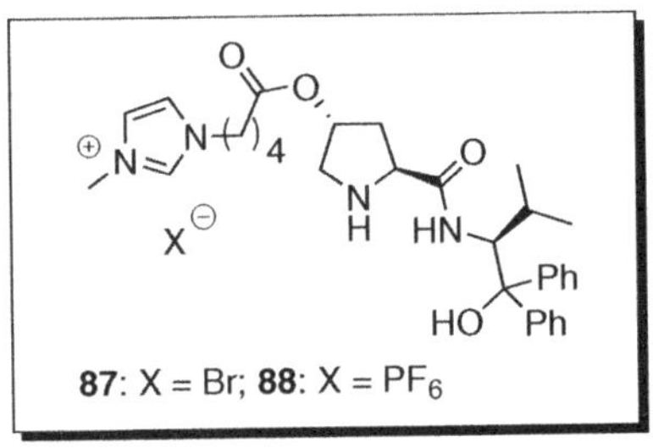

Scheme 10.33. Aldol reactions catalyzed by **88**.

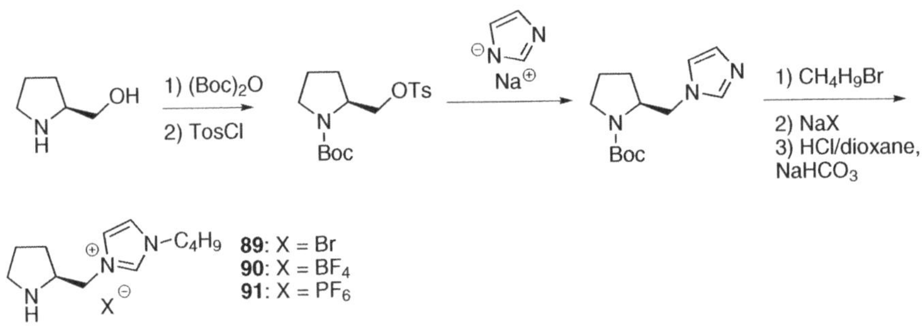

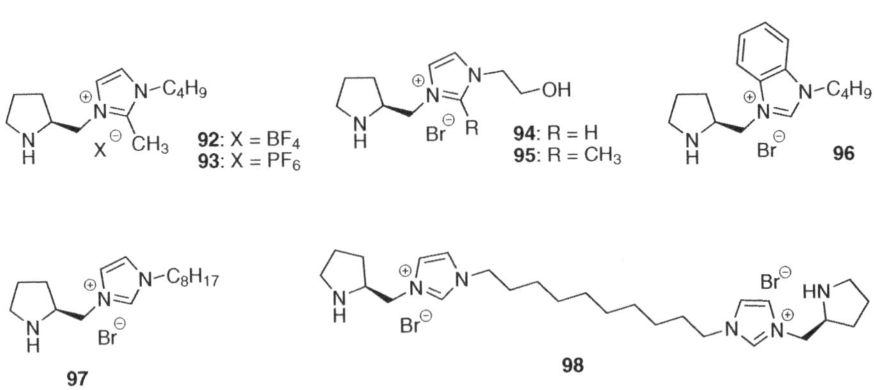

Scheme 10.34. Synthesis of ionic liquid-anchored pyrrolidines **89–98**.

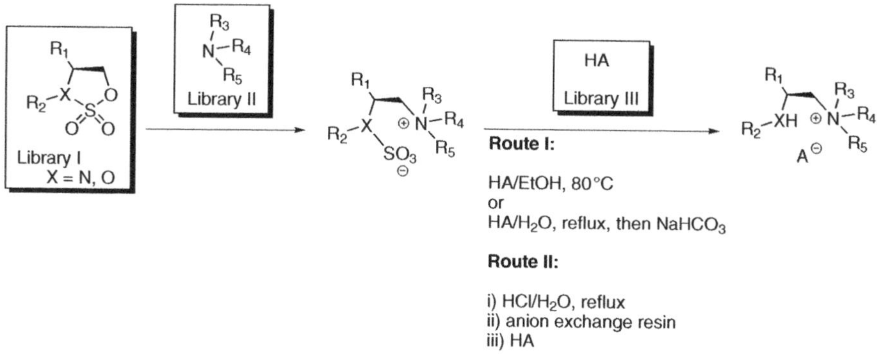

Figure 10.13. Strategy for the combinatorial synthesis of ionic liquid-anchored organocatalysts.

catalyst **95** afforded the aldol product in good yield (61%), it suggested a synergistic effect of the protic group on the aldol pathway. Then, several additives were tested and the best conditions were found when catalyst **90** (20 mol%) was used in the presence of water (100 mol%) and acetic acid (5 mol%). Several aldol reactions with acyclic and cyclic ketones were carried out, but even if yields were good, stereoselectivities were poor. The recyclability of catalyst **90** was also checked. After six cycles the activity decreased. Indeed, in order to obtain the same yield of the first cycle (92%), the reaction time increased from 2 hours (first cycle) to 17 hours. Because of the poor stereoselectivities obtained by the ionic liquid-anchored pyrrolidine catalysts, a further evolution of this type of catalyst was necessary.[57]

In order to test a large number of ionic liquid-based organocatalysts, a combinatorial strategy was developed for the synthesis of libraries of such molecules (Figure 10.13).

Catalysts were synthesized by ring opening of cyclic sulfates or sulfamidates, then the obtained zwitterions were transformed into the final chiral ionic liquid following routes I or II (see structures in Figures 10.14 and 10.15, respectively). Then, some of these molecules were tested as organocatalysts.

Particularly, attention was devoted to the double chiral ionic liquid-anchored molecules such as **117–124** and molecules **125–129** containing nonchiral anions obtained from route II (Figure 10.15). All the molecules obtained from route II displayed an additional functional group, namely a hydrogen-bonding group, in the anion. These molecules were screened as catalysts in the aldol reaction between cyclohexanone and 4-nitrobenzaldehyde; however, stereoselectivities were still poor. Noticeably, organocatalyst **127** (Figure 10.16) gave aldol products (three examples) with enhanced stereoselectivity compared with organocatalyst **90**, highlighting the role of the carboxylic group noncovalently linked to the cation, which plays the role of a noncovalently attached bifunctional ionic liquid.

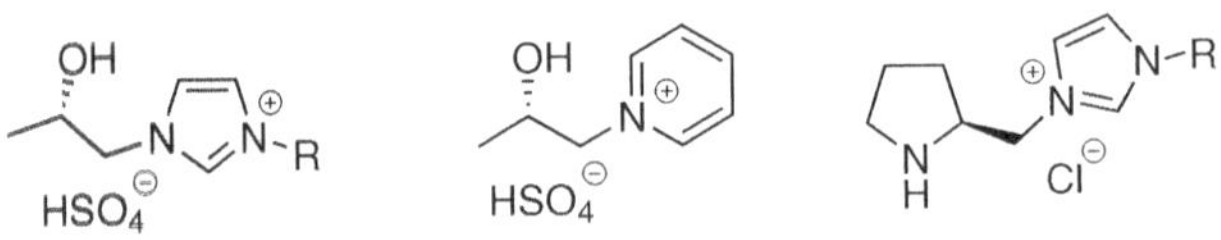

99: R = CH_3
100: R = C_4H_9
101: R = C_8H_{17}
102: R = $C_{12}H_{25}$

103

104: R = CH_3
105: R = C_4H_9
106: R = C_8H_{17}

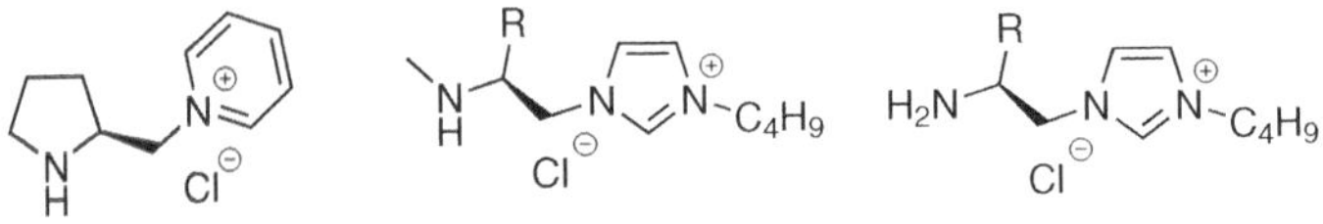

107

108: R = *i*-Pr
109: R = *i*-Bu
110: R = Bn

111: R = Me
112: R = *i*-Bu

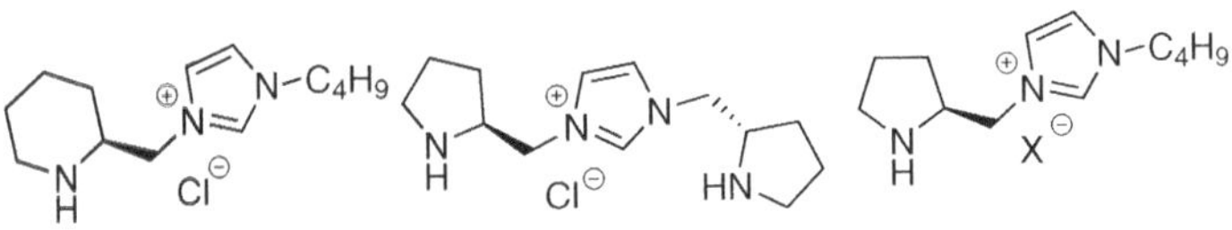

113

114

90: X = BF_4
115: X = ClO_4
116: X = CF_3SO_3

Figure 10.14. Library via Route I.

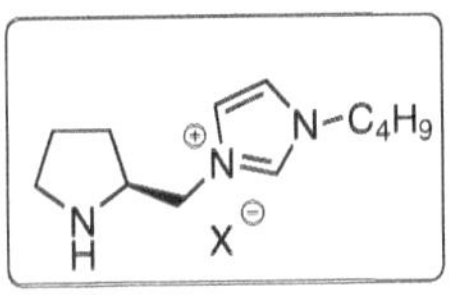

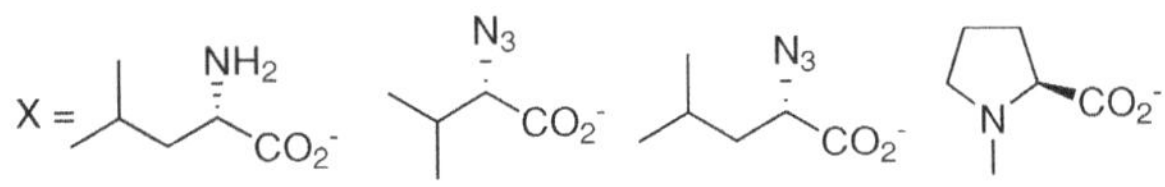

117 **118** **119** **120**

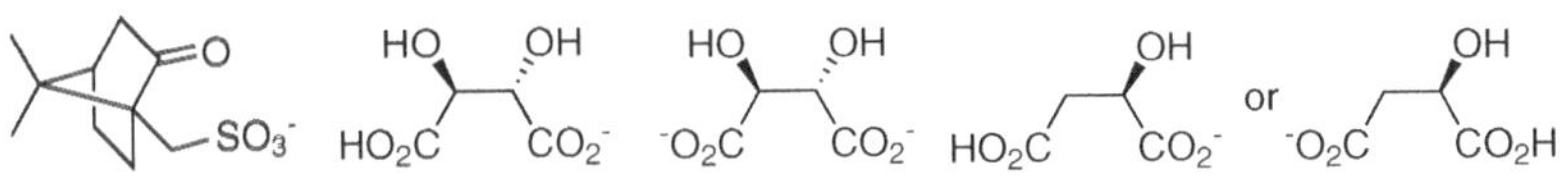

121 **122** **123** **124**

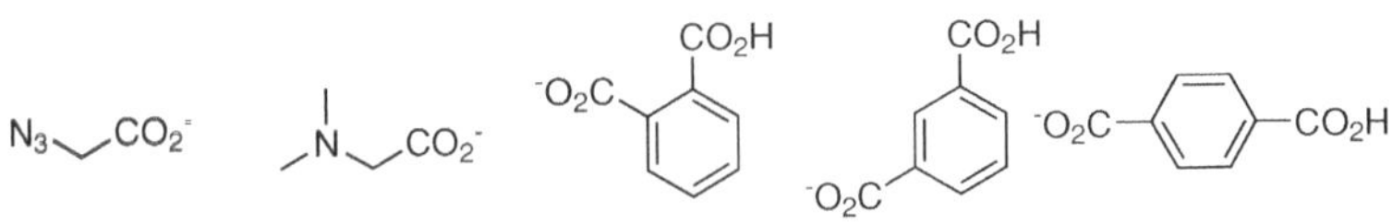

125 **126** **127** **128** **129**

Figure 10.15. Library via Route II.

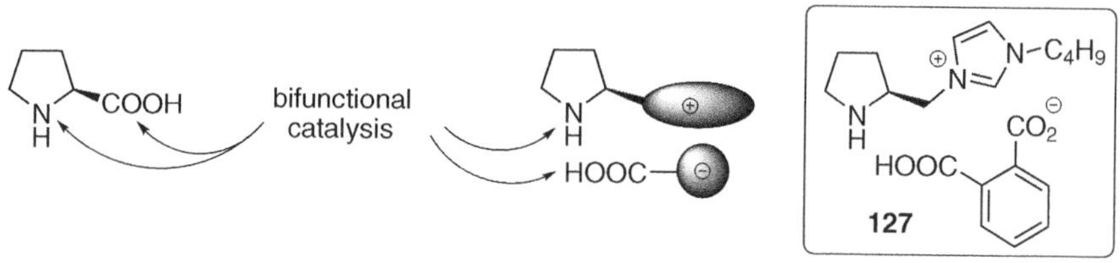

Figure 10.16. Bifunctional catalysts: comparison between proline and compound **127**.

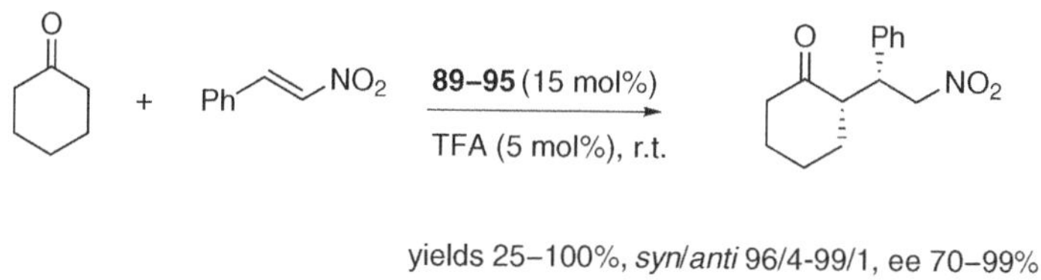

Scheme 10.35. Michael reactions catalyzed by **89–95**.

10.3.2 Michael Reactions

Ionic liquid-anchored organocatalysts have also been successfully employed in the asymmetric Michael addition reactions. Before their first use, in 2005 the application of chiral ionic liquids, synthesized from chiral pool, as reaction media in the enantioselective Michael addition of diethyl malonate to 1,3-diphenyl-prop-2-en-1-one, was also reported. This approach afforded the products in high yields (90–96%) but low enantioselectivity (10–25% ee).[58]

In 2006 the first example of ionic liquid-anchored organocatalysts for Michael addition reactions was reported.[59] Several pyrrolidine derivatives (**89–95**) were tested in the Michael reaction between cyclohexanone and *trans*-β-nitrostyrene (Scheme 10.35). Organocatalysts were used in 15 mol% in the presence of trifluoroacetic acid (TFA) as additive (5 mol%). Catalysts **89** and **90** gave better performances compared with the other catalysts. Recycling experiments, carried out using catalyst **90**, showed no loss in stereoselectivity after four cycles but a loss in activity.

Organocatalysts **89** and **90** were then employed in several Michael addition reactions affording the final products in excellent stereoselectivity in the case of cyclohexanone. Cyclopentanone, acetone, isobutyraldehyde, and isovaleraldehyde gave adducts with lower stereoselectivity (Scheme 10.36).

The high stereoselectivity observed, when cyclohexanone was used, was explained with model **130** (Figure 10.17) in which the ionic liquid moiety would effectively shield the *Si* face of the enamine double bond in the ketone donor and the reaction would

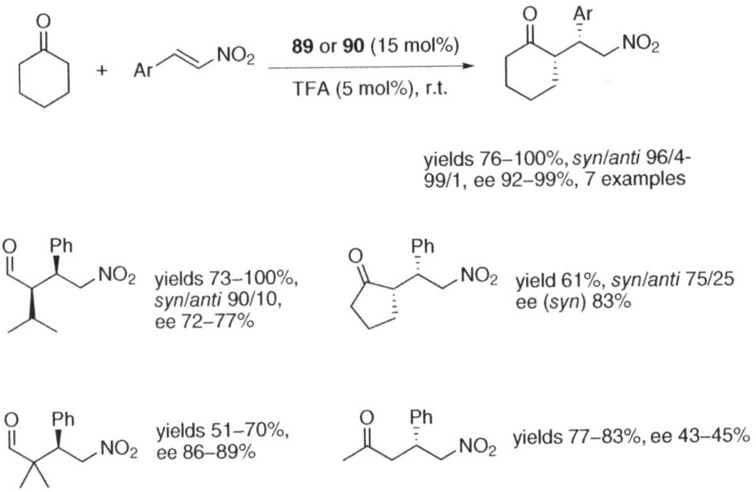

Scheme 10.36. Michael reactions catalyzed by **89–99**.

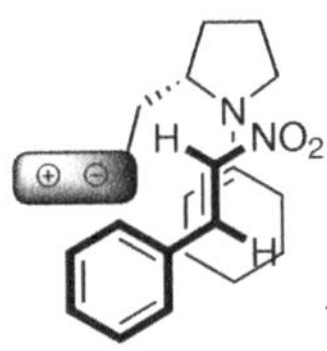

130

Figure 10.17. Transition state model for ketone donors.

occur through a *Re–Re* approach. For aldehyde donors, *anti*-enamine would be formed and the reactions would occur through a *Si–Si* approach.

Surfactant-type ionic liquid-anchored organocatalysts were employed in the Michael addition to nitrostyrenes in water.[60] These organocatalysts were obtained from chiral pyrrolidine imidazolium ionic liquids by replacing the anions with surfactant sulfate or sulfonate (Figure 10.18). This kind of catalysts would simultaneously function as an asymmetric catalyst to promote the reactions and as a surfactant to assist in solubilizing the organic substrates. Following the approaches described in Figure 10.18, catalysts **89**, **96–98**, and **131–136** were prepared and tested in the Michael addition of cyclohexanone to *trans*-β-nitrostyrene in water.

Despite of the hydrophobicity of the cations, use of organocatalysts **97**, **98**, and **136** was unsuccessful. The reactions catalyzed by **89** or **96** afforded no desired product, or poor yield due to polymerization of nitrostyrene in water. Further screening showed catalyst **135** as the optimal catalyst. Interestingly, the latter catalyst was inefficient when used under neat conditions, indicating the unique properties of **135** as a highly efficient organocatalyst in water. Noticeably, vigorous stirring should be maintained for a better outcome of the reaction, suggesting that the reaction should occur through interfacial catalysis.

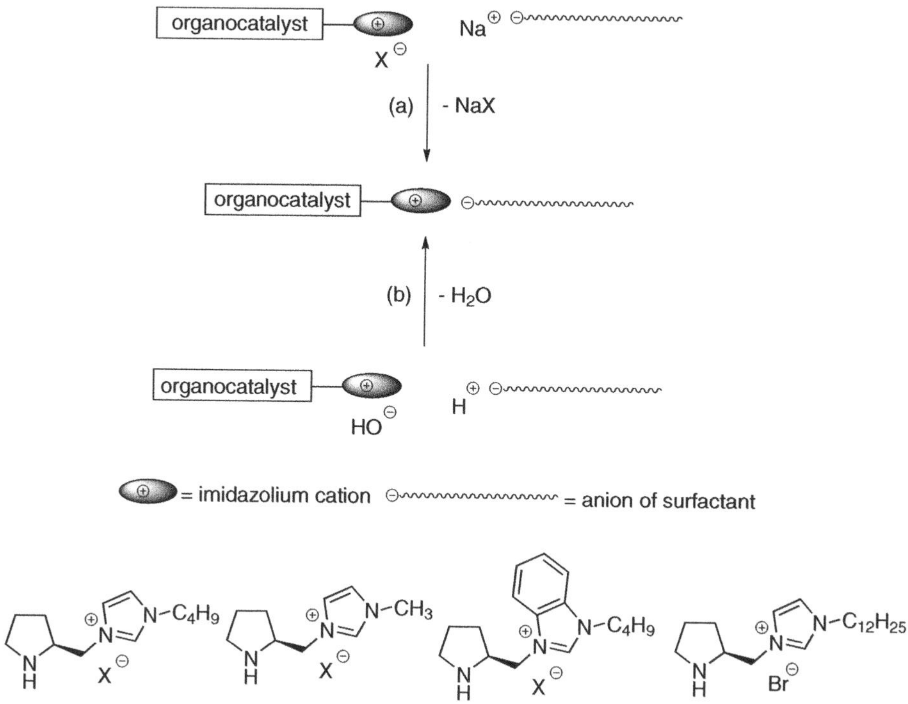

131: X = DS
132: X = DBS

133: X = DS

134: X = DS
135: X = DBS

136

Figure 10.18. Strategy for the synthesis of surfactant-type asymmetric organocatalysts (STAOs) **131–135** by (a) anion metathesis and (b) neutralization.

Catalyst **135** (20 mol%) was then employed in water in the addition of cyclohexanone with a set of nitrostyrenes affording the Michael adducts in good yields (64% to >99%), excellent diastereoselectivity (≥94/6, *syn/anti*), and enantioselectivity (≥91% ee) (Scheme 10.37). Acyclic ketone donors such acetone were tested with little success, while addition of isovaleraldehyde to nitrostyrene gave the product in good yield, high diastereoselectivity (*syn/anti* 97/3), and moderate enantioselectivity (61% ee). Unfortunately, no recycling experiments were carried out.

Ionic liquid-anchored pyrrolidine organocatalysts were also employed for the enantioselective desymmetrizations of prochiral ketones via asymmetric Michael addition reactions to nitrostyrenes.[61] In addition to several known organocatalysts (**89–91, 94, 96–98**), two new organocatalysts (**137** and **138**, Figure 10.19) were also examined.

Screening experiments revealed that the thiazolidine-containing organocatalyst **138** was inactive. The catalytic activity of ionic liquid-anchored organocatalysts with longer

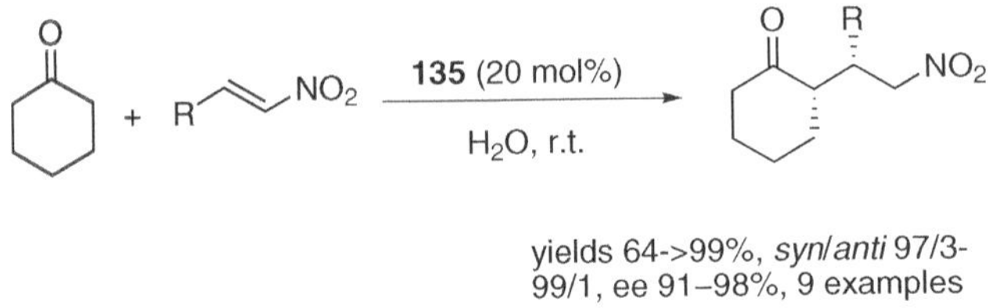

Scheme 10.37. Michael reactions catalyzed by **135**.

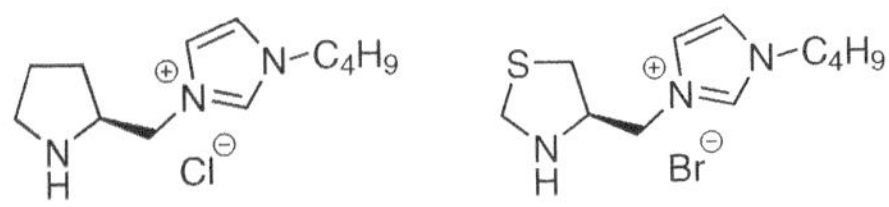

137 **138**

Figure 10.19. Structures of compounds **137–138**.

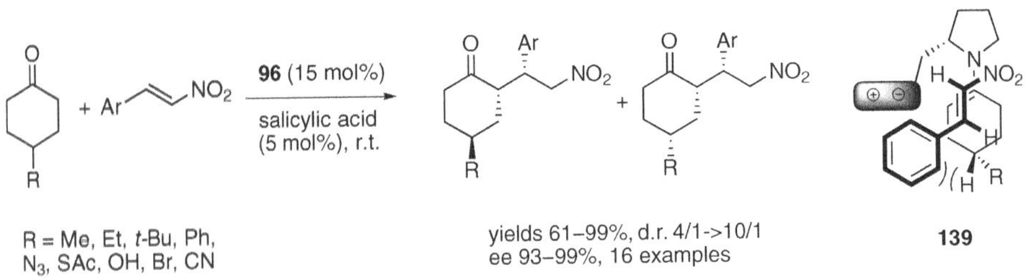

Scheme 10.38. Michael reactions catalyzed by **96**.

(**97**) or protic-group-containing chains (**94**) gave lower yields, while the bis-cation **98** was inactive. Change of anions Cl and Br with BF_4, PF_6 (**137, 89–91**) led to comparable stereoselectivity but increased activity. Better performances were displayed by organocatalyst **96** (Scheme 10.38). Moreover, screening of additives revealed that salicylic acid was the best one.

Under these conditions (**96** 10 mol%, salicylic acid 5 mol%, no added solvent) several Michael addition reactions between prochiral ketones and nitrostyrenes were carried out. Apart from ketones carrying OH, Br, and CN groups, which were unreactive, in all the other examples, high yields and stereoselectivities were observed. These latter were explained by invoking the plausible transition state **139**. Organocatalyst **96** was recovered by precipitation and reused up to four cycles with unchanged stereoselectivity but diminished activity.

In the reported examples, pyrrolidine-based ionic liquid-anchored catalysts, employed for Michael addition reactions, did not contain acidic hydrogens able to form hydrogen bond with substrates (nitrostyrenes). In order to investigate this type of

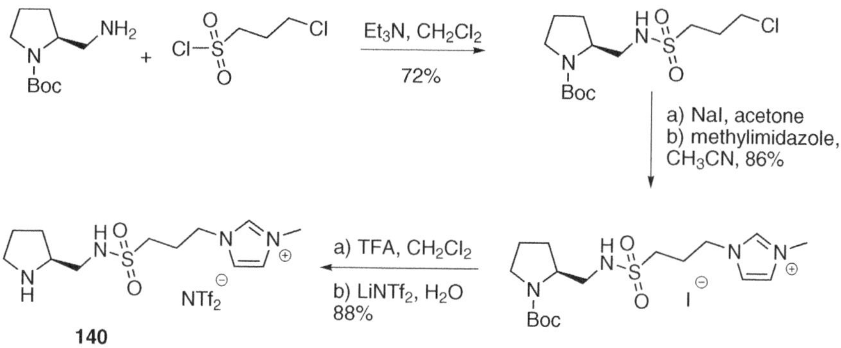

140

Scheme 10.39. Synthesis of ionic liquid-anchored catalyst **140**.

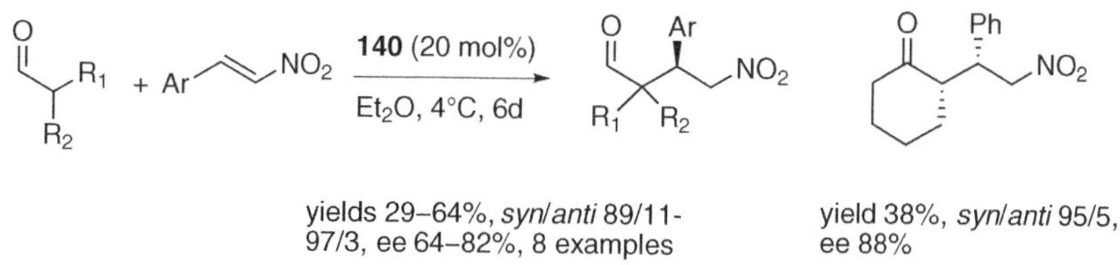

Scheme 10.40. Michael reactions catalyzed by **140**.

organocatalysts, a chiral pyrrolidine moiety including a protonic functionality, linked to an imidazolum ionic liquid, was designed.[62] The introduction of the ionic liquid moiety would help in the recovery of the organocatalyst. Compound **140** was designed with this aim and synthesized starting from (*S*)-2-amino-1-*N*-Boc-pyrrolidine and 3-chloropropanesulfonyl chloride (Scheme 10.39). The pyrrolidine sulfonamide obtained was converted into the final ionic liquid-anchored organocatalyst **140** by reaction with methylimidazole, followed by removal of the Boc group and anion exchange.

Preliminary experiments were carried out on the Michael reaction of isobutyraldehyde and *trans*-β-nitrostyrene. Various solvents were examined at room temperature and the best results were obtained when catalyst **140** (20 mol%) was used in less polar solvents such as Et_2O or $CHCl_3$. Moreover, at a lower temperature (4°C) a small increase in enantioselectivity was observed (82% ee at 4°C; 78% ee at room temperature), although reaction times were quite long (6 days). The catalyst was recovered by precipitation and reused twice with unchanged yield and enantioselectivity.

Then, under the optimized reaction conditions several Michael additions of aldehydes to *trans*-β-nitrostyrenes were carried out (Scheme 10.40). Adducts were obtained in moderate yields (29–64%) good enantioselectivity (64–82% ee), and high diastereoselectivity (89/11-97/3 *syn/anti*). Adduct with cyclohexanone was obtained in low yield (38%) and good enantioselectivity (88% ee) although lower than that obtained with catalyst **90**.

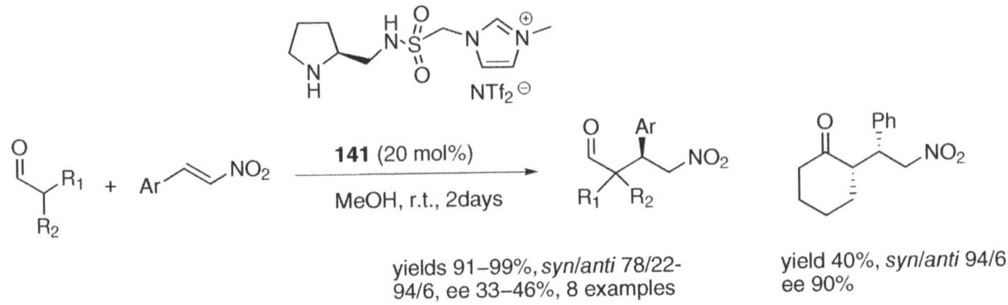

Scheme 10.41. Michael reactions catalyzed by **141**.

A comparison between the catalytic systems **90** (15 mol%)/TFA (5 mol%) and **140** (20 mol%) indicated that the former was more active and selective. Probably, the shielding role of the imidazolium moiety plays a major role compared with the coordinating role of the acidic N–H hydrogen bond.

A more active organocatalyst compared with **140** was later reported by the same authors.[63] Following a similar synthetic strategy for organocatalyst **140**, ionic liquid-anchored organocatalyst **141** was prepared and used for the Michael addition of aldehydes to *trans*-β-nitrostyrenes (Scheme 10.41) and the results were compared with those obtained using catalyst **140**. Michael reactions catalyzed by **141** were carried out in methanol at room temperature, instead of diethyl ether at 4°C for catalyst **140**. Reaction times were significantly shortened (2 days for **141**, 6 days for **140**) and yields markedly increased; however, enantioselectivities observed with **141** were lower than those observed with **140**. Catalyst **141** gave better results compared with **140** when used for the addition of α,β-disubstituted aldehydes. Recycling experiments showed a decreased activity in the fifth cycle.

The stereochemical outcome of the reactions catalyzed by **140** and **141** is believed to be based on the acidity of the N–H bond. In order to increase the N–H acidity, a new type of ionic liquid-anchored pyrrolidine sulfonamide organocatalyst (**142**) was synthesized.[64] The increased acidity is ascribed to the sulfonyl group, which, in turn, is linked to the C-2 position of the electron-withdrawing imidazolium cation. In addition, the C-2 position is more electronegative than the C-4 or the C-5 position. The increased acidity will result in stronger hydrogen bonds that are formed in the transition states of these reactions. As a result, enhancement of the catalytic activity and selectivity, without the need of additives, could be expected. Also, the closer presence of the imidazolium moiety with respect to organocatalysts **140** and **141** could introduce more steric bulk in proximity to the catalytic site, leading to an improved stereoselectivity.

Organocatalyst **142** was prepared as reported in Scheme 10.42 and used, after several preliminary tests, in 10 mol% in *i*-PrOH at room temperature without additives.

Addition reactions of six membered cyclic ketones to nitrostyrenes gave products in high yields and stereoselectivities (Scheme 10.43). Isobutyraldehyde gave good results, while acetone gave good yield but low enantioselectivity. Cyclopentanone gave

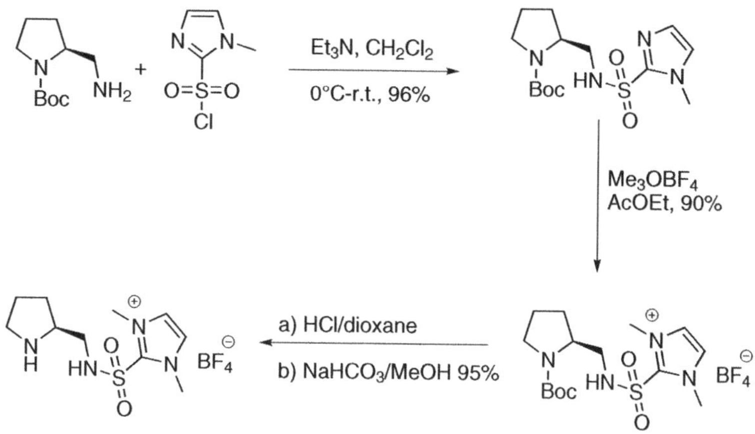

142

Scheme 10.42. Synthesis of ionic liquid-anchored catalyst **142**.

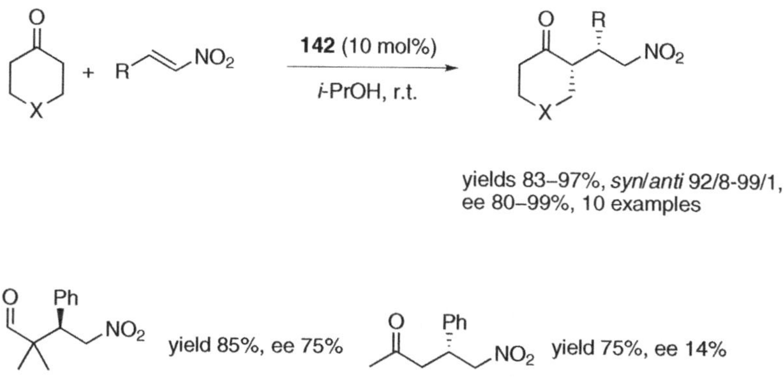

Scheme 10.43. Michael reactions catalyzed by **142**.

low yield. Catalyst **142** was recovered (>90%) by phase separation and reused for five cycles. Stereoselectivity was maintained, but activity decreased in the fourth and fifth cycles.

Pyrrolidine-based organocatalysts of type **143** bearing a 1,2,3-triazole moiety were found good catalysts for the Michael addition of ketones to nitroolefins.[65,66] In particular, compound **144** gave excellent results when used in 10 mol% in chloroform or water without additives[65] or under neat conditions in 10 mol% and in the presence of TFA (2.5 mol%).[66] In both cases no recycling experiments were carried out.

In order to obtain a recyclable catalyst that is able to maintain the high activity of **144**, several ionic liquid-anchored organocatalysts based on the **143** structure were synthesized. The first catalyst of this type was synthesized as reported in Scheme 10.44.[67] Organocatalyst **150** was prepared via the 1,3-dipolar cycloaddition of **146** and

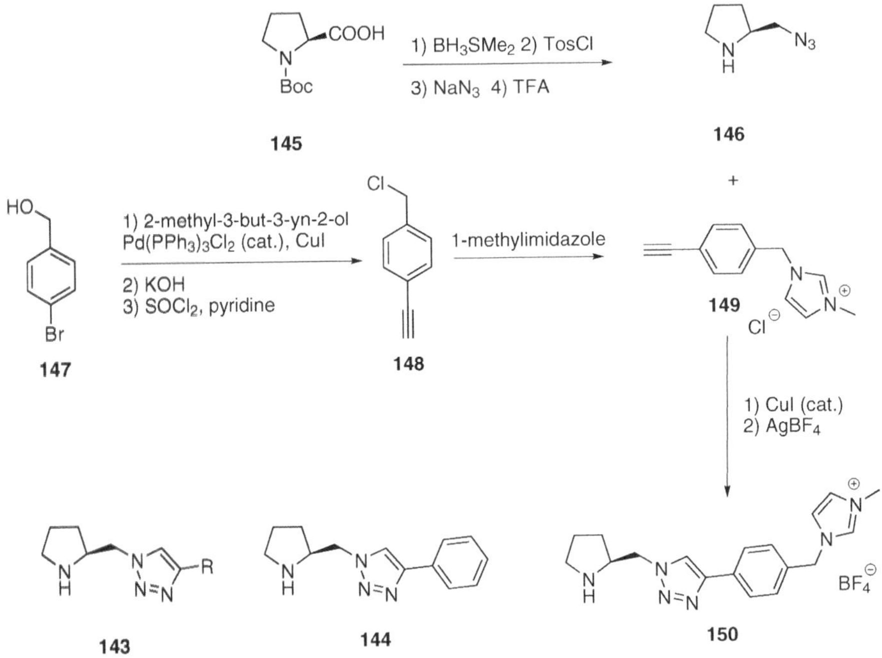

Scheme 10.44. Synthesis of ionic liquid-anchored catalyst **150**.

149 followed by anion exchange. Compound **146** was prepared from Boc-L-proline **145** in four steps, while compound **149** was prepared by reaction of 1-methylimidazole with the chloride **148**. The latter compound was provided, starting from (4-bromophenyl) methanol **147**, via a Sonogashira coupling reaction and decomposition with KOH, followed by treatment with $SOCl_2$.

When catalyst **150** was used in chloroform or water without additives, poor results were obtained, showing the different activity, due to the presence of the imidazolium tetrafluoroborate moiety, compared with catalyst **144**. Other reaction conditions were screened. Finally, it was found that catalyst **150** could be successfully employed under neat conditions (15 mol%) in the presence of TFA (5 mol%) (Scheme 10.45). These reaction conditions led to excellent results in the addition of cyclohexanone to nitroolefins. Lower stereoselectivities were observed when cyclopentanone and acetone were employed.

Differently from catalyst **144**, which was not recovered, catalyst **150** was recovered by precipitation with diethyl ether and reused for four consecutive runs without loss in activity and stereoselectivity.

In catalyst **150** the 1,2,3-triazole unit serves as linker while the imidazolium unit acts as phase tag for recycling. Other authors synthesized the ionic liquid-anchored organocatalysts **152–153** (Scheme 10.46) in which the 1,2,3-triazolium unit does not play the role of linker. In this case the 1,2,3-triazolium unit could act as in the analogue imidazolium compounds **89–90**.[68]

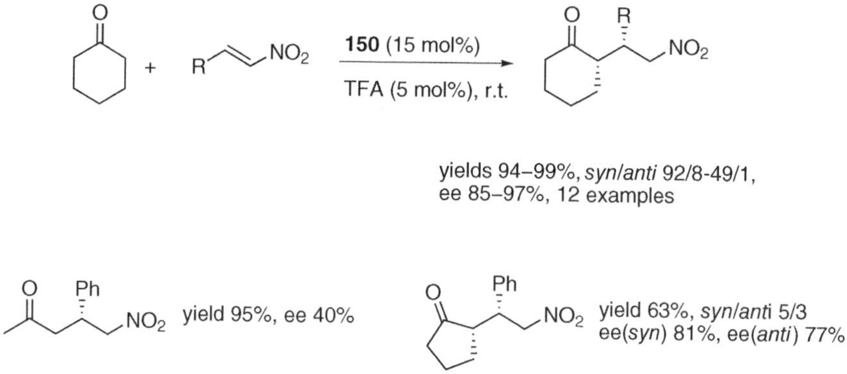

Scheme 10.45. Michael reactions catalyzed by **150**.

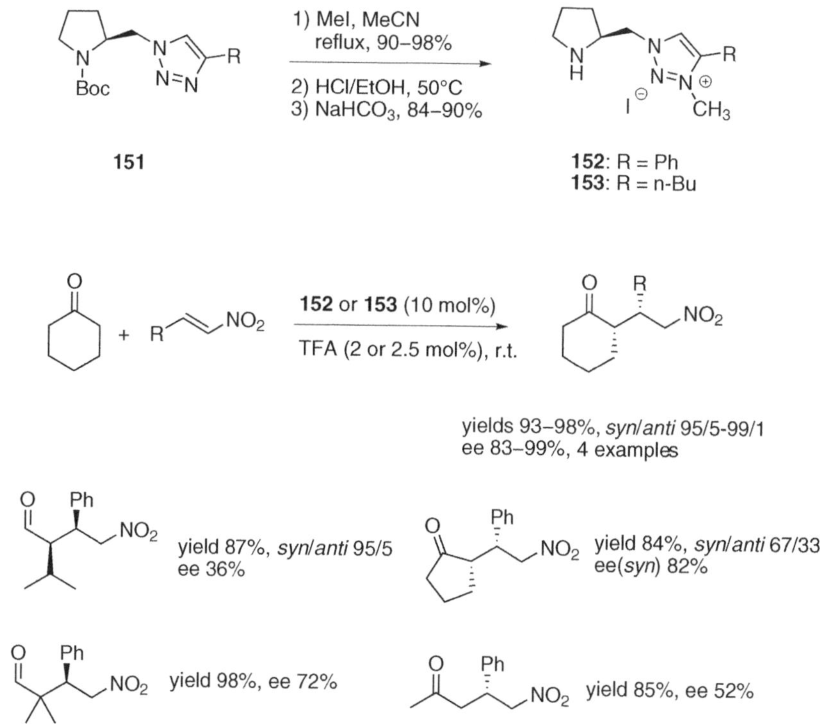

Scheme 10.46. Synthesis of ionic liquid-anchored catalysts **152–153** and Michael reactions catalyzed by **152-153**.

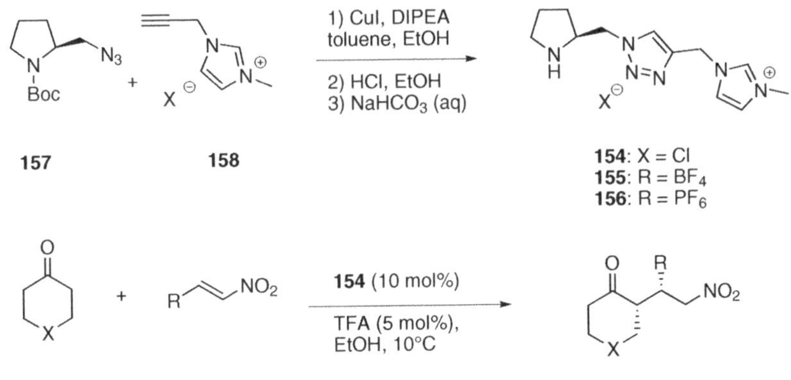

Scheme 10.47. Synthesis of ionic liquid-anchored catalysts **154–156** and Michael reactions catalyzed by **154**.

Organocatalysts **152-153** were synthesized from triazoles **151** by alkylation with iodomethane and subsequent deprotection (Scheme 10.46).

Several Michael addition reactions of ketones to nitrostyrenes were carried out with good results. Recycling investigations (four cycles) with organocatalyst **152** showed that enantioselectivities were unsatisfactory in the last two cycles (Scheme 10.46). Disappointingly, organocatalysts **152-153** are not directly comparable with **89-90** since the anions are different; however, on the whole, organocatalysts **89-90** showed better performances.

Excellent results were observed in the Michael addition of six membered ketones to nitrostyrenes catalyzed by organocatalyst **154** (Scheme 10.47).[69] Organocatalysts **154–156** were prepared through "click chemistry" procedure, starting from **157** and ionic liquids **158**.

Preliminary tests using catalysts **154–156** were carried out investigating the effect of solvent, temperature and reaction times on the reaction between cyclohexanone and *trans*-β-nitrostyrene. Such studies indicated organocatalyst **154** as the most promising when used in 10 mol% in addition to TFA (5 mol%) in EtOH at 10°C for 36 hours. Products were obtained in excellent yields and stereoselectivities (Scheme 10.47). Moreover, organocatalyst **154** was used for eight consecutive cycles without loss in activity and stereoselectivity.

Simple pyridinium ionic liquid-anchored pyrrolidine organocatalysts were easily synthesized starting from (*S*)-(2-aminomethyl)-1-*N*-Boc-pyrrolidine and Zincke's salt, followed by deprotection and anion exchange.[70] Following this procedure four organocatalysts (**157–160**) were prepared (Scheme 10.48).

All organocatalysts **157–160** (15 mol%) catalyzed the asymmetric Michael addition of cyclohexanone to *trans*-β-nitrostyrene, but, overall, compounds **157** and **158** containing Cl and BF_4 anions gave the best performances with high yields as well as

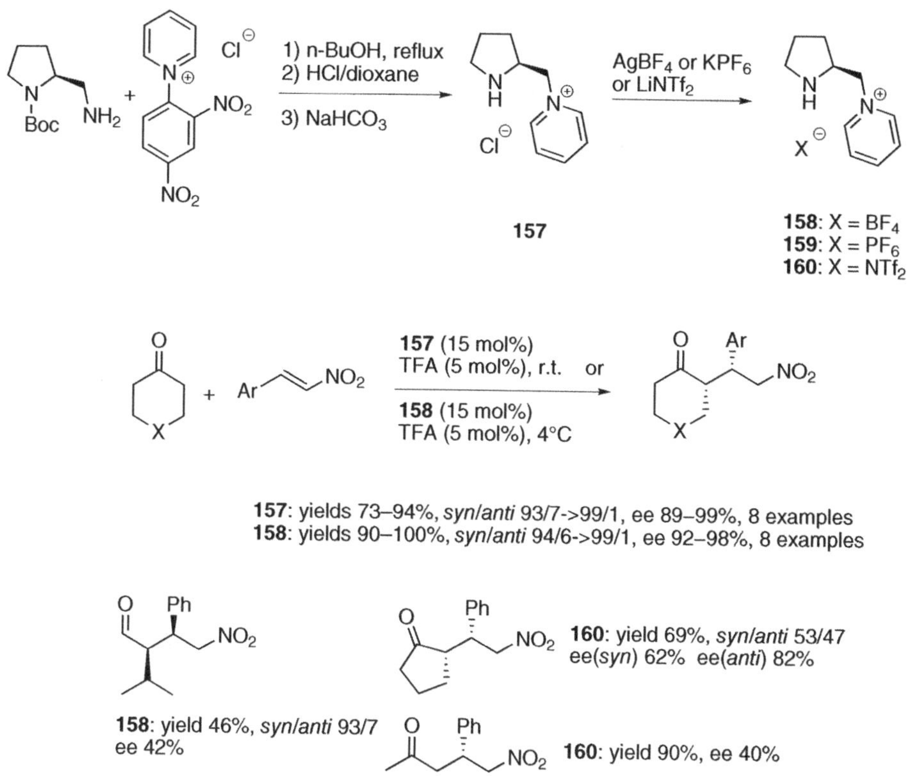

Scheme 10.48. Synthesis of ionic liquid-anchored catalysts **158–160** and Michael reactions catalyzed by **158–160**.

high diastereoselectivity, and enantioselectivity. Addition of TFA (5 mol%) increased the reaction rate without decreasing enantioselectivity. Then, organocatalysts **157** and **158** were used for the reaction between cyclohexanone, or dihydro-2H-pyran-4(3H)-one, and nitrostyrenes to afford the final adducts in high yields and stereoselectivities (Scheme 10.48).

Other donors (isobutyraldehyde, acetone, and cyclopentanone) were also used for the reaction with *trans*-β-nitrostyrene affording the product with modest stereoselectivities. Catalysts were recovered by precipitation, and recycling experiments (three cycles) carried out with **158** showed a decreased yield and steroselectivity in the third cycle. As reported above, the stereoselectivities observed with cyclic ketones were explained by invoking the transition state model **130** (Figure 10.20). Other authors invoked also the occurrence of the transition state model **161** (Figure 10.20) in which the possible attraction between the ionic liquid moiety and nitro group of the substrate should also contribute to the enantioselectivity observed. Alternatively, for organocatalysts bearing N–H acidic hydrogens, the transition state model **162** (Figure 10.20) can also explain the observed stereoselectivity.

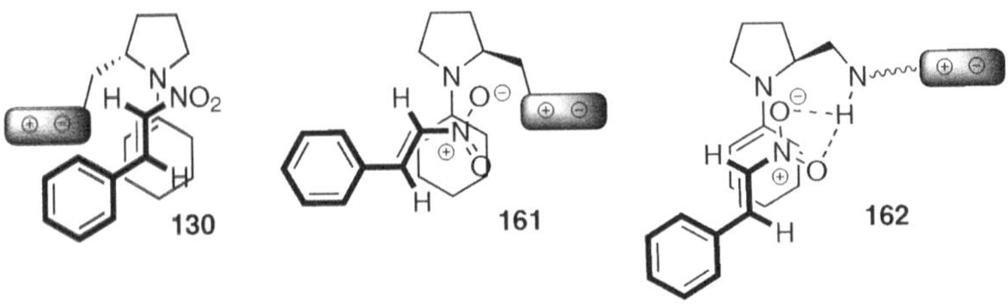

Figure 10.20. Transition state model for ketone donors in the Michael reactions.

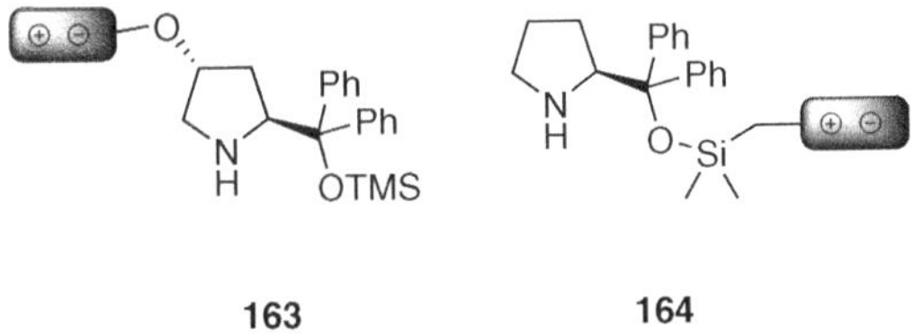

Figure 10.21. General structure of compounds **163** and **164**.

Michael reactions were also carried out using ionic liquid-anchored α,α-diphenyl-(*S*)-prolinol derivatives. Two main examples were reported, the first one carrying the ionic liquid tag in the 4-position of the proline moiety (type **163**) and the second one carrying the ionic liquid tag as substituent in the OTMS group (type **164**) (Figure 10.21).

Three ionic liquid-anchored α,α-diphenyl-(*S*)-prolinols **165–167** of type **163** were synthesized as reported in Scheme 10.49 and used as catalysts in the Michael reaction between α,α-enals and dialkyl malonates.[71]

Catalyst screening in the reaction between dimethyl malonate and *trans*-cinnamaldehyde indicated the higher performances of the OTMS derivative **167**. This catalyst worked well in 96% EtOH at 4°C in 10 mol% loading (Scheme 10.50). Use of water (50 equiv.) or neat conditions gave slightly decreased performances. Then, several Michael reactions were carried out affording the final product in high yields and enantioselectivities. The catalyst was recovered after solvent evaporation and phase separation and used for six consecutive cycles. Although enantioselectivity was unaffected, activity decreased in the fifth and sixth cycles.

α,α-Diphenyl-(*S*)-prolinol **168** of type **164** was synthesized as shown in Scheme 10.51 and used in the Michael addition of aliphatic aldehydes to nitroalkenes.[72] Catalyst **168** was tested in the reaction between propanal and nitrostyrene in 16 different solvents affording the adduct always in excellent enantioselectivity (99% ee) and good diastereoselectivity (85/15 d.r.). Dichloromethane and water were selected as solvents of choice because of the higher activity in these media. Catalyst loading was tested in dichloromethane, down to 0.25 mol% with improved stereoselectivity (>99.5% ee, 96/4 d.r.) and high yield (88%).

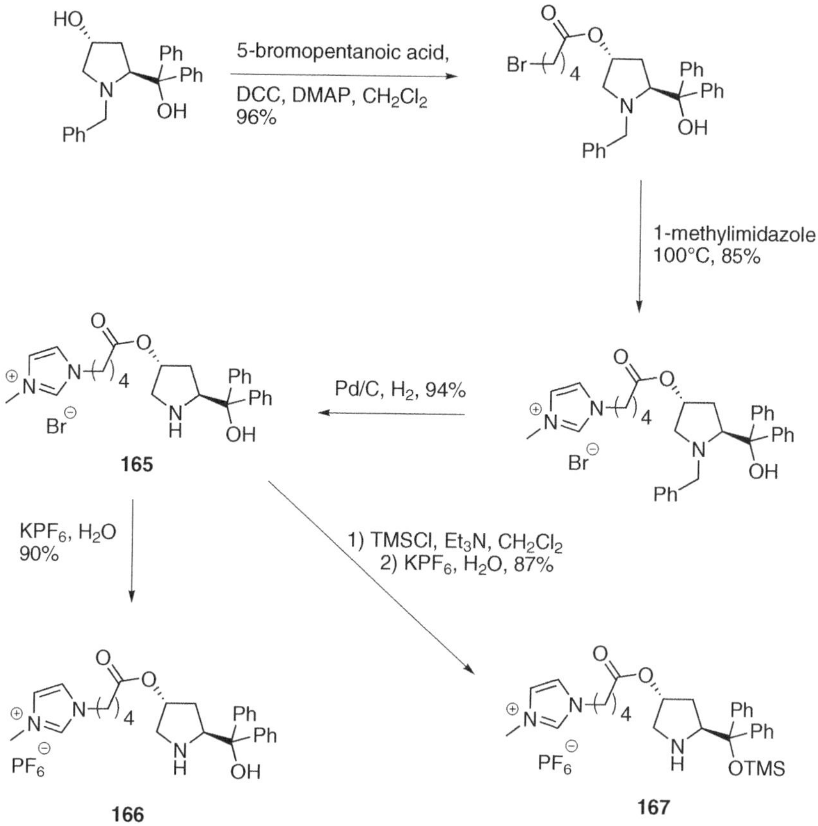

Scheme 10.49. Synthesis of ionic liquid-anchored prolinols **165–167**.

Then several Michael addition reactions were carried out using catalyst **168** in 1–2 mol% in four different conditions (dichloromethane or water with or without benzoic acid as additive). Adducts were obtained in high yields and diastereoselectivities and excellent enantioselectivities (Scheme 10.52).

10.3.3 Morita–Baylis–Hillman Reaction and Claisen–Schmidt Reaction

In addition to aldol reactions and Michael additions, the Morita–Baylis–Hillman (MBH) reaction is one of the most useful carbon–carbon bond-forming reaction and many efforts have been devoted to its development. Ionic liquids have been used as solvents in the MBH reaction. The next step was the design and synthesis of ionic liquid-anchored organocatalysts for this reaction. Again, the main idea was to use the ionic liquid moiety in order to apply the usual biphasic strategy of homogeneous reaction

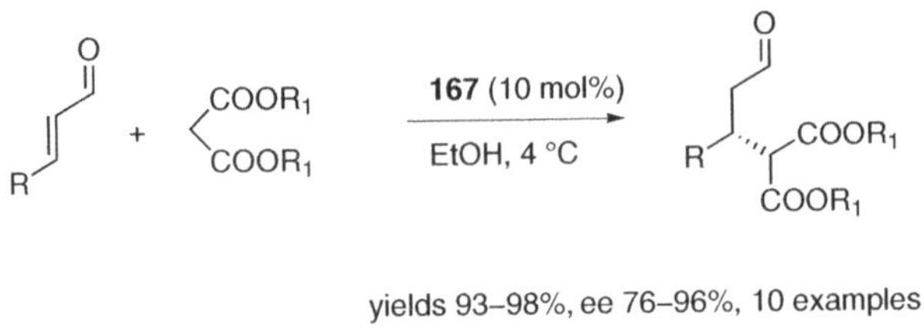

Scheme 10.50. Michael reactions catalyzed by **167**.

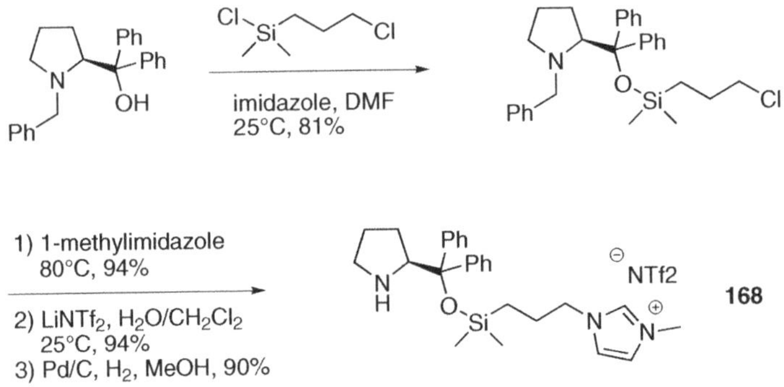

168

Scheme 10.51. Synthesis of ionic liquid-anchored prolinol **168**.

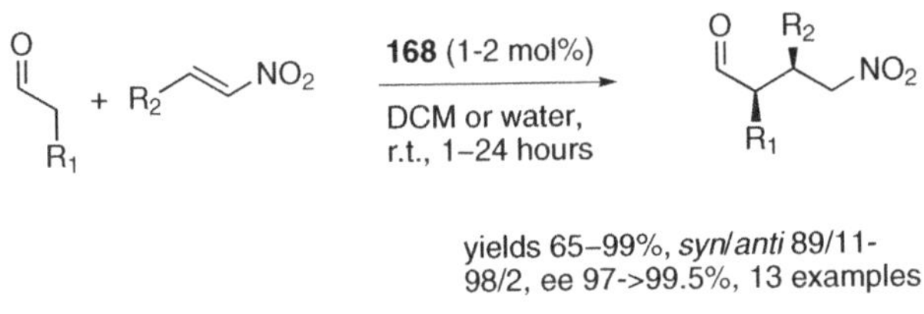

Scheme 10.52. Michael reactions catalyzed by **168**.

and heterogeneous separation. However, imidazolium ionic liquids should be carefully used because they could not be inert under the reaction conditions.[73]

The first examples of ionic liquid-anchored organocatalysts for the MBH reaction were compounds **169–171** (Scheme 10.53).[74] These molecules were synthesized from amino ionic liquids and 3-quinuclidone followed by reduction with sodium borohydride.

After several preliminary studies on catalysts **169–171** and solvents, the optimal reaction conditions were found. The best performances were obtained with catalyst **170** (30 mol%) and methanol (2 equiv.) (Scheme 10.53). These results were in agreement with previous findings that MBH reactions generally proceed faster in protic solvents such as methanol and water than in aprotic solvents. However, the amount of methanol

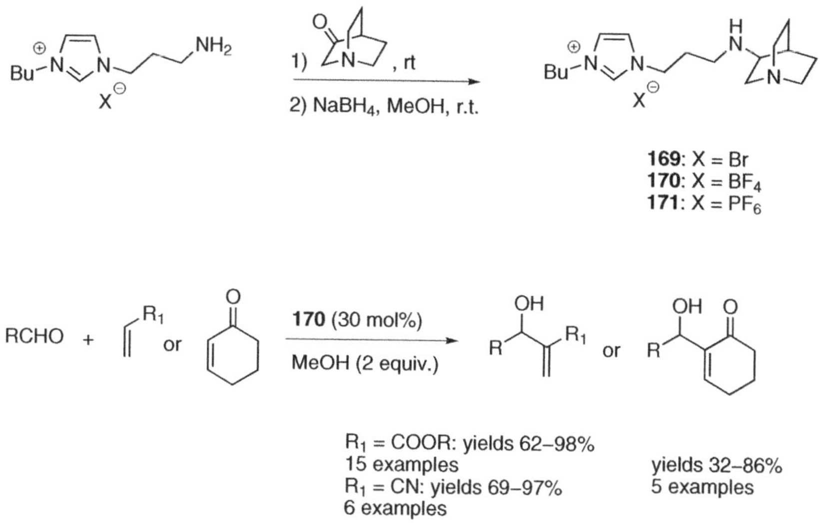

Scheme 10.53. Synthesis of compounds **169–171** and Morita–Baylis–Hillman reactions catalyzed by **170**.

was important to obtain a better yield. Also several ionic liquids were tested as solvents, but yields were lower compared with methanol, indicating again the role that the protic solvent plays in accelerating the reaction, probably via hydrogen bonding. Catalyst **170** was as active as DABCO and slightly less active than 3-benzylaminoquinuclidine, a nonbounded analogue of **170**. Recycling studies, carried out for six consecutive runs, showed a marked decrease in activity, probably due to the reactivity of the ionic liquid moiety.

Ionic liquid-anchored quinuclidine **170** was tested in several MBH reactions with good results. Reactions between aromatic or aliphatic aldehydes gave adducts in 62–98%, and reactions between aromatic aldehydes and acrylonitrile gave adducts in 69–97%. Also good yields were obtained with the less reactive cyclohexenone (32–86%).

Since protic solvents have a beneficial effect on the rate of MBH reactions, the use of a hydroxyl containing ionic liquid could exert a similar accelerating effect. Moreover, it was speculated that this kind of ionic liquid-anchored organocatalyst might offer other advantages such as elimination of solvent or lower catalytic loading. In order to develop this idea, compounds **172–177** were synthesized as reported in the Scheme 10.54.[75]

Catalysts **175–177**, obtained as mixture of diastereoisomers, were inseparable and used directly. Preliminary tests established catalyst **172** as the most useful since it provided slightly better yields compared with the other catalysts. Catalysts **175–177**, afforded the MBH adducts with low enantioselectivity (<10% ee). A comparison between **172** and **169** showed that the latter was a much better catalyst regardless of the solvent used. Noticeably, catalyst **172** gave a higher yield under neat conditions with respect to methanol and a much higher yield with respect to **169** in both reaction conditions. These

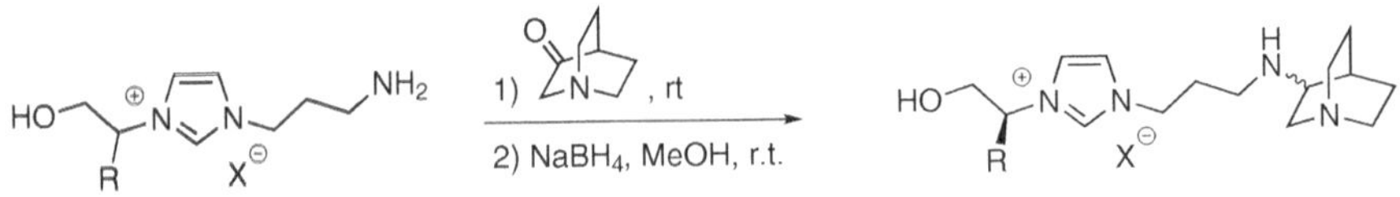

172: R = H, X = Br
173: R = H, X = BF_4
174: R = H, X = PF_6
175: R = CH_3, X = Br
176: R = CH_3, X = BF_4
177: R = CH_3, X = PF_6

Scheme 10.54. Synthesis of compounds **172–177**.

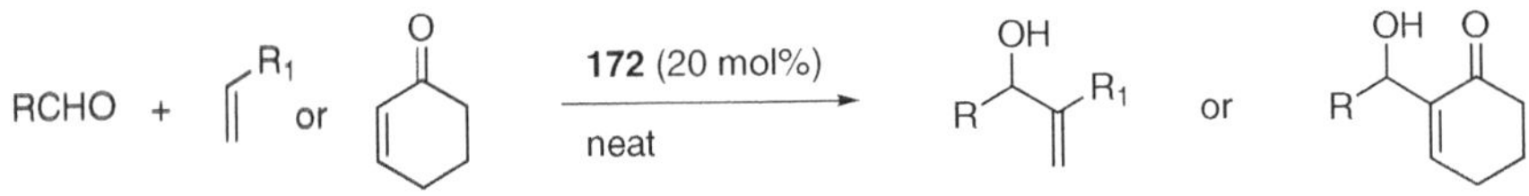

R_1 = COOR: yields 55–98%
17 examples
R_1 = CN: yields 83–97%
8 examples

yields 51–84%
5 examples

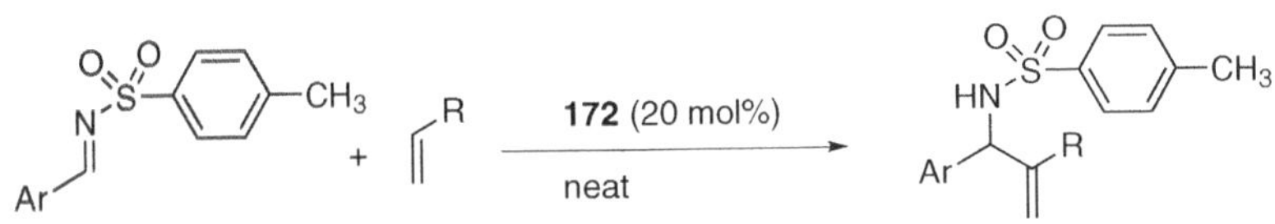

R = $COOCH_3$: yields 76–96%
5 examples
R = CN: yields 83–97%
5 examples

Scheme 10.55. Morita–Baylis–Hillman reactions catalyzed by **172**.

results highlighted the role of the hydroxyl group on the ionic liquid support in activating the aldehyde carbonyl and/or promoting an intramolecular proton transfer. Catalyst **172** was also recovered and reused up to six cycles with better results than catalyst **169**. Employing catalyst **172** in 20 mol%, several MBH and aza-MBH reactions were carried out affording products in good to excellent yields (Scheme 10.55).

Starting from prolinamide **178**, prepared in one step from *N*-Boc-proline, ionic liquid-anchored prolinamide **179** was synthesized (Scheme 10.56).[76]

When organocatalyst **179** was employed in the aldol reaction between acetone and 4-nitrobenzaldehyde, aldol product was isolated in low yield or not detected, regardless of the solvent. The major product of the reaction was the corresponding α,β-unsaturated carbonyl compound (Claisen–Schmidt reaction). In particular, under neat conditions, the condensation product was isolated in high yield.

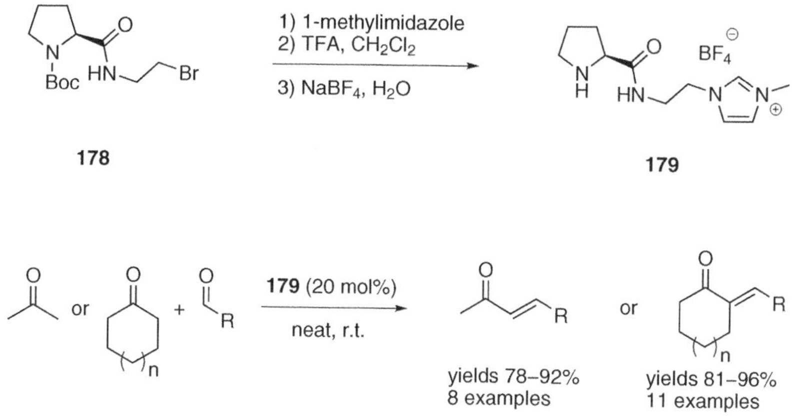

Scheme 10.56. Synthesis of ionic liquid-anchored prolinamide **179** and Claisen-Schmidt reactions catalyzed by **179**.

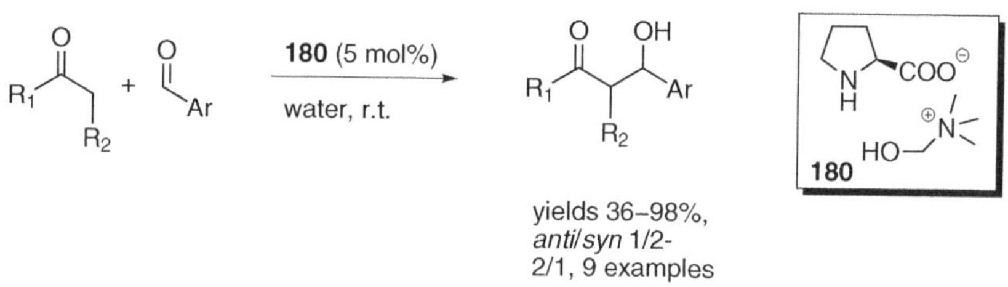

Scheme 10.57. Aldol reactions catalyzed by **180**.

Then, using optimized conditions, ionic liquid prolinamide **179** catalyzed the above reaction between acetone, cyclohexanone, or cyclopentanone with both aromatic and aliphatic aldehydes to give the corresponding α,β-unsaturated products in high yields (Scheme 10.56). The catalyst was recovered and used for seven consecutive cycles with a small loss in activity.

10.4 IONIC LIQUIDS AS ORGANOCATALYSTS

Several simple ionic liquids may act as organocatalysts without the need to be anchored to a catalytic center. These "nonsolvent" ionic liquids have found several synthetic applications. The catalytic center can be both the cation and the anion. Proline and its derivatives have found many applications when linked to the cationic moiety of the "ionic liquid" structure (see previous section). In addition, ionic liquids containing proline anion or cation have been successfully employed.

In the ionic liquid [Choline][Pro] [(2-hydroxyethyl)-trimethylammonium (*S*)-2-pyrrolidinecarboxylic acid salt] **180** (Scheme 10.57), the proline moiety is the anion.[77]

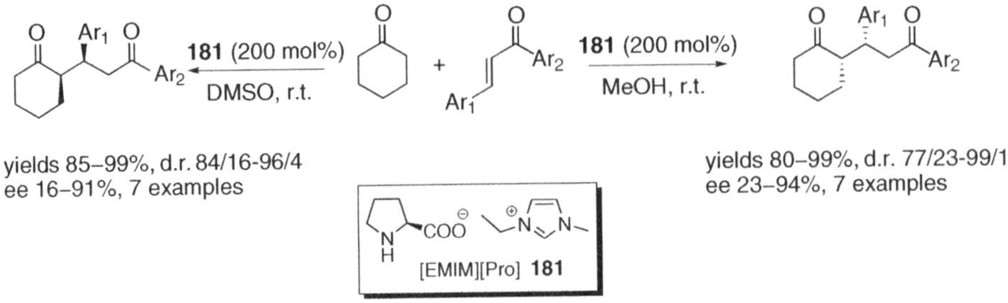

Scheme 10.58. Aldol reactions catalyzed by **181**.

Ionic liquid **180** was prepared from biorenewable materials in two steps (see also Chapter 1). Choline chloride was converted to a choline hydroxide aqueous solution by anion exchange resin, then the solution was neutralized with proline. Catalyst **180** was used in the aldol reactions between ketones and substituted benzaldehydes (Scheme 10.57). Aldol products were obtained in 36–98% yield, but very low enantioselectivity (<10% ee) was observed. The use of water as reaction medium increased the yield of the aldol products. Since catalyst **180** was soluble in the water phase, it was recovered and reused four times with unchanged results.

The use of 1-ethyl-3-methylimidazolium-(*S*)-2-pyrrolidinecarboxylic acid salt [EMIM][Pro] **181** as catalyst for the asymmetric Michael additions of cyclohexanone to chalcones was also reported (Scheme 10.58).[78] Catalyst **181** can be easily prepared in two steps from [EMIM][Br] by anion exchange and neutralization with L-proline.

This catalyst was employed in high loading (200 mol%) to afford the Michael adducts in high yields and variable enantioselectivities (16–94% ee). Screening of solvents for this reaction indicated the methanol was the best solvent; moreover, an interesting solvent-dependent effect was observed when the reaction was carried out in DMSO. Indeed, under the latter conditions the opposite enantiomer was obtained (Scheme 10.58). Since ionic liquid **181** was used in high catalytic loading, its recovery and reuse is of interest. The ionic liquid was recovered as residue after evaporation and extraction of the products. After four recycles yield was maintained, but a small decrease in enantioselectivity was observed.

The chiral ionic liquid catalyst L-prolinium sulfate **182** was an efficient catalyst for the stereoselective synthesis of polyfunctionalized perhydropyrimidines via the three-component Biginelli reaction (Scheme 10.59).[79] By reacting a mixture of aromatic aldehydes, urea or thiourea and the proper active methylene compounds in THF, the corresponding perhydropyrimidines were obtained in high yields and useful levels of enantio- and diastereoselectivities.

Also guanidine-based ionic liquids **183** were synthesized and used as catalysts for aldol reactions without solvent (Scheme 10.60).[80] In this case the effect of the anion was investigated. Several anions were screened ($RCOO^-$ in **183**: lactate, acetate, trifluoroacetate, propionate, *n*-butyrate, and *i*-butyrate), and the best results were obtained when acetate was used.

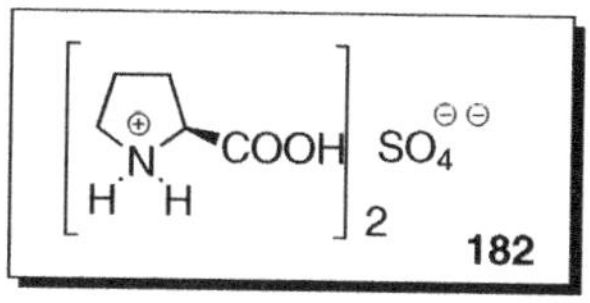

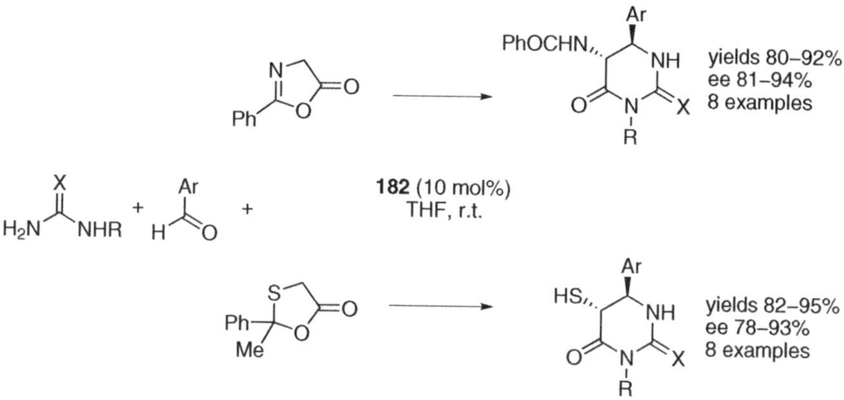

Scheme 10.59. Asymmetric Biginelli reactions catalyzed by **182**.

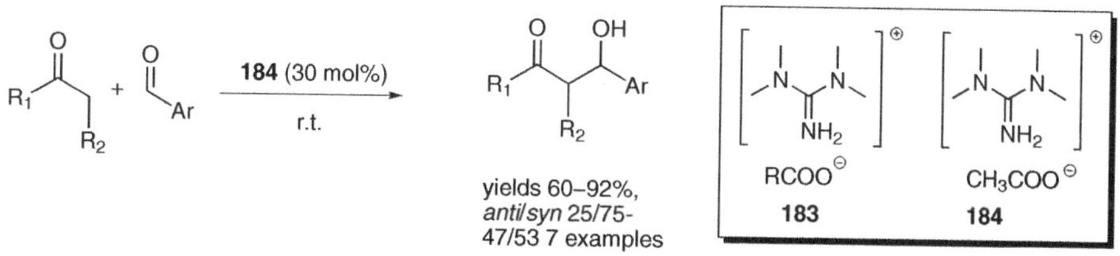

Scheme 10.60. Aldol reactions catalyzed by **184**.

Catalyst **184** was used for several aldol reactions affording the final products in 60–92% yield and low diastereoselectivity. However, the diastereoselectivity was reversed compared with the reactions catalyzed by proline and its derivatives.

Three-component Mannich reactions were investigated by using imidazolium-based or triphenylphosphine-based ionic liquids that acted both as solvents and as catalysts.[81] In the search for a halogen-free and less expensive ionic liquids, with respect to imidazolium-based or triphenylphosphine-based ionic liquids, the functionalized ionic liquid 3-(*N,N,N*-dimethyldodecylammonium)propanesulfonic acid hydrogen sulfate [Ddpa][HSO_4] **185** was prepared (Scheme 10.61).[82]

Ionic liquid **185** was synthesized by sulfonation of the biodegradable surfactant *N,N*-dimethyl-*N*-dodecylammonium)propane sulfonate. This molecule was used as Brønsted acid–surfactant-combined catalyst (10 mol%) for a one-pot three-component Mannich reaction, employing water as solvent. Also other polar solvents such as

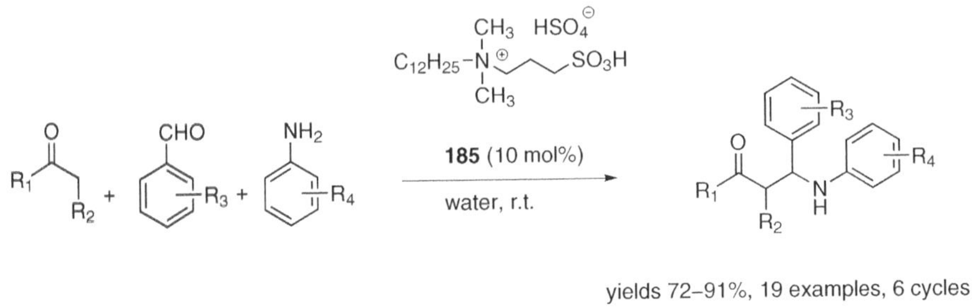

Scheme 10.61. Mannich reactions catalyzed by **185**.

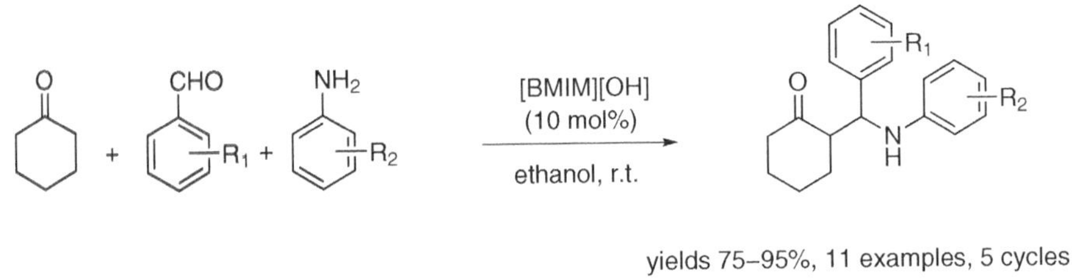

Scheme 10.62. Mannich reactions catalyzed by [BMIM][OH].

ethanol, methanol, and acetonitrile gave comparable or slightly better yields. The activity of catalyst **185** was maintained after six cycles.

Three-component Mannich reactions between cyclohexanone, aromatic aldehydes, and aromatic amines were also carried out in the presence of the Brønsted basic ionic liquid [BMIM][OH] as catalyst (10 mol%) (Scheme 10.62).[83] In this case water gave a lower yield compared with ethanol or acetonitrile. Several Mannich reactions were performed affording the products in high yields. Products were isolated from ethanol by filtration and the filtrate was used in the next cycle without further purification. After five cycles, only a minor decrease in activity was observed.

Amino-functionalized ionic liquid **186** was successfully employed as catalyst in only 0.8 mol% for Knoevenagel reactions in water (Scheme 10.63).[84] Products were obtained in high yields and catalyst was used for six cycles with good reproducibility.

Unsupported ionic liquids were also employed as organocatalysts for carbon dioxide fixation to yield cyclic carbonates. In the first report, ionic liquids [BMIM][BF_4], [BPy][BF_4], [BMIM][Cl] and [BMIM][PF_6] were employed as catalysts in the reaction between propylene oxide and CO_2.[85] At 110°C for 6 hours in the presence of 2.5 mol% of [BMIM][BF_4], propylene carbonate was obtained quantitatively (Scheme 10.64). After the reaction, the propylene carbonate was distilled from the reaction mixture and the ionic liquid catalyst was recycled up to four times with almost unchanged activity. The activity decreased in the order imidazolium > pyridinium and in the order of BF_4 > Cl > PF_6. Later, 1-octyl-3-methylimidazolium tetrafluoroborate [C_8-mim][BF_4] was used in 1.7 mol% at 100°C, working at much higher CO_2 pressure (14 MPa)

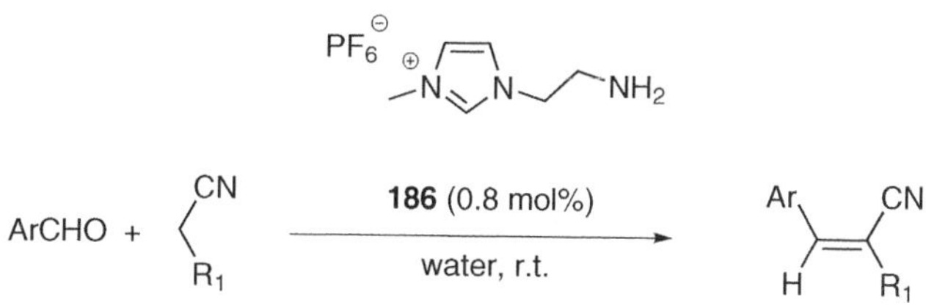

186: R_1 = COOEt: yields 91–98%, 7 examples
R_1 = CN: yields 85–95%, 8 examples

Scheme 10.63. Knoevenagel reactions catalyzed by **186**.

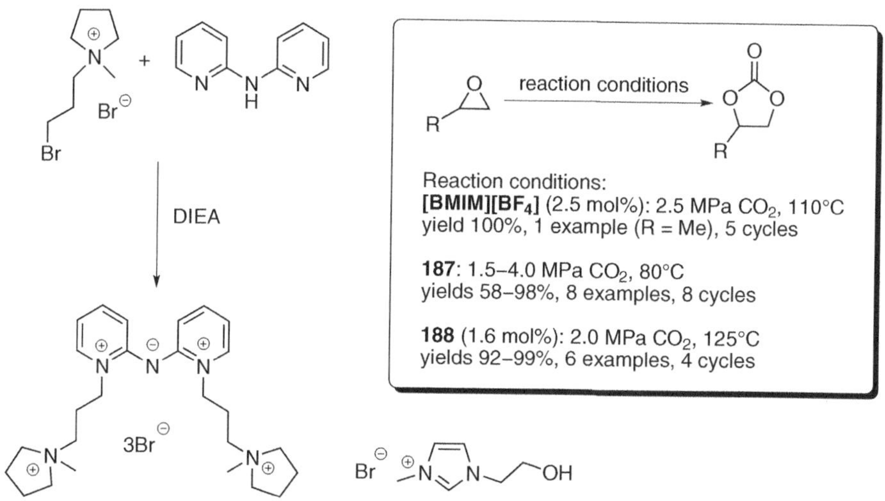

187 **188**

Scheme 10.64. Cycloaddition reactions of epoxides with CO_2 catalyzed by ionic liquid catalysts [BMIM][BF_4], **187-188**.

compared with the previous report. Under these conditions, the turnover frequency (TOF) value for the synthesis of propylene carbonate was 77 times larger than the previous, reaching approximately 100% of yield.[86]

Addition of CO_2 to allyl glycidyl ether gave the corresponding carbonate in high yield (97%) when 1-hexyl-3-methylimidazolium chloride was used in 5 mol% at 100°C and 140 psi CO_2 pressure.[87] Ionic liquid **187** was obtained by reaction between 2,2'dipyridylamine and 1-(3-bromopropyl)-1-methylpyrrolidinium bromide in the presence of *N,N*-diisopropylethylamine (DIEA) (Scheme 10.64). This ionic liquid was used both as a medium and as a catalyst, affording cyclic carbonates in good to high yields, and displayed useful levels of recyclability (eight cycles).[88]

The same reaction was catalyzed by ionic liquid **188** in 1.6 mol% loading (Scheme 10.64).[89] This catalyst was more active than the corresponding chloride ionic liquid and the 1-ethyl-3-methylimidazolium bromide salt [EMIM][Br]. The latter compound, employed in the presence of water or ethanol, gave slightly lower yields, thus showing the cooperative effect of the OH group in activating the epoxide ring opening. Excellent yields were observed in the examples reported and in every case yields were higher than those obtained with [EMIM][Br].

Simple ionic liquids such as [BMIM][BF_4], [BMIM][PF_6], [BBIM][BF_4], and [BMIM][Br] were employed as phase transfer catalysts in the dialkylation and cycloalkylation of active methylene compounds. Results were comparable to those obtained using the known phase transfer catalysts TBAB and TBAI.[90]

10.5 CONCLUSIONS

As can be seen from the literature reported in this chapter, the "nonsolvent" use of ionic liquids in organocatalysis has increased in importance and number of applications. Several useful synthetic processes, such as C–C bond-forming reactions and others such as CO_2 cycloaddition to epoxides, Bayer–Villiger oxidation, or epoxidation have been realized. These reactions can be carried out with the scope of obtaining both chiral and achiral compounds. When these reactions have been realized under asymmetric conditions, chiral compounds have been obtained in very high optical purity. The examples reported highlight how ionic liquids can be considered smart molecules that play several roles in addition to the classical solvent role. These "nonsolvent" roles regard supported and nonsupported ionic liquids, and this chapter has shed some light on how the presence of ionic liquid moieties convey to the materials new important features. Moreover, the "nonsolvent" use of ionic liquids allows the recovery and reuse of the active organocatalytic species, thereby increasing the synthetic usefulness of this approach. Finally, much more research needs to be carried out in order to extend these appealing systems to continuous flow reactions.

REFERENCES

1 B. List, R. A. Lerner, C. F. III Barbas, *J. Am. Chem. Soc.* **2000**, *122*, 2395–2396.

2 K. A. Ahrendt, C. J. Borths, D. W. C. MacMillan, *J. Am. Chem. Soc.* **2000**, *122*, 4243–4244.

3 S. Toma, M. Mečiarova, R. Šebesta, *Eur. J. Org. Chem.* **2009**, 321–327.

4 M. Gruttadauria, F. Giacalone, R. Noto, *Chem. Soc. Rev.* **2007**, *37*, 1666–1688.; (b) F. Cozzi, *Adv. Synth. Catal.* **2006**, *348*, 1367–1390.; (c) M. Benaglia, A. Puglisi, F. Cozzi, *Chem. Rev.* **2003**, *103*, 3401–3430.

5 C. P. Mehnert, R. A. Cook, N. C. Dispensiere, M. Afeworki, *J. Am. Chem. Soc.* **2002**, *124*, 12932–12933.

6 P. Kotrusz, I. Kmentová, B. Gotov, Š. Toma, E. Solčániová, *Chem. Commun.* **2002**, 2510–2511.

7 T.-P. Loh, L.-C. Feng, H.-Y. Yang, J.-Y. Yang, *Tetrahedron Lett.* **2002**, *43*, 8741–8743.

8 M. Gruttadauria, S. Riela, P. Lo Meo, F. D'Anna, R. Noto, *Tetrahedron Lett.* **2004**, *45*, 6113–6116.

9 M. Gruttadauria, S. Riela, C. Aprile, P. Lo Meo, F. D'Anna, R. Noto, *Adv. Synth. Catal.* **2006**, *348*, 82–92.

10 C. Aprile, F. Giacalone, M. Gruttadauria, A. Mossuto Marculescu, R. Noto, J. D. Revell, H. Wennemers, *Green Chem.* **2007**, *9*, 1328–1334.

11 P. Krattiger, R. Kovasy, J. D. Revell, S. Ivan, H. Wennemers, *Org. Lett.* **2005**, *7*, 1101–1103.

12 P. Lozano, E. García-Verdugo, R. Piamtongkam, N. Karbass, T. D. Diego, M. I. Burguete, S. V. Luis, J. L. Iborra, *Adv. Synth. Catal.* **2007**, *349*, 1077–1084.

13 P. Lozano, T. de Diego, D. Carrie , M. Vaultier, J. L. Iborra, *Chem. Commun.* **2002**, 692–693.; (b) S. V. Dzyuba, R. A. Bartsch, *Angew. Chem. Int. Ed.* **2003**, *42*, 148–150.; (c) M. T. Reetz, W. Wiesenhöfer, G. Francio, W. Leitner, *Adv. Synth. Catal.* **2003**, *345*, 1221–1228.; (d) P. Lozano, T. de Diego, S. Gmouh, M. Vaultier, J. L. Iborra, *Biotechnol. Prog.* **2004**, *20*, 661–669.

14 M. I. Burguete, F. Galindo, E. García-Verdugo, N. Karbass, S. V. Luis, *Chem. Commun.* **2007**, 3086–3088.

15 Y. Jiang, C. Guo, H. Xia, I. Mahmood, C. Liu, H. Liu, *J. Mol. Catal. B Enzym.* **2009**, *58*, 103–109.

16 P. Li, L. Wang, M. Wang, Y. Zhang, *Eur. J. Org. Chem.* **2008**, 1157–1160.

17 P. Li, L. Wang, Y. Zhang, G. Wang, *Tetrahedron* **2008**, *64*, 7633–7638.

18 Y. Jiang, C. Guo, H. Xia, I. Mahmood, H. Liu, *Ind. Eng. Chem. Res.* **2008**, *47*, 9628–9635.

19 G. Lai, J. Peng, J. Li, H. Qiu, J. Jiang, K. Jiang, Y. Shen, *Tetrahedron Lett.* **2006**, *47*, 6951–6953.

20 Y. Liu, J. Peng, S. Zhai, J. Li, J. Mao, M. Li, H. Qiu, G. Lai, *Eur. J. Inorg. Chem.* **2006**, 2947–2949.

21 L.-W. Xu, M.-S. Yang, J.-X. Jiang, H.-Y. Qiu, G.-Q. Lai, *Cent. Eur. J. Chem.* **2007**, *5*, 1073–1083.

22 M. Trilla, R. Pleixats, M. W. C. Man, C. Bied, *Green Chem.* **2009**, *11*, 1815–1820.

23 K. Yamaguchi, T. Imago, Y. Ogasawara, J. Kasai, M. Kotani, N. Mizuno, *Adv. Synth. Catal.* **2006**, *348*, 1516–1520.

24 Y. Zhang, Y. Zhao, C. Xia, *J. Mol. Catal. A Chem.* **2009**, *306*, 107–112.

25 L.-F. Xiao, Q.-F. Yue, C.-G. Xia, L.-W. Xu, *J. Mol. Catal. A Chem.* **2008**, *279*, 230–234.

26 P. R. Likhar, S. Roy, M. Roy, M. S. Subhas, M. Lakshmi Kantam, *Catal. Commun.* **2009**, *10*, 728–731.

27 S. Abelló, F. Medina, X. Rodríguez, Y. Cesteros, P. Salagre, J. E. Sueiras, D. Tichit, B. Coq, *Chem. Commun.* **2004**, 1096–1097.

28 Y. Xie, Z. Zhang, T. Jiang, J. He, B. Han, T. Wu, K. Ding, *Angew. Chem. Int. Ed.* **2007**, *46*, 7255–7258.

29 J. Sun, W. Cheng, W. Fan, Y. Wang, Z. Meng, S. Zhang, *Catal. Today* **2009**, *148*, 361–367.

30 S. S. Shinde, B. S. Lee, D. Y. Chi, *Tetrahedron Lett.* **2008**, *49*, 4245–4248.

31 L. Han, S.-W. Park, D.-W. Park, *Energy Environ. Sci.* **2009**, *2*, 1286–1292.

32 S. Udayakumar, V. Raman, H.-L. Shim, D.-W. Park, *Appl. Catal. A* **2009**, *368*, 97–104.

33 J.-Q. Wang, X.-D. Yue, F. Cai, L.-N. He, *Catal. Commun.* **2007**, *8*, 167–172.

34 X. Zheng, S. Luo, L. Zhang, J.-P. Cheng, *Green Chem.* **2009**, *11*, 455–458.

35 K. Motokura, S. Itagaki, Y. Iwasawa, A. Miyaji, T. Baba, *Green Chem.* **2009**, *11*, 1876–1880.

36 K. Qiao, H. Hagiwara, C. Yokoyama, *J. Mol. Catal. A Chem.* **2006**, *246*, 65–69.

37 A. Chrobok, S. Baj, W. Pudło, A. Jarzębski, *Appl. Catal. A* **2009**, *366*, 22–28.

38 H. Zhao, S. V. Malhotra, *Aldrichim. Acta* **2002**, *35*, 75–83.; (b) F. D'Anna, S. Marullo, R. Noto, *J. Org. Chem.* **2008**, *73*, 6224–6228.

39 J. Howarth, K. Hanlon, D. Fayne, P. McCormac, *Tetrahedron Lett.* **1997**, *38*, 3097–3100.

40 N. Jain, A. Kumar, S. Chauhan, S. M. S. Chauhan, *Tetrahedron* **2005**, *61*, 1015–1060.; (b) S. T. Handy, *Curr. Org. Chem.* **2005**, *9*, 959–988.; (c) J. Habermann, S. Ponzi, S. V. Ley, *Mini-Rev. Org. Chem.* **2005**, *2*, 125–137.

41 W. Miao, T. H. Chan, *Adv. Synth. Catal.* **2006**, *348*, 1711–1718.

42 C. Zhuo, L. Yong, X. Hui, H. Chang-gang, D. Xian, *Russ. J. Org. Chem.* **2008**, *44*, 1807–1810.

43 L. Zhou, L. Wang, *Chem. Lett.* **2007**, *36*, 628–629.

44 M. Lombardo, F. Pasi, S. Easwar, C. Trombini, *Adv. Synth. Catal.* **2007**, *349*, 2061–2065.

45 M. Lombardo, F. Pasi, S. Easwar, C. Trombini, *Synlett* **2008**, 2471–2474.

46 M. Gruttadauria, F. Giacalone, R. Noto, *Adv. Synth. Catal.* **2009**, *351*, 33–57.

47 B. List, L. Hoang, H. J. Martin, *Proc. Natl. Acad. Sci. U.S.A.* **2004**, *101*, 5839–5842.

48 N. Zotova, A. Franzke, A. Armstrong, D. G. Blackmond, *J. Am. Chem. Soc.* **2007**, *129*, 15100–15101.

49 S. Aratake, T. Itoh, T. Okano, N. Nagae, T. Sumiya, M. Shoji, Y. Hayashi, *Chem. Eur. J.* **2007**, *13*, 10246–10256.; (b) F. Giacalone, M. Gruttadauria, P. Lo Meo, S. Riela, R. Noto, *Adv. Synth. Catal.* **2008**, *350*, 2747–2760.

50 M. Gruttadauria, F. Giacalone, A. Mossuto Marculescu, R. Noto, *Adv. Synth. Catal.* **2008**, *350*, 1397–1405.; (b) M. Gruttadauria, F. Giacalone, A. Mossuto Marculescu, A. M. P. Salvo, R. Noto, *ARKIVOC* **2009**, (viii), 5–15.; (c) F. Giacalone, M. Gruttadauria, A. Mossuto Marculescu, R. Noto, *Tetrahedron Lett.* **2007**, *48*, 255–259.; (d) M. Gruttadauria, F. Giacalone, A. Mossuto Marculescu, P. L. Meo, S. Riela, R. Noto, *Eur. J. Org. Chem.* **2007**, 4688–4698.

51 N. Mase, Y. Nakai, N. Ohara, H. Yoda, K. Takabe, F. Tanaka, C. F. III Barbas, *J. Am. Chem. Soc.* **2006**, *128*, 734–735.

52 M. Lombardo, S. Easwar, F. Pasi, C. Trombini, *Adv. Synth. Catal.* **2009**, *351*, 276–282.

53 D. E. Siyutkin, A. S. Kucherenko, M. I. Struchkova, S. G. Zlotin, *Tetrahedron Lett.* **2008**, *49*, 1212–1216.

54 D. E. Siyutkin, A. S. Kucherenko, S. G. Zlotin, *Tetrahedron* **2009**, *65*, 1366–1372.

55 M. Lombardo, S. Easwar, A. De Marco, F. Pasi, C. Trombini, *Org. Biomol. Chem.* **2008**, *6*, 4224–4229.

56 S. Luo, X. Mi, L. Zhang, S. Liu, H. Xu, J.-P. Cheng, *Tetrahedron* **2007**, *63*, 1923–1930.

57 L. Zhang, S. Luo, X. Mi, S. Liu, Y. Qiao, H. Xu, J.-P. Cheng, *Org. Biomol. Chem.* **2008**, *6*, 567–576.

58 Z. Wang, Q. Wang, Y. Zhang, W. Bao, *Tetrahedron Lett.* **2005**, *46*, 4657–4660.

59 S. Luo, X. Mi, L. Zhang, S. Liu, H. Xu, J.-P. Cheng, *Angew. Chem. Int. Ed.* **2006**, *45*, 3093–3097.

60 S. Luo, X. Mi, L. Zhang, S. Liu, H. Xu, J.-P. Cheng, *Chem. Commun.* **2006**, 3687–3689.

61 S. Luo, L. Zhang, X. Mi, Y. Qiao, J.-P. Cheng, *J. Org. Chem.* **2007**, *72*, 9350–9352.

62 B. Ni, Q. Zhang, A. D. Headley, *Green Chem.* **2007**, *9*, 737–739.

63 Q. Zhang, B. Ni, A. D. Headley, *Tetrahedron* **2008**, *64*, 5091–5097.

64 B. Ni, Q. Zhang, K. Dhungana, A. D. Headley, *Org. Lett.* **2009**, *11*, 1037–1040.

65 Z.-Y. Yan, Y.-N. Niu, H.-L. Wei, L.-Y. Wu, Y.-B. Zhao, Y.-M. Liang, *Tetrahedron Asymmetry* **2006**, *17*, 3288–3293.

66 S. Luo, H. Xu, X. Mi, J. Li, X. Zheng, J.-P. Cheng, *J. Org. Chem.* **2006**, *71*, 9244–9247.

67 L.-Y. Wu, Z.-Y. Yan, Y.-X. Xie, Y.-N. Niu, Y.-M. Liang, *Tetrahedron Asymmetry* **2007**, *18*, 2086–2090.

68 Z. Yacob, J. Shah, J. Leistner, J. Liebscher, *Synlett* **2008**, 2342–2344.

69 T. Miao, L. Wang, P. Li, J. Yan, *Synthesis* **2008**, 3828–3834.

70 B. Ni, Q. Zhang, A. D. Headley, *Tetrahedron Lett.* **2008**, *49*, 1249–1252.

71 O. V. Maltsev, A. S. Kucherenko, S. G. Zlotin, *Eur. J. Org. Chem.* **2009**, 5134–5137.

72 M. Lombardo, M. Chiarucci, A. Quintavalle, C. Trombini, *Adv. Synth. Catal.* **2009**, *351*, 2801–2806.

73 V. K. Aggarwal, I. Emme, A. Mereu, *Chem. Commun.* **2002**, 1612–1613.

74 X. Mi, S. Luo, J.-P. Cheng, *J. Org. Chem.* **2005**, *70*, 2338–2341.

75 X. Mi, S. Luo, H. Xu, L. Zhang, J.-P. Cheng, *Tetrahedron* **2006**, *62*, 2537–2544.

76 S.-D. Yang, L.-Y. Wu, Z.-Y. Yan, Z.-L. Pan, Y.-M. Liang, *J. Mol. Catal. A Chem.* **2007**, *268*, 107–111.

77 S. Hu, T. Jiang, Z. Zhang, A. Zhu, B. Han, J. Song, Y. Xie, W. Li, *Tetrahedron Lett.* **2007**, *48*, 5613–5617.

78 Y. Qian, S. Xiao, L. Liu, Y. Wang, *Tetrahedron Lett.* **2008**, *49*, 1515–1518.

79 L. D. S. Yadav, A. Rai, V. K. Rai, C. Awasthi, *Tetrahedron* **2008**, *64*, 1420–1429.

80 A. Zhu, T. Jiang, B. Han, J. Huang, J. Zhang, X. Ma, *New J. Chem.* **2006**, *30*, 736–740.

81 G. Zhao, T. Jiang, H. Gao, B. Han, J. Huang, D. Sun, *Green Chem.* **2004**, *6*, 75–77.; (b) S. Sahoo, T. Joseph, S. B. Halligudi, *J. Mol. Catal. A Chem.* **2006**, *244*, 179–182.

82 F. Dong, F. Zhenghao, L. Zuliang, *Catal. Commun.* **2009**, *10*, 1267–1270.

83 K. Gong, D. Fang, H.-L. Wang, Z.-L. Liu, *Monatsh. Chem.* **2007**, *138*, 1195–1198.

84 Y. Cai, Y. Peng, G. Song, *Catal. Lett.* **2006**, *109*, 61–64.

85 J. Peng, Y. Deng, *New J. Chem.* **2001**, *25*, 639–641.

86 H. Kawanami, A. Sasaki, K. Matsui, Y. Ikushima, *Chem. Commun.* **2003**, 896–897.

87 D.-W. Park, N.-Y. Mun, K.-H. Kim, I. Kim, S.-W. Park, *Catal. Today* **2006**, *115*, 130–133.

88 W.-L. Wong, P.-H. Chan, Z.-Y. Zhou, K.-H. Lee, K.-C. Cheung, K.-Y. Wong, *ChemSusChem* **2008**, *1*, 67–70.

89 J. Sun, S. Zhang, W. Cheng, J. Ren, *Tetrahedron Lett.* **2008**, *49*, 3588–3591.

90 S. Muthusamy, B. Gnanaprakasam, *Tetrahedron Lett.* **2005**, *46*, 635–638.

INDEX

Ionic Liquids in Biotransformations and Organocatalysis: Solvents and Beyond, First Edition. Edited by Pablo Domínguez de María.

Printed and bound by CPI Group (UK) Ltd, Croydon, CR0 4YY
16/09/2026
14963096-0002